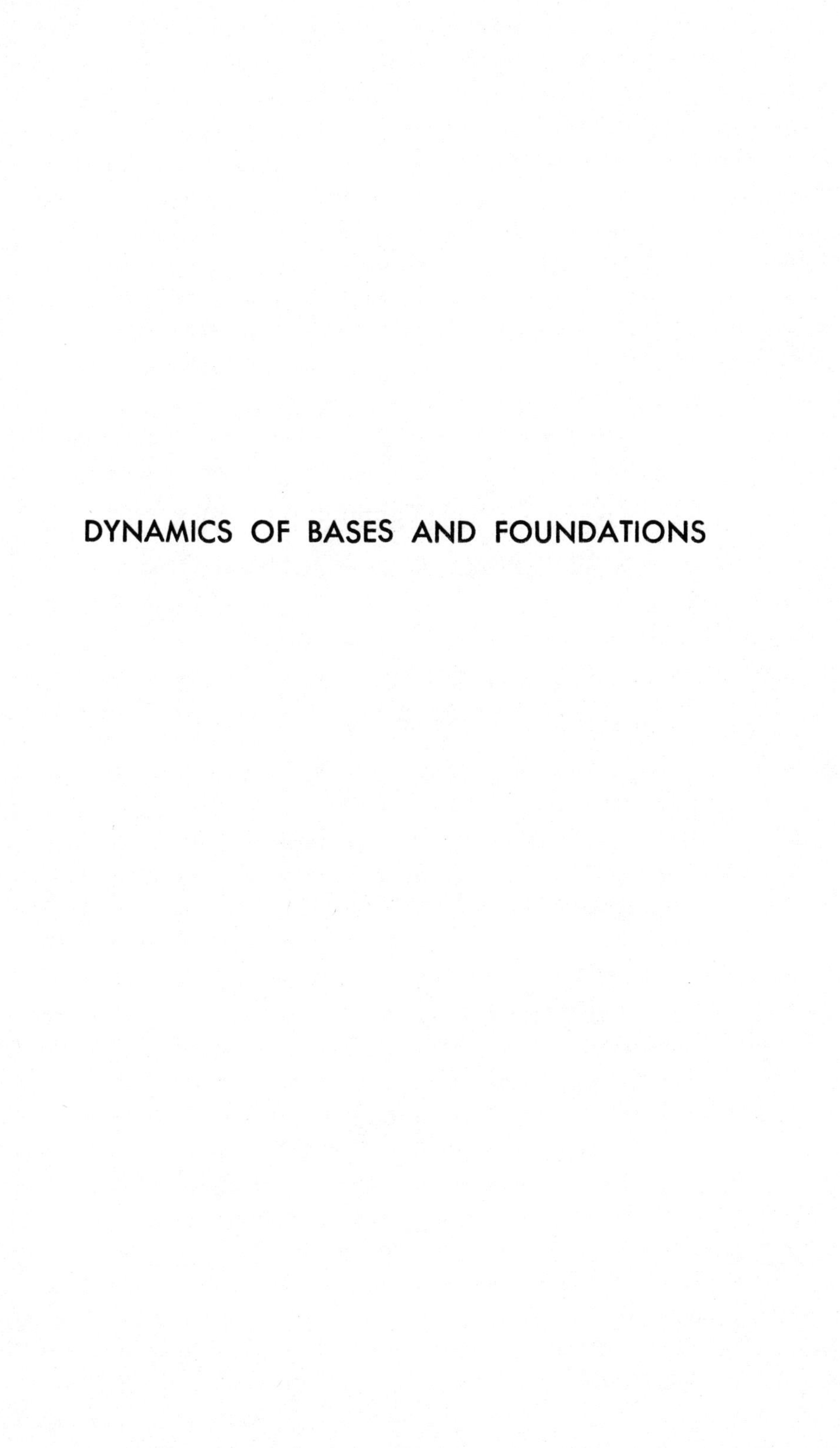

DYNAMICS OF BASES AND FOUNDATIONS

McGRAW-HILL SERIES IN SOILS ENGINEERING AND FOUNDATIONS

BARKAN, Dynamics of Bases and Foundations

LEONARDS, Foundation Engineering

DYNAMICS OF BASES AND FOUNDATIONS

D. D. BARKAN

DOCTOR OF TECHNICAL SCIENCES

TRANSLATED FROM THE RUSSIAN by L. DRASHEVSKA

TRANSLATION EDITED by G. P. TSCHEBOTARIOFF

McGRAW-HILL BOOK COMPANY, INC.

New York San Francisco Toronto London

DYNAMICS OF BASES AND FOUNDATIONS

 Library of Congress Catalog Card Number 62-15743

03650

Revised for English translation by the author in 1959 and 1960.

First published in the Russian language under the title
DYNAMIKA OSNOVANIY I FUNDAMENTOV
and copyright 1948 by Stroyvoenmorizdat, Moscow.

EDITOR'S FOREWORD TO THE ENGLISH TRANSLATION

Since the Fourth International Conference on Soil Mechanics and Foundation Engineering in London (1957), Dr. D. D. Barkan has become well known in the West for his work of recent years on the development of vibratory driving equipment for piles, sheet piles, and cylindrical bridge caissons. However, the remarkable work on machine-foundation problems of the Soil Dynamics Laboratory which he heads at the Institute of Foundations and Underground Structures remains little known in the West.

The development of rational and effective approaches to these problems has been greatly hampered throughout the world by a lack of continuity and organized coordination of efforts in the several disciplines involved. Since its creation some 35 years ago Dr. Barkan's laboratory has made an impressive start in remedying this situation.

I fully endorse the views expressed on the matter in the preface to the Russian text by the late Professor N. M. Gersevanov, who had sponsored the creation of the laboratory. Therefore I am very happy to have the privilege of presenting to the attention of English-speaking engineers the translation of Dr. Barkan's book "Dynamics of Bases and Foundations."† In many respects this book is unique.

The relevant basic premises of soil mechanics, the theory of elasticity, and the dynamics of vibratory motion are presented in a coordinated manner, and the resulting findings are examined in the light of experimental evidence; the use of complex mathematical expressions is simplified by the provision of auxiliary charts and tables; relevant coefficients obtained from theoretical analyses and from laboratory and model tests are checked against and corrected by the *results of numerous measurements*

† The term "base" which appears in the title of the book is used in Russian in the sense of the mass of soil which supports the foundation of any type of structure.

on full-scale industrial installations; design recommendations are made in the light of practical considerations of civil and mechanical engineering; numerical examples illustrate the points made.

This book should therefore be of immediate use to practicing engineers —both civil and mechanical—who have to deal with machine foundations and the vibrations which machines cause; it should also stimulate further creative thought and research in the field.

After a formal agreement on the matter between the Soviet publishing agency Mezhdunarodnaya Kniga (International Book) and the McGraw-Hill Book Company, Inc., Dr. Dominik D. Barkan in 1959 and 1960 reviewed and brought up to date the text of his original 1948 book; the English translation was made from this revised Russian text.

Acknowledgments are due to my wife, Florence Bill Tschebotarioff, who helped me edit the English translation, and to several Princeton colleagues who helped me in the selection of proper mechanical terms—especially to Dr. E. W. Suppiger, Professor of Mechanical Engineering, who read the manuscript.

Gregory P. Tschebotarioff
Professor of Civil Engineering
Princeton University

EDITORS' FOREWORD TO THE ORIGINAL RUSSIAN TEXT

This book is devoted to the elastic and vibratory properties of soil bases, the theory and practice of the design of foundations under machines with dynamic loads, questions of wave propagation from industrial sources, and problems of the effect of these waves on structures.

This book was compiled mainly from results of investigations which were carried out by the author and workers of the laboratory headed by him. Included are data gained from experience in the performance of foundations under machinery at industrial plants of the U.S.S.R.

This book is intended for civil engineers and scientific workers.

THE EDITORS *G. I. Berdichevsky*
D. M. Tumarkin

PREFACE TO THE ORIGINAL RUSSIAN TEXT

The practice of modern industrial construction requires the solution of many problems concerning the performance of foundations subjected to dynamic loads, the propagation of waves in soil, and the effects of these waves on structures.

Some twenty years ago, on the initiative of the undersigned, a special laboratory of dynamics of soil bases and foundations was organized for the study of these problems. The laboratory was headed by Professor D. D. Barkan.

The laboratory work concentrated chiefly on the investigation of elastic properties of soils and changes in their physicomechanical characteristics (in particular, internal friction, cohesion, and void ratio) under the influence of vibrations, on the study of vibrations of massive foundations, and on the study of exciting loads created by various machines.

In addition, much of the laboratory work was devoted to study of the propagation in soils of elastic waves from industrial sources, the screening of these waves by various measures, and the investigation of the influence of several other factors on wave propagation.

Considerable attention was paid by the laboratory to the study of foundation vibrations and the action on structures of waves emanating from foundations.

The extensive research work carried out by the laboratory in the field of dynamics of soil bases and foundations made it possible for the first time to compile standard specifications for the computation, design, and erection of foundations under machinery with dynamic loads.

The use of these standard specifications in the practical design of industrial projects was very effective, in both the engineering and economic aspects of construction.

The book, "Dynamics of Bases and Foundations," by Doctor of Technical Sciences D. D. Barkan, is compiled mainly from materials derived from investigations conducted by his laboratory. This is the first guiding monograph in the field concerned.

The originality of the material and the subjects covered make this book interesting not only for civil engineers who are designers but also for wide circles of scientific and engineering workers who are studying the dynamics of machines and structures.

Professor N. M. Gersevanov
Corresponding Member of the
Academy of Sciences of the U.S.S.R.

PREFACE TO THE ENGLISH TRANSLATION

Technical and industrial progress is accompanied by an increase in the number of machines in use and the growth of their power, as well as by the introduction of blasting operations into construction practice and the increasing speed, intensity, and tonnage of various kinds of transport. All these factors lead to an increase in the influence of shocks and vibrations, the study of which is the subject of a new branch of mechanics designated as industrial seismology. One of the most important divisions of industrial seismology is the dynamics of bases and foundations, which studies the problems of shocks on and vibrations of foundations under industrial installations, as well as the laws governing the propagation of waves from these foundations through soil.

The problems solved by study of the dynamics of bases and foundations are of great importance in engineering practice.

Vibrations of machine foundations are harmful to the operation of the machines themselves. The following case may be quoted as an example: due to an unsatisfactory design, the foundations of several piston gas compressors installed in the same building underwent considerable vibrations. The compressors, being rigidly connected to the foundations, also underwent vibrations of large amplitude which were transmitted to the pipes leading from them. Vibrations frequently caused damage to and even rupture of these pipes near the vibrating compressors. This caused the seepage of explosive gas into the compressor house; as a result the house blew up.

Measurements of vibrations of foundations and anvils under operating hammers show that some 10 per cent of the work of the dropping parts is spent in exciting vibrations of the foundation and anvil.

Vibrations induced by a vibrating foundation often cause poor performance of machine bearings, which may be heated up or may wear dissymmetrically.

In addition, foundation vibrations under high-speed machines cause premature fatigue of machine parts.

Vibrations and shocks also have a harmful effect on the foundations themselves. Although very few cases have been recorded in which machine foundations have been destroyed by their vibrations, some instances are known where vibrations or shocks led to a rapid enlargement of cracks which were initially caused by other factors such as stresses imposed by differential settlements of the foundation. This led to the destruction of the foundations.

Vibrations of and shocks to foundations under machinery lead to considerable nonuniform settlements which necessitate periodic adjustments and gauging of machines.

A foundation subjected to vibrations becomes a source of waves propagated through the soil which have a harmful effect on structures and people within them. These waves also hinder normal operation of plants.

Several examples are presented in Chap. IX showing how waves propagated from foundations under hammers caused damage to structures or fully destroyed them.

Waves from hammer foundations fairly frequently cause differential settlements of columns or walls of forge shops. Shocks on hammer foundations lead also to skewing of locomotive crane tracks, distortions of window sashes, damage of some connections of structural elements, etc.

Waves from machine foundations affect not only structures, but also various technological processes and the operation of precision machines and devices. Cases are known from industrial practice in which machine-foundation shocks hindered the operation of various devices and caused an increase in the amount of rejects in molding and machine shops.

A harmful effect on structures, and especially on technological processes within these structures, may be produced by waves propagated from foundations under low-frequency machines with a stable regime of operation such as compressors, diesels, and saw frames. These machines usually have comparatively low speed, of the same order of magnitude as the principal values characterizing the natural vibrations of structures. Consequently, the frequency of waves propagated from the foundation under a low-frequency machine sometimes coincides with the frequency of natural vibrations of the structure. Considerable vibrations then develop in the superstructure of the building subjected to resonance.

If a building undergoes vibrations of considerable amplitude, many technological processes within this building become difficult or even impossible. Such vibrations have an especially harmful effect on men if living quarters are located in a vibrating building. Interviews with people who lived in such a vibrating building showed that persons subjected day and night to the action of vibrations often suffered from headaches, insomnia, nervousness, etc. A thorough study of the dynamics of bases and foundations is needed to successfully overcome the harmful effects of such vibratory processes.

Problems of the dynamics of bases and foundations are important from an economic standpoint as well. Because the bearing capacity of soils under machinery foundations was usually taken as 50 to 60 per cent of the value corresponding to the action of static loads only, in many cases the bases were reinforced by piles. This raised considerably the price of the foundation. The clarification in recent years of some questions pertaining to the dynamics of bases under structures made it possible to increase design values of the bearing capacity of soils under machinery foundations up to 80 to 100 per cent of the static-load values. Thus the installation of a pile base could be avoided in many cases. The cost of such a base under powerful industrial machinery may reach hundreds of thousands, if not millions, of rubles for a single project.

Until recently it was believed by designers that the larger the mass of a machine foundation, the smaller the amplitude of its vibrations. Therefore foundations under machines were designed as massive blocks attaining sometimes a weight of several thousand tons and placed at depths up to 10 to 12 m.

The erection of such foundations involved the expenditure of considerable funds, materials, and time. The study of questions of the rational design of foundations led to the conclusion that, from the standpoint of dynamic stability, it is often expedient to make the foundations as light as possible.

Owing to this fact, the cost of industrial projects has been considerably reduced in many cases. Let us mention, for example, that foundations under powerful horizontal compressors were formerly built with weights of 2,000 to 2,500 tons; at present they are designed for weights reduced to one-half the above figure. As a result of the development of methods of dynamic computations of foundations, the cost of a foundation under one powerful horizontal compressor is now 150,000 to 200,000 rubles less than it was previously.

Only a few years ago, the weight of a foundation under a hammer was selected to be approximately 100 to 120 times the weight of dropping parts; in addition, it was held necessary to increase the mass of the foundation not by enlarging its area in plan, but by increasing its depth. As a result, foundations under hammers were built as heavy massive blocks sometimes placed at depths up to 10 m. The study of the behavior of foundations under the action of impacts led to the conclusion that it is possible, without increasing the permissible amplitude of vibrations, to design foundations with weights exceeding by not more than 10 to 50 times the weights of the dropping parts. Thus the construction of one foundation under a 3-ton hammer is now cheaper by 25,000 to 30,000 rubles than what it cost previously.

The great economic importance of questions of the dynamics of bases and foundations will be still more evident if one realizes that annually thousands of machines which require the construction of individual foundations are built or moved from one place to another.

The vibrations of foundations under machines and the propagation of waves from industrial sources depend on the elastic properties of soil. It is therefore natural that the first chapter of this book is devoted to the description of these properties. The experimental data which it presents on the dependence of elastic deformations of soil on stresses show that the application of Hooke's law to soils may be considered as a first approximation only. Deviations from Hooke's law are caused by the influence of initial stresses in the soil (which are present because of friction forces, capillary forces, and other factors) on the elastic deformations induced by external loads. Consequently, the elastic constants of soil, particularly its modulus of shear, depend on the values of normal pressure.

The modern theory of massive foundation vibrations is based on the assumption that the hypothesis of linear relationship between settlements and loads is valid. Therefore, for the computation of forced or natural vibrations of foundations it is necessary to know the so-called coefficients of elasticity of soils, i.e., the coefficients of proportionality between stress in the soil along the foundation area in contact with the soil and the settlement or sliding of the foundation. The first chapter presents theoretical and experimental data which make it possible to establish the limiting values of the above coefficients for different soils and the dependence of these coefficients on several factors. At the end of the chapter, experimental data are given concerning the elastic properties of pile foundations. A considerable part of the first chapter is devoted to the results of experimental investigations of elastic properties of soil bases under structures.

The second chapter treats questions related to the formation of irreversible (residual) settlements under the simultaneous action on the soil of static pressure and vibrations. This question is important, first of all, for the estimation of the bearing value of the soil under the combined action of vibrations and static pressure and the clarification of factors affecting residual settlements under the influence of vibrations; second, it is important for the scientific justification of new methods for the provision of artificial bases under structures by means of vibrations. The results of experiments presented in the second chapter show that basically all soil properties (including those which determine the resistance of soils to localized loading) may change considerably under the action of vibrations. Also very strongly affected are the compressive properties of soils, especially sandy soils. Computational problems of residual (plastic) settlements of foundations subjected to vibrations or shocks acquire considerable importance in this connection. The basic theory for the computation of such settlements is given in the second chapter, as are methods to decrease residual settlements.

Data presented in the first two chapters were obtained from investigations carried out by the author or his associates in the laboratory headed by him; a considerable part of these data is being published here for the first time.

The third chapter treats the general theory and results of experimental investigations of vibrations of massive foundations. This chapter concludes the first part of the book, which covers the main general principles of the dynamics of foundations.

Chapters IV to VII form the second part of the book and treat the problems of dynamic computations and design of foundations for special types of machines such as reciprocating engines, machinery with impact loads, rolling mills, turbodynamos, and crushing equipment. The table of contents shows the scope of the subject matter presented. We shall merely mention that for these chapters data are widely used from numerous large-scale instrumental observations of vibrations of existing foundations which have been carried out by the author since 1939.

Finally, Chaps. VIII and IX are devoted to the question of dynamics of bases, i.e., to the propagation through soil of waves from machinery and the action of these waves on structures. These chapters form the third part of the book. A special feature of this part lies in the fact that theoretical material related to the propagation of waves has been corrected here on the basis of numerous experimental data. The latter show that in the solution of some problems (for example, when establishing a relationship between the amplitudes of soil vibrations and distance), theoretical conclusions based on the assumption that soil is an ideal elastic medium are not confirmed experimentally. In such cases the author has tried to introduce the necessary corrections based on experimental results.

In compiling this book the author widely used material obtained from investigations conducted in the Soil Dynamics Laboratory headed by himself.

The author expresses his cordial thanks to Prof. G. P. Tschebotarioff, who spent much time and energy on the editing of the translation and on its preparation for publication in the United States. The author also thanks O. Ya. Shekhter, who helped him in the revision of the text.

D. D. Barkan

CONTENTS

NOTATION

Many symbols used in the original Russian text were also translated and were replaced by symbols customary in the United States. Both sets of symbols are given in the list which follows. This list does not include all subscripts to the main symbols which are defined in the text.

The metric system was used in the Russian text and has been kept in the English translation. The following conversion factors may be helpful to the reader:

$$1 \text{ ton} = 1 \times 10^3 \text{ kg} = 2{,}205 \text{ lb}$$
$$1 \text{ m} = 100 \text{ cm} = 3.28 \text{ ft} = 39.4 \text{ in.}$$
$$1 \text{ cm} = 0.395 \text{ in.}$$
$$1 \text{ m}^2 = 1 \times 10^4 \text{ cm}^2 = 10.35 \text{ ft}^2$$
$$1 \text{ m}^3 = 1 \times 10^6 \text{ cm}^3 = 35.3 \text{ ft}^3$$
$$1 \text{ ton} \times \text{m} = 1 \times 10^5 \text{ kg} \times \text{cm} = 7.25 \times 10^3 \text{ ft} \times \text{lb} = 8.7 \times 10^4 \text{ in.} \times \text{lb}$$
$$1 \text{ kg} \times \text{cm} = 5.61 \text{ lb} \times \text{in.}$$
$$1 \text{ ton/m}^2 = 0.1 \text{ kg/cm}^2 = 206 \text{ lb/ft}^2 \text{ (or} \simeq 0.1 \text{ U.S. ton/ft}^2\text{)} = 1.42 \text{ psi}$$
$$1 \text{ ton/m}^3 = 1 \times 10^{-3} \text{ kg/cm}^3 = 56.6 \text{ lb/ft}^3 = 3.28 \times 10^{-2} \text{ lb/in.}^3$$
$$1 \text{ ton} \times \text{sec}^2\text{/m} = 675 \text{ lb} \times \text{sec}^2\text{/ft}$$
$$1 \text{ kw} = 1.36 \text{ hp}$$
$$1g = 9.81 \text{ m/sec}^2 = 32.2 \text{ ft/sec}^2$$

Symbols

Used in English translation	Definition	Used in Russian text
A	Area of foundation or attached slab in contact with soil; cross-sectional area of hammer piston in Eq. (V-2-2)	F; f
A	Amplitude of vibrations in Eq. (III-I-7); constant; coefficient in Eq. (VIII-4-1)	A
A_f	Amplitude of soil vibrations where source of waves is placed at a depth f below soil surface in Eq. (VIII-5-13)	A_f

SYMBOLS (*Continued*)

Used in English translation	Definition	Used in Russian text
A_0	Permissible amplitude of vibrations in Eq. (V-5-3)	$A_{\text{доп}}$
A_0	Amplitude of soil vibrations where source of waves is placed on soil surface in Eq. (VIII-5-13); ratio in Eq. (VIII-3-24); amplitude of soil vibration at distance r_0 from source	A_0
A_r	Amplitude of vertical soil vibrations at distance r from source in Eq. (VIII-3-42)	A_r
A_{st}	Static vertical displacement of foundation in Eq. (III-1-15)	A_{cm}
A_x	Amplitude of horizontal vibrations in shear in Eq. (III-3-3)	A_x
A_y	Amplitude of horizontal vibrations in shear of center of gravity of whole system in Example IV-4-1	A_y
A_z	Amplitude of vertical vibrations	A_z
A_z^*	Amplitude of forced vertical vibrations in Eq. (III-1-21)	A_z^*
A_ω	Amplitude of vertical soil vibrations in Eq. (VIII-4-9)	A_ω
A_φ	Amplitude of rocking vibrations in Eqs. (III-2-7) and (IV-5-3)	A_φ
$A_{x\varphi}$	Combined amplitude produced by rocking and sliding vibrations in Example IV-4-2	$A_{x\varphi}$
a	Half length of a foundation; full length of a foundation in Art. III-2; coefficient in Eqs. (II-1-5), (VIII-1-8), and (VIII-4-8); amplitude of forced vibrations of a foundation in Eq. (III-1-29); ratio of vibration amplitudes in Eqs. (IV-5-7) and (IV-5-8); distance in Eq. (VI-2-17); velocity of propagation of longitudinal waves in soil in Table VIII-2	a
a_a	Amplitude of anvil vibrations in Eq. (V-4-7)	A_m
a_r	Amplitude of forced vibrations at resonance in Eq. (III-1-34)	a_r
B	Constant in Eq. (II-3-14); amplitude coefficient in Eq. (III-1-7); constant in Art. IV-2-a and Eq. (VIII-4-1); constant of elastic aftereffect in Art. VIII-3-d	B
b	Half width of a foundation; full width of a foundation in Art. III-2; dimensionless value in Eq. (III-1-31); thickness of pad under anvil in Art. V-3; distance in Fig. VI-8 and Eq. (VI-2-17); ratio in Eq. (VIII-1-7); velocity of propagation of transverse waves in soil in Eq. (VIII-6-22)	b
C	Coefficient in Eq. (I-2-13); arbitrary constant in Eqs. (II-3-14), (III-2-5) and (VIII-6-3); total rigidity in Eq. (VI-2-14); constant in Eq. (VIII-3-21)	C
C_a	Arbitrary constant in Eq. (III-4-17)	C_a
C_k	Coefficient of elastic resistance of one pile in a group	C_k
c	Damping constant in Eq. (III-1-17)	n
c	Coefficient in Eqs. (III-4-16), (IV-6-2), and (V-3-5); velocity of surface wave propagation	c
c_a	Velocity of surface waves when there is another soil layer above that surface in Fig. VIII-23	c_a
c_b	Velocity of transverse soil waves	c_b
$\bar{c}$	Reduced coefficient of elastic resistance of a pile in Eq. (I-6-3)	c

SYMBOLS (*Continued*)

Used in English translation	Definition	Used in Russian text
$c_{(A)}$	Coefficient of proportionality variable along the base of a bearing plate	$c_{(F)}$
c_p	Coefficient of proportionality; also called modulus of subgrade reaction	c_p
c_r	Coefficient of rigidity of soil base in Eq. (III-1-2)	c
c_s	Coefficient in Eqs. (I-2-12) and (I-2-14) and Table I-3 for flexible rectangular footing	k_z
c_s'	Coefficient in Eq. (I-2-12) and Table I-3 for rigid rectangular footing	k_s'
c_{sp}	Rigidity of one spring in example in Art. IV-6	$c_{\text{пр}}$
c_u	Coefficient of elastic uniform vertical compression of soil	c_z
c_u'	Coefficient of elastic uniform vertical compression of soil under hammer foundations in Eq. (V-4-5)	c_z'
c_x	$c_\tau A$ in Art. III-4-d	c_x
c_z	$c_u A$ in Eq. (III-4-16)	c_z
c_ψ	Coefficient of elastic nonuniform shear of soil	c_ψ
c_τ	Coefficient of elastic uniform shear of soil	c_φ
c_φ	Coefficient of elastic nonuniform compression of soil	C
C_δ	Coefficient of elastic resistance of a pile in Eq. (I-6-1)	C
D	Arbitrary constant in Eqs. (II-3-14) and (VIII-6-3); diameter of spring coil in Eq. (IV-6-3); expressions in Eqs. (VIII-3-29)	D
d	Diameter of a pile; diameter of a spring in Eq. (IV-6-3)	d
E	Young's modulus; arbitrary constant in Eq. (VIII-6-3)	E
E_c	Young's modulus for concrete	E
e	Void ratio; coefficient of restitution in Eq. (V-2-5)	ϵ
e	Base of natural logarithms	e
$e_{\min}$	Minimum value of void ratio corresponding to densest possible state in Art. II-2	ϵ_∞
$e_{\max}$	Maximum value of void ratio corresponding to loosest possible state	ϵ_0
$e=\epsilon_x+\epsilon_y+\epsilon_z$	Volume expansion	Θ
e^0	Initial value of void ratio prior to start of vibrations in Art. II-2	ϵ^0
F	A function; centrifugal force in Eq. (VI-2-1); arbitrary constant in Eq. (VIII-6-3)	F
F_x, F_z	Horizontal, vertical components of centrifugal force in Eq. (VI-2-2)	P_x, P_z
f	Depth below the soil surface of the source of waves in Eq. (VIII-5-1)	f
f_a	Frequency of natural vibrations of foundation above the springs in example of Art. V-7	λ_p
f_b	Frequency of natural vibrations of foundation below the springs in example of Art. V-7	λ_z
f_n	Natural frequency of vibrations $= \omega/2\pi$	λ

Symbols (*Continued*)

Used in English translation	Definition	Used in Russian text
f_{na}	Frequency of natural vibrations of attached masses in Eq. (IV-5-9); frequency of natural vibrations of the anvil in Eq. (V-3-6)	λ_p
f_{nd}	Natural frequency of damped vibrations in Eq. (III-1-18)	λ_1
f_{nx}	Natural frequency of horizontal vibrations in Fig. I-31 and Eq. (III-3-2)	λ_x
f_{nz}	Natural frequency of vertical vibrations in Eq. (III-1-5)	λ_z
$f_{n\varphi}$	Natural frequency of free rocking vibrations of a foundation in Eq. (III-2-6)	λ_φ
$f_{n\psi}$	Natural frequency of torsional vibrations around a vertical axis in Eq. (III-3-5)	λ_ψ
f_1, f_2	Functions in Eqs. (III-1-26) and (VIII-1-13); frequencies of forced vibrations in example of Art. VII-1	f_1, f_2
f_{n1}, f_{n2}	First two main natural frequencies of a foundation, equal to the two positive roots of Eq. (III-4-8) [see Table (III-7)]	λ_1, λ_2
$\bar{f}_{n1}$, $\bar{f}_{n2}$	Main natural frequencies of a foundation for limiting case $\epsilon = 0$ in Eq. (III-4-20)	λ_1, λ_2
G, μ	Modulus of elasticity in shear, i.e., modulus of rigidity	μ, G
g	Acceleration of gravity	g
H	Height; distance; thickness; coefficient (with various subscripts)	H
h	Height; ratio ω/a in Eq. (VIII-2-5)	h
I	Moment of inertia of a bearing plate with respect to one of its axes; moment of inertia of a mass in Eq. (VII-3-6)	J
i	$\sqrt{-1}$; radius of gyration of mass of foundation and machine in Art. III-4; moment of inertia per unit of mass in Eq. (III-4-21)	i
J_1, J_2	Bessel's functions for solution of Eq. (III-1-26); used in Eq. (VIII-5-7)	J_1, J_2
$\bar{J}$	Expressions in Eq. (VIII-3-16)	$\bar{J}$
J_z	Polar moment of inertia of the base contact area of a foundation	J_z
j	Coefficient in equation defining v_0^* in Art. II-3	j
K	Coefficient of lateral earth pressure (ratio between lateral and vertical pressures); coefficient in Eq. (I-3-1)	ξ
K_{in}	Kinetic energy in Art. V-7	T
K_z	Coefficient of elastic resistance of a pile group	K_z
K_φ	Coefficient of elastic nonuniform resistance of pile foundations in Eq. (I-6-11)	K_φ
k	Constant in Eq. (III-1-34); number of engine cylinders in Eq. (IV-2-9); dynamic correction coefficient in Eq. (V-4-5); ratio in Eq. (VI-2-9); ratio ω/b in Eq. (VIII-2-5); number in Eq. (VIII-6-28)	k
k_φ	Coefficient in Eq. (I-3-5)	k_φ
k_τ	Coefficient in Eq. (I-4-6)	K_x

Symbols (*Continued*)

Used in English translation	Definition	Used in Russian text
L	Distance between a pile and the axis of rotation of the foundation in Eq. (I-6-10); distance; length	ξ
L	Distance between the axis of rotation and the center of gravity of the vibrating mass in Fig. III-9	h
L	Wavelength in Eq. (VIII-6-27)	L
L_b	Length of transverse waves in Eq. (VIII-5-15)	L_b
L_c	Wavelength in Eq. (VIII-2-25)	L_c
L_s	Length of shear waves propagated through soil from foundation [see solution of Eq. (III-1-26)]	L_b
l	Length; depth of pile penetration in Fig. I-28; constant in Eq. (III-1-34)	l
l	Distance in Fig. III-9	ξ
l_0	Depth of pile fixation in soil; span of beam in Eq. (VI-2-4a)	l_0
M	External moment; mass; moment of inertia in Eq. (VI-2-13)	M
M_i	External exciting moment in Eqs. (III-2-1) and (III-4-12)	M
M_m	Moment of inertia of the mass in respect to the y axis, which passes through the center of gravity of foundation and machine in Fig. III-11 and Eq. (III-4-2)	Θ
M_{m0}	Moment of inertia of the total vibrating mass (foundation and machine) with respect to axis through center of gravity of base contact area of foundation in Art. III-4	Θ_0
M_r	Moment of soil reactions in Eq. (III-2-2)	M_r
$M_{(t)}$	Moment of external exciting forces in Eq. (III-4-4)	$M_{(t)}$
M_z	Exciting moment acting in a horizontal plane in Eq. (III-3-4)	M_z
M_1	Mass of the crank in Eq. (IV-2-3)	M_1
M_2	Mass of the rod, crosshead, and piston in Eq. (IV-2-3)	M_2
M_3	Mass of the connecting rod in Eq. (IV-2-3)	M_3
m	Mass; mass of foundation in Eq. (V-2-3); empirical coefficient in Eq. (VIII-5-14)	m
m_0	Mass of falling ram in Eq. (V-2-3)	m_0
m_1	Mass of foundation in Eqs. (V-3-5)	m_1
m_2	Mass of anvil in Eqs. (V-3-5)	m_a
m_s	Mass of soil in Eq. (III-1-37)	$m_{\text{гр}}$
m_t	Total mass of soil and foundation	M
N	Number of oscillations per minute (speed) $= 60\omega/2\pi = 60n$	N
n	Number of piles in Eq. (I-6-4); number of oscillations per second (in hertz) $= \omega/2\pi$; number of springs in example of Art. IV-6	n
n	Coefficient of vibroviscosity in Eq. (II-1-2)	ν
n_a	Ratio of weight of anvil and frame W_a to weight of dropping parts of hammer W_0 in Eq. (V-5-9)	n_m
n_f	Ratio of weight of foundation W_F to weight of dropping parts of hammer W_0 in Eq. (V-5-9)	n_ϕ
P	Exciting force	P
P_{act}	Actual load on a spring in example of Art. IV-6	$P_{\text{действ.}}$

Symbols (*Continued*)

Used in English translation	Definition	Used in Russian text
P_c	Centrifugal force in Art. IV-2-a	W_1
P_p	Permissible load on a spring in examples of Arts. IV-6 and V-7	$P_{доп}$
P_r	Peripheral stress transmitted to pulley by belt in Eq. (IV-3-2)	P_r
P_T	Total horizontal tangential pressure on soil along foundation base; resultant force of pull on belt in Eq. (IV-3-1); total horizontal exciting force in Eq. (IV-5-1) and example of Art. VII-1	P_x
$P(t)$	Exciting force which changes with time	$P(t)$
P_x	Horizontal force; horizontal projection of a force	P_x
P_z	Vertical force; vertical projection of a force; exciting load in Eq. (VIII-3-9)	P_z
p	Exciting force $P(t)$ per unit of mass in Eqs. (III-1-11) and (IV-5-9); pressure on piston of hammer in Eq. (V-2-2)	p
p_{dy}	Dynamic pressure on soil beneath a foundation in Eqs. (II-3-3) and (V-5-1)	σ_d
p_l	Limit pressure on soil beneath a foundation in Art. II-3	σ_l
p_0	Permissible pressure on soil if only static load is acting in Eq. (V-5-1)	p_{cm}
p_{st}	Static pressure on soil beneath a foundation in Eqs. (II-3-2), (III-1-35), and (V-5-1)	σ_{cm}
p_z	Vertical pressure on horizontal soil surface	p_z
q	Coefficient of required pressure decrease on soil in the presence of vibrations in Eq. (II-3-3)	q
q_0	Radial component of soil surface displacement in Eq. (VIII-5-2)	q_0
R	Total reaction of soil against a bearing plate in Eqs. (I-2-3) and (III-1-25); radius of a plate in Eq. (I-2-6); force acting anywhere in Eq. (II-1-2); distance between a center of gravity and an axis of rotation in Art. IV-2-a; ratio in Eq. (VI-2-19); radius in Eq. (VIII-4-3)	R
R_i	Elastic force in Eq. (VI-2-13)	R_i
R_{dy}	Maximum dynamic pressure which, if exceeded, causes large settlements in Eq. (II-3-3)	R_d
R_{st}	Maximum static pressure which, if exceeded, causes large settlements in Eq. (II-3-2)	R_s
R_p	Permissible torsional stress of the spring material in example of Art. IV-6	$R_{доп}$
r	Radius of a sphere in Eq. (II-1-2); ratio in Eq. (II-3-11); eccentricity of impact in Eq. (V-2-4); radius of gyration in Eq. (VI-2-13); radius in Eqs. (VIII-1-13)	r
r_0	Radius of a circle in Eq. (III-1-26); eccentricity of machine rotor in Eq. (VI-2-2)	r_0
S	Settlement; perimeter of a pile in Eq. (I-6-3); load coefficient in Eq. (II-1-3)	z, s

SYMBOLS (*Continued*)

Used in English translation	Definition	Used in Russian text
S	Coefficient defining slope of settlement-time curve during vibration in Eq. (II-3-13)	S
S_{av}	Average settlement of a flexible plate in Eq. (I-2-10)	Z_{cp}
S'_{av}	Average value of horizontal sliding of a foundation	u_{cp}
S_e	Total elastic settlement of a bearing plate	z
S'_e	Total elastic sliding of a bearing plate	u_x
S_t	Total settlement of a bearing plate in Eq. (I-2-1)	Z_n
S'_{tot}	Total sliding of a bearing plate in Eq. (I-4-2)	u
s	Shearing strength of soil; distance in Fig. VIII-12	S
T	Period of vibration; tension in a machine belt in Eq. (IV-3-1)	T
t	Time; distance between piles	t
u	Slope of a curve in Art. II-2	u
u, v, w	Components of displacements	u, v, w
u_a, v_a	Displacements on soil surface in Chap. VIII	u_z, v_h
u_r	Radial displacement of soil in Eq. (VIII-1-15)	u_r
V	Volume	V
v	Velocity; peripheral speed of engine belt in Art. IV-3	v
v_0	Initial velocity	v_0
v_0^*	Maximum value of initial velocity of vibratory foundation motion at which no residual settlements will occur in Art. II-3	v_0^*
v_s	Velocity of shear waves in soil [see solution of Eq. (III-1-26)]	v
W	Work; engine power	W
W	Weight	Q
W_m	Moment of inertia of the mass of the whole system with respect to an axis passing through common center of gravity perpendicular to the plane of vibrations	Θ_m
W_0	Moment of inertia of the mass of foundation and machine with respect to horizontal axis of rotation through the center of gravity of foundation in Eq. (III-2-1)	Θ_0
W_z	Moment of inertia of the mass of foundation and machine with respect to vertical axis of rotation in Eq. (III-3-4)	Θ_z
w	Water content (by weight)	p or w or q
w_0	Water content in Eq. (I-1-19)	p_0
X, Y, Z	Components of mass forces in Eq. (VIII-1-1)	X, Y, Z
X_i	Projection on x axis of all external and inertial forces acting on foundation in Eq. (III-1-4)	X_i
x	Horizontal displacement of the center of gravity of foundation in Eq. (III-3-1); horizontal displacement of the common center of gravity of foundation and machine in Art. III-4; projection on a horizontal axis of the displacement of the center of gravity of foundation in Eq. (V-8-1)	x
x_0	Horizontal displacement of center of gravity of contact area of foundation in Art. III-4	x_0
Y_0, Y_1, Y_u	Bessel functions in Eqs. (VIII-3-27) and (VIII-4-4) (see also symbol J)	Y_0, Y_1, Y_u

SYMBOLS (*Continued*)

Used in English translation	Definition	Used in Russian text
Z_i	Projection on z axis of all external and inertial forces acting on foundation in Eq. (III-4-1)	Z_i
x, y, z	Rectangular coordinates	x, y, z
x_i, y_i, z_i	Coordinates of a point i of moving machine parts in Eq. (IV-2-1); coordinates of centers of gravity of separate elements of a vibrating system in Example IV-4-1	x_i, y_i, z_i
x_0, y_0, z_0	Coordinates of the common center of gravity of the vibrating system in Example IV-4-1	x_0, y_0, z_0
z	Vertical displacement of foundation	z
$\dot{z}$	First derivative of vertical displacement	$\dot{z}$
$\ddot{z}$	Second derivative of vertical displacement	$\ddot{z}$
z_{st}	Elastic settlement of foundation under the action of the weight of foundation and machinery	Z_s
α	Ratio of half length a to half width b of a foundation in Eq. (I-2-11); coefficient in Eqs. (I-1-16) and (I-1-18a); coefficient of vibratory compaction in Eq. (II-2-1); angle of phase shift between displacement and soil reaction = $\varphi - \epsilon$ in Eq. (III-1-30); coefficient defining c in Eqs. (III-1-17) and (III-1-37); arbitrary constant in solution of Eq. (III-4-5); characteristic number of crank mechanisms in Eq. (IV-2-7); pressure reduction coefficient in Eq. (V-5-1); hammer coefficient in Eq. (V-7-2); coefficient in Eq. (VI-2-4a) for frame foundations; coefficient in Eq. (VI-2-24); charge coefficients in Eq. (VIII-1-19); coefficient in Eq. (VIII-2-20); coefficient in Eq. (VIII-6-1); coefficient in Eq. (VIII-6-13); coefficient of absorption of wave energy in Eq. (VIII-3-43)	α
β	Coefficient in Eqs. (I-1-11) and (II-1-1); coefficient of increase of foundation mass due to participation of soil in Eq. (III-1-40); ratio of foundation frequencies in Art. III-4, Fig. III-13, and Eq. (IV-5-14); angle between cranks of different cylinders in Eq. (IV-2-9); ratio of dynamic displacements of foundation and anvil in Eq. (V-4-2); charge coefficient in Eq. (VIII-1-19); coefficient in Eqs. (VIII-2-6), (VIII-2-20), and (VIII-6-26)	β
γ	Unit weight in Eqs. (II-1-4) and (VIII-1-1); angle of phase shift between exciting force and the displacement it induces in Eq. (III-1-22); ratio of moments of inertia following Eq. (III-4-7); angle with the horizontal of straight line through axes of rotation of pulleys in Art. IV-3; coefficient depending on parameters of the engine in Eqs. (IV-6-9) and (VIII-3-26); ratio of frequencies in Eq. (V-4-3); moment causing rotation of the beam of a frame foundation by an angle equaling unity in Eq. (VI-2-17)	γ
$\gamma_{xy}, \gamma_{xz}, \gamma_{yz}$	Shearing-strain components in rectangular coordinates	$\epsilon_{xy}, \epsilon_{xz}, \epsilon_{yz}$

SYMBOLS (*Continued*)

Used in English translation	Definition	Used in Russian text
Δ	Ratio of damping coefficient c to natural frequency f_{nz}; relative change in volume in Eq. (VIII-1-1)	Δ
$\Delta(\omega^2)$	Coefficient in Eq. (IV-6-8)	$\Delta(\omega^2)$
δ	Elastic settlement of a pile; effect of vibrations on decrease of internal friction in soil in Art. II-1-a; ratio in Eq. (IV-5-7); lateral displacement of frame foundation in Eq. (VI-2-15)	δ
∇^2	Laplace operator	∇^2
ϵ	Phase shift between exciting force and soil reaction in Eq. (III-1-25); eccentricity in distribution of the mass of foundation and machine in Eq. (III-4-15); coefficient of nonuniformity of engine speed in Eq. (IV-5-14); distance in Eq. (VI-2-14) and Fig. VI-8	ϵ
ϵ_x, ϵ_y, ϵ_z	Unit elongations in x, y, z directions	ϵ_{xx}, ϵ_{yy}, ϵ_{zz}
ϵ_e	Elastic unit deformations	e
ϵ'_x	Horizontal elastic displacement of a pile head	e_x
η	Element of width of a foundation base area in Eq. (I-4-3); ratio of the acceleration of vibrations to the acceleration of gravity in Eq. (II-1-1); dynamic modulus (magnification factor) of undamped vibrations in Eq. (III-1-15); ratio of amplitudes (degree of vibration absorption) in Eq. (IV-6-11); hammer coefficient in Eq. (V-2-1) and Art. V-7	η
η	Coefficient of viscosity in Art. VIII-3-d	γ
η^*	Dynamic modulus of damped vibrations in Eq. (III-1-23)	η^*
η_0	Threshold of vibratory compaction in Eq. (II-2-5)	η_0
Θ	Angle in cylindrical system of coordinates in Eq. (VIII-3-16)	Θ
$\varkappa$	Coefficient in an Eq. of Art. III-1-d and Fig. III-5; root of Eq. (VIII-2-17) defined by Eq. (VIII-2-18)	$\varkappa$
ϑ	Ratio in Eq. (VIII-3-16)	ϑ
λ	Lamé's constant $= E/[(1 + \nu)(1 - 2\nu)]$	λ
μ	Coefficient in Eq. (I-6-5); ratio of limiting frequencies in Art. III-4; ratio of natural horizontal and rocking frequencies in Art. III-13; coefficient of friction of belt on pulley in Eq. (IV-3-4); ratio of masses of damper and foundation in Eq. (IV-5-14); ratio of masses above and below springs in Eq. (IV-6-8); ratio of masses of foundation and ram in Eq. (V-2-6); ratio of masses of anvil and foundation in Eqs. (V-3-8) and (V-4-4); dynamic coefficient in Eq. (VII-3-5)	μ
μ	Lamé's constant ($= G$)	μ
$\mu_0 = G_0$	Modulus of rigidity of soil in its initial state when no external stresses are acting thereon	μ_0
ν	Poisson's ratio	σ
ξ	Element of length of a foundation base area in Eq. (I-4-3); reduced coefficient of damping in Art. II-1-c and Eq. (III-1-39); ratio of actual to natural frequencies (frequency	ξ

SYMBOLS (*Continued*)

Used in English Translation	Definition	Used in Russian text
	ratio) of vibration in Eq. (III-1-16); reduced damping coefficient in Eq. (III-1-39); ratio of the limit natural frequency of a foundation to the exciting frequency in Eq. (IV-6-9); constant in Eq. (VIII-2-6)	
ρ	c/f_{nd}, coefficient in Eq. (II-3-7); ratio of constants expressed by Eq. (III-4-9); ratio of amplitudes in Eq. (III-4-13); square root of the ratio of moment of hammer and foundation mass inertia to the mass of hammer in Eq. (V-2-6); density of soil in Table VIII-1; kr in Eq. (VIII-3-16)	ρ
ρ	Radius vector of a point within the area of a bearing plate	ρ
σ	Normal stress	p
$\sigma_x, \sigma_y, \sigma_z$	Normal components of stress parallel to x, y, z axes	X, Y, Z
τ	Shearing stress; relative time in Eq. (II-3-7)	ρ, τ
τ_{av}	Average shear stress in soil along plane of contact with foundation	p_x
$\tau_{xy}, \tau_{xz}, \tau_{yz}$	Shear-stress components in rectangular coordinates	Y_x, Z_x, Z_y
Φ	Function in Eq. (VIII-6-3)	Φ
ϕ	Angle of internal friction of soil; tan ϕ = coefficient of internal friction of soil	f
ϕ_{st}	Angle of internal friction without vibrations; tan ϕ_{st} = coefficient of internal friction without vibrations	f_{cm}
φ	Slope angle of a deflected plate along the line $x = 0$ in Eq. (I-3-1); angle of rotation of foundation in Eqs. (I-6-10) and (III-2-2) and Fig. III-9; phase shift between exciting force and displacement in Eq. (III-1-30); angle of rotation of crankshaft; angle of belt contact in Fig. IV-6; function in Eq. (VIII-3-1)	φ
φ_0	Arbitrary constant in Eq. (III-2-5); initial velocity of rotation of foundation in Eq. (V-2-4)	φ_0
Ψ	Function in Eq. (VIII-6-1)	Ψ
ψ	Angle of rotation of a foundation around its vertical axis in Eq. (III-3-5); coefficient of energy absorption in Art. II-1; angle in Fig. VIII-12	ψ
Ω	Function in Eq. (VIII-6-9)	Ω
ω	Frequency of vibrations; frequency of exciting force; angular velocity of machine rotation [in Eq. (IV-2-2)] $= 2\pi N/60 = 2\pi n$	ω

I

ELASTIC PROPERTIES OF SOIL

I-1. Elastic Deformations of Soil and Its Elastic Constants

a. Generalized Hooke's Law. There are two kinds of external forces acting on a deformable body: body forces and surface forces.

Body forces act on volume elements. They comprise gravity forces, inertial forces, and forces of mutual attraction. The value of the body force per unit volume is called the mass force.

Surface forces act on surface areas. Classical examples are hydrostatic pressure and forces of internal and external friction. Forces of interaction between the parts of a body belong to the category of internal surface forces.

The magnitude of the surface force per unit area of the surface is called stress. Stresses induced by internal surface forces acting in deformable bodies are called elastic stresses.

The magnitude of the stress at a point depends on the orientation of the stress plane through the point. The stress acting on any inclined plane through a point can be obtained from stresses acting on three planes passing through the point and perpendicular to the coordinate axes x, y, and z. Thus a stress acting on a plane perpendicular to the x axis may be resolved into three components: σ_x, τ_{xy}, and τ_{xz} (Fig. I-1); σ_x will be the normal component of the stress and τ_{xy} and τ_{xz} will be shearing components parallel to the y and z axes. Correspondingly, stresses acting on planes perpendicular to the y and z axes may be resolved into the components τ_{yx}, σ_y, τ_{yz} and τ_{zx}, τ_{zy}, σ_z; σ_y and σ_z will be the normal components, and the rest components of the shearing stress. The stress conditions of

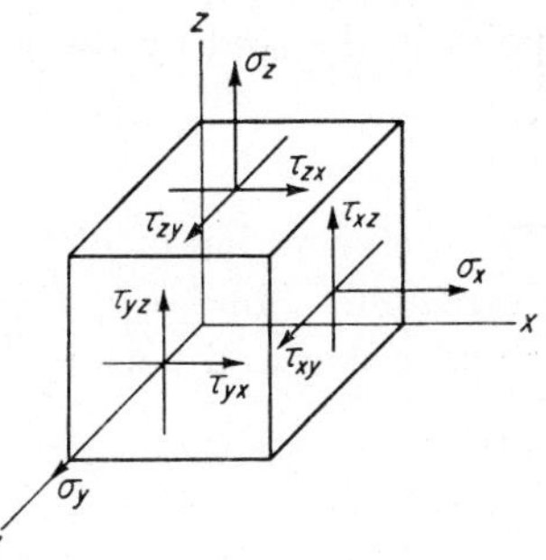

FIG. I-1. Components of stress.

the body at a point are then determined by the totality of nine values:

$$\begin{matrix} \sigma_x & \tau_{xy} & \tau_{xz} \\ \tau_{yx} & \sigma_y & \tau_{yz} \\ \tau_{zx} & \tau_{zy} & \sigma_z \end{matrix}$$

where $\tau_{zy} = \tau_{yz} \qquad \tau_{zx} = \tau_{xz} \qquad \tau_{xy} = \tau_{yx}$

Only six of the nine components of stress are not interdependent.

If a body undergoes deformations, then under the action of stresses the distances between points of the body (i.e., its form) will change. Let u, v, and w be the components of displacement along the coordinate axes x, y, and z, respectively. In the general case the values of these components depend both on the coordinates of the point under consideration and on time.

Assuming that deformations are small, we may represent the linear strain of an element of length parallel to the x axis as

$$\epsilon_x = \frac{\partial u}{\partial x}$$

Similarly the linear strains of elements parallel to the y and z axes are

$$\epsilon_y = \frac{\partial v}{\partial y} \qquad \epsilon_z = \frac{\partial w}{\partial z}$$

However, the state of strain of a body is determined not only by linear strains but also by the rotation of some linear elements in respect to other elements. As a result of deformations, a right angle between two linear elements parallel to the x and y axes and originating at the same point is changed by the value

$$\gamma_{xy} = \frac{\partial v}{\partial x} + \frac{\partial u}{\partial y}$$

This deformation or distortion of the right angle is called shear strain or detrusion. Similarly, the distortions of the right angles between the x and z axes and between the y and z axes will equal

$$\gamma_{xz} = \frac{\partial w}{\partial x} + \frac{\partial u}{\partial z}$$

$$\gamma_{yz} = \frac{\partial v}{\partial z} + \frac{\partial w}{\partial y}$$

The quantities ϵ_x, ϵ_y, ϵ_z and γ_{xy}, γ_{xz}, γ_{yz}, which determine the deformation of the body, are called components of strain.

If the stresses in the deformed body are of such magnitudes that $\gamma_{xy} = \gamma_{xz} = \gamma_{yz} = 0$ (that is, no shearing occurs), then the deformation is accompanied only by a change in volume. The relative change in volume for small deformations equals

$$e = \epsilon_x + \epsilon_y + \epsilon_z$$

If $\epsilon_x = \epsilon_y = \epsilon_z = 0$, then the deformation leads to the sliding of some elements of the body relative to other elements, and the volume of the body will not change. According to the theory of elasticity, any small deformation can be resolved into a sum of deformations accompanied only by change in volume and of deformations accompanied only by shear.

If after the removal of the load a body resumes its initial shape (i.e., the one it had before the load was applied), then this body is elastic. A truly solid body, as a rule, possesses only a partial elasticity, because, after unloading, the body does not exactly resume its initial shape (i.e., residual deformations have taken place). We shall consider here only elastic deformations which can always be calculated by subtracting the value of the residual deformation from the total deformation corresponding to the given loading.

It is known from experience that elastic deformations increase proportionally to increases in external loads. Assuming that internal stresses in a body are proportional to external loads, we may reach the conclusion that deformations of a body depend on the stresses which act within it. In general form this interrelationship between the stresses and deformations can be analytically expressed by six functions, as follows:

$$\sigma_x = F_x(\epsilon_x,\epsilon_y,\epsilon_z,\gamma_{xy},\gamma_{xz},\gamma_{yz})$$
$$\cdots\cdots\cdots\cdots$$
$$\tau_{yz} = F_{yz}(\epsilon_x,\epsilon_y,\epsilon_z,\gamma_{xy},\gamma_{xz},\gamma_{yz})$$

Let us assume that before external forces were applied to a body, it was in such a state that no internal stresses were present (it was in an undeformed state). Under these conditions the functions F must be such that if

$$\epsilon_x = \epsilon_y = \cdots = 0$$

then $$F_x = F_y = \cdots = 0$$

We assume that deformations are small and therefore in expanding the functions F_x, F_y, . . . into Taylor's series we restrict ourselves to terms of the first order only. We obtain

$$\begin{aligned} \sigma_x &= C_{11}\epsilon_x + C_{12}\epsilon_y + C_{13}\epsilon_z + C_{14}\gamma_{xy} + C_{15}\gamma_{xz} + C_{16}\gamma_{yz} \\ &\cdots\cdots\cdots\cdots\cdots\cdots \\ \tau_{yz} &= C_{61}\epsilon_x + C_{62}\epsilon_y + C_{63}\epsilon_z + C_{64}\gamma_{xy} + C_{65}\gamma_{xz} + C_{66}\gamma_{yz} \end{aligned} \qquad \text{(I-1-1)}$$

Equations (I-1-1) present the analytical expression of the generalized Hooke's law, which can be formulated as follows: each of the six components of the *stress* is a *linear function* of the six components of the *strain.*

The coefficients C_{ij} are constants which depend on the elastic properties of the solid body. There are 36 constants in the general case. However, it is proved in the theory of elasticity that only 21 are independent; the remaining 15 are fixed by the interrelationships $C_{ij} = C_{ji}$.

The smallest number of independent constants is necessary to characterize an isotropic homogeneous body, i.e., one whose elastic properties are identical in all directions and at all points. Two elastic constants describe such a body.

The generalized Hooke's law for homogeneous isotropic bodies may be written as follows:

$$\begin{aligned} \sigma_x &= \lambda e + 2\mu\epsilon_x \\ \sigma_y &= \lambda e + 2\mu\epsilon_y \\ \sigma_z &= \lambda e + 2\mu\epsilon_z \\ \tau_{yx} &= \mu\gamma_{xy} \\ \tau_{zx} &= \mu\gamma_{xz} \\ \tau_{zy} &= \mu\gamma_{yz} \end{aligned} \tag{I-1-2}$$

where λ and μ are elastic constants, called Lamé's constants.

More than two elastic constants characterize anisotropic bodies. For example, the elastic properties of a pyrite crystal are determined by three constants, those of fluorite and rock salt by six, those of barite and topaz by nine.

Equations (I-1-2) show directly that the elastic constant μ is the coefficient of proportionality between the stress and the shearing strain; therefore μ is called the modulus of elasticity in shear. This modulus is usually measured by observing the torsion of a sample or by the study of pure shear.

It is difficult to create simple experimental conditions for which the relationships between the stresses and the strains would be expressed only by λ. Therefore, the coefficient of proportionality between the tensile or compressive stress and the tensile or compressive strain is commonly used as the second elastic constant. Its magnitude is determined directly from the testing of prismatic or cylindrical bars subjected to elongation or compression. Assuming that elongation induced by a uniform stress σ occurs in the direction of the z axis, we obtain for the stress components

$$\sigma_x = 0 \qquad \sigma_y = 0 \qquad \sigma_z = \sigma$$

Since

$$\tau_{yx} = \tau_{zx} = \tau_{zy} = 0$$

we have

$$\gamma_{xy} = \gamma_{xz} = \gamma_{yz} = 0$$

Substituting the expressions for the normal stresses into the first three of Eqs. (I-1-2), we obtain

$$\begin{aligned} \lambda e + 2\mu\epsilon_x &= 0 \\ \lambda e + 2\mu\epsilon_y &= 0 \\ \lambda e + 2\mu\epsilon_z &= \sigma \end{aligned} \tag{I-1-3}$$

From these equations we find directly

$$\epsilon_z = \frac{\lambda + \mu}{\mu(3\lambda + 2\mu)} \sigma \tag{I-1-4}$$

$$\epsilon_x = \epsilon_y = - \frac{\lambda}{2\mu(3\lambda + 2\mu)} \sigma \tag{I-1-5}$$

The quantity

$$E = \frac{\mu(3\lambda + 2\mu)}{\lambda + \mu} \tag{I-1-6}$$

which defines the relationship between a tensile (or compressive) stress and the elongation (or compression) caused by this stress, is called the *normal modulus of elasticity* or *Young's modulus.*

Since the expressions for ϵ_x and ϵ_z have different signs, it follows that axial elongation of a bar will be accompanied by lateral contraction. From (I-1-4) and (I-1-5) we obtain

$$\nu = \frac{\epsilon_x}{\epsilon_z} = \frac{\epsilon_y}{\epsilon_z} = \frac{\lambda}{2(\lambda + \mu)} \tag{I-1-7}$$

Hence it follows that the ratio between the relative lateral contraction and the relative axial elongation does not depend on the shape of the cross section and is a constant for a given material.

The quantity ν is called Poisson's ratio and lies within the limits $0 < \nu < \frac{1}{2}$. Young's modulus and Poisson's ratio are the two principal quantities defining the elastic properties of materials. They are usually applied in engineering calculations.

Solving Eqs. (I-1-6) and (I-1-7) for λ and μ, we obtain their expressions in terms of Young's modulus E and Poisson's ratio ν.

$$\lambda = \frac{\nu E}{(1 + \nu)(1 - 2\nu)} \tag{I-1-8}$$

$$\mu = \frac{E}{2(1 + \nu)} = G \tag{I-1-9}$$

b. Applicability of Hooke's Law to Soil. Since Eqs. (I-1-2) are valid only for a homogeneous isotropic body in which no initial stresses are present, it is evident that Young's modulus and Poisson's ratio may be applied as elastic constants only to bodies satisfying these conditions. Therefore it is necessary to clarify whether it is possible to consider that soil satisfies the conditions of homogeneity and isotropy.

Some soils, such as sands, clays with sand, and clays, consist of particles of different materials surrounded by air or by a film of capillary water containing solutions of various salts and gases. The rigidity of separate particles is much higher than that of the soil as a whole. Therefore in the investigation of elastic (and total) deformations of soils, the particles composing the soil may be considered to be absolutely rigid.

The dimensions of the particles of the soil skeleton vary and range between 5 mm (gravelly soils) and several microns (i.e., thousandths of a millimeter). Suppose we have a small cubic element of soil, several millimeters on a side. It is clear that such a cube, if of clay, can be considered, with a large degree of precision, to be homogeneous. A cube of similar dimensions consisting of gravelly soil must be considered to be nonhomogeneous. However, if the sides of a cube of gravelly material measure several tens of centimeters, then it is evident that it may be assumed to consist of homogeneous material. Hence it follows that soil may be considered homogeneous only in such volume elements as have dimensions which are large in comparison with the dimensions of the soil particles. Therefore, by a stress or deformation at a selected point in the soil is meant an average stress over an area or a volumetric deformation whose dimensions (i.e., of both the area and the volume) are large in comparison with the dimensions of the soil particles. With these limitations, any discrete system, including soil, which consists of statistically uniform particles, may be considered a homogeneous body. The elastic properties of a volume element whose dimensions are large in comparison with the dimensions of soil particles will be constant throughout the element.

Soil particles are distributed more or less at random. Therefore the elastic properties of soil are the same in all directions, and soil may be considered an isotropic body. The only exception is presented by layered systems of soils in which separate layers are characterized by different properties. In such cases a soil may be considered to be isotropic only within the boundaries of a layer.

In addition to the forces of cohesion which are present to a lesser or greater degree in all solid bodies, soil is characterized by the presence of the forces of internal friction and capillary forces which induce the appearance of initial internal stresses.

The generalized Hooke's law, formulated in the general case by Eqs. (I-1-1), is based on the assumption that the stresses and deformations in the initial state (i.e., before an external load is applied) are equal to zero. In reality, initial stresses are present in all solid bodies. However, their action can be neglected if they are small in comparison with the induced stresses. Then Eqs. (I-1-2) for a homogeneous isotropic body are valid.

Sometimes the initial stresses in a body may attain magnitudes comparable with working stresses or even surpassing them. An example is the earth, the interior of which is characterized by considerable stresses caused by the mutual attraction of large masses. It is known that most severe earthquakes are caused by internal stresses in the crust of the earth. In such cases the use of Hooke's law in the form (I-1-2) without considering initial stresses may lead to large errors.

If considerable initial stresses are present, then initial deformations cannot be considered to be small, and the principle of superposition ceases to be valid in relation to deformations. Therefore it is not possible to apply conventional methods, which take the state before deformation as the unstressed state, when computing deformations.

Initial stresses in soil are induced by the action of friction, by capillary forces, and by processes leading to the swelling and consolidation of soils. In sandy soils, initial stresses result mostly from friction; in cohesive soils the initial state of stress may be influenced by a number of physicochemical processes, the natures of which have not as yet been sufficiently investigated.

The following method can be employed to detect the presence of initial stresses in sandy soil: a load is applied axially to a soil sample filling a vessel with perfectly elastic walls. The sample is unloaded and a special device is used to measure radial deformations of the vessel. The readings taken after the unloading differ from those taken before the loading. This shows that after the load which acted on the sand was removed, stresses developed in the soil and induced elastic radial deformations of the vessel. These initial stresses are explained by the wedging action of grains and can exist only if forces of friction are acting between sand particles.

The influence of initial stresses on elastic deformations of a soil may be found only by comparing with experimental data some deductions which follow from Eqs. (I-1-2).

Let us indicate some contradictions between theoretical conclusions based on Eqs. (I-1-2) and experimental data.

Equations (I-1-2) were derived on the basis of the assumption that in the absence of external forces the stresses equal zero, and thus deformations also equal zero. Figure I-2 presents typical shear-test results on sand (Fig. I-2*a*) and on an artificially prepared sample of clay (Fig. I-2*b*).

Figure I-3 shows a device which was used to investigate elastic shearing in sand. The device consists of two metal walls *1* which rotate freely around their supports. The sharpened horizontal edges of the walls rest on the upper and lower plates *2* and *3*. The load *4* was placed on the plate *3*. This load pressed on the walls and thus increased the stability of their sharp edges in grooves of the upper and lower plates.

The test sand *5* was placed between the two rigid walls when they were in a vertical position. Tamping was used to make the sand as dense as possible. The vertical edges of the walls were bent around a wooden block to prevent sand from trickling out.

The application of the load *6* induced a horizontal displacement of the upper plate and resulted in a distortion of the sample. These shear-test conditions are very close to those of pure shear.

Shear tests on clay samples were performed by applying a horizontal force to a metallic plate glued to the upper surface of the sample. The lower surface of the sample was glued to the base of the device. These shear-test conditions are also close to those of pure shear when the

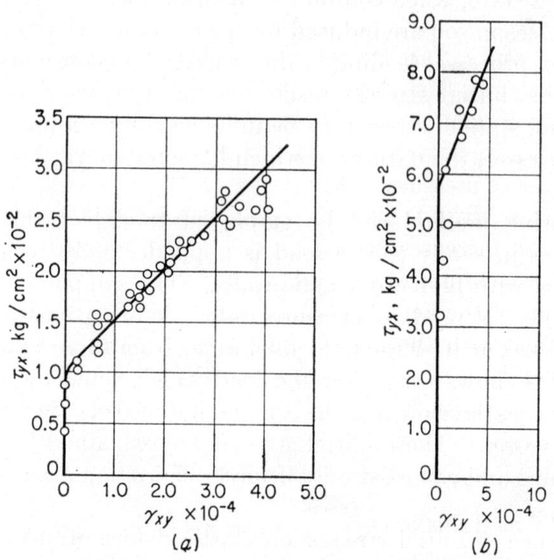

FIG. I-2. Shearing stress-strain relationships: (*a*) sand; (*b*) artificially compacted clay.

relationship between shear stresses and shear strains is determined by one of the last three of Eqs. (I-1-2).

The shear-test results as plotted in Fig. I-2 indicate that, within limits of test errors, there exists a linear relationship between shear stress and strain. However, the curves show that shearing deformations develop only in cases in which the stress attains a certain value. Apparently this value of the stress corresponds to the frictional force per unit area of the sand sample.

Shear-test results may be in agreement with Hooke's law if the influence of initial friction stresses acting between soil particles is taken into account in the last three of Eqs. (I-1-2). Then the relationship between the components of

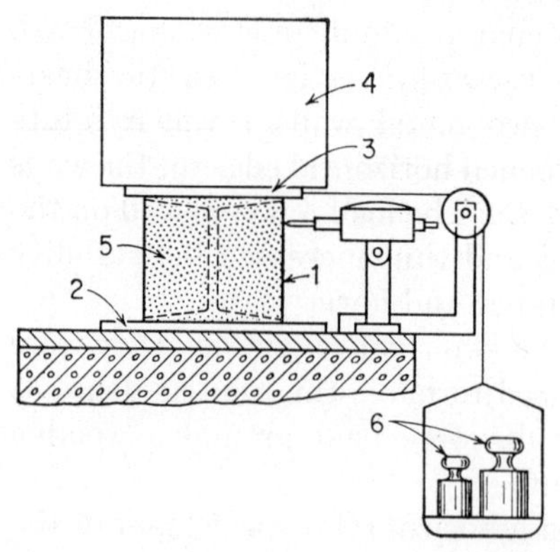

FIG. I-3. Device for testing sand in shear.

shear stress and the corresponding strains is determined by the equations

$$\begin{aligned} \tau_{yx} - \tau_{yx}{}^0 &= 2\mu\gamma_{xy} \\ \tau_{zx} - \tau_{zx}{}^0 &= 2\mu\gamma_{yz} \\ \tau_{zy} - \tau_{zy}{}^0 &= 2\mu\gamma_{xz} \end{aligned} \qquad \text{(I-1-10)}$$

where $\tau_{yx}{}^0$, $\tau_{zx}{}^0$, and $\tau_{zy}{}^0$ are the components of the initial shearing stresses.

It follows directly from Eqs. (I-1-2) that the shearing stresses do not depend on the normal stresses. Results of shear tests on a clay sample under different normal stresses show that the modulus of rigidity increases with an increase of pressure on the sample. The results of field investigations of soils (see Art. I-4) also show that the coefficient of elastic uniform shear of the soil increases proportionally to the normal pressure on the soil. Thus the modulus of rigidity is not a constant, but depends upon the magnitudes of the acting normal stresses.

In accordance with the above-mentioned experiments, let us assume that the dependence of the modulus of rigidity of soil μ on the normal components of stress is determined by the expression

$$\mu = \mu_0 - \beta(\sigma_x + \sigma_y + \sigma_z) \qquad \text{(I-1-11)}$$

where μ_0 is the modulus of rigidity of the soil in its initial state (when no external stresses are acting on it).

Substituting Eq. (I-1-11) into the right-hand parts of Eqs. (I-1-2), we obtain the following stress-strain relationships:

$$\begin{aligned} \sigma_x &= \lambda e + 2[\mu_0 - \beta(\sigma_x + \sigma_y + \sigma_z)]\epsilon_x \\ &\cdots\cdots\cdots\cdots\cdots\cdots \\ \tau_{xy} &= 2[\mu_0 - \beta(\sigma_x + \sigma_y + \sigma_z)]\gamma_{xy} \\ &\cdots\cdots\cdots\cdots\cdots\cdots \end{aligned} \qquad \text{(I-1-12)}$$

In these expressions, the stress components represent the differences between the stresses induced by the acting loads and the initial stresses. Equations (I-1-12) are not linear, although Eqs. (I-1-2) are.

Let us use Eqs. (I-1-12) to consider the problem of the compression of a cube along the z axis. Let $\sigma_z = -\sigma$, and

$$\sigma_x = \sigma_y = 0$$

Substituting these expressions for normal stresses into Eqs. (I-1-12), we obtain the following relations:

$$\begin{aligned} \lambda e + 2(\mu_0 + \beta\sigma)\epsilon_x &= 0 \\ \lambda e + 2(\mu_0 + \beta\sigma)\epsilon_y &= 0 \\ \lambda e + 2(\mu_0 + \beta\sigma)\epsilon_z &= -\sigma \end{aligned} \qquad \text{(I-1-13)}$$

Adding these equations,

$$[3\lambda + 2(\mu_0 + \beta\sigma)]e = -\sigma$$

Then

$$e = -\frac{\sigma}{3\lambda + 2(\mu_0 + \beta\sigma)}$$

Substituting this expression for e into the left-hand parts of Eqs. (I-1-13), we find

$$\epsilon_x = \epsilon_y = \frac{\lambda\sigma}{2[3\lambda + 2(\mu_0 + \beta\sigma)](\mu_0 + \beta\sigma)} \tag{I-1-14}$$

$$\epsilon_z = -\frac{\lambda + \mu_0 + \beta\sigma}{[3\lambda + 2(\mu_0 + \beta\sigma)](\mu_0 + \beta\sigma)}\sigma \tag{I-1-15}$$

The modulus of elasticity will be

$$E = \frac{\sigma}{\epsilon_z} = \frac{[3\lambda + 2(\mu_0 + \beta\sigma)](\mu_0 + \beta\sigma)}{(\lambda + \mu_0 + \beta\sigma)} = \alpha E_0 \tag{I-1-16}$$

where E_0 is Young's modulus determined by Eq. (I-1-6), and α is a coefficient which takes into account the influence of normal stresses on the shearing strength:

$$\alpha = \frac{[3\lambda + 2(\mu_0 + \beta\sigma)](\mu_0 + \beta\sigma)(\lambda + \mu_0)}{\mu_0(3\lambda + 2\mu_0)(\lambda + \mu_0 + \beta\sigma)} \tag{I-1-17}$$

Thus it is seen that the modulus of elasticity depends on the normal pressure σ. For small values of σ, the value of α is close to unity, and the value of E is close to the value of E_0. With an increase in pressure, the modulus of elasticity also increases.

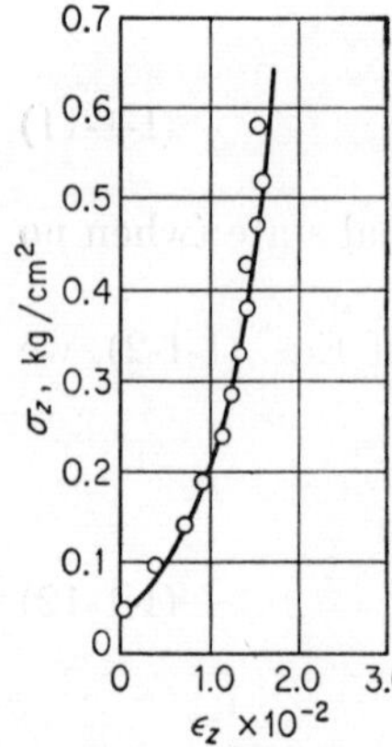

FIG. I-4. Relationship between compressive stresses σ_z and vertical elastic unit deformations ϵ_z of clay and silt cubes.

These conclusions are confirmed by compression tests on cubes of cohesive soils. The results of all tests on clays and silty clays with some sand reveal an increase in the rigidity of the soil with an increase of normal pressure. Figure I-4 presents a typical curve of interrelationship between compressive elastic deformations and normal pressure acting on the sample. Analysis of this curve leads to the conclusion that elastic properties of samples change considerably, even within the small range of pressures applied in testing.

Dividing (I-1-14) by (I-1-15), we obtain the following expression for Poisson's ratio:

$$\nu = \frac{\lambda}{2(\lambda + \mu_0 + \beta\sigma)} \tag{I-1-18}$$

This expression shows that the magnitude of Poisson's ratio is not constant. If the normal stresses (including their initial magnitudes) are small, then the value of ν is close to the value given by Eq. (I-1-7) for elastic bodies for which Hooke's law is valid. With an increase in the normal pressure, Poisson's ratio decreases in soils for which Eqs. (I-1-12) are valid.

In bodies for which the conditions expressed by Eqs. (I-1-12) are valid, the modulus of elasticity E in tension will decrease with an increase in the tensile stresses; at the same time Poisson's ratio ν will increase. Thus the hypothesis concerning the dependence of the modulus of shear rigidity on the normal stresses leads to the conclusion that the elastic properties of soil are different for compression and for tension.

The elastic deformations of a soil depend on the time period during which the load is applied. The influence of time is particularly marked in cohesive soils. It is known that settlements of structures erected on such soils continue for a long period of time, sometimes reaching decades. Similarly, elastic deformations do not disappear immediately after unloading, but do so over an extended period. This phenomenon of elastic aftereffect is observed in soils to a much greater degree than in other materials such as metals and concrete.

Owing to the influence of the elastic aftereffect, elastic deformations depend on the rate of load application. This is the reason why values of elastic constants found in some soils under conditions of slow changes in loads may differ from corresponding values obtained as a result of investigations of elastic dynamic processes (for example, in studies of wave propagation or natural frequencies of vibration of soil samples).

c. Young's Modulus and Poisson's Ratio for Soils. It was shown above that even when only elastic deformations of soil are considered, stress-strain relations in soil are much more complicated than those stipulated by Hooke's law. However, the assumption of a relationship more complicated than Hooke's law will lead to the necessity of employing a nonlinear theory of elasticity operating with nonlinear differential equations. The solution of these equations, even in the simplest problems, leads to considerable difficulties, and the application of relations analogous to Eqs. (I-1-12) becomes practically impossible. Thus it is necessary to restrict the analysis by the assumption that the soil strictly follows Hooke's law. Then linear differential equations may be used when considering the equilibrium or movement of soils, and solutions may be found for many problems. It should, however, be kept in mind that the numerical values of elastic soil constants should be selected with due consideration of the influence of the simplifying assumptions.

In computations of settlements of structures under the action of static nonrepetitional loads, only residual settlements are usually determined, although elastic settlements under the structures are also observed. However, elastic settlements are small in comparison with residual and are not taken into account in computations. Therefore when computing settlements or deformations resulting from the action of nonrepetitional loads, one does not employ elastic constants, but rather constants which determine the relationship between total deformations of soil (residual and elastic) and stresses acting therein. Thus instead of Young's modulus, the modulus of total deformation or the compressibility coefficient is used, sometimes determined from the first branch of the compression curve. The testing of strength properties of soils in field laboratories usually is limited to the determination of the above-mentioned soil properties. Up to the present time, the elastic constants of soils have seldom been investigated in soil laboratories, and so no reliable data are available concerning the interrelationship of these properties with the physicomechanical properties of soils (grain size, porosity, moisture content, etc.).

N. N. Ivanov and T. P. Ponomarev,[21,†] in their investigations of compression of clay cubes, did not find any lateral expansion. Hence they came to the conclusion that Poisson's ratio equals zero for these soils. K. Terzaghi[45] investigated the coefficient of lateral earth pressure in different soils by comparing the magnitudes of forces which had to be applied to steel bands, inserted in vertical and horizontal positions into a consolidometer, in order to remove them. As a result of these investigations he came to the conclusion that the coefficient of lateral earth pressure, i.e., the ratio σ_x/σ_z, has a constant value for each soil; for sands it equals 0.42, for clays 0.70 to 0.75. The coefficient of earth pressure is related to Poisson's ratio by the formula

$$K = \frac{\nu}{1 - \nu}$$

from which

$$\nu = \frac{K}{1 + K}$$

From the above-quoted values of the lateral-earth-pressure coefficient we obtain the values of Poisson's ratio: for sands, 0.3; for clays, 0.41–0.43. G. I. Pokrovsky and associates[34] investigated the coefficient of lateral earth pressure by means of a specially designed device. They found that the coefficient depends on the normal stress σ_z; with an increase in σ_z, the value of K increases, asymptotically approaching a value characterizing a certain soil. N. M. Gersevanov[17] showed that this dependence of K on the value of σ_z is explained by the influence of initial stresses acting in soils.

For samples of an artificially prepared clay which they studied, G. I. Pokrovsky and his associates obtained an asymptotic value of the lateral-

† Superscript numbers refer to the References at the end of the book.

earth-pressure coefficient equaling 0.62 to 0.65. Correspondingly, the value of Poisson's ratio was 0.38 to 0.40.

Ramspeck[36] determined values of Poisson's ratio on the basis of measurements of the rate of propagation of longitudinal and shear waves. For moist clay, he obtained a value of Poisson's ratio of 0.5, for loess, 0.44. For sandy soils, in three cases Poisson's ratio was found to equal 0.42 to 0.47; in one case, 0.31. Finally, N. A. Tsytovich[46] recommends for sandy soils $\nu = 0.15$ to 0.25; for clay with some sand and silt, 0.30 to 0.35; for clays, 0.35 to 0.40.

This author and R. Z. Katsenelenbogen investigated in the laboratory the effects of the moisture content of clay and of sand admixed thereto on Poisson's ratio, determined from data obtained from a sample tested both in compression and in a consolidometer. It was found that Poisson's ratio does not depend on the moisture content, but an increase in the sand admixture results in a decrease of Poisson's ratio. The average value of ν for pure clay was found to be 0.50; for clay with the admixture of 30 per cent sand it was 0.42.

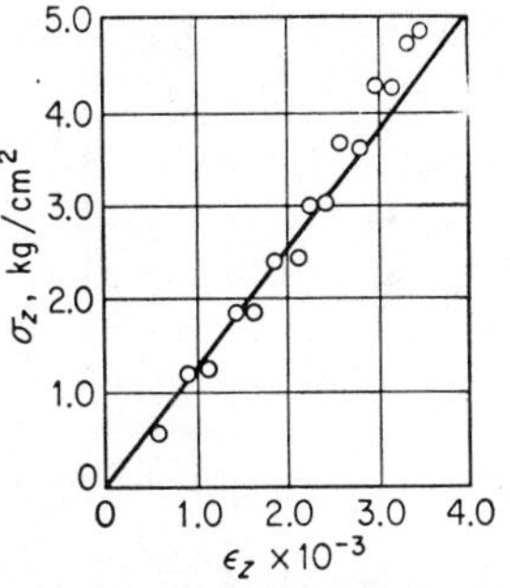

FIG. I-5. Relationship between compressive stresses σ_z and elastic unit deformations ϵ_z obtained from hysteresis loop of loading and unloading.

In spite of variations in the values of ν obtained by authors who employed different methods and investigated various soils, it can be held as an established fact that Poisson's ratio is smaller for sands than for clays. All experimental data show that the value of Poisson's ratio for clays lies close to 0.5. Most of the results show that for sands it is about 0.30 to 0.35.

The author and R. Z. Katsenelenbogen performed laboratory investigations of Young's modulus for artificially prepared soils. Since sand cannot be tested by means of cube compression, the sand samples were placed in a consolidometer. By means of repeated loading and unloading, elastic deformations corresponding to certain values of vertical pressure were found. On the basis of these test results, curves showing the interrelationship between elastic deformation and vertical pressure on a soil were plotted for each soil. Figure I-5 presents one of these characteristic diagrams. The slope coefficient of this curve is related to Young's modulus and Poisson's ratio by the following formula:

$$\alpha = \frac{1 - \nu}{(1 + \nu)(1 - 2\nu)} E \tag{I-1-18a}$$

Hence, knowing α and ν, it is easy to find Young's modulus. In accordance with the experimental data quoted above, we assumed ν for all sands investigated to be 0.35.

Figure I-6 presents values of Young's modulus for the Lyuberetsky sand, obtained as a result of investigations of samples with several different moisture contents by means of the method just described. It is seen from this figure that experimental values of E change only slightly with changes in moisture content, within the range of testing errors; hence it is possible to consider that Young's modulus for sand

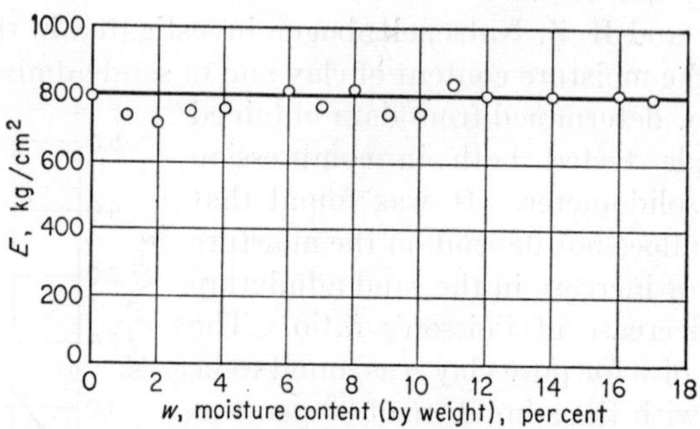

Fig. I-6. Young's modulus for sands with different water contents.

does not depend on moisture content. Analogous results were obtained for other sands.

Experiments also showed that the grain size of sand has comparatively minor influence on Young's modulus. Table I-1 presents values of Young's modulus obtained for sands with different grain sizes.

Table I-1. Values of Young's Modulus for Different Grain Sizes of Sand

Grain size of sand, mm	*E, kg/cm²*
1.25–1.50	450
1.00–1.25	520
0.60–0.80	620
0.35–0.60	480
0.30–0.35	480
0.20–0.30	620

Let us mention finally that, contrary to the opinion expressed in publications,[46] the experiments established that Young's modulus for sand does not depend much on the porosity of the sand.

Thus the results of experiments lead to the conclusion that Young's modulus for a pure sand containing no silt or clay admixtures is a stable characteristic which does not change much with changes in the moisture content, grain size, or porosity of the sand.

However, the Young modulus of a cohesive soil depends to a large degree on the properties mentioned above. Figure I-7 presents values of Young's modulus obtained as a result of tests on artificially prepared clay samples with different moisture contents. The diagram shows that for a clay with moisture content close to 30 per cent, Young's modulus is very small. This clay had the following characteristics: plastic limit = 16.16 per cent; liquid limit = 29.13 per cent; void ratio = 0.70; molecular

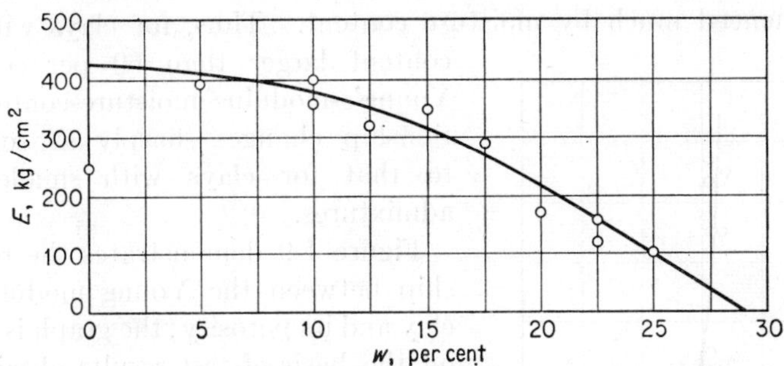

FIG. I-7. Young's modulus for a clay with varying water contents.

water absorption capacity = 13.32. The values of Young's modulus corresponding to certain values of the moisture content were obtained from compression tests of clay cubes.

Figure I-7 also shows the Young's-modulus–moisture-content relationship expressed by the empirical formula

$$E = E_0\left(1 - \frac{w^2}{w_0^2}\right) \tag{I-1-19}$$

where E_0 is the value of Young's modulus for the clay sample with zero moisture content and w_0 is the moisture content of the clay for which Young's modulus is very small (theoretically equal to zero).

If E_0 = 430 kg/cm² and w_0 = 29 per cent, the graph plotted in Fig. I-7 fully agrees with values of E established experimentally.

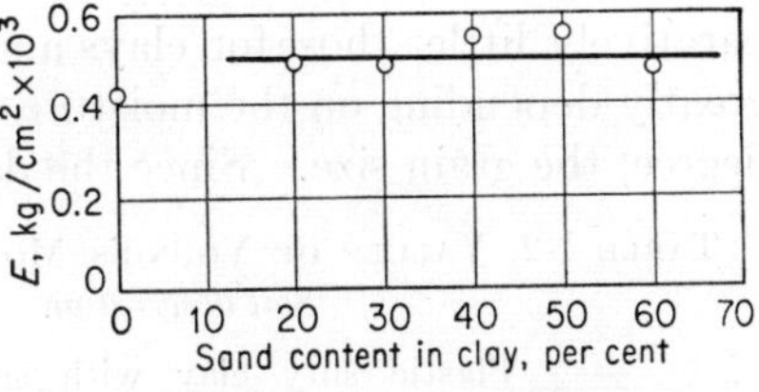

FIG. I-8. Young's modulus for a compacted clay with varying sand contents.

The experiments showed that, within a certain range, the addition of Lyuberetsky white quartz sand to the clay does not induce large changes in the values of E. This is demonstrated by Fig. I-8, giving values of E obtained for clay samples containing differing amounts of sand.

The limit value w_0 of the moisture content depends on the sand admixed to the clay. Experiments established that in soil samples containing less than 50 per cent sand, w_0 decreases with an increase in the sand content. For example, if a pure clay is characterized by $w_0 = 29$ per cent, then clay containing 20 per cent sand is characterized by $w_0 = 24$ per cent, clay containing 30 to 40 per cent sand has $w_0 = 20$ per cent, and finally, in clay containing 50 to 60 per cent sand the ultimate moisture content decreases to 14 to 15 per cent.

The Young modulus for clays containing more than 60 per cent sand is not influenced much by moisture content. Thus, for clays with sand content larger than 60 per cent, the Young's-modulus–moisture-content relationship changes sharply as compared to that for clays with smaller sand admixtures.

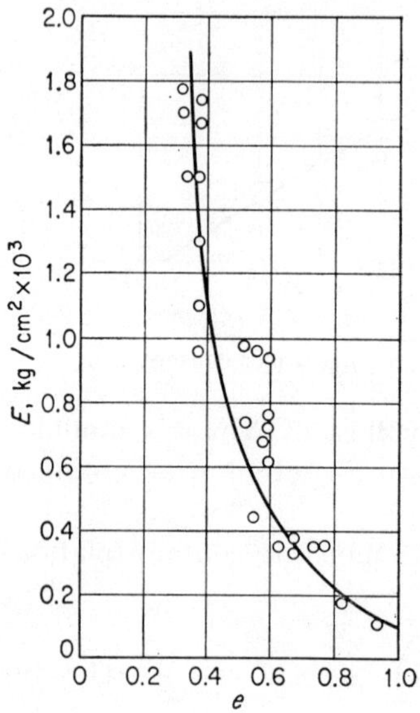

FIG. I-9. Young's modulus for a clay with varying void ratios e.

Figure I-9 demonstrates the relationship between the Young modulus of a clay and its porosity; the graph is plotted on the basis of test results obtained by the author and R. Z. Katzenelenbogen. Values of the void ratio are plotted along the horizontal axis, values of E along the vertical axis. While the void ratio changed from 0.32 to 0.93, the value of Young's modulus decreased from 1,700 to 120 kg/cm², i.e., fourteenfold.

On the basis of these data, it is possible to conclude that Young's modulus for clayey soils depends to a large degree on physicomechanical properties. While the values of E for sands change comparatively little, those for clays and clays with silt and sand may change greatly depending on the moisture content, the void ratio, and, to a lesser degree, the grain size. Since the data now available are not sufficient for

TABLE I-2. VALUES OF YOUNG'S MODULUS FOR DIFFERENT TYPES OF SOILS

Soil description	*E, kg/cm²*
Plastic silty clay with sand and organic silt	310
Brown saturated silty clay with sand	440
Dense silty clay with some sand	2,950
Medium moist sand	540
Gray sand with gravel	540
Fine saturated sand	850
Medium sand	830
Loess	1,000–1,300
Loessial soil	1,200

generalizations concerning the effects of these characteristics on Young's modulus, it is not possible to recommend any values of E for cohesive soils for use in design computations. It is necessary to determine the design values of E in each case by direct testing of undisturbed soil taken from the construction site.

In conclusion, Table I-2 is presented, giving Young's modulus computed from the coefficients of elastic uniform compression for different soils; the values of E were found as a result of field testing (see Art. I-2) and from laboratory investigations.

I-2. Coefficient of Elastic Uniform Compression of Soil c_u

a. The Coefficient of Subgrade Reaction. To determine the bearing capacity of a soil, load tests are used during which a concentrated load is transferred to the soil through a rigid bearing plate. Except for special detailed soil investigations, plates not larger than 5,000 cm² are used.

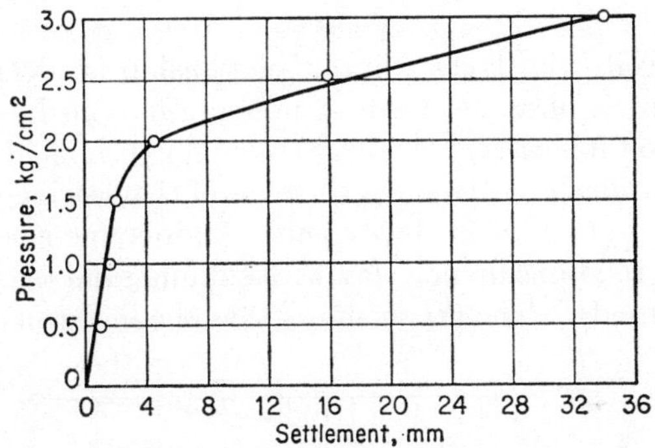

FIG. I-10. Results of load test of a 1.0-m² plate on sand.

By increasing in steps the load formed by the weight of the bearing plate and the whole installation, and by measuring the settlement corresponding to each load increment, one may derive data for a curve showing the dependence of plate settlement on the magnitude of external pressure on the soil.

The results of load tests make it possible to draw some general conclusions concerning the relationship between settlement and load.

Figure I-10 gives the relationship between vertical normal pressures and settlements obtained from load tests on very moist medium sand (a rigid bearing plate of 1 m² area was used). Figure I-11 shows the curve obtained from tests on loessial soils; a plate of 8 m² area was used.

Two distinct sections are observed in each of the graphs presented. The first section reveals a linear settlement-pressure relationship. For medium sand, this relationship is observed for pressures smaller than

1.5 kg/cm² (Fig. I-10). The second section, which corresponds to higher pressures, is characterized by a nonlinear settlement-pressure relationship expressed by a convex curve. In this section, settlement increases at a greater rate than pressure.

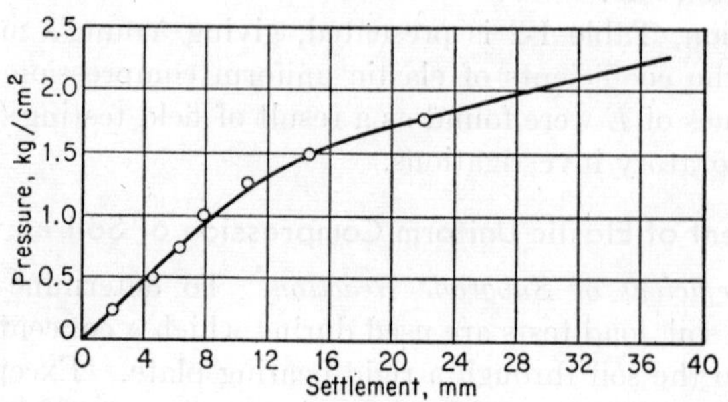

FIG. I-11. Results of load test of a 8.0-m² plate on loess.

For loessial soil (Fig. I-11) a linear relationship between pressure and plate settlement is observed up to a pressure of 1.25 kg/cm²; then the nonlinear section follows.

A similar relationship between settlement of the bearing plate and pressure on the soil is observed in clayey soils. This is confirmed by Fig. I-12, showing such a relationship in a clay with silt and sand (a bearing plate of 8 m² area was used). The graph shows that the linear plate-settlement–

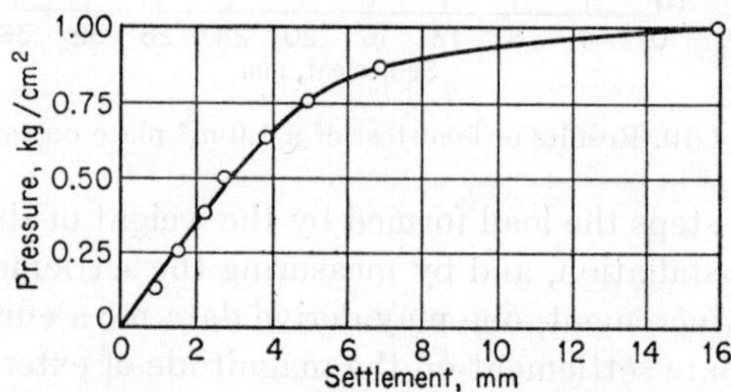

FIG. I-12. Results of load test of an 8.0-m² plate on clay.

load relationship exists up to 0.75 kg/cm²; then settlement increases at a greater rate than pressure.

This characteristic settlement-pressure relationship may be generalized for various soils. A proportionality limit may be established, below which the pressure-settlement relationship is linear; i.e., below a certain point,

$$p_z = c_p S_t \qquad \text{(I-2-1)}$$

where p_z = normal pressure on soil
c_p = coefficient of proportionality, also called modulus of subgrade reaction
S_t = total settlement of bearing plate resulting from external pressure

The proportionality limit is a criterion in the selection of the permissible bearing value of soil under static load, often taken as being equal to or somewhat lower than the proportionality limit.

A question arises as to whether the proportionality limit is a constant physical property of a soil or whether it depends on such test conditions as the bearing-plate area or shape. The experimental data available are not sufficient to allow a fairly positive answer to this question. The proportionality limit apparently does not depend on the area of the bearing plate used in the investigations and *apparently* depends only on soil properties. This has been confirmed, for example, by D. E. Polshin's tests performed on loessial soils. He studied settlements of six square plates having areas of 0.25, 0.50, 1.0, 2.0, 4.0, and 8.0 m². The experiments did not reveal any definite relationship between the bearing-plate area and the proportionality limit. This leads to an *assumption* that the proportionality limit is a characteristic constant property for a given soil.

If the validity of the law of linear relationship between bearing-plate settlement and pressure thereon is confirmed by experiment, then, in the range of pressures smaller than the proportionality limit, there exists also a linear relationship between deformation at any point in the soil and stress induced at that point by external forces acting on a limited area of the soil surface. Therefore, for stresses not exceeding a certain limit, the soil is considered to be a linearly deformable body. The theory of linearly deformable bodies is applied in studies of deformations and stress conditions in soil.

b. Coefficient of Elastic Uniform Compression c_u. Figure I-13 presents the graph of the results of tests performed by the author and associates, who investigated the settlement of a 1.4-m² bearing plate on a loessial soil under conditions of repeated loading and unloading. The maximum stress transferred to the soil was increased in each loop. In the first loop, maximum stress reached only 0.34 kg/cm²; i.e., it was much smaller than the proportionality limit, which in loess is approximately equal to 1.25 to 1.5 kg/cm². In spite of the relatively small value of maximum stress in this loop, residual settlement was about 90 per cent of total settlement. If one considers that the elastic limit corresponds to the magnitude of stress at which residual settlement forms only a negligible part (around 2 to 3 per cent) of total settlement, then, in the case concerned, the elastic limit of the soil evidently is much smaller than the proportionality limit. This conclusion apparently may also be applied to other soils. The experimental results are not sufficient to permit conclusions concerning numerical values of the elastic limit for various soils. Probably for most soils (not for preloaded soils and not for rocks) this limit equals only several tenths of a kilogram per square centimeter; i.e., it is much lower than the usually permissible working pressures on soil. Thus, for small pressures only, the assumption is possible that soils of the clay and sand types satisfy the condition of reversibility of deformations.

A graph of the footing settlement-pressure relationship under conditions of repeated loading and unloading permits easy separation of the

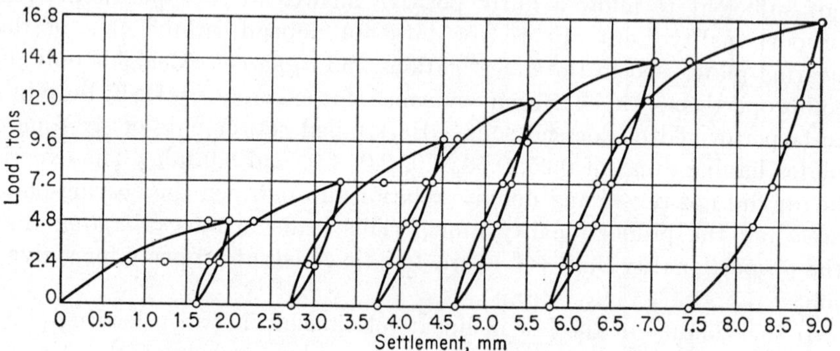

FIG. I-13. Results of load test of a 1.4-m² plate on loess.

elastic part of the deformation from the total settlement. Figure I-14 indicates the result of an analysis performed on Fig. I-13. The circles indicate the magnitudes of elastic settlement for the maximum pressure in each loop. The graph shows that within a range of pressures up to 1.0 kg/cm², elastic settlements change proportionally to pressure.

Figure I-15 presents graphs of elastic settlements obtained by the author for 1,600- and 3,600-cm² bearing plates on medium sand of natural moisture content. Similar graphs showing interrelationships between settlement and pressure on the plate were obtained for other soils. On the basis of these investigations, the statement can be made that within a certain range there is a proportional relationship between elastic settlements of footings and external uniform pressure on the soil; i.e.,

$$p_z = c_u S_e \tag{I-2-2}$$

where c_u is the coefficient of proportionality, called the coefficient of elastic uniform compression of soil, and S_e is the elastic settlement of the bearing plate due to the external pressure.

The coefficient of elastic uniform compression c_u of a soil differs from the coefficient of proportionality c_p [in Eq. (I-2-1)], but the difference is not often taken into consideration. While c_p relates the *total settlement* to the external pressure, c_u relates *only the elastic part of the total settlement* to the pressure. Since the total settlement of a foundation is always larger than its elastic settlement, c_u is larger than c_p in all soils, without exception.

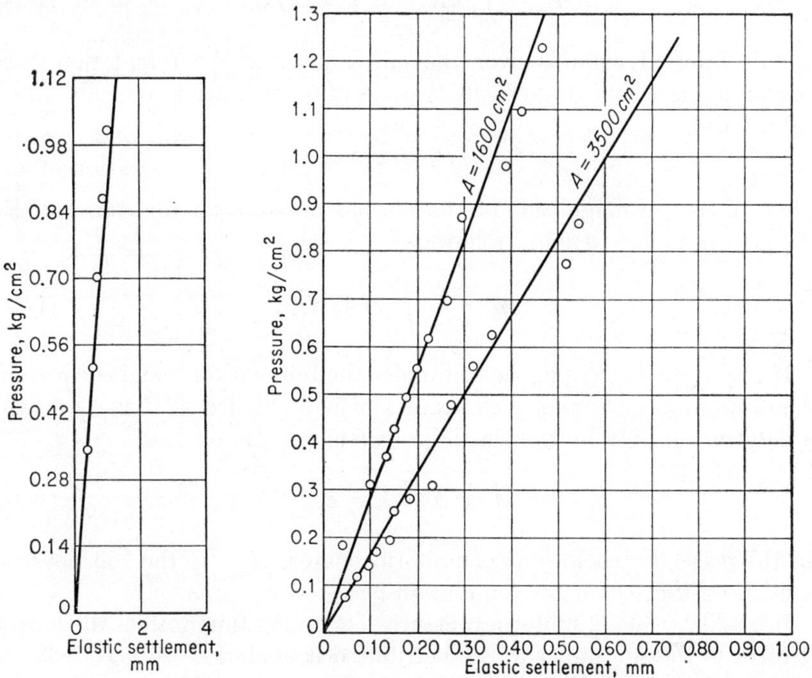

FIG. I-14. Evaluation of the hysteresis loops of Fig. I-13.

FIG. I-15. Evaluation of hysteresis loops obtained during load tests on footings of two sizes.

c. The Effect of the Area and Shape of a Foundation Base on the Coefficient of Elastic Uniform Compression. The coefficient of elastic uniform compression and the coefficient of proportionality would be constant values (characteristic of certain soils and independent of test conditions and area and shape of the foundation base) if the stresses in the soil under a uniformly loaded foundation remained constant at any point. Otherwise, both coefficients would depend on the area of the footing base and its shape.

If dR is the soil reaction on the element dA of the foundation base contact area, then, assuming that at any point directly under dA the settlement S of the soil is proportional to the stress, we obtain

$$dR = c(A)S\,dA$$

where $c(A)$ is a variable coefficient of proportionality whose magnitude depends on the elastic constants of the soil and the coordinates of the element of area dA. The total reaction of the soil equals the load P_z imposed on it. Therefore,

$$P_z = \int_A dR = S \int_A c(A)\, dA \qquad \text{(I-2-3)}$$

This integral extends over the entire area of the foundation base. Settlement is assumed to be constant, and the foundation to be absolutely rigid. But

$$P_z = p_z A$$

where p_z is the magnitude of the normal pressure on the soil, and Eq. (I-2-3) can be rewritten as follows:

$$p_z = \frac{S}{A} \int_A c(A)\, dA \qquad \text{(I-2-4)}$$

If the normal stress in the soil under the foundation remains constant, then $c(A)$ will also remain constant and will equal c_u. Hence its value is determined only by the elastic properties of the soil, and

$$\int_A c(A)\, dA = c_u A \qquad \text{(I-2-5)}$$

In this case the coefficient of uniform compression of the soil does not depend on the size of the foundation base contact area.

In reality, when a uniform pressure acts on the foundation, the normal stresses in the soil under it are distributed irregularly.

Since machinery foundations consist of either a rigid block placed directly on the soil or a framed elastic system supported by a rigid plate placed on the soil, the problem of distribution of stresses under the base contact area of a rigid plate which has undergone uniform settlement of a certain magnitude is of great interest.

Sadovsky[38] gave a solution to this problem for a circular base contact area of a rigid plate. For soil directly under the base contact area the expression for $c(A)$ may be written in the form

$$c(A) = \frac{E}{1 - \nu^2} \frac{1}{\pi \sqrt{R^2 - \rho^2}} \qquad \text{(I-2-6)}$$

where E = Young's modulus of soil
ν = Poisson's ratio
R = radius of plate
ρ = radius vector of point under consideration in soil under bearing plate ($\rho < R$)

Equation (I-2-6) shows that the stress under the base contact area of a rigid bearing plate is smallest at the center of the plate. The stresses at other points grow proportionally with distance from the center; the maximum value is reached near the edges of the foundation, where $\rho = R$. Theoretically this maximum equals infinity. However, no actual material can stand infinitely large stresses, and the soil at the foundation edges will undergo plastic deformations; the stresses in these zones will be of finite magnitudes, although they will be much larger than the stresses at points near the bearing-plate center.

Multiplying both parts of Eq. (I-2-6) by the area element dA and integrating over the entire foundation base contact area, we obtain, according to Eq. (I-2-4),

$$
\begin{aligned}
p_z &= S \frac{E}{1-\nu^2} \frac{1}{A\pi} \int_A \frac{\rho \, d\rho \, d\varphi}{\sqrt{R^2 - \rho^2}} \\
&= S \frac{E}{1-\nu^2} \frac{1}{A} \int_0^{2\pi} d\varphi \int_0^R \frac{\rho \, d\rho}{\sqrt{R^2 - \rho^2}} \qquad \text{(I-2-7)}
\end{aligned}
$$

By integration we find:

$$
p_z = 1.13 \frac{E}{1-\nu^2} \frac{S}{\sqrt{A}} \qquad \text{(I-2-8)}
$$

Comparing the right-hand parts of Eqs. (I-2-2) and (I-2-8), we find the following expression for c_u:

$$
c_u = 1.13 \frac{E}{1-\nu^2} \frac{1}{\sqrt{A}} \qquad \text{(I-2-9)}
$$

Thus the equation for the coefficient of elastic uniform compression of a soil, obtained from the solution of the theory-of-elasticity problem concerning the distribution of normal stresses in the soil under the base contact area of a rigid plate, leads to the conclusion that if the settlement of the foundation is uniform, the stresses under its base are not distributed uniformly. The coefficient of elastic compression c_u (or, if the solution is given for a proportionally deformed soil, the coefficient of proportionality c_p) depends not only on the elastic constants E and ν, which define the elastic or linear properties of a soil, but also on the size of the base contact area of the foundation. The coefficient c_u changes in inverse proportion to the square root of the base area of the foundation.

If the foundation consists of an absolutely flexible plate uniformly loaded by a vertical pressure, then stresses in the soil under the foundation will be distributed uniformly, but settlement under the foundation will vary. For an absolutely flexible foundation, the coefficient of elastic uniform compression is the ratio of the uniform pressure to the average value of settlement

$$c_u = \frac{p_z}{S_{av}} \tag{I-2-10}$$

Schleicher[39] gave a solution for an absolutely flexible footing with rectangular base. According to his solution, the average settlement value is determined by the equation

$$S_{av} = \frac{1-\nu^2}{E}\sqrt{A}\,\frac{1}{\pi\sqrt{\alpha}}\left[\ln\frac{\sqrt{1+\alpha^2}+\alpha}{\sqrt{1+\alpha^2}-\alpha} + \alpha\ln\frac{\sqrt{1+\alpha^2}+1}{\sqrt{1+\alpha^2}-1} - \frac{2}{3}\,\frac{(1+\alpha^2)^{3/2}-(1+\alpha^3)}{\alpha}\right]p_z \tag{I-2-11}$$

where $\alpha = 2a/2b$

$2a$, $2b$ = length, width of foundation

From Eq. (I-2-2) we have:

$$S = p_z c_u$$

Assuming $S = S_{av}$, we obtain for c_u:

$$c_u = \frac{c_s C}{A} \tag{I-2-12}$$

where

$$C = \frac{E}{1-\nu^2} \tag{I-2-13}$$

c_s is a coefficient which depends only on the ratio α between the length and the width of the foundation base. It is equal to

$$c_s = \frac{\pi\sqrt{\alpha}}{\ln\dfrac{\sqrt{1+\alpha^2}+\alpha}{\sqrt{1+\alpha^2}-\alpha} + \alpha\ln\dfrac{\sqrt{1+\alpha^2}+1}{\sqrt{1+\alpha^2}-1} - \frac{2}{3}\left[(1+\alpha^2)^{3/2} - \dfrac{1+\alpha^3}{\alpha}\right]} \tag{I-2-14}$$

Table I-3 gives values of c_s for various values of α, computed from Eq. (I-2-14).

Table I-3. Auxiliary Values for the Computation of Settlements of Rigid and Flexible Bearing Plates

α	c_s	c_s'
1	1.06	1.08
1.5	1.07	
2	1.09	1.10
3	1.13	1.15
5	1.22	1.24
10	1.41	1.41

The same table gives also values of c_s' computed by M. I. Gorbunov-Posadov[20] for an absolutely rigid rectangular foundation. The difference between c_s and c_s' is not larger than 3 per cent. Hence, there will be little difference between the values of the coefficient of elastic uniform compression as computed for absolutely rigid and absolutely flexible rectangular plates. Table I-3 indicates that if the base area of a foundation remains constant, the coefficient of elastic uniform compression increases with an increase in α. Foundations for machinery are seldom constructed with a large ratio between the side lengths; in most cases

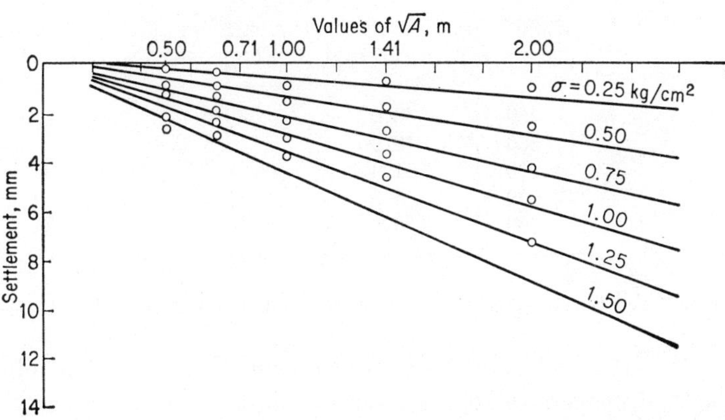

Fig. I-16. Dependence of settlements on the loaded area under different pressures.

it does not exceed 3. Hence, it can be held that, in practice, the coefficient of elastic uniform compression for a machine foundation does not depend on α.

The foregoing conclusions concerning the effect of the area of the foundation base on the settlement and on the coefficient of elastic uniform compression will not change if we assume that the soil under the plate behaves, not as an elastic body, but as a proportionally deformable body.

d. Experimental Investigations of the Coefficient of Elastic Uniform Compression. The conclusions of the theory discussed above are fully confirmed by experimental investigations performed within the range of pressures smaller than the proportionality limit; then, residual settlements appear in addition to elastic settlements, and the soil does not behave like an elastic body. Figure I-16 presents graphs obtained by D. E. Polshin which show the settlement of rigid bearing plates on loess plotted against the areas of the plates. The square roots of the bearing-plate areas are plotted as abscissas, the settlements as ordinates. The test footings were square. Within the range of test errors, the relationship between settlement and the square root of base area may be considered to be directly proportional, which fully agrees with theoretical conclusions.

A number of tests on various soils give us a chance to verify the validity of the theoretical conclusion that the coefficient of elastic compression increases in inverse proportion to the square root of bearing-

TABLE I-4. COMPUTED AND MEASURED VALUES OF THE COEFFICIENT OF ELASTIC UNIFORM COMPRESSION c_u OF A MEDIUM SAND FOR DIFFERENT AREAS OF LOADING

A, cm^2	c_u, kg/cm^3	
	Experimental	Computed
200	30.0	30.0
500	18.0	18.8
1,000	11.4	13.4
3,000	8.9	7.7
7,500	6.6	4.9

plate area. All tests were performed on rigid bearing plates with varying areas.

Gerner[16] performed laboratory tests on medium sands and obtained values of c_u for several plate areas (see Table I-4).

Figure I-17 gives curves, plotted on the basis of the data of Table I-4, which show changes in c_u plotted against changes in plate area. The

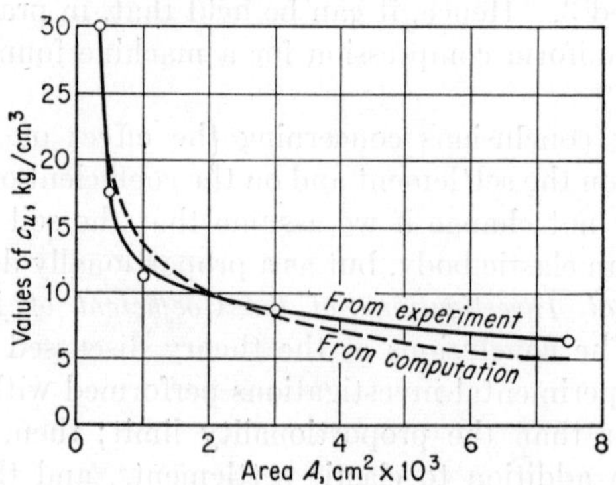

FIG. I-17. Dependence of the coefficient of elastic uniform compression of soil c_u on the loaded area A.

values of c_u were obtained from experiments and from theoretical considerations. The latter values were computed as follows: according to Eq. (I-2-12), the coefficients of elastic compression c_{u1} and c_{u2} for two foundations with different areas A_1 and A_2 are related by the formula

$$c_{u2} = c_{u1}\sqrt{\frac{A_1}{A_2}} \tag{I-2-15}$$

As a result of field tests performed by the author and the late Ya. N. Smolikov on gray soft saturated clays with sand and silt, the following values of c_u were obtained for foundations with areas of 1.5, 1.0, and 0.5 m²: 2.1, 2.52, and 3.50 kg/cm³. Inserting the values $A_1 = 0.5$ m² and $c_{u1} = 3.5$ kg/cm³, we obtain from Eq. (I-2-15) the following values of c_u for the other two foundations investigated: 2.45 kg/cm³ for $A = 1.0$ m², and 2.02 kg/cm³ for $A = 1.5$ m².

Table I-5 gives values of c_u for foundations with three different base areas obtained from tests performed by the author and A. I. Mikhalchuk on saturated brown clays with silt and sand. When the values presented in this table were computed, a value of c_{u1} equal to 4.4 kg/cm³ and corresponding to $A = 2.0$ m² was inserted in Eq. (I-2-15). The computed values of c_u for areas of 4 and 8 m² were larger than those found experimentally. This difference was apparently caused by the diversity of soil properties at the site of field investigations.

Table I-5. Computed and Measured Values of the Coefficient of Elastic Uniform Compression c_u of a Saturated Clay for Different Areas of Loading

A, m²	c_u, kg/cm³	
	Experimental	Computed
8	2.05	2.2
4	2.5	3.12
2	4.4	4.4

The author, P. A. Saichev, and Ya. N. Smolikov performed tests on loess at its natural moisture content, using foundations with square bases of different sizes. The results of these investigations are presented in Table I-6.

All the investigations confirm that the coefficient of elastic uniform compression of a soil decreases with an increase in the area of the foundation base. The same experiments indicate, however, that values of c_u computed from Eq. (I-2-15) decrease at a higher rate than those established experimentally. Experimental studies of forced and free vibrations of actual foundations with bases attaining 100 m² area indicate that experimentally established values of c_u are much larger than those obtained from computations in which values of c_u, found by testing a foundation of small area, were adjusted to a larger area. The author studied free vibrations of the foundation for a horizontal compressor. The foundation was erected on loessial soil and had a base area of 90 m². From the results of the studies the coefficient of elastic uniform compression of the soil was found to be 4.7 kg/cm³. Another foundation had a base area of approximately 1.5 m² and an experimentally determined c_u of 10.8 kg/cm³. Using Eq. (I-2-15) and this data, we can find the value of c_u for $A_2 = 90$ m².

$$c_u = 10.8 \frac{1.5}{90} = 1.4 \text{ kg/cm}^3$$

This computed value of c_u is almost 3.6 times smaller than that established experimentally.

Similar results were obtained in other studies of actual machinery foundations. They indicate that, for foundations with large base areas,

TABLE I-6. COMPUTED AND MEASURED VALUES OF THE COEFFICIENT OF ELASTIC UNIFORM COMPRESSION c_u OF LOESS FOR DIFFERENT AREAS OF LOADING

A, m²	c_u, kg/cm³	
	Experimental	Computed
0.81	14.2	14.2
1.4	10.8	10.6
2.0	10.2	9.0
4.0	8.0	6.4

the coefficient c_u changes at a much lower rate than is indicated by Eq. (I-2-15).

This disagreement between theory and experimental data can be explained by the apparent dependence of the elastic constants of a soil, particularly the modulus of rigidity, on the normal vertical pressure, as discussed in Art. I-1. The properties of a uniform soil change with depth. With an increase in base contact area, a greater depth of soil is affected by the weight of the foundation, and the influence of deeper soil layers on foundation settlement increases.

e. Values of the Coefficient of Elastic Uniform Compression for Soils Used in Design Computations. The true values of the coefficients of elastic uniform compression, just as do Young's modulus and Poisson's ratio, depend on a number of factors whose influence in each separate case is very hard to evaluate. Therefore, in construction practice involving the erection of a sufficiently large number of foundations for machinery subjected to dynamic loads, special investigations of the elastic properties of soils are necessary on the construction site. Sometimes these investigations cannot be carried out. Therefore it often happens that designers of machinery foundations assume tentative values of the coefficients of elastic uniform compression selected on the basis of data secured from tests performed on similar soils at other construction sites.

TABLE I-7. SUMMARY OF AVERAGE VALUES OF THE COEFFICIENT OF ELASTIC UNIFORM COMPRESSION c_u OF DIFFERENT SOILS OBTAINED UNDER VARYING CONDITIONS

Description of soil	Tentative value of permissible load on soil, kg/cm²	Coefficient of elastic uniform compression c_u, kg/cm³	Contact areas of foundation bases, m²; types of tests (D = dynamic; S = static)
Gray plastic silty clay with sand and organic silt	1.0	1.4	1.5, 1.0, 0.5 (S; D)
Brown saturated silty clay with sand	1.5	2.0	8.0, 4.0, 2.0 (S; D)
Dense silty clay with some sand (above ground-water level)	Up to 5	10.7	8.9 (D)
Medium moist sand	2	2.0	1.5 (D)
Dry sand with gravel	2	2.0	0.25 (D)
Fine saturated sand	2.5	3.0–3.5	11.6 (D)
Medium sand	2.5	3.1	8.75 (D)
Gray fine dense saturated sand	2.5	3.4	up to 15 (S; D)
Loess with natural moisture content	3	4.5	0.81, 1.4, 2.0, 4.0 (S; D)
Moist loessial soil	3	4.7	90 (a foundation for compressor) (D)

Table I-7 gives a number of values of the coefficient of elastic uniform compression c_u established by special tests carried out under the supervision of the author. These tests dealt with foundations having several base areas built on various soils.

The values of the coefficient of elastic uniform compression, obtained by testing foundations with base areas smaller than 10 m², were later recalculated by means of Eq. (I-2-15) for an area of 10 m².

From the data of Table I-7, the author compiled Table I-8, which gives tentative values of c_u for four types of soil.

This table is included in the official *Instructions for the Design and Construction of Machinery Foundations.*

TABLE I-8. RECOMMENDED DESIGN VALUES OF THE COEFFICIENT OF ELASTIC UNIFORM COMPRESSION c_u

Soil group category	Soil group	Permissible load on soil under action of static load only, kg/cm²	Coefficient of elastic uniform compression c_u, kg/cm³
I	Weak soils (clays and silty clays with sand, in a plastic state; clayey and silty sands; also soils of categories II and III with laminae of organic silt and of peat)........................	Up to 1.5	Up to 3
II	Soils of medium strength (clays and silty clays with sand, close to the plastic limit; sand)...............	1.5–3.5	3–5
III	Strong soils (clays and silty clays with sand, of hard consistency; gravels and gravelly sands; loess and loessial soils)...........................	3.5–5	5–10
IV	Rocks..................................	greater than 5	greater than 10

I-3. Coefficient of Elastic Nonuniform Compression of Soil c_φ

Let us consider the bending of a flexible rectangular plate with sides of length $2a$ and $2b$ acted upon by the moment M, which bends the plate with respect to the y axis (Fig. I-18). Using the equations for settlement at any point produced by stresses in an element of plate area situated at a certain distance from the point, and applying the method of superposition of settlements induced by several loads, D. E. Polshin[35] derived an equation for the slope angle φ of the deflected plate at any point of its surface and at points along the straight line $x = 0$. Since the plate is assumed to be absolutely flexible, the slope angle of the deflected plate at points along this straight line depends on the y coordinate.

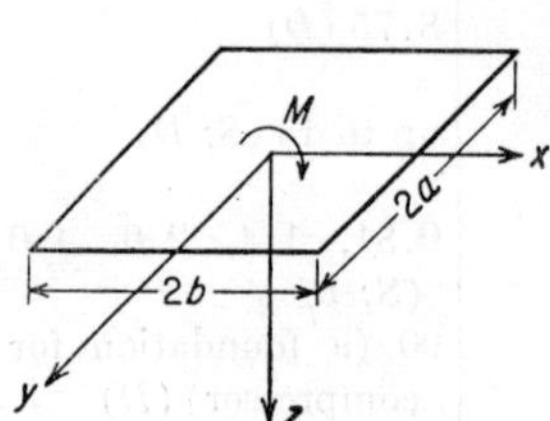

FIG. I-18. Induced tilting of a loaded plate causes nonuniform compression of soil.

Polshin gives the following expression for the angle φ depending on the value of y:

$$\varphi(y) = K\left[(a+y)\ln\frac{\sqrt{(a+y)^2+b^2}+b}{\sqrt{(a+y)^2+b^2}-b} + (a-y)\ln\frac{\sqrt{(a-y)^2+b^2}+b}{\sqrt{(a-y)^2+b^2}-b}\right] \qquad \text{(I-3-1)}$$

where

$$K = \frac{1-\nu^2}{\pi E}\frac{M}{I}$$

M is the external moment acting on the plate in the zx plane and I is the moment of inertia of the plate with respect to the y axis.

From Eq. (I-3-1), the author and B. M. Terenin derived an equation which expresses the effect of both the elastic properties of soil and the size of the foundation area on the coefficient of elasticity of the soil as a result of rotation of the foundation base by the angle φ. It was assumed that the slope angle φ of the tilted base of the foundation (an absolutely rigid plate) differs very little from the average total slope angle φ_{av} of tilt of an absolutely flexible plate.

$$\varphi_{av} = \frac{1}{a} \int_0^a \varphi(y)\, dy$$

Substituting under the integral the expression for the function $\varphi(y)$ from Eq. (I-3-1) and integrating, we obtain

$$\varphi_{av} = \frac{K}{a} \left[2a^2 \ln \frac{\sqrt{4a^2 + b^2} + b}{\sqrt{4a^2 + b^2} - b} + b(\sqrt{4a^2 + b^2} - b) \right] \quad \text{(I-3-2)}$$

On the other hand, we may assume that the normal vertical soil reaction at any point under the base of an absolutely rigid bearing plate depends only on the settlement of this point and is proportional to it. Then we obtain an equation for the value of the soil reaction on the foundation area element dA tilted through an angle φ by external bending moment M. This equation is as follows:

$$dR = c_\varphi L \varphi\, dA$$

where c_φ is the coefficient of elastic nonuniform compression of soil, i.e., the coefficient of proportionality, analogous to c_u, which was introduced when the uniform settlement of foundations was considered. L is the distance between the element dA and the axis of rotation.

The moment of the elementary reaction dR with respect to this axis is

$$dM = c_\varphi L^2 \varphi\, dA$$

The total reactive moment transmitted by the soil to the foundation is

$$M = c_\varphi \varphi \int_A L^2\, dA = c_\varphi I \rho$$

Assuming that $\varphi_{av} = \varphi$, we obtain:

$$\frac{1 - \nu^2}{E} \frac{1}{\pi a} \left[2a^2 \ln \frac{\sqrt{4a^2 + b^2} + b}{\sqrt{4a^2 + b^2} - b} + b(\sqrt{4a^2 + b^2} - b) \right] = \frac{1}{c_\varphi} \quad \text{(I-3-3)}$$

or

$$c_\varphi = \frac{k_\varphi C}{\sqrt{A}} \quad \text{(I-3-4)}$$

where

$$k_\varphi = \frac{2\pi\alpha\sqrt{\alpha}}{2\alpha^2 \ln\left[(\sqrt{4\alpha^2+1}+1)/(\sqrt{4\alpha^2-1}-1)\right] + \sqrt{4\alpha^2+1} - 1} \tag{I-3-5}$$

and φ, as before, equals a/b; C is defined by Eq. (I-2-13).

It follows from Eq. (I-3-4) that c_φ decreases with an increase in the area of the foundation base.

From Eqs. (I-2-12) and (I-3-4) we obtain

$$\frac{c_\varphi}{c_u} = \frac{k_\varphi}{c_s} \tag{I-3-6}$$

Equation (I-3-6) indicates that the coefficient of elasticity of the soil c_u, established by means of tests with uniform loading, is not equal to the coefficient c_φ, established by means of tests with nonuniform loading. Thus, the coefficients of elasticity of a soil, relating the external load to the settlement of a foundation, depend not only on the elastic properties of the soil and the size of the foundation, but also on the character of the load transmitted to the soil.

Table I-9 gives the computed values of the coefficient k_φ and the ratios c_φ/c_u for different values of α. The data of this table lead to the conclusion that, with an increase in α, the coefficient of elastic nonuniform compression c_φ increases faster than c_u.

TABLE I-9. VALUES OF THE COEFFICIENT OF ELASTIC NONUNIFORM COMPRESSION c_φ, EXPRESSED IN TERMS OF c_u, FOR DIFFERENT RATIOS α OF THE LENGTH TO THE WIDTH OF A FOUNDATION

α	k_φ	$\frac{c_\varphi}{c_u}$	α	k_φ	$\frac{c_\varphi}{c_u}$
1.0	1.984	1.87	3	2.955	2.63
1.5	2.254	2.11	5	3.700	3.04
2.0	2.510	2.31	10	4.981	3.53

The author and A. I. Mikhalchuk were the first to verify experimentally the theoretical findings related to c_φ. This was done during the experimental studies of foundations on saturated brown clays with silt and sand described above. By means of these investigations, it was for the first time established experimentally that the coefficient of elastic nonuniform compression c_φ differs from the coefficient of elastic uniform compression c_u. Prior to these experiments, the coefficients were considered equal in all design computations.

Values of c_φ obtained experimentally for brown clays with silt and sand and for other soils are given in Table I-10. The tests were carried out with square foundations; both static and dynamic methods were employed for determining the values of c_φ.

The data of Table I-10 confirm experimentally the basic theoretical findings related to c_φ. The test results agree with the theoretical considerations indicating a decrease in the value of the coefficient c_φ with an increase in area. The test results also indicate that the ratio c_φ/c_u

TABLE I-10. VALUES OF THE COEFFICIENT OF ELASTIC NONUNIFORM COMPRESSION c_φ FOR DIFFERENT SOILS AND SIZES OF FOUNDATIONS

Description of soil	Base contact area of foundation, m^2	c_u, kg/cm^3	c_φ, kg/cm^3	$\frac{c_\varphi}{c_u}$
Saturated brown silty clays with some sand	2.0	4.40	12.0	2.73
	4.0	2.50	4.0	1.60
	8.0	2.05	3.0	1.46
Saturated gray soft silty clays with some sand	0.5	3.5	3.55	1.02
	1.0	2.52	3.61	1.44
	1.5	2.11	3.79	1.80
Loess at natural moisture content	0.81	14.2	25.0	1.76
	1.4	10.8	17.6	1.63
	2.0	10.2	15.5	1.51
	4.0	8.0	12.9	1.61
Saturated gray fine dense sands	4.0	7.5	14.5	1.92
	8.0	5.6	9.5	1.71
	15.0	4.0	9.2	2.30

changes within the limits established by the theory for foundations with square bases; experiments established the average value of this ratio to be 1.73, which differs by only 8 per cent from that found on the basis of theoretical considerations. Thus it is possible, if necessary, to limit experimental investigations of the coefficients of elasticity of soil c_u and c_φ on the site of foundation construction. After one of these coefficients is established experimentally, the second may be computed by means of theoretically derived formulas. Figure I-19 illustrates the results of tests on elastic nonuniform compression performed on loessial soil. It presents one of the graphs of relationship between the slope angle of the deflected foundation and the value of the moment, in cases where the foundation was loaded on both sides of the axis of tilting.

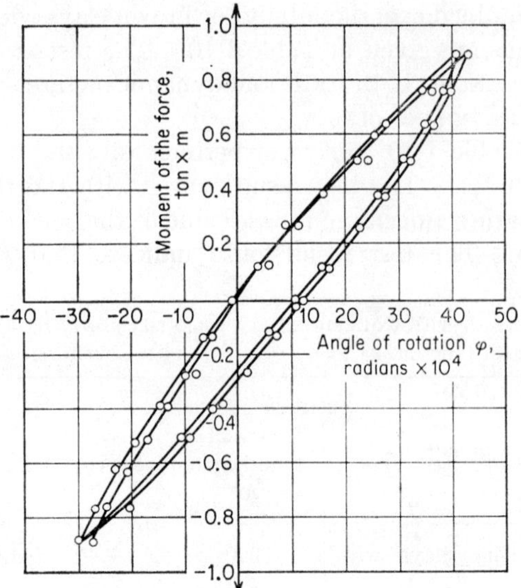

FIG. I-19. Angle of rotation (tilting) φ of a test plate as a function of the applied moment M.

I-4. Coefficient of Elastic Uniform Shear of Soil c_τ

If a test foundation is subjected to the action of a horizontal force applied at the level of its base contact area, it will slide in the direction of that force. The displacement at any time depends on the magnitude of the force at that time. However, after the force drops to zero, there remains a residual displacement of the foundation. The sliding of a foundation subjected to a horizontal force in principle does not in any way differ from settlement. As shown by experience, under conditions corresponding exactly to those established for soil compression (see Art. I-2), it is possible to consider that there is a linear relationship between the sliding of a foundation and the average shearing sliding stress developed along the foundation base contact area, i.e.,

$$\tau_{av} = c_\tau S'_e \tag{I-4-1}$$

where τ_{av} is the average shearing stress in the soil at the plane of contact with the foundation, and S'_e is the elastic part of the total horizontal sliding of the foundation base under the action of τ_{av}.

c_τ, by analogy with the coefficient of elastic compression c_u, is called the coefficient of elastic uniform shear of soil.

In order to clarify the dependence of c_τ on Young's modulus, Poisson's ratio, and the size and shape of the foundation base, let us use the existing solution[31] of the problem of deformation of a semi-infinite solid under the action of a horizontal force P_t applied to its surface (the point of application of P_t is made the origin of the coordinate system):

$$S'_{\text{tot}} = \frac{P_t}{\pi}\frac{1-\nu^2}{E}\left[\frac{1}{\sqrt{x^2+y^2}} + \frac{\nu}{1-\nu}\frac{x^2}{(x^2+y^2)^{3/2}}\right] \tag{I-4-2}$$

In order to obtain the solution for a load uniformly distributed over a rectangular area with sides $2a$ and $2b$, we make the following substitutions in (I-4-2):

$$P_t = \tau_{\text{av}}\, d\eta\, d\xi$$
$$x = x^* - \eta$$
$$y = y^* - \xi$$

We then integrate Eq. (I-4-2) over the entire loaded area. Thus we obtain:

$$S'_{\text{tot}} = \frac{\tau_{\text{av}}}{\pi}\frac{1-\nu^2}{E}\left[\int_{-b}^{b}\int_{-a}^{a}\frac{d\eta\, d\xi}{\sqrt{(\eta-x^*)^2+(\xi-y^*)^2}} + \frac{\nu}{1-\nu}\int_{-b}^{b}\int_{-a}^{a}\frac{(\eta-x^*)^2\, d\eta\, d\xi}{\sqrt{[(\eta-x^*)^2+(\xi-y^*)^2]^3}}\right] \tag{I-4-3}$$

The average value of the horizontal sliding along the surface area $2a$ by $2b$ equals

$$S'_{\text{av}} = \frac{1}{ab}\int_0^b\int_0^a S'_{\text{tot}}\, dx^*\, dy^*$$

Substituting the formula for S'_{tot} from (I-4-3) into (I-4-4) and integrating, we obtain:

$$S'_{\text{av}} = \frac{4a}{\pi}\frac{1-\nu}{E}\left(\frac{1}{\alpha}\sinh^{-1}\alpha + \sinh^{-1}\frac{1}{\alpha} - \frac{1}{3}\left[\frac{1}{\alpha^2}(\sqrt{1+\alpha^2}-1) + \sqrt{1+\alpha^2} - \alpha\right] + \frac{\nu}{1-\nu}\left\{\frac{1}{\alpha}\sinh^{-1}\alpha + \frac{1}{3}\left[\sqrt{1+\alpha^2} - \alpha - \frac{2}{\alpha^2}(\sqrt{1+\alpha^2}-1)\right]\right\}\right)\tau_{\text{av}} \tag{I-4-4}$$

where $\alpha = a/b$.

On the other hand, according to Eq. (I-4-1), the value of the elastic sliding of the foundation base equals

$$S'_e = \frac{\tau_{\text{av}}}{C_\tau}$$

Assuming $S'_e = S'_{av}$, we obtain the following expression for the coefficient of elastic uniform shear of soil:

$$c_\tau = \frac{k_\tau C}{\sqrt{A}} \tag{I-4-5}$$

where the coefficient k_τ equals

$$k_\tau = \frac{\pi}{2\sqrt{\alpha}\,((1/\alpha)\sinh^{-1}\alpha + \sinh^{-1}(1/\alpha) - \tfrac{1}{3}[(1/\alpha^2)(\sqrt{1+\alpha^2} - 1)} + \frac{\pi}{\sqrt{1+\alpha^2} - \alpha] + [\nu/(1-\nu)]\{(1/\alpha)\sinh^{-1}\alpha} + \frac{\pi}{\tfrac{1}{3}[\sqrt{1+\alpha^2} - \alpha - (2/\alpha^2)(\sqrt{1+\alpha^2} - 1)]\})} \tag{I-4-6}$$

Equation (I-4-5) leads to the conclusion that the coefficient of shear c_τ, just as do the coefficients of compression c_u and c_φ, depends on the foundation base area, decreasing proportionally to $1/\sqrt{A}$.

Unlike c_s and k_φ [see Eqs. (I-2-14) and (I-3-5)], k_τ depends not only on the ratio α between the lengths of the sides, but also on the value of Poisson's ratio. Table I-11 gives values of the coefficient k_τ computed for different values of ν.

Table I-11. Values of the Coefficient k_τ [Eq. (I-4-6)] for Varying Values of the Poisson Ratio ν and of the Ratio α of the Length to the Width of a Foundation

ν	α						
	0.5	1.0	1.5	2.0	3.0	5.0	10.0
0.1	1.040	1.000	1.010	1.020	1.050	1.150	1.250
0.2	0.990	0.938	0.942	0.945	0.975	1.050	1.160
0.3	0.926	0.868	0.864	0.870	0.906	0.950	1.040
0.4	0.844	0.792	0.770	0.784	0.806	0.850	0.940
0.5	0.770	0.704	0.692	0.686	0.700	0.732	0.940

The ratio between c_u and c_τ, equaling

$$\frac{c_u}{c_\tau} = \frac{c_s}{k_\tau}$$

is governed not only by the value of α, but also by the value of ν. Since the value of ν is larger in cohesive soils than in sandy soils (see Art. I-1), it follows that under otherwise equal conditions the relative value of the coefficient of elastic shear c_τ must be smaller in clayey soils than in sandy soils.

Experimental investigations of the coefficient of elastic shear c_τ were carried out on the same soils for which c_u and c_φ were studied (see Arts. I-2 and I-3). Experiments were performed on gray saturated plastic silty clays with some sand; several test foundations of reinforced concrete with base areas of 1.5, 1.0, and 0.5 m² were used. Some footings were square, others were rectangular, but all had the same area.

The following method was employed during the tests: hooks were embedded in the footings 5 to 10 cm above the base level. Wire ropes were tied to the hooks and were connected by a turnbuckle to a dynamometer. Twisting the turnbuckle produced a horizontal force which acted along the foundation base and induced its sliding. The shearing stress was gradually increased and was measured by the dynamometer; the

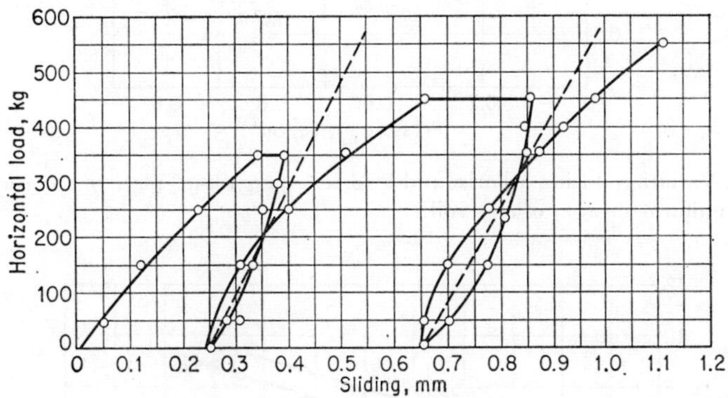

FIG. I-20. Results of a horizontal load test on saturated silt.

displacement of the foundation was measured by a gauge and was recorded for certain values of stress. In each test, loading-unloading cycles were obtained. Figure I-20 presents curves of the relationship between sliding and horizontal stress, plotted on the basis of data obtained in one of the tests.

Experiments established that with an increase in normal pressure on the soil, c_τ increases. It was also found that the value of c_τ for a given soil depends not only on the normal pressure, but also on the duration of its action. Thus, immediately after application of the vertical load, the value of the coefficient of elastic shear is smaller than the corresponding value established a little later. Under constant vertical pressure, the rate of increase of c_τ gradually decreases with time, and after several hours changes in c_τ become so small that its value is practically constant.

For illustration let us analyze Fig. I-21, which gives curves of the relationship between c_τ and normal pressure, plotted on the basis of data obtained in tests with footings having 1 m² base area. After c_τ was determined under a normal vertical pressure of 0.53 kg/cm², the test was interrupted, and the footing remained under this load for 20 hr. Then c_τ was determined for the second time under the same pressure and was found to be 30 per cent larger than before the interruption: c_τ was 1.10 kg/cm³ before the interruption, and 1.45 after. Figure I-21 reveals that, due to the interruption in the test, the curve of relationship between c_τ

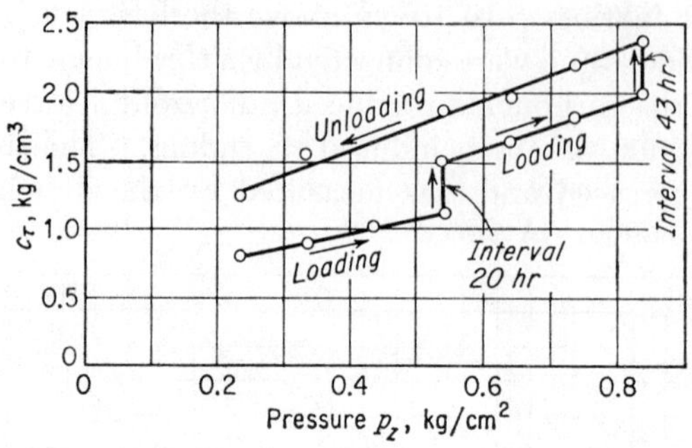

FIG. I-21. Variation during a horizontal load test on a 1.0-m² plate of the coefficient of elastic uniform shear c_τ of clay soil.

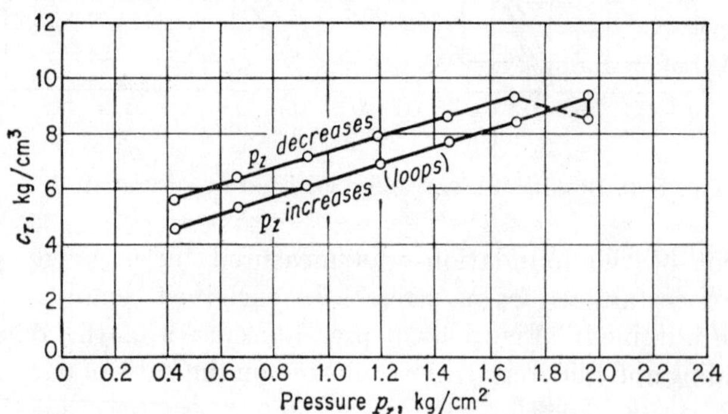

FIG. I-22. Variation during a horizontal load test of the coefficient of elastic uniform shear c_τ of loessial soil.

and p_z was broken at the point $p_z = 0.53$ kg/cm². Later p_z was increased again at a regular rate, and c_τ also grew smoothly.

An interruption of 43 hr was made once more, after the normal pressure attained 0.83 kg/cm². It was found again that c_τ was larger after the interruption than before. A regular decrease in the normal pressure caused a smooth decrease in c_τ. For unloading, the curve relating c_τ and normal pressure turned out to be parallel to the corresponding curve for loading.

A similar increase in c_τ with the time of normal pressure action was found in tests on loess. This is illustrated by Fig. I-22, which presents the curve of relationship between c_τ and normal pressure. Test conditions were the same as described for saturated plastic silty clays with some sand. The normal pressure on soil was increased in approximately equal time intervals. The coefficient c_τ was determined for each new value of the normal pressure p_z. After the maximum pressure of 2.0 kg/cm² had been attained, it was found that under this pressure the horizontal force induced large settlements of the foundation.

Ya. N. Smolikov and the author investigated the relationship between the coefficient of elastic uniform shear and normal pressure for gray saturated fine sands. A footing with an area of 4 m² was used. The

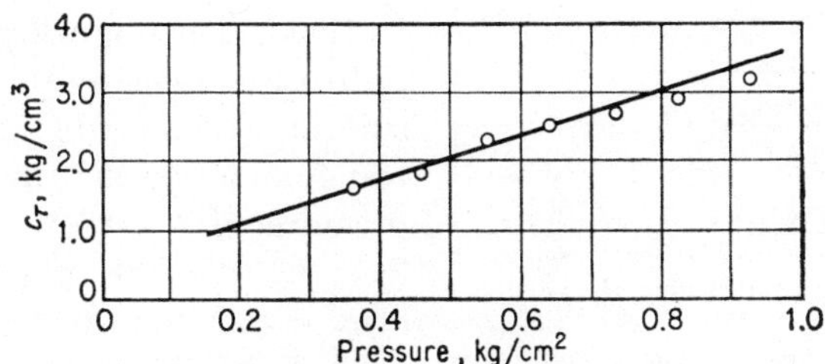

Fig. I-23. Variation during a horizontal load test on a 4.0-m² plate of the coefficient of elastic uniform shear c_τ of fine saturated sand.

results of these investigations, shown in Fig. I-23, indicate that in sands, as in plastic silty clays with some sand and in loess, the coefficient of elastic uniform shear increases proportionally to the increase in normal pressure on the soil.

The tests under consideration demonstrate that the elastic resistance of a soil to shear depends on the normal pressure. The fact that the magnitude of elastic shear depends not only on the shearing stress, but also on the normal pressure, indicates that the law of independent action of forces cannot be applied in cases of wide ranges of changes in the magnitudes of external loads, even when considering the elastic deformations of soils. This deviation of soil from the properties of linear systems is apparently explained by the fact, already discussed in Art. I-1, that the modulus of elasticity in shear of a soil depends on the normal pressure. Therefore, as noted before, there are limitations to the application of Hooke's law to soils.

Investigations of test foundations with different base contact areas indicate that according to Eq. (I-4-5) the coefficient of elastic uniform shear of a soil decreases in inverse proportion to the square root of the foundation base area. Figure I-24 shows values of this coefficient established experimentally for test foundations on loess.

The dotted line is plotted on the basis of the assumption that c_τ changes in inverse proportion to the square root of the foundation base contact area.

The conclusion can be reached, as it was in relation to the coefficient of elastic uniform compression, that the relationship expressed by Eq. (I-4-5) is valid only for foundations with comparatively small contact areas. It can be held that for foundations with base areas larger than 10 to 12 m², c_τ does not depend on the area.

According to the data of Table I-11, for foundations with square bases and for cases in which Poisson's ratio lies in the range 0.3 to 0.5, the ratio c_u/c_τ changes within the range 1.22 to 1.50. The following values of the ratio c_u/c_τ were established as a result of experiments with the same test foundation and under a normal pressure of 0.4 to 0.5 kg/cm²: for loess, $c_u/c_\tau = 1.9$; for gray fine sand, $c_u/c_\tau = 2.20$ to 2.40. Since theoretically derived values of the ratio c_u/c_τ do not take into account the influence of normal pressure on the magnitude of c_u, it is advisable in design computations to use values of this ratio close to those established experimentally. It can be tentatively assumed that $c_u/c_\tau = 2$.

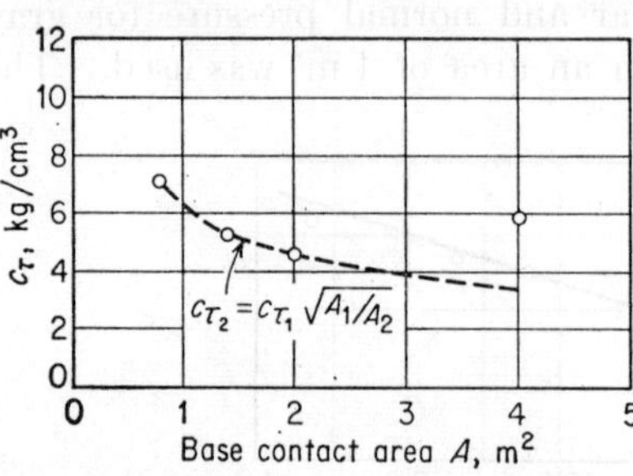

Fig. I-24. Dependence of the coefficient of elastic uniform shear c_τ of loessial soils on the loaded area.

I-5. Coefficient of Elastic Nonuniform Shear of Soil c_ψ

If a foundation is acted upon by a moment with respect to the vertical axis, it will rotate around this axis. Tests show that the angle of rotation ψ of a foundation is proportional to the external moment. Therefore we may write:

$$M_z = c_\psi J_z \psi \tag{I-5-1}$$

where M_z = external moment producing rotation of base of foundation around a vertical axis to angle ψ

J_z = polar moment of inertia of contact base area of foundation

c_ψ = coefficient of elastic nonuniform shear

In the rotation of a foundation around a vertical axis, the base of the foundation undergoes nonuniform sliding; hence the term "coefficient of elastic nonuniform shear" may be applied to the coefficient c_ψ. Experiments show that its magnitude is somewhat larger than that of the coefficient of elastic uniform shear c_τ. Table I-12 gives values of c_τ and c_ψ established by tests performed on the same foundations on plastic silty clay with some sand and on gray sands.

TABLE I-12. VALUES OF THE COEFFICIENTS OF ELASTIC UNIFORM SHEAR c_τ AND OF ELASTIC NONUNIFORM SHEAR c_ψ FOR DIFFERENT SOILS AND FOUNDATION AREAS

Soil	A, m²	c_τ, kg/cm³	c_ψ, kg/cm³	$\frac{c_\psi}{c_\tau}$
Gray plastic saturated silty clays with some sand	1.5	1.27	2.72	2.14
	1.0	1.64	2.33	1.42
	0.5	1.88	2.08	1.01
Gray fine saturated sands	1.0	2.54	3.90	1.54
	4.0	2.20	2.85	1.30
	15.0	1.90	2.20	1.16

As established by the results of all tests, c_ψ is larger than c_τ. The average value of c_ψ/c_τ was in all cases found to equal 1.43. Therefore, it can be assumed that a value of $c_\psi = 1.5\ c_\tau$, when used for design computations, will not differ greatly from the actual value.

I-6. Elastic Resistance of Pile Foundations

a. Elastic Resistance of Piles under a Vertical Load. The use of piles in foundations subjected to vibrations and shocks may be necessary in four cases: (1) if the total pressure on the soil, both static and dynamic, is larger than the bearing capacity of the soil, considering the dynamic action of the foundation on the soil; (2) if it is necessary to increase the natural frequency of vibrations of the foundation, which is considered to be a solid body resting on an elastic basis; (3) if it is necessary to decrease the amplitude of natural or forced vibrations; (4) if it is necessary to decrease the residual dynamic settlement of the foundation.

The second and third cases refer mainly to foundations for machines with dynamic loads; the first and fourth cases are more common in the design of foundations for structures subjected to the dynamic influence of waves emanating from machine foundations.

If a pile foundation is to be used because the total pressure on the soil is larger than its bearing capacity (taking into account the dynamic action of the foundation on its base), then conventional methods of design, such as are generally used in the design and computations of pile foundations producing only static pressure on soil, are applied. In such cases the practical procedure of pile-foundation design consists in the determination of the number of piles needed from the known value of the bearing capacity of a single pile. This value is determined by load tests, and the length of the piles is selected on the basis of test pile driving.

If the use of a pile foundation is due to the need for increasing the natural frequency of vibrations and decreasing the amplitude of free or

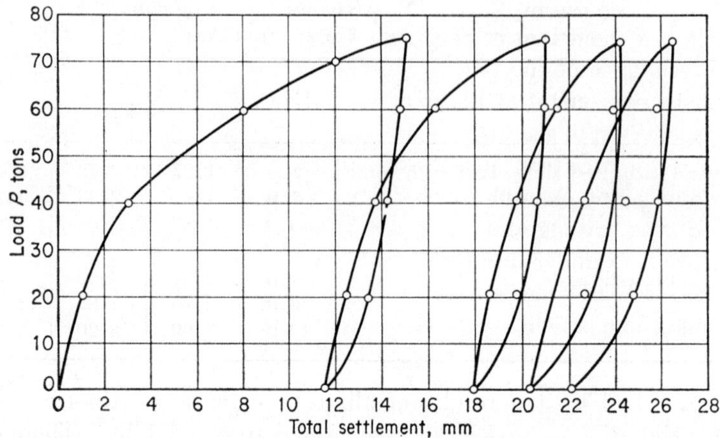

FIG. I-25. Results of a vertical load test on a single reinforced concrete pile.

forced vibrations of the foundation, then the practical design procedure has some peculiarities which will be examined in this article.

A pile foundation, like the natural soil base beneath it, is characterized by elastic properties.

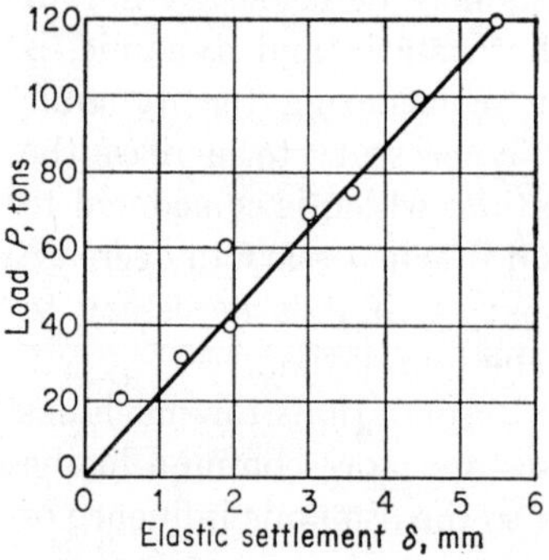

FIG. I-26. Evaluation of hysteresis loops of Fig. I-25.

Figure I-25 shows the results of a test in which a static vertical load was applied to a single reinforced-concrete pile. It is clear from the graph that the settlement of the pile decreases with a decrease in load. It is also seen that if the maximum value of load is constant, then the decrease in the elastic settlement also remains constant. Thus after the first unloading, the elastic settlement of the pile was 3.5 mm; after the second unloading it was 3.10 mm; after the subsequent ones, 3.85 and 3.3 mm. Within the range of measurement errors, which for many reasons can be fairly significant in pile tests of this type, it is possible to consider that the elastic settlement of the pile up to a load of 75 tons remained constant and on the average was equal to 3.4 mm. Figure I-26 gives a graph showing the relationship between the elastic settlement of the pile and the magnitude of the load thereon. This graph shows that the elastic settlement of the pile is proportional to the magnitude of the load. Thus, denoting by P the load acting on the pile, and by δ the elastic settlement, we obtain

$$P = C_\delta \delta \tag{I-6-1}$$

where C_δ is a coefficient of proportionality which we call the coefficient of elastic resistance of the pile; it represents the load required to induce a unit elastic settlement of the pile.

The coefficient of elastic resistance of the pile depends on soil properties, pile characteristics (e.g., length), and the length of time the pile has been in the soil. For example, the elastic resistance of a pile has different values during driving and some time later.

The problem under discussion was a subject of the author's investigations, performed by measuring vertical vibrations of the soil in the vicinity of the pile being driven. Soil vibrations induced by pile driving have the same period as the vertical vibrations of the pile itself, the latter being considered a solid body resting on an elastic base. Therefore, by measuring soil vibrations, it is possible to estimate the vibrations of the pile. The relationship between the period of vibrations T and the coefficient of elastic resistance C_δ is given by the expression

$$T = 2\pi \sqrt{\frac{m}{C_\delta}} \tag{I-6-2}$$

where m is the mass of the pile and the pile-driving mechanism. By means of Eq. (I-6-2) it is possible to establish the coefficient of elastic resistance of the pile from the measured period.

Figure I-27 shows the soil profile at the site of the test pile and gives diagrams illustrating the effect of the number of blows on the periods and the depth of pile penetration. The reinforced-concrete pile was 10 m long with cross section 30 by 30 cm. The driving was performed by means of a 3-ton steam hammer; the height of the drop was the same at each stroke—around 1.0 m.

A study of the diagrams illustrating the effect of depth of pile penetration on changes in periods of vibrations leads to the conclusion that an increase in depth of penetration does not produce a decrease in period of vibrations, as could be anticipated on the basis of the assumption that the elastic resistance of a soil acts mostly along the lateral pile surfaces; on the contrary, the period of vibrations even increases somewhat.

A pause of 10 days was made in the pile driving; when it was resumed, the average refusal of piles was found to be 10.5 mm per 11 blows. During this additional pile driving, the periods of vibration were measured. The results of these measurements are also given in Fig. I-27;

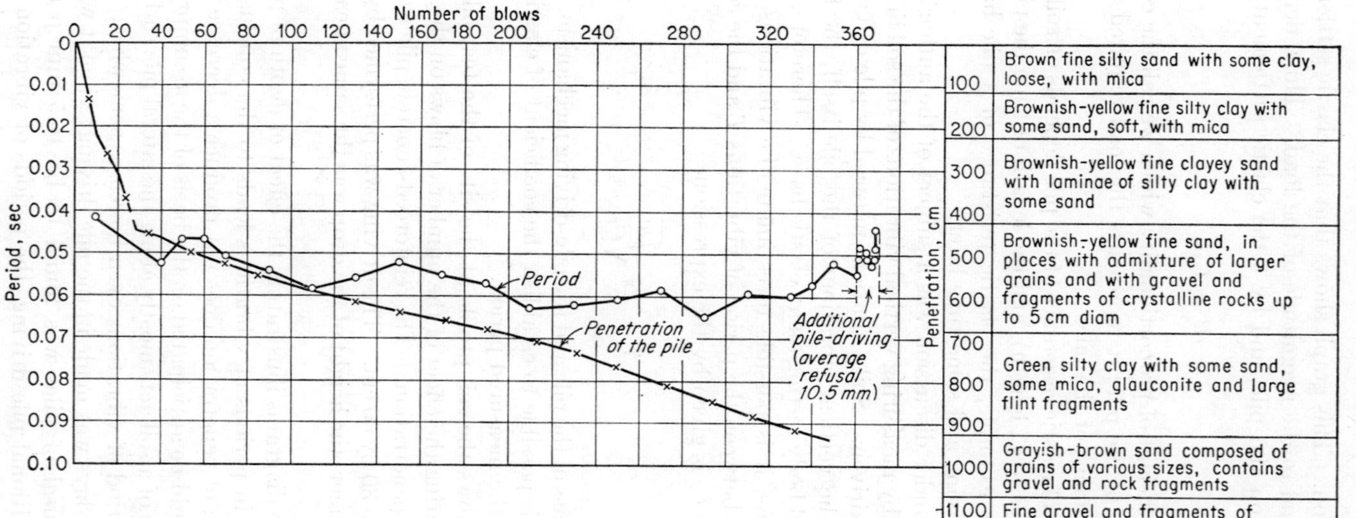

FIG. I-27. Driving record of a pile.

they indicate that, after the "rest," the period of natural vibrations of the pile decreases somewhat. During pile driving, the average period (at a penetration depth of 6 to 9 m) was 0.060 sec; during the additional driving it decreased to 0.048 sec. Since squares of periods are inversely proportional to the coefficients of elastic resistance, this decrease in the period after the pile "rest" attests to the fact that the elastic resistance of the soil increased after the "rest" by approximately 45 per cent.

A decrease in the period of natural vertical vibrations of the pile resulting from a "rest" may occur only on account of an increase in the elastic resistance of the soil acting along the lateral surface of the pile.

This conclusion fully agrees with the experience derived from construction practice: it is known that a few days after the completion of pile driving the dynamic resistance of the pile increases. Because of this increase, the refusal of piles is determined, not immediately after pile driving, but some time later. The author found in several cases that during driving the period of vibrations of a pile is constant. This point has also been confirmed by the experiments of F. A. Kirillov and S. V. Puchkov,[23] who measured periods of soil vibrations during the driving of two piles. When the first pile was driven, the period of vibration did not change with an increase in the depth of penetration, but remained equal to 0.14 sec; observations on the second pile showed a decrease in the period of vibrations from 0.17 sec at a penetration depth of 4.75 m to 0.13 sec at a depth of 8.00 m. In this case, the decrease in period is explained by changes in soil properties.

These experimental data lead to the conclusion that as the pile penetrates into the soil, the depth of its penetration has little effect on the elastic resistance of the soil; therefore it should be assumed that elastic soil reacts mostly on the pile tips.

As noted before, this conclusion is valid only when piles remain in the soil for a short time. When piles are in the soil for a long time, the cohesion between the soil and the lateral side surfaces of the piles increases, resulting in an increase in elastic resistance. The magnitude of elastic resistance resulting from soil action along the lateral pile surfaces may be considerable as compared to the magnitude of elastic resistance caused by soil reactions against the tips of piles. This is confirmed by experimental pullout tests of piles subjected to static loads. Figure I-28 gives values of C_δ for several wooden piles, established experimentally for pile driving and for pulling out. The experiments were carried out by Ya. N. Smolikov on saturated fine gray sands. Comparison of these values reveals that the elastic resistance of piles is approximately 30 per cent higher during driving than it is during pullout. Hence it follows that the elastic resistance of a pile which has been in the soil for a long time is conditioned mainly by soil reactions along the side surfaces of the pile.

Since the magnitude of the soil reactions on these surfaces increases with the depth of pile penetration, C_δ should increase with an increase in pile length. This is confirmed by testing piles of different lengths (see Fig. I-28, which gives diagrams which illustrate the effect of pile length on the coefficient of elastic resistance of the pile).

On the basis of the results of experimental investigations, the coefficient of elastic resistance of a pile C_δ may be considered to be approximately

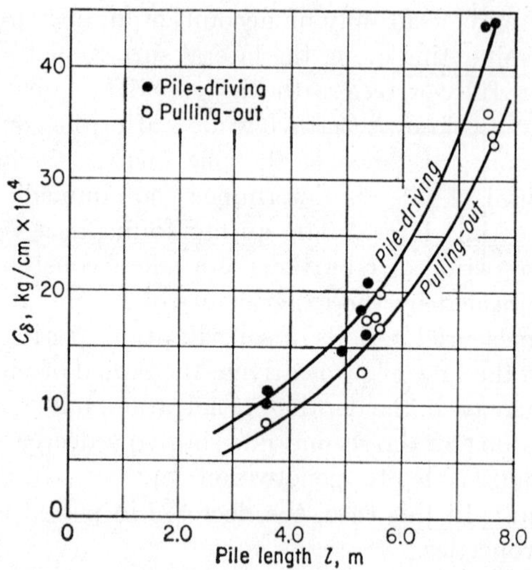

FIG. I-28. Variation of the coefficient C_δ of a pile during a driving and pullout test.

proportional to the square of the depth l of pile penetration and to its perimeter S:

$$C_\delta = \bar{c}Sl^2 \tag{I-6-3}$$

The reduced coefficient $\bar{c}$ has the dimensions tons per meter^4.

Let us assume that the coefficient of elastic resistance of a pile group is proportional to the number of piles in the group. Thus if n denotes the number of piles, the coefficient of elastic resistance of the group equals

$$K_z = nC_\delta \tag{I-6-4}$$

An increase in the number of piles under the foundation leads to an increase in its area of contact with the soil; therefore the coefficient of elastic uniform compression of the pile group increases proportionally to the contact area of the foundation. This conclusion is confirmed by experimental data. Figure I-29 gives a diagram of the dependence of the coefficient of elastic uniform compression K_z for pile foundations on the contact areas of these foundations, and consequently on the number of piles under the foundation. The distance between the piles in all cases was 0.81 m; the pile length was also constant, equaling 5.5 m. Tests were performed on saturated gray fine sands.

Curve 1 shows the relationship between K_z and the foundation contact area (or the number of piles). It reveals that, unlike the case of spread foundations resting on natural soils (curve 2), the coefficient of elastic uniform compression for pile foundations increases proportionally to the

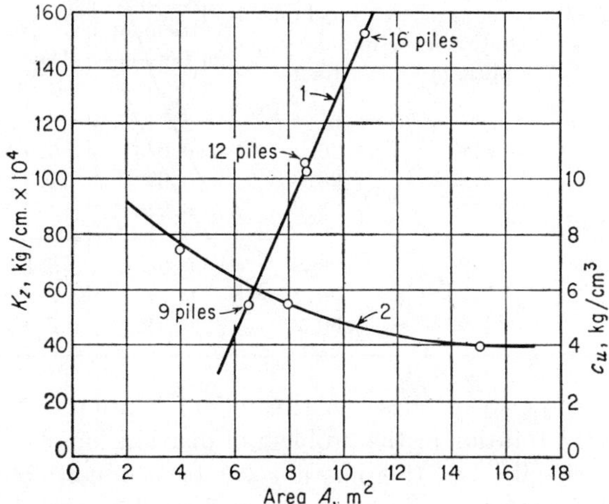

FIG. I-29. Variation with size of footing of the coefficient of elastic resistance K_z of a pile group (curve 1) and of the coefficient of elastic uniform vertical compression c_u of soil (curve 2) under a footing without piles.

increase in the foundation contact area, or, other conditions being equal, proportionally to the number of piles.

Table I-13 gives results of vertical load tests performed on six similar foundations, each resting on nine piles. The lengths of the piles were identical, but the distances between them were different. The perimeters of all piles were approximately the same and equal to $S = 0.9$ m. Tests were carried out on saturated gray fine sands.

It follows from Table I-13 that the distance between piles in a group affects the value of the coefficient of elastic resistance C_δ of the piles in the group. When the distance between piles increases, the elastic resistance of each pile in the group increases; when this distance is sufficiently large, the resistance of each pile in the group approaches the resistance of a single pile. Thus, in order to evaluate the elastic resistance of each pile in the group, a corrective influence coefficient should be introduced to take into account the distance between piles. Then the coefficient of elastic resistance of one pile in the group will equal:

$$C_k = \mu C_\delta \tag{I-6-5}$$

The experimental data available are insufficient to evaluate with an adequate degree of precision the value of the influence coefficient μ for

TABLE I-13. VALUES OF COEFFICIENT C_δ OF ELASTIC RESISTANCE OF PILES

Pile length, m	Distance between piles, m	Load on foundation, tons	Elastic settlement of foundation, cm	C_δ, kg/cm
	0.81	30.2	0.049	6.9×10^4
5.5	1.22	28.8	0.028	11.5×10^4
	1.62	28.8	0.026	12.4×10^4
	0.81	28.8	0.063	5.1×10^4
3.5	1.22	30.2	0.047	7.2×10^4
	1.62	30.2	0.048	7.0×10^4

various distances between piles. A theoretical solution of this problem would require the solution of the problem of distribution of soil deformation around each pile, but this has not yet been done. Therefore the influence coefficient necessarily has to be evaluated on the basis of those insufficient experimental data which were obtained as a result of testing of pile groups with different distances between piles. The values of the coefficient of elastic resistance of piles of various lengths presented in Fig. I-28 give a basis for the determination of the value of the coefficient of resistance of a single pile. With $\mu = 1$ for this pile, it is possible, by using the data of Table I-13, to compute the value of the influence coefficient for various ratios of the distance t between piles to the pile diameter d. The values of influence coefficient established in this way are given in Table I-14.

TABLE I-14. CORRECTION COEFFICIENT μ (FOR USE WITH THE C_δ VALUES OF TABLE I-13) EXPRESSED AS A FUNCTION OF THE RATIO OF t (SPACING) TO d (DIAMETER) OF PILES

$\frac{t}{d}$	μ
	1.0
6	0.65
4.5	0.64
3	0.41

It follows from Table I-14 that the mutual influence among piles in the group is significant, even when the distance between the piles is six times larger than the pile diameter. If this ratio is three instead of six, then the elastic resistance of one pile in the group is only 40 per cent of the elastic resistance of a separate pile.

In the design of pile foundations, a value of t/d is usually selected within the range 5 to 6. According to the data of Table I-14, the design value of μ corresponding to such distances is about 0.65.

As noted before, the value of the coefficient of elastic resistance C_k of one pile in the group is the main characteristic of the elastic properties of the pile base under a foundation. This value may be established by

TABLE I-15. SUMMARY OF VALUES OF COEFFICIENT C_k OF ELASTIC RESISTANCE OF A PILE IN A GROUP

Type of engine	Description of soil beneath foundation	Foundation contact area, m²	Number of piles	Distance between piles t, m	Pile length, m	Perimeter of piles, m	C_k, kg/cm
Vibrator	Saturated soft gray silty clays with some sand	1.5	5	0.55	3.0	0.6	0.1×10^4
Log-sawing frame	Soft plastic clay underlaid by saturated medium sand	17.4	40	0.5	6.5	0.8	1.3×10^4
Log-sawing frame	Saturated sand with admixture of organic silt and blue plastic clay	24.8	30	0.7	6.0	0.8	3.4×10^4
Log-sawing frame	Saturated grayish-brown fine sand with laminae of organic silt	24.8	20	0.8	6.0	0.8	7.7×10^4
Log-sawing frame	Saturated fine sand with admixture of organic silt	12.0	35	0.4	3.0	0.8	5.3×10^4
Log-sawing frame	Saturated fine sand with admixture of organic silt	18.0	35	0.5	3.0	0.8	37.8×10^4
Vibrator	Loess and loessial soil	4.0	9	0.75	6.0	0.8	1.1×10^4

means of experimental investigations of the elastic properties of single piles. There have not been many experiments of this kind. C_k may be also determined from results of testing of pile groups or pile foundations. Table I-15 gives several values of C_k obtained by the author and Ya. N. Smolikov as a result of investigations of test foundations and foundations for log-sawing frames.

The data of Table I-15 make it possible to select a tentative design value of C_k for the computation of vertical vibrations of machine foundations under soil conditions similar to those described in the table. If several pile foundations are to be built on the same construction site, then design values of C_k should be established from special investigations of test foundations.

b. Elastic Resistance of Pile Foundations in Cases of Eccentric Vertical Loading. Let us assume that under the action of an external moment M acting in one of its main vertical planes, the foundation rotates by a small angle φ. Then it will be under the action of the reactive moment of a pile located at a distance L from the axis of rotation; the value of this moment is $C_\delta L^2\varphi$. By equating the external moment M and the sum of the moments transmitted to the foundation by the piles, we obtain

$$M = C_\delta\varphi \sum_{i=1}^{n} L_i^2 \tag{I-6-6}$$

Here, as before, C_δ is the coefficient of elastic resistance of one pile [Eq. (I-6-3)]. Equation (I-6-6) can be rewritten as

$$M = K_\varphi \varphi$$

where

$$K_\varphi = C_\varphi \sum_{i=1}^{n} L_i^2 \tag{I-6-7}$$

may be called, as is done with a natural soil base, the coefficient of elastic nonuniform resistance of pile foundations; its units are kilograms × centimeters or tons × meters.

Table I-16 gives the results of an investigation of three test foundations under the action of nonuniform vertical loading; nine piles were driven under each of the foundations.

TABLE I-16. VALUES OF COEFFICIENT K_φ OF ELASTIC NONUNIFORM RESISTANCE OF PILE FOUNDATIONS

Pile length, m	Perimeter of piles, m	Distance between piles, m	ΣL^2, m²	External moment, tons × m	Angle of rotation of the foundation, radians	K_φ, tons/m	C_δ, kg/cm
3.6	0.9	1.62	15.2	14.4	1.17×10^{-4}	12.3×10^4	8.1×10^4
3.6	0.9	1.22	9.0	12.6	1.92×10^{-4}	6.6×10^4	7.3×10^4
7.2	0.9	1.22	9.0	18.0	1.10×10^{-4}	16.3×10^4	18.2×10^4

Comparison of the values of C_δ in Table I-16 with the values obtained from testing foundations under a uniform vertical load (Table I-13) shows that there is comparatively little difference between the values of the reduced coefficient of elastic resistance obtained by these two methods.

c. Elastic Resistance of Pile Foundations to Horizontal Forces. If the main exciting forces act on a foundation in a horizontal direction, then in some cases pile foundations are provided in order to increase the natural frequencies or decrease the amplitudes of excited horizontal vibrations.

When a pile is designed to take into account the action of a horizontal force, it is assumed that the pile is fixed in the soil at a depth greater than a certain value l_0. The depth of fixation is taken on the order of 1 to 1.5 m. The influence of soil reactions on pile deformation is disregarded, and it is assumed that the resistance of the pile to the horizontal force is determined only by the properties and depth of fixation of the pile.

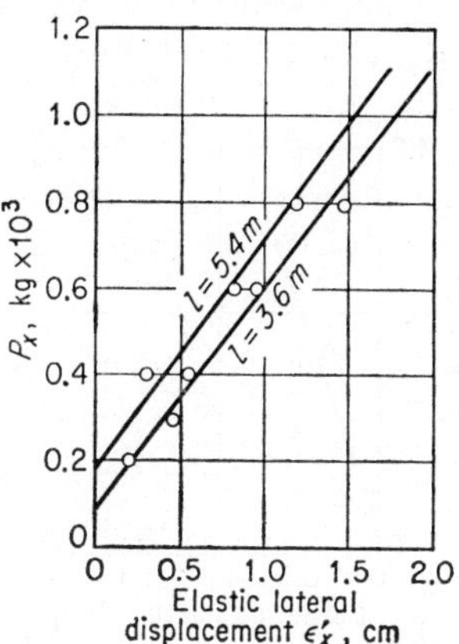

FIG. I-30. Dependence of elastic lateral displacement $\varepsilon'x$ of pile head on the lateral load and length of pile.

On the basis of these assumptions, it can be stated that an increase in length to a value greater than the depth of fixation will have no effect on the elastic resistance of the pile. This is confirmed by experiments. Figure I-30 gives two graphs of the relationship between the elastic displacement of a pile head and the magnitude of the horizontal force acting on piles having respective lengths of 5.4 and 3.6 m. The coefficients of elastic resistance of the piles to horizontal forces are equal. They are defined by the relation P_x/ε_x'. Figure I-31 furnishes data on experimentally established frequencies of natural horizontal vibrations

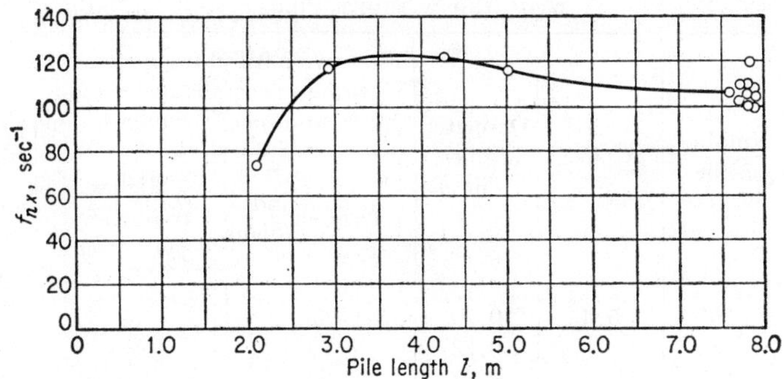

FIG. I-31. Natural frequency of horizontal vibrations of single reinforced-concrete piles of varying lengths.

of single reinforced-concrete piles of different lengths but of equal cross sections (30 by 30 cm). The distribution of points over the diagram confirms the conclusion concerning the small influence of the depth of pile driving on the elastic resistance of a pile against horizontal forces. Thus, it can be held that the employment of long piles will not contribute to the elastic resistance of pile foundations to horizontal forces.

If one assumes, as was done on the preceding pages, that the elastic resistance of a pile to horizontal forces is determined mostly by its size and the elastic properties of its material, then one must conclude that the pile cross section will have great effect on its coefficient of elastic resistance to horizontal forces. This is confirmed by testing piles with different cross sections. Figure I-32 gives graphs of the relationship between the elastic deformation of a pile and the value of the horizontal force. Tests were performed on wooden piles of 29 to 33 cm diameter. For piles with a diameter of 29 cm, the coefficient of elastic resistance was found to be 1.0×10^4 kg/cm; for piles with a diameter of 23 cm, it was 0.41×10^4 kg/cm, i.e., 2.5 times smaller.

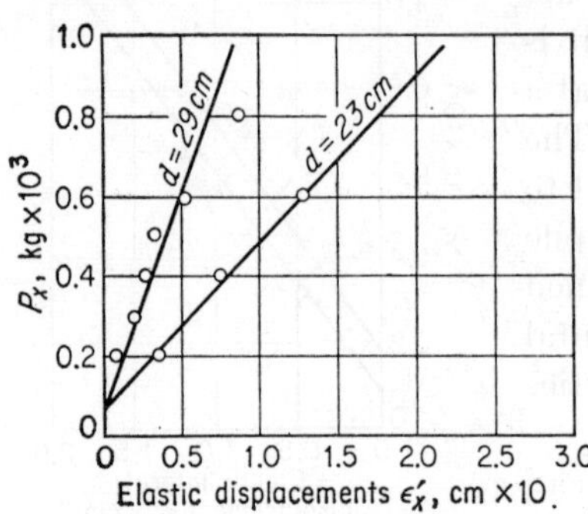

FIG. I-32. Elastic horizontal displacements of the heads of two piles of different diameter.

Accordingly, it can be anticipated that for reinforced-concrete piles the coefficient of elastic resistance will be much larger than for wooden piles.

TABLE I-17. NATURAL FREQUENCIES OF FOUNDATIONS WITH AND WITHOUT PILES

Foundation base contact area, m²	Number of piles	Length of piles, m	Diameter of piles, m	First two natural frequencies of vibrations accompanied by sliding & rotation of the foundation		Type of foundation base
				Lowest frequency, cycles/sec	Higher frequency, cycles/sec	
10.5	16	5.4	0.29	82	266	Wooden piles
8.6	12		0.23	75	240	
8.3	12		0.28	86	270	
6.5	9		0.23	55	270	
10.6	9		0.28	60	253	
4	...	...	....	86	158	Natural soil base
4	...	...	....	92	222	
15	...	...	....	64	143	

These peculiarities in the resistance of piles to horizontal forces put limitations on their employment for the reinforcement of foundations in order to increase the lowest natural frequencies of vibrations. A comparison of test foundations constructed on wooden piles with those placed directly on soil indicates that the use of wooden piles has no effect on the increase of natural frequencies when the frequencies depend on the resistance of the base to vibrations in a horizontal direction. This is shown by Table I-17, giving values of the first two main natural frequencies experimentally established for foundations placed on piles and for foundations placed directly on natural soil.

Comparison of the lowest natural frequency of a foundation erected on piles with that of a foundation placed directly on soil reveals that the piles do not increase the lowest natural frequency; on the contrary, the piles decrease its value somewhat. Thus, pile foundations are characterized by a value of elastic resistance to shear which is not larger than in foundations placed directly on natural soil. The fact that the higher frequencies are larger in pile foundations than in foundations placed directly on natural soil shows that the value of the elastic resistance to rotation is larger in pile foundations.

II

EFFECTS OF VIBRATIONS ON RESIDUAL SOIL SETTLEMENTS

GENERAL

Studies and observations of models and existing foundations subjected to shocks and vibrations show that they may undergo settlements many times larger than those imposed by static loads. Therefore it often happens that settlements of foundations undergoing vibrations greatly endanger the safety of structures. An extremely adverse effect is produced by the vibrations of foundations under machinery—in particular, under forge hammers. These vibrations lead to settlements not only of the hammer foundations, but also of the footings under the columns or walls of a building where hammers are located. Considerable deformations of the superstructure result, often endangering the safety and stability of the whole building.

Several cases illustrating the above points are presented in Chap. IX.

Considerable residual settlements caused by a joint action of static loads and vibrations appear particularly often when foundations rest on noncohesive soils such as water-saturated sands, even if the sands are characterized by low values of porosity. Static investigations of these soils either by means of load tests or in a consolidometer show that they possess high values of resistance, sometimes larger than corresponding values of cohesive soils such as plastic clays. Therefore sands having medium and higher-than-medium densities are considered good natural bases. This is true, however, only for static loads. If a sand or other noncohesive soil is subjected simultaneously to both static loads and shocks (or vibrations), then, as shown by experience with industrial construction, resistance to external loads decreases considerably. This is verified experimentally by plunging into the soil subsurface vibrators, vibrating piles, pipes, or cylinders with considerable diameters. The settlements of vibrating foundations, as well as the penetration of the above structural elements into the soil to a considerable depth, may be observed not only in noncohesive soils, but also in cohesive soils, for example, in heavy clays. Of course, the quantitative effect of vibrations and shocks on cohesive soils may be much smaller than on noncohesive soils, especially sands.

The physical processes which cause changes in soil properties during vibrations (such as the decrease in resistance against external loads leading to the development of residual settlements) are not clarified as yet.

Experiments show that vibrations cause changes in the dissipative properties of a soil, i.e., in the forces of internal dry and viscous friction, in the forces of cohesion (which determine the initial resistance of soil to shear), in the forces of external friction, in the hydrodynamic properties (such as in the coefficient of permeability and in the pore water pressure), and in elastic and plastic characteristics such as Young's modulus, the modulus of shear, and the limits of elasticity and of plasticity.

In addition, it was established that if subjected to intensive vibrations, some soils (especially sandy fills and loose silty and clayey soils) lose their resistance to shear to such a degree that their mechanical properties are closer to those of viscous liquids than of solids.

It follows that in the process of vibration, all properties of a soil may undergo basic changes, including those properties which govern the soil resistance to localized loading.

It is not possible at present to determine with a sufficient degree of precision the value of the residual settlement of a foundation subjected to vibrations or shocks. However, the data available from experimental and theoretical investigations make it possible to evaluate the influence of certain factors on the settlement of a foundation and consequently to take suitable measures to decrease settlement.

II-1. Effect of Vibrations on the Dissipative Properties of Soils

a. Internal Friction and Cohesion of Soil during Vibrations. G. I. Pokrovsky and associates were the first to investigate experimentally the influence of vibrations on the coefficient of internal friction. The results of these experiments were published in 1934. They show that the coefficient of internal friction depends on the kinetic energy of vibrations; as the energy increases, the coefficient decreases, approaching a value 25 to 30 per cent smaller than that observed before vibrations. At the same time, the author investigated experimentally the quantitative dependence of the coefficient of internal friction of sand on the amplitude of vibrations. Some results of these experiments are given in Fig. II-1. The curves show an increase in displacement with an increase in magnitude of the shearing force.

Test 1 was conducted on a freshly filled sand without vibrations. Test 2 involved vibrations with an amplitude of 0.50 mm and a frequency of 140 sec^{-1}. Test 3 was conducted after vibrations ceased. Test 4 involved vibrations of the same frequency as test 2, but an amplitude of vibrations of only 0.15 mm.

A comparison of the curves based on data obtained with an without vibrations permits the conclusion that vibrations have considerable effect on the resistance of soil to shear.

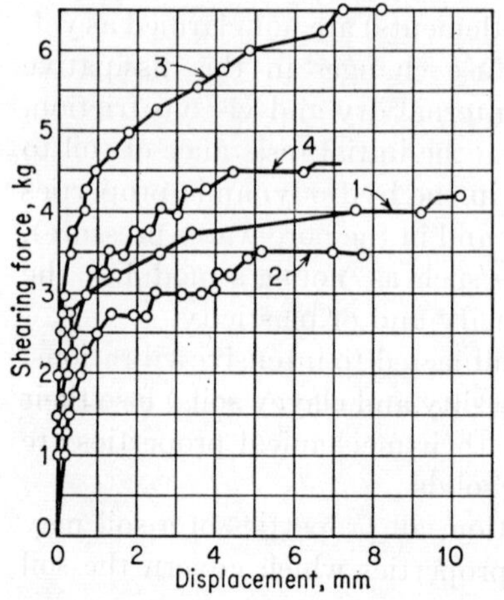

FIG. II-1. Effect of shearing force on displacement, with and without vibrations.

Without exception, the slope of the shearing-force–shearing-deformation curve is smaller with vibrations than without. Since the coefficient of elastic shear is proportional to the tangent of that slope angle, it is clear that the elastic resistance of soil to shear is lower during vibrations than in their absence.

The experiments also established that the coefficient of internal friction is lower during vibrations than in the static state. With an increase in acceleration, the value of the coefficient of internal friction decreases, approaching asymptotically a limit which depends on the properties of a soil.

In addition to experiments with dry soils, analogous experiments were conducted with sand having a moisture content of 10 to 12 per cent. It was established that vibrations cause a smaller decrease in the coefficient of friction in moist soils than in dry soils. Apparently this may be attributed to the fact that the absolute value of the forces of cohesion between particles of sand having a moisture content up to 10 to 12 per cent is much higher than the corresponding value in air-dried sand. As the forces of cohesion in a soil increase, the influence of vibrations on changes in physicomechanical properties decreases.

Comprehensive experimental investigations of the effect of vibrations on the coefficient of internal friction of sand were performed by I. A. Savchenko under the supervision of the author. Figure II-2 shows the installation used for these experiments. It consists of a device for the determination of the shearing resistance of a soil either subjected or not subjected to vibrations, a vibrator 1 driven by a direct-current motor, and a recording instrument 2.

The device for measuring the shearing resistance of a soil consists of three principal parts: a box 3, in which shearing is produced, a spring jack 4, to apply a vertical pressure on the soil sample, and the setup 5 for the application of shearing stresses to the soil, consisting of a cable spanned over a pulley and a loading platform. The shear box 3 is rigidly attached to the vibrating platform 6 of the device; it consists of a lower cylinder closed at its bottom and two rings inserted into a socket. The calibrated spring placed into the guides of the cylinder is the principal part of the spring jack 4. The supports 7 of the jack are rigidly attached to the vibratory platform 6 of the device.

In order to ensure the static nature of the action of the shearing load on the soil, the loading platform 5 is mounted on vibration absorbers.

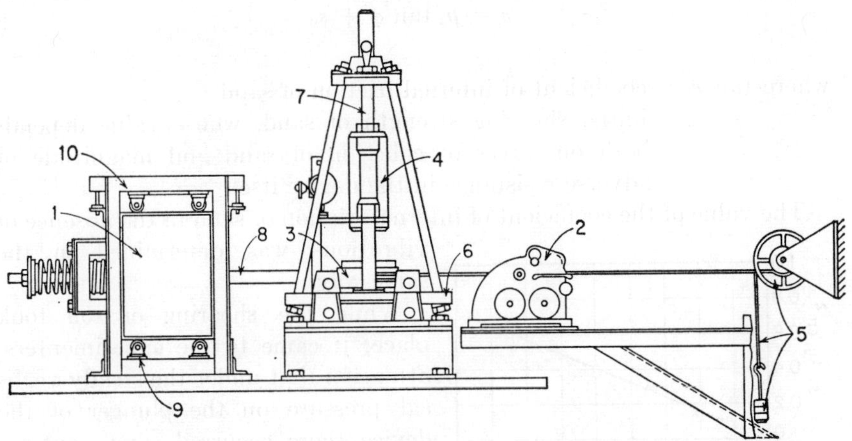

FIG. II-2. Installation used for investigation of the shearing resistance of soil.

Horizontal vibrations in the direction of shearing are induced by means of a vibrator of the usual two-shaft type. The vibrator is connected to the vibratory platform 6 by means of a turnbuckle 8 and can easily slide on the rollers 9 mounted on the guide frame 10, which is rigidly connected to the foundation under the installation.

During tests, the speed of the vibrator could be increased to 3,000 rpm. The amplitude of vibrations changed up to 1.7 mm, depending on the magnitude of the moment of the eccentrics of the vibrator.

The influence of vibrations on the coefficient of internal friction was evaluated by comparing its value determined in the absence of vibrations with the value obtained during soil vibrations, all other conditions remaining equal.

The coefficient of internal friction was determined as follows: the shear box was filled with sand and was placed on the vibratory platform. Then the sand was compacted by means of vibrations under a normal pressure of 0.5 kg/cm². After that the pressure on the soil was increased to a certain magnitude and shearing of the soil was induced while the box was vibrated at selected amplitudes and frequencies.

The regime of vibrations was kept constant during a test, but shear was produced under varying normal pressures. The results of each experiment were presented as graphs of relationship between shearing strength s and normal pressure p_z. An example of such a graph is shown in Fig. II-3. Analogous graphs secured from other experiments show that a linear relationship between the shearing strength of a soil and the normal pressure on it is observed with and without vibrations. This relationship is described as follows:

$$s = p_z \tan \phi + s_0$$

where $\tan \phi$ = coefficient of internal friction of sand
s_0 = initial shearing strength of sand, whose value depends both on forces of cohesion of sand and magnitude of adverse resistance in the device itself

The value of the coefficient of internal friction of sand in the absence of vibrations was determined in the same way.

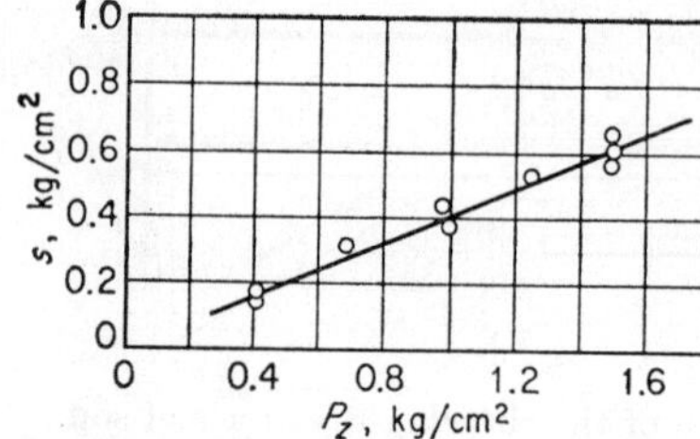

FIG. II-3. Relationship between the shearing strength of a soil and normal pressure.

While the shearing of soil took place, it came to the experimenters' attention that under the steady vertical pressure on the plunger of the device there occurred some contraction and then some loosening of the sand in the device. To investigate this phenomenon, measurements of the settlement of the plunger were taken. These measurements showed that with an increase in horizontal pressure the soil at first contracted; then began a process of expansion of the soil, which continued until shear was completed.

The process of initial contraction and subsequent expansion of the soil was observed both when the soil was subjected to vibrations and when it was not. In the latter case this process develops more intensively, especially the contraction phase. The tests also showed that as normal pressure grows, the changes of porosity, produced by shear, decrease.

The relative change in porosity during shear is small; therefore it does not substantially affect the shearing strength of a soil as determined by its coefficient of internal friction.

Figure II-4 illustrates the relationship between the coefficient of internal friction tan ϕ and the amplitude of vibrations of a dry medium-grained sand while the frequency remains constant. These graphs show that the coefficient of internal friction of sand decreases continuously as the amplitude increases.

The dependence of tan ϕ for the same sand on the angular frequency of vibrations ω is more complicated (Fig. II-5). As frequency increased up

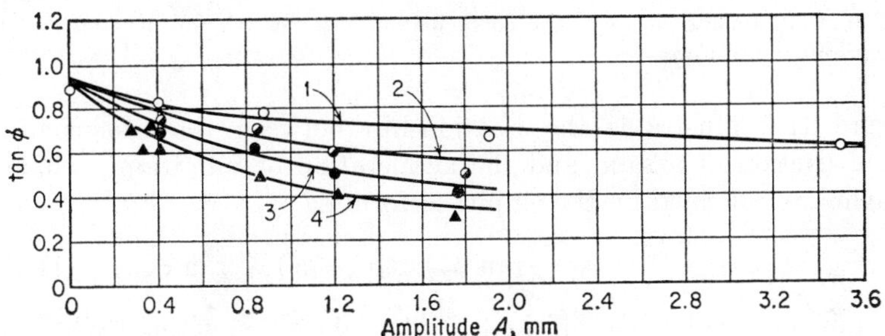

FIG. II-4. Relationship between the coefficient of internal friction of dry medium-grained sand and the amplitude of vibrations: curve 1, $\omega = 25 \text{ sec}^{-1}$; curve 2, $\omega = 144 \text{ sec}^{-1}$; curve 3, $\omega = 177 \text{ sec}^{-1}$; curve 4, $\omega = 208 \text{ sec}^{-1}$ (ω = angular frequency of vibrations).

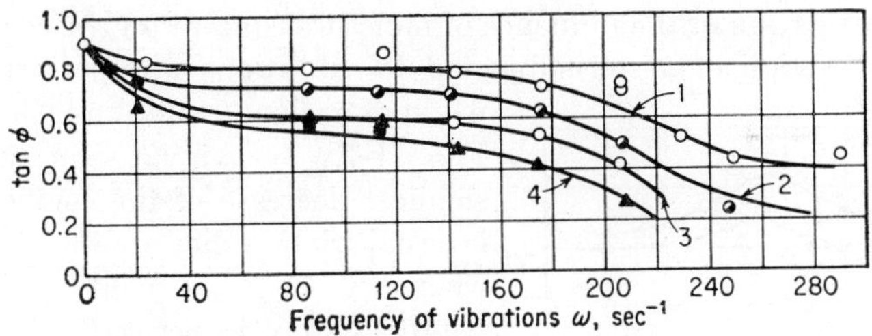

FIG. II-5. Relationship between the coefficient of internal friction of sand and the frequency of vibrations: curve 1, $A = 0.35$ mm; curve 2, $A = 0.85$ mm; curve 3, $A = 1.2$ mm; curve 4, $A = 1.6$ mm.

to 180 sec^{-1}, the coefficient of internal friction slowly decreased; then, as the frequency grew from 180 to 250 sec^{-1}, the coefficient of internal friction sharply decreased; subsequent increases in the value of the frequency had almost no influence on the coefficient of internal friction.

These data permit the conclusion that there exists a range of frequencies which corresponds to small changes in the coefficient of internal friction of sand.

It follows also that tan ϕ for a sand depends on both the amplitude and the frequency of vibrations.

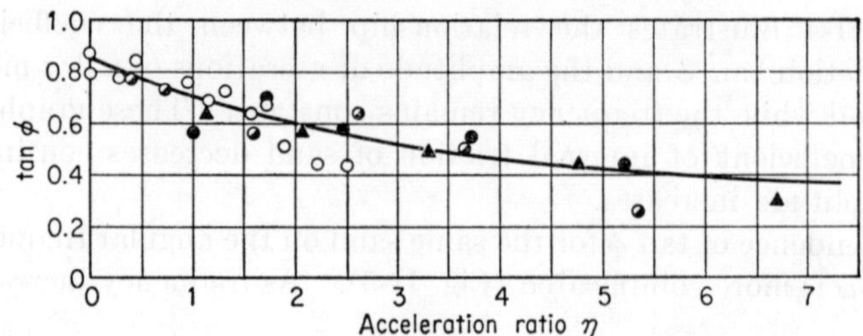

FIG. II-6. Relationship between the coefficient of internal friction of sand and the acceleration of vibrations.

Figure II-6 illustrates the relationship between the coefficient of internal friction of a sand and the acceleration of vibrations, which is approximately defined by the empirical formula

$$\tan \phi = (\tan \phi_{st} - \tan \phi_\infty) \exp(-\beta\eta) + \tan \phi_\infty \qquad \text{(II-1-1)}$$

where $\tan \phi_{st}$ = value of coefficient of internal friction without vibrations
$\tan \phi_\infty$ = limit value of coefficient of internal friction
η = ratio of acceleration of vibrations to acceleration of gravity
β = coefficient determining effect of vibrations (for dry medium-grained sand, $\beta = 0.23$)

The investigation of the influence of moisture content on the coefficient of internal friction of a soil subjected to vibrations was performed on medium-grained sand, as shown in Fig. II-7. It can be seen that the smallest decrease of the coefficient of internal friction due to vibrations was when the moisture content equaled approximately 13 per cent.

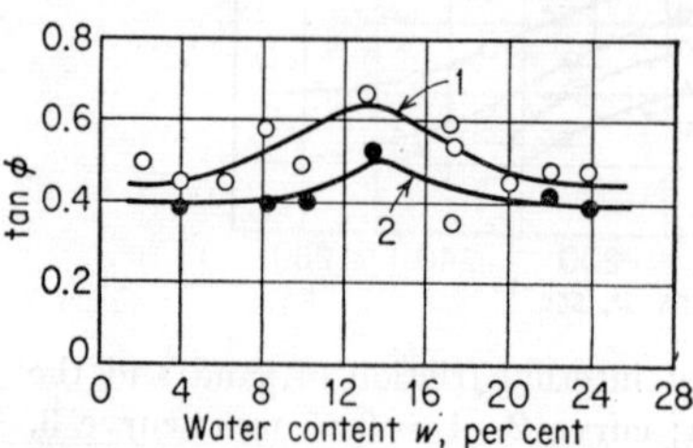

FIG. II-7. Relationship between the coefficient of internal friction of sand subjected to vibrations and its moisture content: curve 1, $\omega = 144\ \text{sec}^{-1}$; curve 2, $\omega = 250\ \text{sec}^{-1}$; $A = 0.35$ mm for both curves.

The study of the influence of grain size on the effect of vibrations was performed on four varieties of sand under two regimes of vibrations, as shown in Fig. II-8. The values of the coefficients of internal friction of these sands are different in the absence of vibrations; therefore it appeared advisable to compare, not the absolute values of these coefficients, but the values δ characterizing the effect of vibrations thereon:

$$\delta = \frac{\tan \phi_{st} - \tan \phi}{\tan \phi_{st}}$$

where tan ϕ_{st} and tan ϕ are coefficients of internal friction respectively without and with vibrations.

The value δ shows the degree of decrease in the value of the coefficient of internal friction under the influence of vibrations.

The data graphically presented in Fig. II-8 show that the effect of vibrations increases in proportion to sand-grain diameter.

b. Vibroviscous Soil Resistance. As indicated above, the effect of viscous forces of friction is most clearly displayed in noncohesive (freshly filled) sand undergoing intensive vibrations. Tests show that in such a sand the forces of internal dry friction between particles may completely

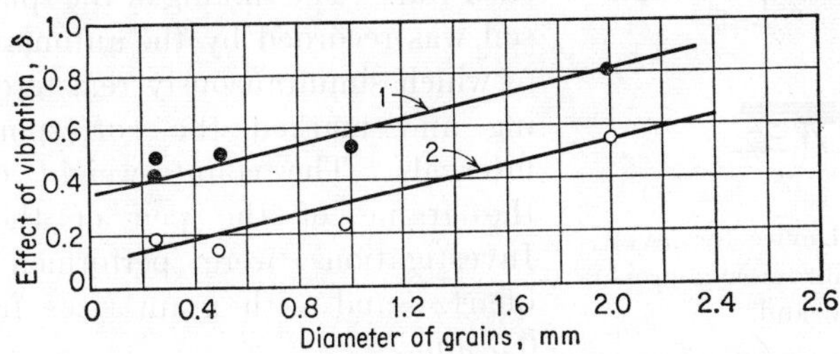

FIG. II-8. Relationship between the effect of vibrations and the diameter of sand grains: curve 1, ω = 144 sec^{-1}; curve 2, ω = 250 sec^{-1}; A = 0.35 mm for both curves.

disappear under the action of vibrations, and the sand may acquire the mechanical properties of a viscous fluid. Then objects placed on this sand will sink into it with a certain velocity if their unit weight exceeds the unit weight of the sand and will float if their unit weight is lower than that of the sand.

This soil property may be characterized by the coefficient of vibroviscosity. For the determination of this coefficient and for the investigation of its dependence on the acceleration of vibrations, the moisture content, and other factors, the method of a falling sphere may be used, as in the study of viscous fluids. This method is based on the well-known Stokes' law, which establishes the dependence of velocity v of the motion of a sphere in a viscous fluid on the force R acting thereon, the radius r of the sphere, and the coefficient of viscosity n of the fluid:

$$R = 6\pi nvr \tag{II-1-2}$$

If the sphere is under the action of weight only, then, designating by γ_1 the reduced unit weight of the sphere (i.e., the weight corresponding to the unit volume of the sphere) and by γ_2 the unit weight of soil, we may rewrite Eq. (II-1-2) as follows:

$$S = nv \tag{II-1-3}$$

where

$$S = \tfrac{2}{9}r^2(\gamma_1 - \gamma_2) \tag{II-1-4}$$

The device used by the author for the investigation of the coefficient of vibroviscosity is shown in Fig. II-9. A container 1, 30 by 30 cm in cross section, 40 cm high, and filled with sand to the height of 30 to 35 cm, was placed on the vibratory platform 2, which could be subjected to vertical vibrations of a selected amplitude and frequency. A metallic sphere 3 of 28 mm diameter was placed on the soil surface. A load 4 was imposed on this sphere by means of a thin (2.5 mm diameter) steel rod. The sinking of the sphere into the soil was recorded by the automatic recorder 5, which simultaneously registered the sinking and marked the corresponding time intervals. The counterweight 6 balanced the frame of the pen of the recorder. Investigations were performed on white quartz sand with grain sizes from 0.2 to 0.5 mm.

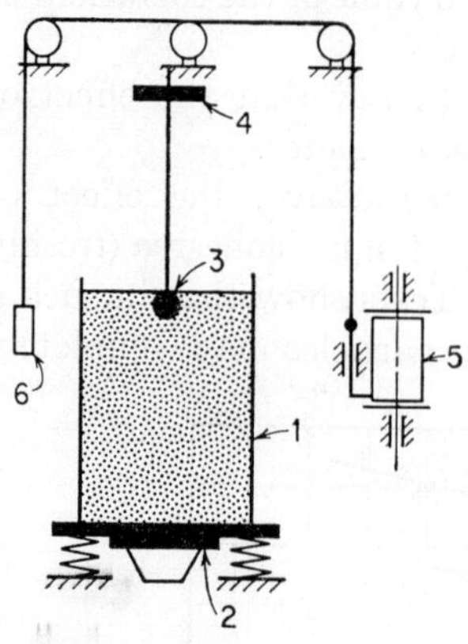

FIG. II-9. Device for investigation of vibroviscous properties of filled sand.

The following method was used in the experimental investigation of the relationship between the coefficient of vibroviscosity and the acceleration of vibrations.

In the course of all the experiments the moisture content of the sand

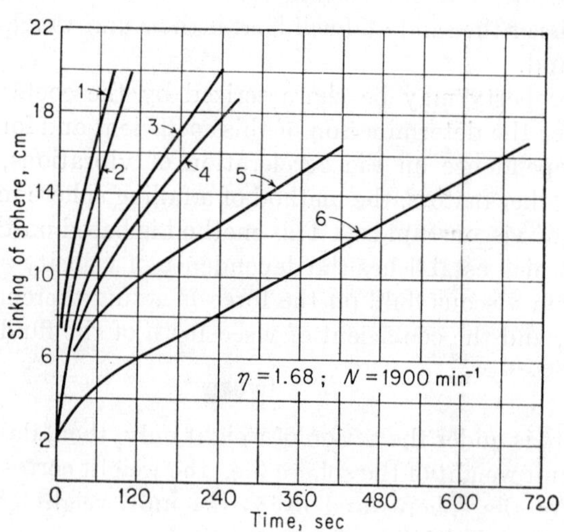

FIG. II-10. Relationship between the depth of sinking of sphere into vibrating sand and time: curve 1, load = 1.9 kg; curve 2, load = 1.5 kg; curve 3, load = 1.2 kg; curve 4, load = 1.0 kg; curve 5, load = 0.6 kg; curve 6, load = 0.3 kg.

was equal to zero. Prior to testing the sand was compacted by means of vibration so that the value of its void ratio was close to its minimum (approximately 0.5). The unit weight of soil corresponding to this porosity was approximately 1.77 g/cm³.

While the value of the acceleration of vibrations remained the same, the sinking of the sphere was recorded under variable loading.

After the completion of a series of tests with a constant acceleration of vibrations and selected values of loading, the acceleration was changed and a new series of experiments was conducted, in order to record the sinking of the sphere which corresponded to various values of loading. This procedure was repeated successively.

The graphs of relationship between time and the depth of sinking of the sphere into vibrated sand are shown in Fig. II-10. These curves are plotted for constant acceleration of vibrations and variable loads. It is first seen from the graphs that the velocity of sinking varies: it decreases with the depth of sinking. However, as the sphere sinks, its acceleration approaches zero, and its velocity attains a more or less stable value depending on the magnitude of the load and the acceleration of vibrations.

Equation (II-1-3) may be applied only for the time interval corresponding to steady sinking velocity, when, consequently, the influence of inertia may be neglected.

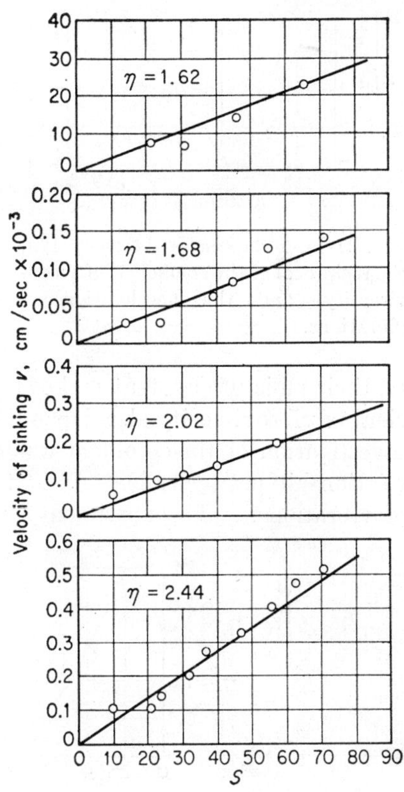

FIG. II-11. Relationship between the velocity of sinking of sphere into vibrating sand and load S.

Graphs in Fig. II-11 present relationships—for constant values η of the acceleration of vibrations—between the sinking velocity of the sphere and S (i.e., the load acting on the sphere). These graphs confirm the linear relationship, established by Eq. (II-1-3), between the steady sinking velocity and the load. It is directly seen from these graphs that the coefficient of proportionality between the two, i.e., the coefficient of vibro-viscosity, depends essentially on the acceleration of vibrations.

Figure II-12 presents a graph of relationship between the value of $1/n$ (that is, the reciprocal of the coefficient of vibroviscosity n) and the ratio η between the acceleration of vibrations and the acceleration of gravity. This graph shows that when accelerations of vibrations are lower than $\eta = 1.5g$, vibrations practically do not affect the value of the coefficient of vibroviscosity. A sharp decrease in n, and consequently an increase in $1/n$, is found only when $\eta > 1.5$. The relation between $1/n$ and η may be approximately described by the equation

$$\frac{1}{n} = a(\eta - \eta_0) \qquad \text{(II-1-5)}$$

where η_0 may be designated as the threshold of the vibroviscose state of soil.

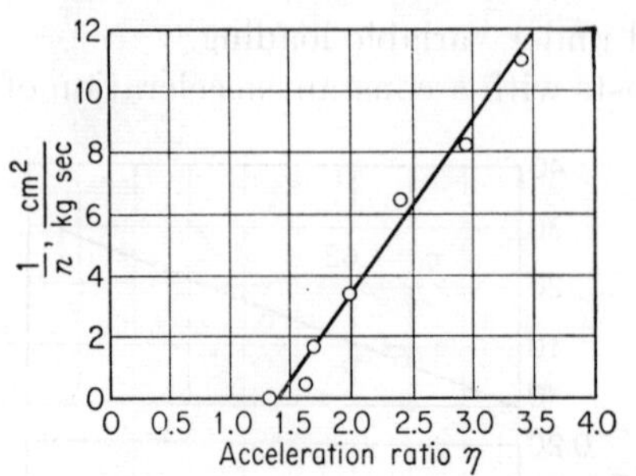

FIG. II-12. Relationship between $1/n$, the reciprocal of the coefficient of vibroviscosity, and the acceleration of vibrations.

The fact that the magnitude of the forces of cohesion in soils depends on their moisture content makes possible the assumption that the coefficient of vibroviscosity also depends on moisture content. The method of investigation of this problem was analogous to the investigation of the relationship between the coefficient of vibroviscosity and the acceleration of vibrations. The experiments were performed with the same sphere

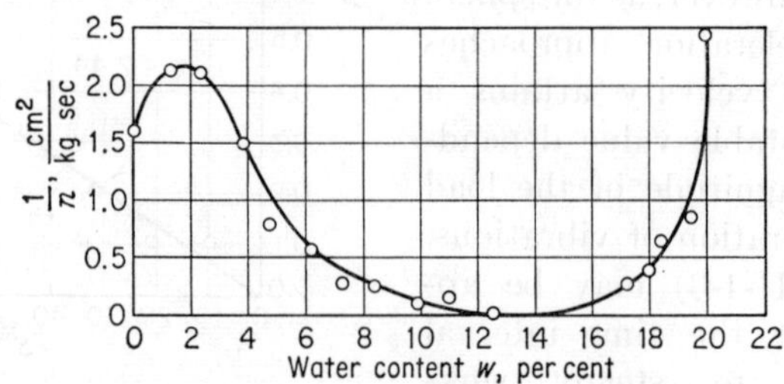

FIG. II-13. Relationship between $1/n$, the reciprocal of the coefficient of vibroviscosity, and the moisture content of sand fill.

and with the same sand, but with variable moisture content; the acceleration of vibrations and the total load imposed on the sphere (including its weight) remained the same in all the experiments.

Figure II-13 shows the relationship between the value $1/n$ and the moisture content of a soil. With increase in moisture content to approximately 13 per cent, the coefficient of vibroviscosity increases approximately 220 times. A further increase in moisture content leads to a sharp decrease in n. When moisture content equals some 20 per cent, n attains about the same value as in dry sand. If one is to assume that the coefficient of vibroviscosity of a soil is proportional to the forces of friction and cohesion acting between its particles, then the graph in Fig. II-13 illustrates how these forces change with moisture content.

The results of these experiments lead to the assumption that, other conditions remaining equal, the residual settlements of foundations (or the sinking into a soil as a result of vibrations) will have the largest velocity when the sand either is dry or has a moisture content approaching the largest possible moisture content for the given soil.

An analogous dependence of the coefficient $1/n$ on the moisture content was obtained for artificially prepared clayey soils.

c. Damping Properties of Soils. The study of the viscous properties of soil by means of a falling sphere is evidently possible if the intensity of vibrations is so large that the soil loses a considerable part of its shearing resistance and behaves as a viscous body in the sense of Newton and Bingham. Otherwise, the viscous properties of soils are manifested to a much smaller degree than the elastic and plastic properties, so that the method of a falling sphere is no longer applicable. However, although the viscous properties of a soil may be manifested very slightly, they may have a significant influence on the following factors: free vibrations, forced vibrations under conditions of resonance or close to resonance, the magnitude of energy spent for the maintenance of vibrations, the propagation of waves, and others.

Therefore the study of these properties of soils, which are known in the literature as damping properties, has a practical interest.

Damping properties of soils cause stresses which depend not only on deformation but also on rates of deformation. Therefore soil reactions against foundations depend not only on settlement, but also on rate of settlement.

To simplify computations, it is generally assumed that soil presents a linearly deformable elastic-viscous body and that stresses are linearly related to the deformations and the rates thereof.

The determination of the damping properties of materials, including soils, may be performed by several methods: from observations of the damping of free vibrations of soil samples, from the amplitude of forced vibrations under conditions of resonance, from the phase shift between the existing periodic force and the soil deformation, and, finally, from hysteresis loops.

The last method is the simplest; therefore we shall dwell on it in some detail.

Experiments show that even when stresses are small and are considerably lower than the elastic limit, the relationship between stresses and deformations is not linear and single valued: to the same value of stress there correspond values of deformation which are different under conditions of loading and of unloading. Therefore, a cycle of successive loading and unloading is graphically presented, not by a straight line, but by a closed curve which is usually called a loop of elastic hysteresis.

A greater degree of nonhomogeneity of a discrete material is reflected by a greater intensity of display of the damping properties of that material and by a larger area of the hysteresis loop. Since soils (especially sandy soils) are more micrononhomogeneous and discrete than steel and concrete, for instance, the phenomena of hysteresis are displayed in soils much more intensively than in these other materials.

The larger the loop area, the more intensive are the damping properties of the material, i.e., the larger is the ability to absorb mechanical work in an irreversible form by transforming it into heat. Quantitatively this work ΔW equals the area of the hysteresis loop. If W is the total work performed during loading up to the maximum deformation, then the ratio

$$\psi = \frac{\Delta W}{W}$$

is the coefficient of absorption which determines the amount of energy absorbed by the material per unit of energy spent for deformation per cycle. This coefficient may be related to the coefficient of resistance to vibrations, which is the coefficient of proportionality between the soil reaction and the velocity of vibrations.

Let us assume that the diagram of free vibrations of a foundation will have the shape of a periodic curve, analogous to that shown in Fig. II-22. If we take two successive maximum deflections of the foundation from the equilibrium state with amplitudes of, for example, A_1 and A_2, the corresponding energies of the foundation vibrations will equal

$$W_1 = \frac{k}{2} A_1^2 \qquad W_2 = \frac{k}{2} A_2^2$$

where k is the coefficient of rigidity of soil.

The energy absorbed by the soil per period of vibration equals

$$\Delta W = \frac{kA_1^2}{2}\left(1 - \frac{A_2}{A_1}\right)$$

Hence,

$$\psi = 1 - \left(\frac{A_2}{A_1}\right)^2$$

On the other hand (with general reference to Chap. III), for free vibrations of a foundation with one degree of freedom,

$$A_1 = A \exp\left(-\frac{cT}{4}\right) \qquad A_2 = A \exp\left(-\frac{5cT}{4}\right)$$

where c = damping constant
T = period of vibrations
A = amplitude of vibrations without taking into account damping properties of soil†

Therefore,
$$\psi = 1 - \exp(-2cT)$$
hence
$$\xi = \frac{c}{f_n} + -\frac{1}{4\pi} \ln(1 - \psi)$$

Here f_n is the natural frequency of vibrations of the foundation. Coefficient ξ is designated as the coefficient of damping of vibrations.

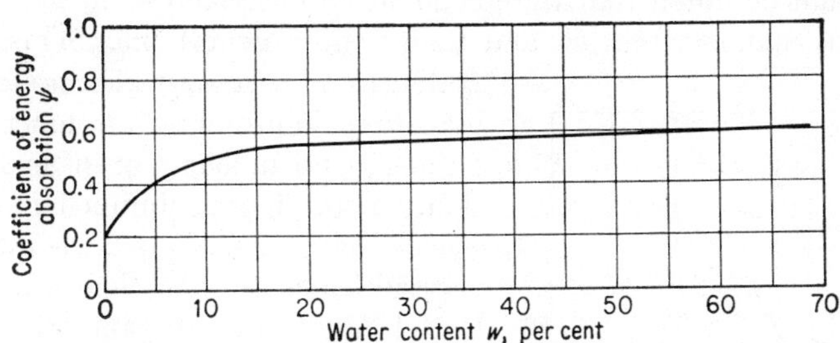

FIG. II-14. Relationship between the coefficient of energy absorption of a clayey soil and its moisture content.

The values of ψ were determined by I. A. Savchenko under laboratory conditions. His data show that this coefficient is a comparatively stable soil characteristic.

The experiments established that the value of the coefficient of absorption ψ both for sandy and clayey soils does not depend on the rate of load application, or on the frequency of changes in the load, or on the maximum stress in the hysteresis loop.

Contrary to existing opinions, it was established experimentally that in soil undergoing shearing deformations the value of the moisture content has little influence on the coefficient of energy absorption ψ of sand.

Investigations of natural clayey soils with some sand and silt content showed that when the soil undergoes compressive deformations, ψ increases proportionally to the increase in the moisture content (see Fig. II-14).

The absolute value of ψ is smaller for compressive deformations than for shearing deformations.

The grain size of sand (Table II-1) has a considerable effect on the coefficient of absorption: the latter increases with increase in grain size.

† These equations for A_1 and A_2 are obtained from Eq. (III-1-18) by setting $B = 0$ (what corresponds to the excitement of free oscillations by a shock) and by setting the time τ equal to one-fourth and five-fourths of the period.

Apparently this can be explained by the growing dispersity and micro-nonhomogeneity of sand with an increase in grain size.

TABLE II-1. RELATIONSHIP BETWEEN THE COEFFICIENT ψ OF ENERGY ABSORPTION OF SAND UNDER CONDITIONS OF SHEAR AND THE GRAIN SIZE

Grain size of sand, mm	ψ
0.10–0.25	0.64
0.25–0.50	0.68
1.00–2.00	0.79

It should be noted that the phenomenon of hysteresis in soil develops differently under vibration and under static conditions. This is illustrated by Fig. II-15, which presents two hysteresis loops obtained for medium-grained sand undergoing shear deformation. Loop 1 was obtained in the absence of vibrations, loop 2 when the sand was subjected to vibrations with a horizontal amplitude (in the direction of shear) of $A_x = 1.2$ mm with frequency $\omega = 155 \text{ sec}^{-1}$. Figure II-15 shows that the area of the hysteresis loop is much smaller when sand vibrates. This is explained by the considerable decrease in the shearing resistance of vibrated sand.

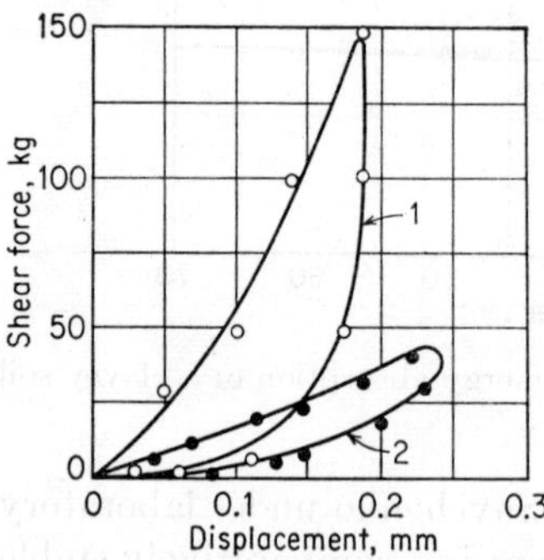

FIG. II-15. Hysteresis loops in sand: (1) in the absence of vibrations; (2) with vibrations.

For the above test conditions, the coefficient of energy absorption of the sand was decreased 25 per cent by the vibrations. The values of the coefficient of absorption permit only the estimation of the influence of certain properties of soil on its coefficient of damping of vibrations ξ. The true value of ξ depends essentially on the transfer of the energy of vibrations to the soil. This transfer depends on the size of the foundation area in contact with soil, the foundation weight, and the properties of the soil as a medium in which elastic waves are propagated. Therefore the design values of the coefficient of damping of vibrations should be taken from results of vibration observations on models or existing foundations (see Chap. III).

II-2. Effect of Vibrations on the Porosity and Hydraulic Properties of Soils

The porosity and the hydraulic properties described by the coefficient of permeability and by pore pressures are basic physicomechanical characteristics of soils. Therefore the study of the effect of vibrations on these properties is of great interest not only from the point of view of the design and erection of foundations subjected to dynamic loads, but also from the point of view of the use of vibrations for the improvement of engineering properties of soils.

Experiments show that the principal vibration parameter which determines the effect of vibrations and shocks on the compaction of soils is the acceleration, or rather the inertial force, which acts on the soil particles during vibration.

The inertial forces are proportional to the density of particles forming a dispersed system. Therefore, other conditions remaining equal, systems containing particles characterized by high specific gravity will

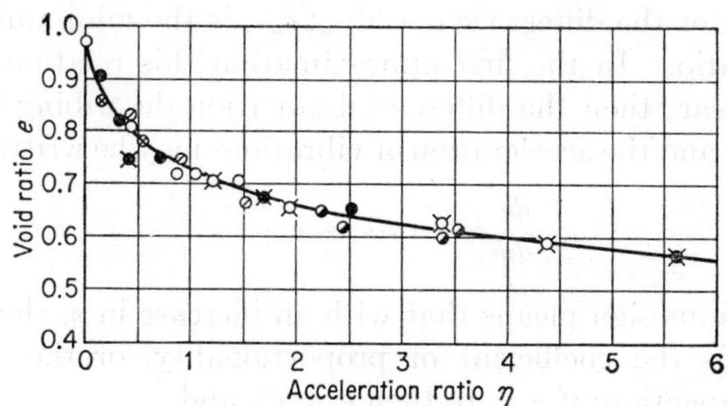

FIG. II-16. Vibratory consolidation curve of sand fill.

undergo a more intensive compaction than systems built up of particles with lower specific gravity. However, there is so little difference between the values of specific gravity of various noncohesive soils that it may be considered that the compaction produced by vibrations is a function only of the acceleration of the vibrations transmitted to the soil.

Once the dependence of the void ratio on the amplitude and frequency of vibration is experimentally established, it is possible to plot graphs showing a relationship between the void ratio and the acceleration of vibrations. Figure II-16 presents such a graph plotted for sand. On the abscissa is plotted the ratio η between the acceleration of vibrations and the acceleration of gravity.

Experimental investigations indicated that the function $e = e(\eta)$ is of the same type for all soils. In the initial state, all sands are characterized by a value of the void ratio approaching the greatest limit value; however, when sands are subjected to vibrations, even to those characterized by low acceleration, they undergo considerable compaction.

As the void ratio decreases, it becomes less susceptible to changes in vibration acceleration. For large accelerations, the relative void-ratio change is small, and its value may be considered to approach the minimum.

This kind of dependence of the void ratio on the acceleration of vibrations is observed not only in sands, but also in other disperse systems capable of undergoing vibratory compaction.

By analogy with the static consolidation curves, the graphs establishing a relationship between the void ratio and the acceleration of vibrations may be named the "vibratory consolidation curves." In highway construction the dependence of the void ratio on the intensity of impacts is sometimes determined. The graphs showing this dependence may be called the "impact consolidation curves."

Experimental investigations indicate the existence of a nonlinear relationship between the void ratio and the acceleration of vibrations (or the ratio η). Therefore the slope of the curve $e(\eta)$ is a variable depending on the values of e or the difference $e - e_{\min}$; $e_{\min}$ is the minimum limit value of the void ratio. In the first approximation this relationship may be considered linear; then the differential equation describing the relationship between e and the acceleration of vibrations may be written as follows:

$$\frac{de}{d\eta} = -\alpha(e - e_{\min}) \qquad \text{(II-2-1)}$$

Here the minus sign means that with an increase in η, the value $de/d\eta$ decreases; α is the coefficient of proportionality, or the coefficient of vibratory compaction; if $\eta = 0$, then $e = e^0$, and

$$\frac{1}{\alpha} = \frac{e^0 - e_{\min}}{\tan u}$$

where u = slope of curve

e^0 = initial value of void ratio when $\eta = 0$

The physical significance of the coefficient $1/\alpha$ may be stated as follows: if there is a linear relationship between e and η, then the value $1/\alpha$ equals the value of the acceleration of vibrations which induces the maximum possible soil compaction, i.e., a void ratio equal to $e_{\min}$.

By integration Eq. (II-2-1), we obtain

$$e = e_{\min} + C \exp(-\alpha\eta)$$

where C is a constant of integration determined from the following condition: If $\eta = 0$, then $e = e^0$ and $C = e^0 - e_{\min}$. Consequently

$$e = e_{\min} + (e^0 - e_{\min}) \exp(-\alpha\eta) \qquad \text{(II-2-2)}$$

Let us assume that prior to compaction by vibration the soil was in its loosest possible state, so that the void ratio was the largest possible, $e = e_{\max}$; the equation of the vibratory consolidation curve in this case will read as follows:

$$e = e_{\min} + (e_{\max} - e_{\min}) \exp(-\alpha\eta) \qquad \text{(II-2-3)}$$

If, as a result of vibrations with the acceleration $\eta = \eta_0$, the void ratio will decrease to e^0, then

$$e^0 = e_{\min} + (e_{\max} - e_{\min}) \exp(-\alpha\eta_0) \tag{II-2-4}$$

Substituting this value of e^0 into the right-hand part of Eq. (II-2-2), we obtain

$$e = e_{\min} + (e_{\max} - e_{\min}) \exp[-\alpha(\eta + \eta_0)]$$

It follows that in the above equation η_0 is the value of the vibration acceleration needed to bring the soil from the state of least compaction (characterized by the void ratio $e_{\max}$) to the natural (initial) state of static compaction (characterized by the void ratio e^0; $e_{\max} > e^0$).

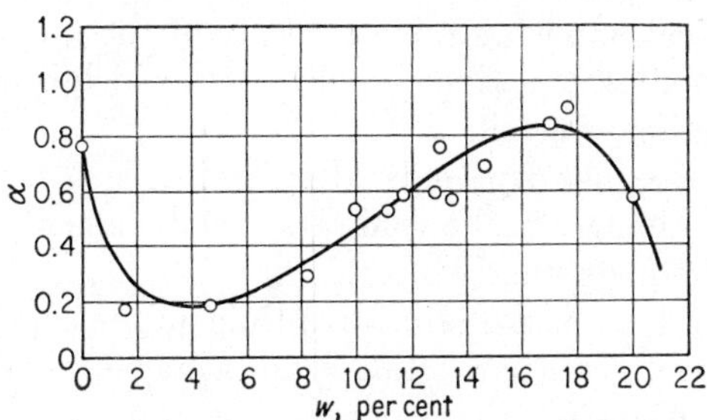

FIG. II-17. Relationship between the coefficient of vibratory compaction α of sand and its moisture content.

Equation (II-2-3) contains three parameters: $e_{\max}$, $e_{\min}$, and α; while the parameters $e_{\max}$ and $e_{\min}$ by definition do not depend on the soil moisture content and are affected only by the grain-size distribution and the size of the largest sand particles, the value of α depends also on the moisture content.

It follows from the graph of Fig. II-17 that the moisture content has an appreciable effect on the coefficient of vibratory compaction α. When the moisture content of a sand is low (less than 5 per cent), α sharply decreases; consequently the capacity of a soil to compact under the action of vibrations also decreases. When the moisture content increases, the coefficient of vibratory compaction gradually grows, and when the moisture content attains some 16 to 17 per cent (corresponding to some 80 per cent of the complete water saturation of voids in sand) it reaches a maximum value of $\alpha = 0.82$ to 0.88.

When the moisture content becomes larger than $w = 16$ to 17 per cent, the coefficient α decreases, and when the sand is fully saturated by water it equals approximately eight-tenths of its largest value, corresponding to the optimum moisture content of the soil.

The existence of an optimum moisture at which the greatest compaction is achieved (other conditions being equal) is characteristic not only for sands, but also for other soils (for example, for clays with sand and for clays with some sand and silt).

The foregoing results of investigations concerning the dependence of the coefficient of vibratory compaction α on the moisture content w of a soil explain to some degree the considerable settlements of foundations bearing dynamic loads when erected on water-saturated sands.

When a soil is subjected to vibrations with a selected acceleration, the process of compaction of the soil does not occur instantly, but develops during a certain period. As the time of action of the vibrations increases, the void ratio decreases at a diminishing rate and after a certain time reaches a constant value, regardless of further vibration.

Of great practical importance is the study of the question of the time interval, corresponding to a given acceleration, which is necessary for the complete compaction of a soil.

Experiments show that dry sands, after having been vibrated for 30 sec, do not reveal any changes in the void ratio. For cement, the corresponding time interval equals some 120 sec.

Investigation of the effect of vibration time on the compaction of moist soils shows that complete compaction is not achieved even after the soil is subjected to vibrations for 3 to 4 min.

If static stresses exceed the elastic limit, then residual deformations appear. In the same way, vibrations result in compaction if the acceleration exceeds a certain value which may be termed the "threshold of vibratory compaction." The existence of this value is confirmed experimentally.

It is also established that when a soil is subjected to vibrations with a selected acceleration, the threshold of vibratory compaction is raised to a value equaling this acceleration. And vice versa, if a soil (or any other disperse body capable of changing its void ratio under the action of vibrations) has a void ratio e^0 corresponding on the vibratory consolidation curve to the value η_0 of the vibration acceleration, then vibrations with an acceleration lower than η_0 will not cause any changes in the void ratio.

Therefore, the condition of the compaction of soil under the action of vibrations may be written in the form

$$\eta > \eta_0 \tag{II-2-5}$$

where η_0 is the threshold of vibratory compaction, i.e., the acceleration of vibrations corresponding to the void ratio prior to vibrations. Its value is determined from Eq. (II-2-4).

The concept "threshold of vibratory compaction" makes it possible to determine the dimensions of the zone of soil compaction caused by a selected vibrator.

Laboratory investigations showed that soil compaction under a given regime of vibrations depends essentially on the magnitude of the load acting on the soil. If a sand already compressed in a consolidometer is subjected to vibrations, the degree of further compaction will be relatively low. Moreover, when a static load is applied to a sand, it will be compacted under the action of vibrations only when the acceleration of vibrations exceeds a certain value, larger than the one needed for compaction if no static load is present. Hence, the threshold of vibratory compaction depends not only on the initial value of the void ratio, but also on the static normal stresses acting within the soil. The larger the magnitude of these stresses, the higher the threshold of vibratory compaction of the soil.

The effect of static load on soil compaction by vibration is attributed to the fact that the forces of friction between particles increase with an increase of pressure on the soil. Vibrations then cause a smaller change in the density of the soil. In other words, under the action of pressure the sand, so to say, acquires a greater cohesion and consequently does not respond as much to vibratory compaction.

Static pressure may have an appreciable influence on the vibratory compaction of soil. In particular, it may be assumed that with an increase in depth the effect of vibrations on soil compaction will decrease.

When elastic waves are produced in a soil by a local source, it is evident that the acceleration of soil vibrations will depend on the coordinates of the point under consideration. Therefore in a soil subjected to vibrations induced, for example, by machinery foundations, the void ratio will not remain constant, but will undergo changes which will be governed by changes in the acceleration within the soil, i.e., by the distance from the foundation.

If a point with the coordinates (x,y,z) has an acceleration defined by the ratio η, whose value depends on these coordinates (η, as before, is the ratio between the acceleration of vibrations and the acceleration of gravity), then the void ratio at this point will be determined by Eq. (II-2-2).

It follows from the definition of the threshold of vibratory compaction that compaction will occur only at those points of the soil where the inequality (II-2-5) is satisfied.

Beyond the zone of vibratory compaction a reverse inequality is valid, and at the boundaries of this zone

$$\eta(x,y,z) = \eta_0$$

Thus the boundary of the zone of vibratory compaction of the soil represents a surface of equal acceleration of vibrations; hence, the determination of the zone of compaction is reduced to the computation of the field of accelerations of soil vibrations excited by an oscillating foundation.

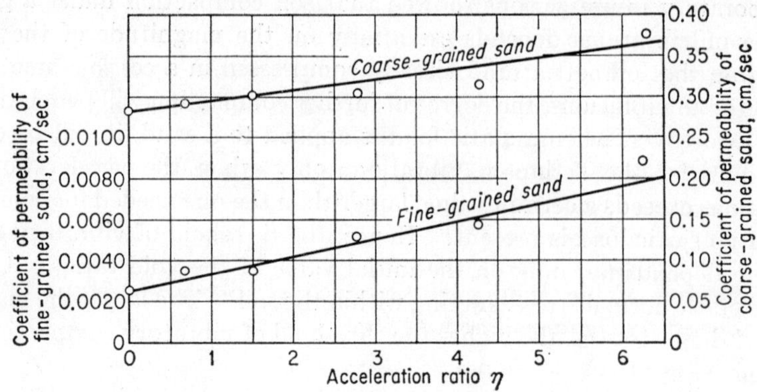

FIG. II-18. Relationship between the coefficient of permeability and the acceleration of vibrations.

Experiments with water-saturated soils show that vibrations lead to changes in the hydraulic state of a soil as determined by the coefficient of permeability and the pore pressures in the ground water.

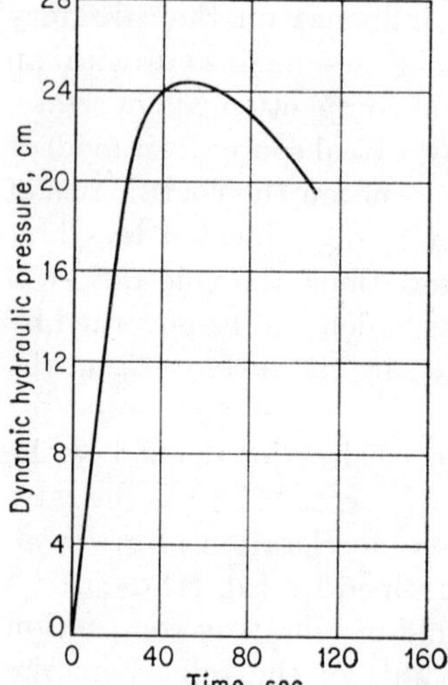

FIG. II-19. Changes with time in dynamic hydraulic pressure in vibrating sand.

Figure II-18 shows two graphs of changes in the coefficient of permeability of sands depending on the intensity of vibrations. These graphs show that the coefficients of permeability increase proportionally to the acceleration. It is also seen that the effect of vibrations is larger in fine-grained than in coarse-grained sands.

The effect of vibrations on the coefficient of permeability may be utilized for the injection of different solutions into a soil, for example, for soil stabilization by chemicals.

Experimental investigations performed by N. N. Maslov, V. A. Florin, M. N. Goldshteyn, and others show that the pore pressure in water-saturated sands changes under the influence of vibrations. This is illustrated by Fig. II-19, which clearly shows changes with time in the pore pressure in sand subjected to vibrations with an acceleration of only 15 cm/sec^2. It is seen from this graph that at first the pressure grows, then it continuously decreases over a long period. When sand is subjected to impulsive vibrations caused by shocks, a different pattern of changes in pressure with time is observed; namely, the time interval corresponding to the increase in pressure is much shorter than in the previous case. The nature of the development of dynamic hydraulic pressures in a soil subjected to vibrations or shocks is insufficiently understood as yet. There are reasons to assume that the development of dynamic pressures in ground water leads to a decrease in the shearing resistance of the water-saturated soil and consequently to the loss of stability of earth masses.

II-3. Elements of the Theory of Residual Settlements of Foundations Induced by Dynamic Loads

The theory of the development of residual settlements of foundations subjected not only to static but also to dynamic loads has not been elaborated so far to a degree sufficient to serve as a basis for the estimation in each particular case of the value of the residual settlement. This is due not only to the mathematical difficulties involved in the theoretical analysis of a foundation subjected to the action of loads which vary with time, but also to the fact that so far there is no sufficiently verified rheological soil model, nor have there been established any numerical values, expressing soil properties, which would permit the determination of the magnitude of the residual dynamic settlement.

In computations of settlements of structures under the action of static loads, soil is usually considered to be a linearly deformable, although inelastic, body.

Variations of settlements with time may be taken into account, for example, by the introduction of stresses which depend on the rate of deformation. Thus in the general case, the computation of settlements of foundations subjected to the action of static loads only is based on the use of a model of a linearly deformable viscous body whose elastic limit equals zero.

It is hardly justifiable, however, to use such a rheological model in the case of dynamic loads, because, as shown by experiments, the development of residual settlements of foundations occurs intensively under conditions in which the total stress (imposed by the static and dynamic loads) exceeds the definition of the limit value.

When stresses are lower than the limit value, settlements of foundations subjected to vibrations and shocks will be of the same order of magnitude as settlements caused by static loads. But since the static pressure imposed on a soil by foundations under machinery is comparatively low (as a rule, it is lower than that imposed by foundations under buildings and structures), settlements caused by this pressure will also be comparatively low. However, if the total pressure on the soil exceeds a limit value p_l, the process of development of settlements is intensified, and they may reach considerable values, often up to several tens of centimeters. The settlements are nonuniform and may lead to an impermissible tilting of the foundation.

Thus the condition for the absence of residual settlements of foundations under machines may be written in the form

$$p_{st} + p_{dy} < p_l \tag{II-3-1}$$

where p_{st} and p_{dy} are the static and dynamic pressures on the soil under the foundation.

Since vibrations affect the internal friction and cohesion of a soil (and consequently its resistance to local loads), the value of the limiting

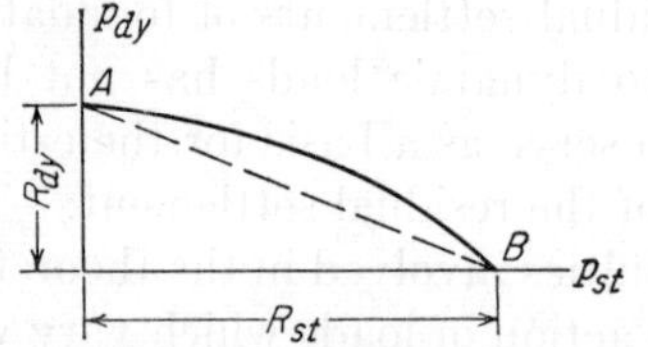

Fig. II-20. Required static pressure decrease in the presence of vibrations.

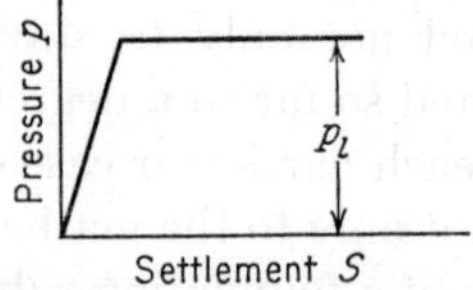

Fig. II-21. Simplified settlement diagram in the presence of a limiting pressure p_l.

pressure p_l for a given soil will depend on the characteristics of the stress cycle, i.e., on the relationship between p_{st} and p_{dy}.

Analogously to the known relationship between the fatigue limit of a material and the characteristic of the stress cycle, it may be assumed that the relationship between the limiting values of the static and dynamic stresses in the cycle will be determined by graphs similar to curve AB in Fig. II-20. Designating by R_{st} the maximum static pressure corresponding to the intensive development of settlements under the action of static pressure only, and by R_{dy} the maximum dynamic pressure corresponding to the intensive development of settlements under the action of dynamic pressure only, and straightening the curve AB, one can establish the following relationship:

$$p_{st} \leq qR_{st} \tag{II-3-2}$$

where

$$q = 1 - \frac{p_{dy}}{R_{dy}} \tag{II-3-3}$$

q is the coefficient of required pressure decrease on soil in the presence of vibrations.

In accordance with the foregoing discussion, let us consider that as long as the normal pressure on the soil is lower than its limit value p_l, settlement of the foundation will be elastic and, consequently, the foundation subjected to the action of dynamic loads will undergo only elastic vibrations.

When the normal pressure on a soil reaches p_l, there appear residual settlements which increase under the action of this pressure, even if its value remains constant. Thus the relationship between the pressure imposed on a soil by the foundation and the settlement of the foundation will be illustrated by a graph similar to the one presented in Fig. II-21.

If we take into account the soil reactions which depend on the rate of settlement of the foundation, we arrive at a rheological model of the soil, which corresponds to an elastoplastic viscous body.

Let us first consider the forced vertical vibrations of a foundation. The amplitudes of vibrations of machinery foundations usually lie within such a small range of values that they undergo vibrations without separating from the soil. Then the maximum dynamic pressure on the soil equals

$$p_{dy} = c_u A_z$$

where A_z is the amplitude of vertical vibrations of the foundation.

Substituting this expression for p_{dy} into inequality (II-3-1) and solving for the amplitude of forced vibrations, we come to the following condition for the absence of residual settlements of the foundation:

$$A_z \leq \frac{p_l - p_{st}}{c_u} \tag{II-3-4}$$

For a great majority of machines whose foundations were designed more or less correctly, the actual values of vibration amplitudes are much lower than the limiting values determined by the above inequality. This explains the fact that foundations under machines with a steady regime of operation (turbogenerators, reciprocating engines, etc.) do not undergo significant residual settlements, even when erected on water-saturated fine-grained sands, which are especially susceptible to changes in properties under the action of vibrations.

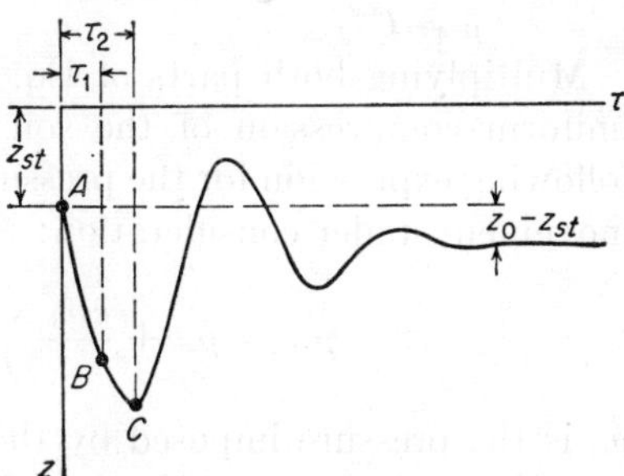

FIG. II-22. Damped vibrations due to sudden impact.

The determination of residual settlements is of special practical interest in foundations undergoing unsteady free vertical vibrations induced by vertical impact or by waves propagated through soil. Amplitudes of such vibrations may attain a relatively large value and may therefore induce relatively large dynamic stresses.

A diagram of such vibrations will be similar to the curve presented in Fig. II-22. Let us assume that section AB of the vibration diagram corresponds only to elastic settlement of the foundation. A differential equation of the settlement may be written in the form:

$$\ddot{z}_1 + 2c\dot{z}_1 + f_{nz}^2 z_1 = g \tag{II-3-5}$$

This equation differs from Eq. (III-1-17) only by the presence of a constant in the right-hand part. This constant is the acceleration of gravity and takes into account the effect of the weight of the foundation and machinery. All symbols are the same as those in Art. III-1.

Let us assume that free vertical vibrations of the foundation are caused by a centered impact. Under the action of this impact, the foundation acquires an initial velocity in a downward direction equal to v_0; its values are determined by Eq. (V-2-8).

Initial conditions of the foundation movement may be written in the form: If $t = 0$, then

$$z = z_{st} \qquad \text{and} \qquad \dot{z}_1 = v_0 \tag{II-3-6}$$

where z_{st} is elastic settlement of the foundation under the action of the weight of foundation and machinery.

The solution of Eq. (II-3-5) with initial conditions expressed by Eqs. (II-3-6) will be

$$z_1 = z_{st} + \frac{v_0}{f_{nd}} \exp(-\rho\tau) \sin \tau \tag{II-3-7}$$

where f_{nd} = frequency of natural vertical vibrations of foundation under consideration of damping: $f_{nd}^2 = f_{nz}^2 - c^2$

τ = relative time: $\tau = f_{nd}t$

$\rho = C/f_{nd}$

Multiplying both parts of Eq. (II-3-7) by the coefficient c_u of elastic uniform compression of the soil under the foundation, we obtain the following expression for the pressure on soil at the phase of the foundation movement under consideration:

$$p_{z1} = p_{st} + \frac{c_u v_0 \sqrt{1 + \rho^2}}{f_{nz}^2} \exp(-\rho\tau) \sin \tau \tag{II-3-8}$$

p_{st} is the pressure imposed by the weight of foundation and machinery.

As long as $p_{z1} > p_l$, the foundation settlements remain elastic. Let us assume that at a certain time $\tau = \tau_1$, to which corresponds the point B on the vibration diagram (Fig. II-22), the pressure on the soil will reach the limiting value p_l. Setting $p_{z1} = p_l$ in Eq. (II-3-8), we obtain the equation

$$\frac{(p_l - p_{st})f_{nz}^2}{c_u v_0 \sqrt{1 + \rho^2}} = \exp(-\rho\tau) \sin \tau \tag{II-3-9}$$

from which we can determine the time τ_1.

If the limiting pressure p_l is reached at the maximum settlement of the foundation, then

$$\frac{dz_1}{dt} = 0$$

from which we obtain

$$\tan \tau_1 = \frac{1}{\rho} \tag{II-3-10}$$

The last equation determines the condition of absence of foundation settlements, in the case of foundation vibrations caused by a vertically centered impact.

If the condition of Eq. (II-3-10) is not satisfied, then residual settlements of the foundation will develop; section BC in the vibration diagram of Fig. II-22 will correspond to these settlements. Along this section, the soil reactions will not depend on the settlement, and therefore the differential equation of the foundation settlement will be

$$\ddot{z}_2 + 2c\dot{z}_2 = g\left(1 - \frac{1}{r}\right) \tag{II-3-11}$$

where

$$r = \frac{p_{st}}{p_l}$$

The solution of Eq. (II-3-11) must satisfy the following conditions: If $\tau = \tau_1$, then

$$z_2 = z_1 = \frac{p_l}{c_u} \tag{II-3-12}$$

$$\frac{dz_2}{d\tau} = \frac{dz_1}{d\tau} = S \tag{II-3-13}$$

The integral of Eq. (II-3-11) will be

$$z_2 = -B\tau + C \exp(-2\rho\tau) + D \tag{II-3-14}$$

where

$$B = \frac{p_l(1 - r)(1 + \rho^2)}{2c_u} \tag{II-3-15}$$

Using the conditions of Eqs. (II-3-12) and (II-3-13), we obtain two equations for the determination of the arbitrary constants C and D. Solving these equations, we find

$$C = -\frac{B + S}{2\rho} \exp(-2\rho\tau) \qquad D = B\tau_1 + \frac{B + S}{2\rho} + \frac{p_t}{c_u}$$

The development of residual settlements will proceed up to the moment $\tau = \tau_2$, at which time

$$\frac{dz_2}{d\tau} = 0$$

From this condition we obtain the equation for the determination of the time τ_2:

$$\exp\left[-2\rho(\tau_2 - \tau_1)\right] = \frac{B + S}{B} \qquad \text{(II-3-16)}$$

Substituting the value τ_2 thus obtained into Eq. (II-3-14), we obtain the maximum total (i.e., elastic and residual) foundation settlement:

$$z_{2,\max} = \frac{p_l}{c_u} - B(\tau_2 - \tau_1) + \frac{S}{2\rho} \qquad \text{(II-3-17)}$$

If $\tau > \tau_2$, then the foundation will move upward under the influence of the elastic reaction of the soil (it will undergo an elastic rebound), and the displacements of the foundation will be elastic only.

Let us consider the motion of the foundation as shown on the third section of the vibration diagram (Fig. II-22). To simplify the computations, let us transfer the initial point of displacements to the point C; we shall also read the time τ from this point.

The differential equation of foundation motion corresponding to the elastic rebound will have the form

$$\ddot{z}_3 + 2c\dot{z}_3 + f_{nz}{}^2 z_3 = g\left(1 - \frac{1}{r}\right) \qquad \text{(II-3-18)}$$

The solution of this equation satisfying the conditions: If $\tau = 0$, then

$$z_3 = 0 \qquad \text{and} \qquad \frac{dz_3}{d\tau} = 0$$

will be

$$z_3 = \frac{1 - r}{r}\frac{p_{st}}{c_u}\left[\exp(-\rho\tau)(\cos\tau + \rho\sin\tau) - 1\right] \qquad \text{(II-3-19)}$$

When $\tau = \infty$, the elastic displacement equals

$$z_{3,\infty} = -\frac{1 - r}{r}\frac{p_{st}}{c_u} \qquad \text{(II-3-20)}$$

Thus the magnitude of the residual settlement only, caused by one impact, equals

$$z_0 = z_2 - z_{3\infty} = \frac{p_{st}}{c_u}\frac{S}{2\rho}(\tau_2 - \tau_1) \qquad \text{(II-3-21)}$$

Assuming in Eq. (II-3-7) that $z_1 = p_l/c_u$ and that it satisfies Eq. (III-3-10), we may determine the maximum value of the initial velocity of vibratory foundation motion, at which no residual settlements will occur; the magnitude of this velocity equals

$$v_0^* = j\frac{p_l - p_{st}}{\sqrt{p_{st}}}\sqrt{\frac{g}{c_u}}$$

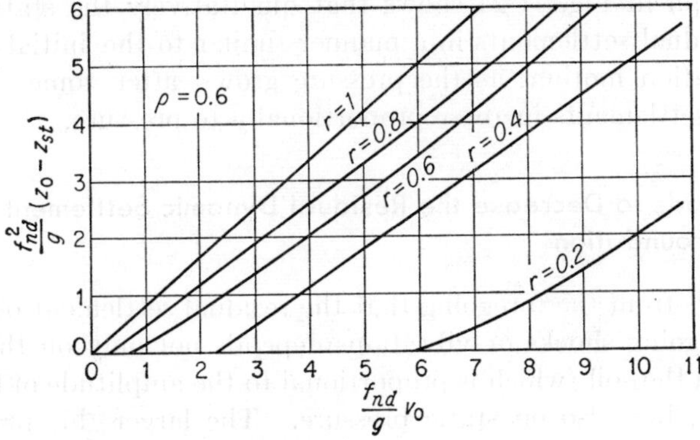

FIG. II-23. Relationship between the residual settlement of a foundation and the initial velocity v_0 of its motion.

The value of coefficient j depends on ρ: if $\rho = 0$, then $j = 1$; if $\rho = 0.3$, then $j = 1.47$; if $\rho = 0.6$, then $j = 1.84$.

Using Eqs. (V-2-8) and (V-3-2), we may select the values of foundation weight and contact area with soil at which no residual settlements of the foundation will occur.

It is not possible to establish directly from Eq. (II-3-21) the effect on the residual settlements of static pressure or the velocity of the initial movement of the foundation. However, by computing z_0 for various specific conditions, it is possible to plot graphs showing the relationship between residual settlement and certain parameters. These graphs make it possible to estimate the effects of these parameters.

Figures II-23 and II-24 present such graphs. The curves in Fig. II-23 show the relationship between residual settlements and initial velocity of foundation motion for various values of the ratio r between the static pressure p_{st} and the limiting pressure p_l [see Eq. (II-3-11)].

It follows from these curves that after the initial velocity of the foundation reaches the value v_0^*, the residual settlements grow approximately in proportion to the initial velociy (especially for large values of the coefficient ρ).

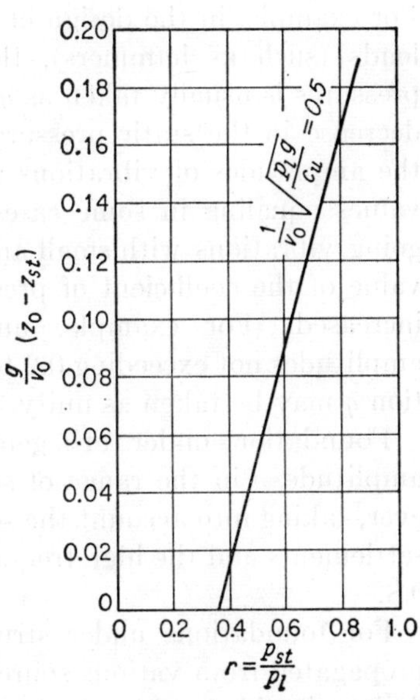

FIG. II-24. Relationship between the residual dynamic settlement of a foundation and the static pressure on the soil.

The graph in Fig. II-24 shows that qualitatively the static pressure affects residual settlements in a manner similar to the initial velocity of the foundation motion: as the pressure grows, after some "threshold" value the settlements increase proportionally to pressure.

II-4. Methods to Decrease the Residual Dynamic Settlement of a Foundation

It follows from the foregoing that the residual settlement of a foundation undergoing shocks or vibrations depends not only on the dynamic pressure on the soil (which is proportional to the amplitude of foundation vibrations), but also on static pressure. The larger this pressure, the larger the settlements. Therefore for foundations undergoing shocks or vibrations (such as foundations under machinery or under buildings subjected to the action of vibrations), the design should provide a smaller static pressure on the soil than for foundations which support only static loads. It follows from Eq. (II-3-3) that the coefficient of reduction of permissible pressure on soil q, in cases where the foundation is subjected to dynamic loads, depends on the value of the dynamic pressure and, consequently, on the amplitude of vibrations. The larger this amplitude, the smaller should be the static pressure on the soil under the foundation For example, in the design of foundations under machinery with impact loads (such as hammers), the coefficient of reduction of permissible pressures is usually taken as $q = 0.4$ to 0.5. This considerable required decrease in the static pressure on the soil is explained by the fact that the amplitudes of vibrations under hammers may attain relatively large values equaling in some cases 1.2 to 1.5 mm. For foundations undergoing vibrations with small amplitudes and not too high frequencies, the value of the coefficient of pressure reduction should be correspondingly increased. For example, under reciprocating engines vibrating at amplitudes not exceeding 0.2 to 0.3 mm, the coefficient of pressure reduction q may be taken as unity.

Foundations under turbogenerators undergo vibrations at much smaller amplitudes—in the range of several microns or tens of microns. However, taking into account the sensitivity of these machines to nonuniform settlements and the high frequency of their vibrations, q is assumed to be 0.8.

For foundations under structures subjected to the action of waves propagated from various sources (including those under machinery), the design should provide a reduced static pressure on the soil as compared to its value when only static loads are to be supported. Disregarding this factor (especially in the construction of forge shops and shops with crushing equipment) may lead to considerable nonuniform settlements of footings under columns or of foundations under some sections of walls, as a result of which a structure may undergo damage (cf. Art. IX-3).

In order to prevent considerable nonuniform settlements of foundations under structures subjected to the action of vibrations or shocks, it is necessary (especially in cases where foundations are to be erected on water-saturated fine-grained sands) to assign varying values of static pressure on the soil; the pressure values under foundation sections located close to the source of waves should be lower than those assigned under sections located at greater distances. The static pressures thus selected should be inversely proportional to the amplitude of vertical vibrations taking place under the action of waves propagated through the soil.

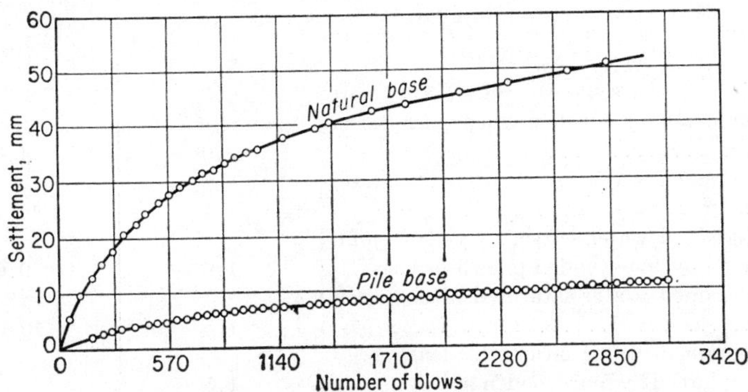

FIG. II-25. Relationship between foundation settlement and number of impacts.

The size of the soil zone in which dynamic foundation settlements may develop is not large either in depth or in plan. Therefore the use of any measures aimed at an increase in the limiting pressure p_l, even at small depths below the foundation contact area, will contribute to the decrease of dynamic settlements due to shocks or vibrations. If the soil under a foundation consists of sands, then chemical stabilization or cementation of the soil to a depth of 3 to 4 m may be possible.

A decrease of the dynamic settlements may be achieved by any measures which increase the soil density and consequently the limiting value p_l (for example, by driving short piles).

Figure II-25 gives a curve of relationship between the settlements of two identical foundations and the number of impacts. One of these foundations was erected on natural soil (medium-grained yellow water-saturated sand) and the other on five piles driven to a depth of some 3 m; four piles were driven at the corners and the fifth in the center. The foundation contact area in both cases equaled 1.5 m²; the foundation weight was 5.7 tons. Natural vertical vibrations of the foundation were induced by the impacts of a ram having a weight of 160 kg and dropped from a height of 1 to 1.1 m.

Similar investigations conducted under different soil conditions showed that a floating pile base had a very favorable effect on dynamic settlements.

The values of foundation settlements (in field experiments) after 1,500 blows on piles, presented in Table II-2, show that in all cases the reinforcement of the base under the foundation by means of short piles resulted in a considerable decrease of settlements induced by vibrations. When the foundation was erected on loess reinforced by wooden piles, the settlement of the foundation under the action of vibrations was reduced to one-ninth its value on natural soil. A foundation erected on short wooden piles driven into dense gray sands had settlements equal to approximately one-thirtieth of the settlement value on natural sand.

TABLE II-2. DATA ON DECREASE OF DYNAMIC SETTLEMENTS BY THE USE OF PILES

Characteristics of the base	Foundation area in contact with soil, m²	Settlement of the foundation after 1,500 blows, mm
Loess in natural state	2.0	8.9
Loess reinforced by seven soil piles 4.5 m long	2.0	3.6
Loess reinforced by seven wooden piles 4.5 m long	2.0	1.0
Medium-grained water-saturated yellow sand of medium density	1.5	45.7
Medium-grained water-saturated yellow sand, reinforced by five wooden piles 3 m long	1.6	9.0
Medium-grained water-saturated dense gray sand	1.5	19.0
Medium-grained water-saturated dense gray sand reinforced by four wooden piles 3 m long	1.5	0.5

A decrease in residual settlements may also be achieved by decreasing the amplitude of vertical vibrations by selecting rational dimensions for the foundation.

III

THEORY OF VIBRATIONS OF MASSIVE MACHINE FOUNDATIONS

III-1. Vertical Vibrations of Foundations

a. Basic Assumptions. In general, the investigation of vibrations of a massive foundation placed on the soil surface can be reduced to the investigation of vibrations of a solid block resting on a semi-infinite elastic solid. To date no solution of this problem has been found. Therefore several simplifying assumptions concerning vibrations of solid blocks placed on soil are necessary.

First of all let us assume that there is a linear relation between the soil reacting on a vibrating foundation and the displacement of this foundation. Then the relation between the displacements and the reactions will be determined in terms of the coefficients of elastic uniform and nonuniform compression, as well as a coefficient of elastic shear. In Chap. I, the dependence of these coefficients on the elastic properties of the soil and on the size of the foundation was established; also, the numerical values of the coefficients for various soils were given. In addition, it is necessary to assume that the soil underlying the foundation does not have inertial properties, but only elastic properties as described by the coefficients. Thus, the foundation is considered to have only inertial properties and to lack elastic properties, while the soil is considered to have only elastic properties and to lack properties of inertia. These assumptions concerning foundation and soil make it possible, in the general case, to analyze foundation vibrations as a problem of a solid body resting on weightless springs, the latter serving as a model for the soil.

Frequently foundations under machinery are embedded into soil to a certain depth. In this case, the elastic reactions of the soil act not only along the horizontal contact surface between soil and foundation, but also along the side surfaces of the foundation. These reactions may have considerable effect on the frequencies of free vibration of the foundation and on the coefficient of damping. Therefore, reactions along the side surfaces of the foundation have considerable effect on amplitudes of free or forced vibration under conditions close to resonance.

It is difficult to evaluate in each case the effect of side reactions on foundation vibration. This effect is tentatively taken into account in design computations by increasing the values of the coefficients of elasticity of the base. For example, this method is applied in computations of foundations for forge hammers. If a foundation undergoes only forced vibrations (as, for example, foundations under reciprocating machinery), and the design values of frequencies of natural vibrations of this foundation are larger than the operational frequency of rotation of the machine, then the effect of the side reactions is relatively small and can be neglected. In these cases, disregarding the above soil reactions is conservative, since it results in a design dynamic stability lower than the actual stability.

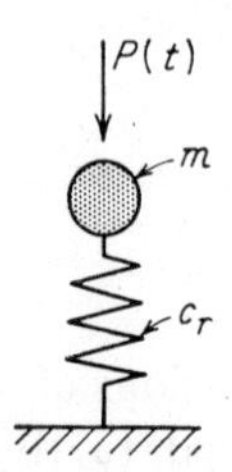

FIG. III-1. Vibration of a centered mass resting on a spring.

b. Vertical Vibrations of Foundations Neglecting the Damping Effect of Soil Reactions. Let us consider vibrations of the foundation caused by a vertical exciting force $P(t)$ which changes with time. We assume that the center of mass of the foundation and machine and the centroid of the area of foundation in contact with soil lie on a vertical line which coincides with the direction of action of the exciting force $P(t)$. In this case, the foundation will undergo only vertical vibrations. Since the foundation is assumed to be an absolutely rigid body, its displacement is determined by the displacement of its center of gravity. As mentioned above, weightless springs serve as a model for the soil. Thus, the problem of vertical vibrations of a foundation is reduced to the investigation of vibrations of a centered mass resting on a spring (Fig. III-1). Let us denote by z the vertical displacement of the foundation computed with respect to the equilibrium position. We shall consider z to be positive in a downward direction. If the displacement of the center of gravity of the foundation equals z, then the reaction of the spring (i.e., the foundation base) will equal

$$R = W + c_r z \qquad \text{(III-1-1)}$$

where W = weight of foundation and machine
c_r = coefficient of rigidity of the base

$$c_r = c_u A \qquad \text{(III-1-2)}$$

c_u = coefficient of elastic uniform compression of soil
A = horizontal contact area of foundation with soil

If c_u has the dimensions tons per cubic meter and A is expressed in square meters, then c_r will evidently have the dimensions tons per meter. Using d'Alembert's principle, we may obtain the differential equation of vertical foundation vibrations. According to this principle, the equation of motion may be written in the same way as the equation of statics if one adds the inertial force to the external forces acting on a moving body. Then the equation of motion for the foundation will be

$$-m\ddot{z} + W + P(t) - R = 0$$

or, using Eq. (III-1-1), we obtain

$$m\ddot{z} + c_r z = P(t) \qquad \text{(III-1-3)}$$

where m = mass of foundation and machine: $m = W/g$

g = acceleration of gravity

Dividing both parts of Eq. (III-1-3) by the mass m, we rewrite this equation as follows:

$$\ddot{z} + f_{nz}^2 z = p(t) \qquad \text{(III-1-4)}$$

where

$$f_{nz}^2 = \frac{c_r}{m} = \frac{c_u A}{m} \qquad \text{(III-1-5)}$$

$$p(t) = \frac{P(t)}{m}$$

Equation (III-1-4) describes the vertical vibrations of a foundation under the action of an exciting force.

Let us consider the case in which no exciting force acts on the foundation, but the motion results from an impact or from an initial displacement of the foundation. Setting $p(t) = 0$ in Eq. (III-1-4), we obtain

$$\ddot{z} + f_{nz}^2 z = 0 \qquad \text{(III-1-6)}$$

This equation corresponds to the case in which the motion occurs only under the action of the inertial forces of the foundation and the elastic reaction of the base. Such vibrations are called natural or free vibrations. For example, foundations under forge hammers may be subjected to such vibrations.

The general solution of the homogeneous differential Eq. (III-1-6) may be written as follows:

$$z = A \sin f_{nz}t + B \cos f_{nz}t \qquad \text{(III-1-7)}$$

Hence it is seen that free vibration under the action of elastic reactions and inertia forces is a harmonic motion with frequency f_{nz}, called the "natural circular frequency of vertical vibrations of the foundation." According to Eq. (III-1-5), this frequency is determined only by the foundation mass and the elasticity of the base and does not depend on the nature or condition of the exciting force.

Since the frequency of vibration is the number of oscillations per second, the period of the natural vibration, i.e., the time for one oscillation, is related to the circular frequency f_{nz} by the following equation:

$$T_{nz} = \frac{2\pi}{f_{nz}}$$

The numbers of oscillations per minute and per second are related to the circular frequency of vibrations by the following simple formulas:

$$N = \frac{60}{2\pi} f_{nz} \qquad n = \frac{f_{nz}}{2\pi}$$

One oscillation per second is called a hertz.

The coefficients A and B in Eq. (III-1-7) represent the amplitudes of natural vibrations of the foundation. Their values depend only on the initial conditions of motion, i.e., on the magnitudes of the velocity (or the displacement) of the foundation at a certain moment of time taken as the initial moment. Natural vibrations of foundations under machines are usually caused by an impact, i.e., the foundations experience a certain initial velocity. Therefore let us consider only this particular case.

Let us assume that at $t = 0$,

$$z = 0 \quad \text{and} \quad \dot{z} = v_0 \tag{III-1-8}$$

Differentiating both parts of the solution (III-1-7) with respect to time, we obtain the following expression for the velocity of the foundation:

$$\dot{z} = Af_{nz} \cos f_{nz}t - Bf_{nz} \sin f_{nz}t \tag{III-1-9}$$

Setting $t = 0$ in Eqs. (III-1-7) and (III-1-9), we obtain the following expressions for constants A and B:

$$A = \frac{v_0}{f_{nz}} \qquad B = 0$$

Thus, when vertical natural vibrations of the foundation are caused by an impact, the displacement is determined by the equation

$$z = \frac{v_0}{f_{nz}} \sin f_{nz}t \tag{III-1-10}$$

While the frequency of natural vibrations of a foundation depends only on the inertia and the elastic properties, the amplitude, i.e., the maximum deflection from the equilibrium position, depends also on the initial conditions of the motion, being proportional to initial velocity.

Returning to Eq. (III-1-4) for forced vertical vibrations of foundations, let us consider the case in which the exciting force $P(t)$ is a harmonic function of the time, for example, $P(t) = p \sin \omega t$ (ω is exciter frequency and

$$p = \frac{P}{m}$$

where P is the exciting force).

An exciting force which changes with time according to $\sin \omega t$ or $\cos \omega t$ is of special interest in the study of forced vibrations of foundations, since in design work, exciting loads imposed by machines are usually harmonic functions of time. Substituting into the right-hand part of Eq. (III-1-4) the expression

$$p(t) = p \sin \omega t$$

we obtain the equation for forced vertical vibrations of foundations:

$$\ddot{z} + f_{nz}^2 z = p \sin \omega t \tag{III-1-11}$$

The general solution of this differential equation presents the sum of two solutions, corresponding to free and to forced vibrations caused by a given exciting force. Due to the action of damping soil reactions, free vibrations are damped out a short time after the beginning of the forced motion of foundations, and there remain only forced vibrations. The solution of Eq. (III-1-11), corresponding only to these steady-state vibrations, is as follows:

$$z = A_z \sin \omega t \tag{III-1-12}$$

We obtain the expression for the amplitude A_z of forced vibrations by substituting the formula for z [Eq. (III-1-12)] into differential Eq. (III-1-11); then we have

$$A_z = \frac{P}{m(f_{nz}^2 - \omega^2)} \tag{III-1-13}$$

The solution (III-1-12) shows that the frequency of forced vibrations is equal to the frequency of the exciting force. Thus, unlike the frequency of natural vibrations, the frequency of forced vibrations does not depend on the inertial and elastic properties of the foundation and its base. Since the exciting loads of machines are usually repeated periodically with every revolution of the machine, in many cases the frequency of the exciting force equals that of rotation of the machine.

In general, this conclusion is valid for all linear mechanical systems not capable of producing self-excited vibrations. Therefore, the identity of the frequency of forced vibrations and the frequency of exciting loads acting on the foundation holds so long as the relationship between elastic foundation displacements and soil reactions is linear. Numerous comparisons of the frequency of forced vibrations of machine foundations with the frequency of exciting forces developed by these machines confirm the coincidence of these frequencies. Thus it is clear that in forced vibrations of a foundation, there really exists a linear relationship between the foundation displacement and the soil reaction.

Figure III-2 presents graphs of the effect of the magnitude of the exciting force on the amplitude of forced vertical vibrations of a test foundation. These graphs substantiate the linear character of the relationship established by Eq. (III-1-13).

It follows from the same equation that the amplitude of forced vibrations depends also on the mass of the foundation and the difference between the frequencies of free and forced vibrations.

In order to better understand the influence of the mass and the natural frequency of the foundation, we transform expression (III-1-13) into

$$A_z = \frac{P}{m f_{nz}^2} \frac{1}{1 - \omega^2 / f_{nz}^2} \qquad \text{(III-1-14)}$$

Since

$$\frac{P}{m f_{nz}^2} = A_{st}$$

where A_{st} is the displacement of the foundation under the action of force P if the latter were applied statically, expression (III-1-14) may be rewritten as

$$A_z = \eta A_{st} \qquad \text{(III-1-15)}$$

where η is a dynamic modulus (or magnification factor)

$$\eta = \frac{1}{1 - \xi^2} \qquad \text{(III-1-16)}$$

and $\xi = \omega / f_{nz}$ is the frequency ratio.

The value of the dynamic modulus depends only on the interrelationship between the frequency of the exciting force and the natural frequency of the foundation.

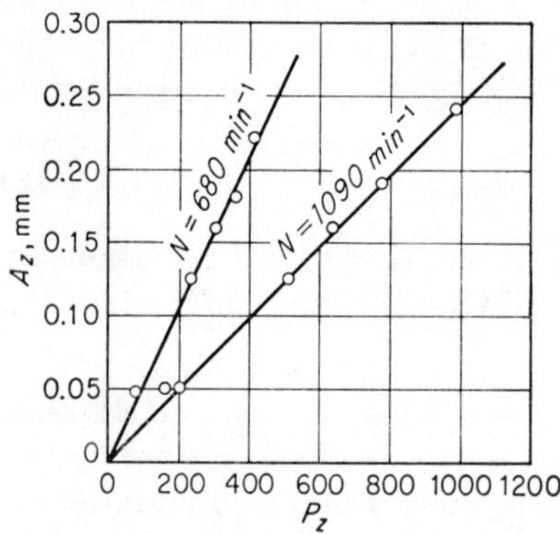

FIG. III-2. Relationship between the amplitude of vertical vibrations A_z and the exciting force P_z.

If the frequency of the excited vibrations is small in comparison with the natural frequency of the foundation, then the value of the dynamic modulus is close to unity and the amplitude of forced vibrations of the foundation does not differ much from A_{st}; the latter represents a static displacement of the foundation under the action of the exciting force P. With an increase in the frequency of the exciting force, ξ also increases; thus the denominator in expression (III-1-16) decreases, leading to an increase in the value of the dynamic modulus. When $\xi = 1$, i.e., when the frequency of the exciting force equals the natural frequency of the foundation, the amplitude of vibrations of the foundation theoretically equals infinity. This corresponds to resonance. With further increase in the frequency of the exciting force, ξ becomes larger than 1; the dynamic modulus continuously decreases with an increase of ξ, and when $\xi = \sqrt{2}$, the dynamic modulus again equals 1. For ranges of frequency of exciting force corresponding to $\xi > \sqrt{2}$, the dynamic modulus uniformly decreases, asymptotically approaching zero. Hence it follows that an exciting force, the frequency of which is large in comparison with the natural frequency, may induce forced vibrations with an amplitude of infinitesimal value. This conclusion is used as the guiding principle for the design of various devices for insulation from vibrations, particularly for insulating machines and engines. When the exciting frequency is given, the design of a device for insulating a machine or an engine should be made in such a way that the frequency of natural vibrations of the device is as small as possible compared to the exciting frequency (for example, the frequency of oscillations caused by traffic).

Foundations under reciprocating machinery are usually designed in such a way that the natural frequency of the foundation is higher than the operating frequency of the machine; i.e., $\xi < 1$. If one increases the foundation mass without changing the foundation area, the frequency of natural vibrations decreases and the value of ξ increases. It follows that the denominator in expression (III-1-16) decreases, causing an increase in the dynamic modulus. Thus, other conditions being equal, an increase in the foundation height is accompanied by an increase in the amplitude of forced vibrations. This is the reason why modern foundations for machines (especially for reciprocating machinery) are designed as blocks with large bases and minimum height.

c. Vertical Vibrations of Foundations Considering the Damping Effect of Soil Reactions. As mentioned in the foregoing discussion, under conditions of resonance the amplitude of forced vibrations theoretically approaches infinity. However, this contradicts experimental data which show that under conditions of resonance, the amplitude of vibrations still remains finite. This contradiction between experience and theory is explained by the fact that amplitudes of vibrations with a frequency close to the natural frequency of the foundation are affected by deviations of the mechanical properties of the soil from those of an ideal elastic body. Like any real body, soil deviates from an idealized model represented by an ideally elastic solid. In reality, irreversible processes, characterized by energy dissipation, occur in soil. This deviation of soil properties from those of an ideally elastic solid may be taken into account if one assumes that the reaction of the soil depends not only on the displacement of the foundation, but also upon its velocity. Since the velocities of foundation vibrations are rather low, it can be taken as a first approximation that the damping reactions of soil are proportional to the first power of the velocity of vibration. Then the equation of free vibrations of foundations may be written as follows:

$$\ddot{z} + 2c\dot{z} + f_{nz}{}^2 z = 0 \tag{III-1-17}$$

This expression differs from Eq. (III-1-6) by the presence of the term $2c\dot{z}$. Here

$$c = \frac{\alpha}{2m} \tag{III-1-17a}$$

is called the damping constant; its double value equals the coefficient of resistance α per unit of foundation mass. Usually $c < f_{nz}$. The solution of Eq. (III-1-17) corresponding to this case is as follows:

$$z = \exp(-ct)(A \sin f_{nd}t + B \cos f_{n1}t) \tag{III-1-18}$$

where f_{nd} is the natural frequency of vertical vibrations of foundations in cases where the reaction of soil depends not only on the displacement, but also on the velocity. Substituting solution (III-1-18) into the left-hand part of differential Eq. (III-1-17), we find that the solution will satisfy this equation for any values of A and B if

$$f_{nd}{}^2 = f_{nz}{}^2 - c^2$$

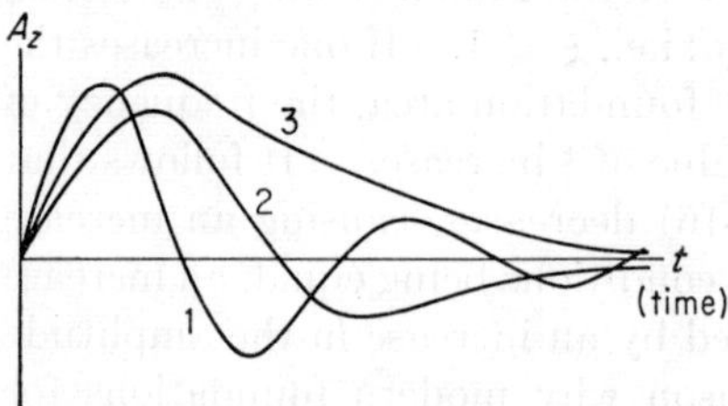

FIG. III-3. Effect of damping on vibration amplitude A_z: curve 1, damping coefficient c is small; curve 2, c approximately equals natural frequency f_{nz}; curve 3, c is larger than f_{nz}.

Hence it follows that the damping properties of a soil decrease the natural frequency of vibration of the foundation. If the damping constant is small in comparison to the natural frequency of the foundation, then the influence of the damping properties of the soil on the natural frequency may be neglected.

However, the effect of the damping reactions of soil on the amplitudes of free vibrations of a foundation is rather considerable, even in cases of small values of c. It follows directly from Eq. (III-1-18) that amplitudes of vibrations decrease exponentially with time. This is illustrated by curve 1 of Fig. III-3.

If $c \cong f_{nz}$, then the character of free vibrations of foundations will correspond to curve 2. For large values of damping constant, when $c > f_{nz}$, free vibrations are not possible, and, under the action of an impact or an initial displacement, the foundation will undergo non-periodic motion, as shown by curve 3 of Fig. III-3.

Thus damping reactions of the soil have considerable effect on free vibrations of foundations and on amplitudes of forced vibrations, especially under conditions close to resonance.

We obtain an equation for forced vibrations of foundations, including the effect of damping reactions of the soil, if we insert into the right-hand part of Eq. (III-1-17) the value of the exciting force; as before, we assume the latter to equal $p \sin \omega t = (P/m) \sin \omega t$; then we have

$$\ddot{z} + 2c\dot{z} + f_{nz}^2 z = p \sin \omega t \tag{III-1-19}$$

The solution of this equation, corresponding only to steady forced vibrations of foundations, will be

$$z = A_z^* \sin (\omega t - \gamma) \tag{III-1-20}$$

Here A_z^* is the amplitude of forced vibrations:

$$A_z^* = \frac{P}{m\sqrt{(f_{nz}^2 - \omega^2)^2 + 4c^2\omega^2}} \tag{III-1-21}$$

The phase shift between the exciting force and the displacement induced by this force equals

$$\tan \gamma = \frac{2c\omega}{f_{nz}^2 - \omega^2} \tag{III-1-22}$$

Similarly, Eq. (III-1-21) may be reduced to the form

$$A_z^* = \eta^* A_{st}$$

The dynamic modulus η^* in this case will equal

$$\eta^* = \frac{1}{\sqrt{(1 - \xi^2)^2 + 4\Delta^2\xi^2}} \tag{III-1-23}$$

where
$$\Delta = \frac{c}{f_{nz}}$$

Figure III-4 presents curves of interrelationship between η^* and ξ; the latter is the ratio of the frequency of forced vibrations to the natural frequency of the foundation. These graphs are plotted for different magnitudes of Δ, proportional to the damping constant.

The particular case $\Delta = 0$ corresponds to the previously discussed case of foundation vibrations where the damping effects of soil reactions were not considered. Only here will the amplitude at resonance increase without limit. At all other times, when $\Delta \neq 0$, the amplitude remains finite. The larger the value of Δ, the smaller the amplitude. Under conditions of damping, the maximum value of the amplitude corresponds to

$$\xi = \sqrt{1 - 2\Delta^2}$$

Thus if damping occurs, the resonance frequency decreases somewhat, and the dynamic modulus at resonance equals

$$\eta^*_{\max} = \frac{1}{2\Delta\sqrt{1 - \Delta^2}} \tag{III-1-24}$$

Hence it follows that the larger the damping constant, the smaller the dynamic modulus at resonance.

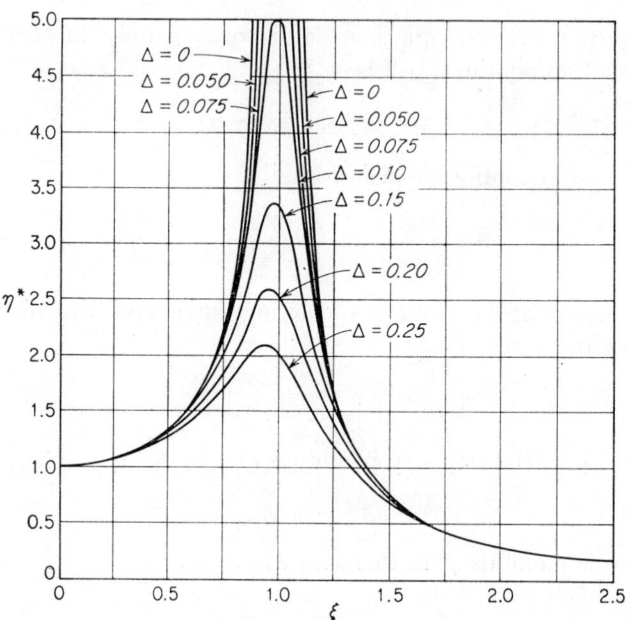

Fig. III-4. Relationship between the dynamic modulus of damped vibrations η^* and the ratio ξ of actual to natural frequencies ω/f_{nz}, for varying values of the reduced damping coefficient Δ.

The graphs presented in Fig. III-4 also show that the greatest effect of damping reactions of soil is observed in the zone of resonance, when the value of ξ is approximately equal to unity. When the difference $1 - \xi^2$ increases, the influence of damping soil properties on amplitudes of forced vibrations decreases; when ξ is large in comparison to unity, this effect may be neglected. Since foundations under machines with a steady regime of work are usually designed in such a way that there is a significant difference between ξ and unity, the effect of soil damping may be neglected in many computations of amplitudes of forced vibrations.

d. The Effect of Soil Inertia on Forced Vertical Vibrations of Foundations. The foregoing theory of vertical vibrations of foundations is based on the assumption that soil reactions may be represented by weightless springs characterized by the coefficient c_u. This model differs considerably from the real properties of soil. Therefore, the results obtained should be considered as a first approximation only. As stated previously, an accurate solution of the problem of vertical vibrations of foundations resting on soil necessitates a consideration of the problem of vibrations of a solid resting on an elastic base, which in the simplest case presents a semi-infinite elastic solid. Limiting our analysis by considering only vertical forced vibrations induced by the force $P \sin \omega t$ which harmonically changes with time, we may write the equation for this case as follows:

$$m\ddot{z} + R \exp (i\omega t) = P \exp [i(\omega t + \epsilon)] \qquad \text{(III-1-25)}$$

where R = magnitude of soil reaction against foundation
ϵ = phase shift between exciting force and soil reaction

In order to solve Eq. (III-1-25) it is necessary to determine the dependence of the value R upon the displacement, the characteristics of the foundation, and the soil properties.

E. Reissner[37] gave an approximate solution of the problem of vibrations of a solid body resting on an elastic semi-infinite mass; for the computation of R, he used the magnitude of the settlement of the soil under the center of a uniformly loaded absolutely flexible circular area. O. Ya. Shekhter[41] showed a mistake involved in Reissner's solution† and solved the same problem taking the magnitude of settlement of soil (needed for the computation of the value R) as an arithmetic mean between the magnitudes of settlement under the center of a flexible circular area and under its edge. As a result of rather complicated computations which are omitted here, the following simple relationship between R and z was established:

$$z = -\frac{R}{r_0 G}(f_1 + if_2) \exp (i\omega t) \qquad \text{(III-1-26)}$$

where G = modulus of elasticity in shear of soil
r_0 = radius of a circle = $\sqrt{A/\pi}$
A = contact area between foundation and soil
f_1, f_2 = functions depending on ratio between radius r_0 of circle and length of shear waves propagated by foundation under machine, and also depending on Poisson's ratio of soil

The following formulas give values of f_1 and f_2 corresponding to a Poisson ratio of $\nu = 0.5$, with a degree of precision sufficient for practical computations:

$$f_1 = -0.130 + 0.0536\varkappa^2 - 0.0078\varkappa^4 + \cdots$$
$$f_2 = 0.0545 J_1(1.047\varkappa)[1 + J_0(1.047\varkappa)] + 0.0474\varkappa - 0.0065\varkappa^3 + \cdots$$

† This mistake was also noted in the paper by R. N. Arnold, G. N. Bycroft, and G. B. Worburton in *J. Appl. Mech.*, vol. 22, no. 3, pp. 391–400, 1955.

where $\varkappa = 2\pi r_0/L_s$
L_s = length of shear waves propagated from foundation
J_1, J_0 = Bessel's functions of first kind, of order one, zero

The length of the shear waves is

$$L_s = 2\pi \frac{v_s}{\omega}$$

where v_s is the velocity of shear waves in soil. For conventional designs of machinery foundations the value $\varkappa$ is considerably smaller than unity.

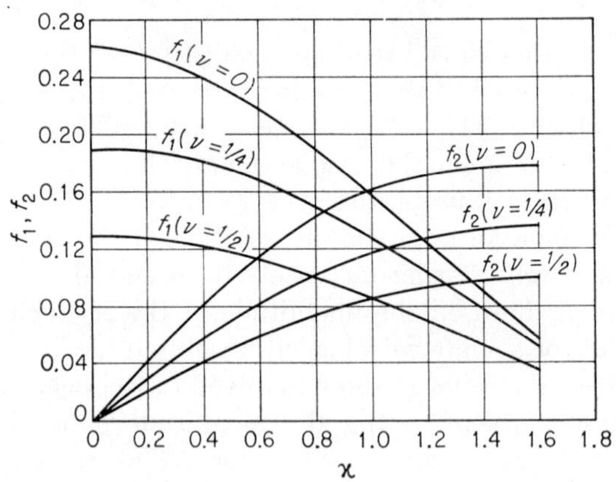

FIG. III-5. Auxiliary diagram for the solution of Eq. (III-1-26).

Figure III-5 presents graphs of f_1 and f_2 depending on magnitudes of the independent variable $\varkappa$; these graphs are plotted for Poisson's ratio equal to 0, 0.25, and 0.5 and make it possible to avoid computations when determining values of f_1 and f_2 corresponding to selected values of $\varkappa$ and of ν.

Substituting the value of z determined by Eq. (III-1-26) into the left-hand part of Eq. (III-1-25), we obtain:

$$\frac{Rm\omega^2}{Gr_0}(f_1 + if_2)\exp(i\omega t) + R\exp(i\omega t) = P\exp[i(\omega t + \epsilon)]$$

From this we obtain two equations for determining R and ϵ:

$$\frac{m\omega^2 f_2}{Gr_0} R = P\sin\epsilon$$

$$\left(\frac{m\omega^2 f_1}{Gr_0} + 1\right) R = P\cos\epsilon$$

Solving this system of equations, we find

$$\tan \epsilon = \frac{(m\omega^2/Gr_0)f_2}{1 + (m\omega^2/Gr_0)f_1} \tag{III-1-27}$$

$$R = \frac{P}{\sqrt{[1 + (m\omega^2/Gr_0)f_1]^2 + [(m\omega^2/Gr_0)f_2]^2}} \tag{III-1-28}$$

The expression thus found for R is substituted into the right-hand part of Eq. (III-1-26); by neglecting the imaginary part, we obtain the following formula for the amplitude a of forced vibrations of the foundation:

$$a = \frac{P}{Gr_0}\sqrt{\frac{f_1^2 + f_2^2}{[1 + (m\omega^2/Gr_0)f_1]^2 + [(m\omega^2/Gr_0)f_2]^2}} \tag{III-1-29}$$

The phase shift between the exciting force P and the displacement z equals $\varphi = \alpha + \epsilon$, where $\tan \alpha = -f_1/f_2$ (phase shift between displacement and reaction of soil). Using Eq. (III-1-27), we obtain

$$\tan \varphi = -\frac{f_2}{f_1 + (m\omega^2/Gr_0)(f_1^2 + f_2^2)} \tag{III-1-30}$$

Let us introduce a dimensionless value b;

$$b = \frac{m}{\gamma r_0^3} \tag{III-1-31}$$

where γ is the soil density. Then

$$\frac{m\omega^2}{Gr_0} = \varkappa^2 b$$

and formulas for the amplitude and phase of vibrations will be rewritten as follows:

$$\frac{aGr_0}{P} = \sqrt{\frac{f_1^2 + f_2^2}{(1 + \varkappa^2 b f_1)^2 + (\varkappa^2 b f_2)^2}} \tag{III-1-32}$$

$$\tan \varphi = -\frac{f_2}{f_1 + b\varkappa^2(f_1^2 + f_2^2)} \tag{III-1-33}$$

Figure III-6 presents graphs of changing aGr_0/P depending on changes of $\varkappa$, or, what is the same thing, changes in the frequency of excitement. These graphs have much in common with resonance curves for a system with one degree of freedom subjected to damping.

Thus, although the initial Eq. (III-1-25) does not take into account damping properties of soil, amplitudes of vibrations never reach infinity with changes in frequencies of excitement as is the case in an ideally elastic system with one degree of freedom. This means that even an ideally elastic soil has a damping effect on the amplitude of foundation vibrations. This is explained by the fact that the energy of a vibrating foundation, due to its propagation into the soil, is continuously dissipated; therefore the vibrations of a foundation, even of one resting on an ideal elastic solid representing a semi-infinite elastic mass, are damped with

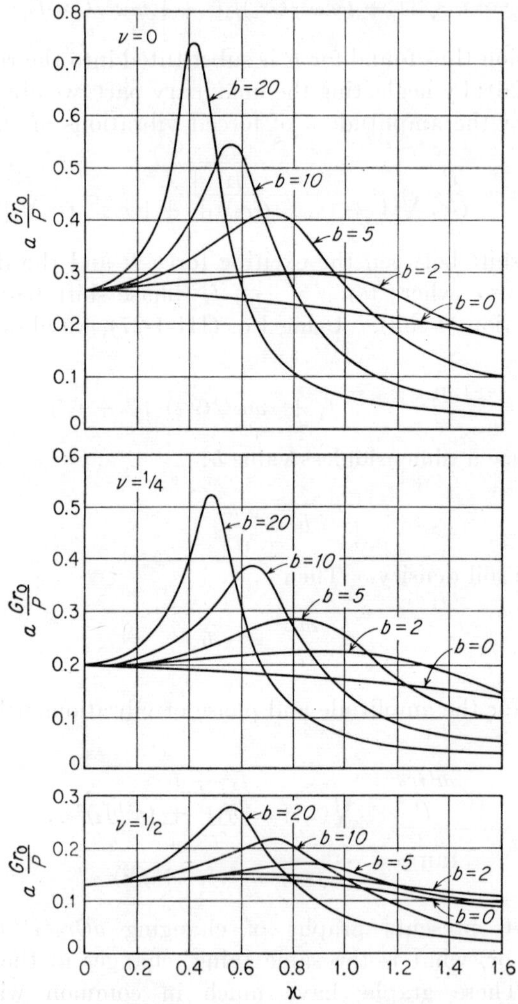

FIG. III-6. Auxiliary diagram for the solution of Eq. (III-1-32).

time. Since, other conditions being equal, amplitudes of maximum vibrations depend upon the value of Poisson's ratio, it follows that the damping of foundation vibrations also depends thereon. A comparison of graphs in Fig. III-6, plotted for several values of Poisson's ratio, shows that an increase in this ratio leads to an increase in dissipation of energy from a vibrating foundation into the soil, and hence to a result equivalent to an increase in the damping properties of the soil.

The fact that amplitudes of vibrations depend considerably on the value of b shows that the damping properties of soil are determined not only by its characteristics (inertia and elastic properties) but also by the size and mass of the foundation.

The interrelationship between resonance values aGr_0/P and the value b is illustrated in Fig. III-7 by dashed lines. It is seen that this relationship is close to a linear one of the type

$$a_r \frac{Gr_0}{P} = k + pb \qquad \text{(III-1-34)}$$

where a_r is the amplitude under conditions of resonance.

The value b is determined by Eq. (III-1-31). Taking into account that

$$m = \frac{W}{g} \qquad r_0 = \sqrt{\frac{A}{\pi}}$$

where W is the weight of the foundation and machine and A is the contact area of the foundation with soil, Eq. (III-1-31) may be rewritten as follows:

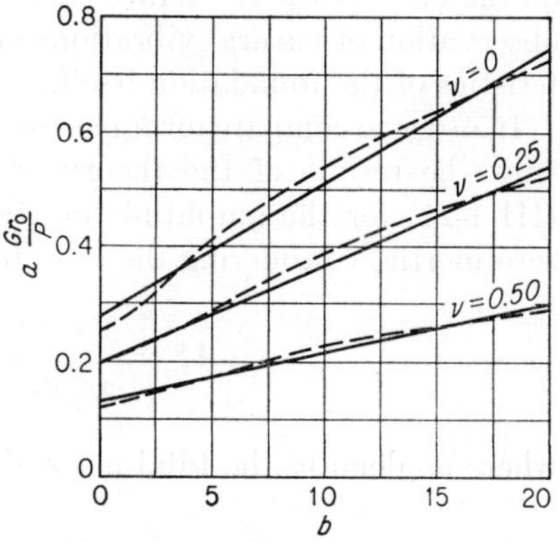

FIG. III-7. Dependence of peak (resonance) values of Fig. III-6 on Poisson's ratio ν.

$$b = \frac{\pi^{3/2}}{\gamma} \frac{p_{st}}{\sqrt{A}} \qquad \text{(III-1-35)}$$

where γ = unit weight of soil

p_{st} = normal static pressure

Substituting this value of b into Eq. (III-1-34), we obtain for the amplitude of vibrations under conditions of resonance (for $P = 1$)

$$a_r = \frac{k\pi^{1/2}}{GA^{1/2}} + \frac{l\pi^2}{G\gamma} \frac{p_{st}}{A} \qquad \text{(III-1-36)}$$

Since the unit weight of soils varies within a comparatively narrow range, the value of γ in Eq. (III-1-36) may be considered to be constant, equaling, for example, 1.70 tons/m³. Hence the conclusion is possible that the inertial characteristics of soil have small effect on amplitudes under conditions of resonance.

Equation (III-1-36) also shows that the resonance amplitude (reduced to unit of exciting force) increases with an increase in normal static pressure on the foundation base and decreases with an increase in the foundation contact area. Thus, under otherwise equal conditions, the damping of foundation vibrations by the soil decreases with an increase of the static pressure beneath the foundation. With an increase in the foundation contact area, resonance amplitudes decrease, i.e., damping increases. Consequently, damping properties of soil depend not only on its physicomechanical properties, but also on the foundation size and mass. Hence it is understandable why a damping constant established on the basis of the resonance curve of forced vibrations or on the basis of observation of natural vibrations of a foundation depends on the characteristics of the foundation itself.

In order to compare formulas for computations (III-1-32) and (III-1-33) with the results of the theory of vibrations, let us rewrite expression (III-1-21) for the amplitude of vibrations of a foundation on a base of zero inertia, considering the base to consist of weightless springs.

$$A_z^* = \frac{P}{\sqrt{(c_r - m_t\omega^2)^2 + (\alpha\omega)^2}} \tag{III-1-37}$$

where m_t denotes the total mass of foundation and soil,

$$m_t = m + m_s$$

Assuming that the foundation contact area is a circular area with radius r_0, we transform Eq. (III-1-37) by introducing $\varkappa$ and b as variables.

Multiplying (III-1-37) by Gr_0, we obtain

$$A_z^* \frac{Gr_0}{P} = \frac{1}{[(c_r/Gr_0) - (m_t\omega^2/Gr_0)]^2 + (\alpha\omega/Gr_0)^2} \tag{III-1-38}$$

According to (III-1-5), if $m = m_t$ we will have

$$c_r = m_t f_{nz}{}^2 = c_u A$$

Further, using (I-1-9), (I-2-12), and (I-2-13), we obtain

$$c_r = c_s \frac{E\sqrt{A}}{(1 - \nu^2)} = c_s \frac{2\sqrt{\pi}\, Gr_0}{1 - \nu}$$

Here, as in (I-2-14), c_s is a coefficient which depends on the geometric form of the foundation contact area; for a circular area it equals 1.083 if one considers settlement as the arithmetic mean of settlements under the center of the circular area and its edge. We denote

$$\frac{\alpha}{2m_t f_{nz}} = \xi \tag{III-1-39}$$

where α refers to Eq. (III-1-17a). Then

$$\left(\frac{\alpha\omega}{Gr_0}\right)^2 = \frac{4m_t^2\xi^2\omega^2 f_{nz}^2}{G^2 r_0^2} = \frac{8\sqrt{\pi}}{1-\nu}\,\frac{\omega^2 m_t \xi^2}{Gr_0}\,c_s$$

Let $m_t = \beta m$, where β is the coefficient of increase in foundation mass due to participation of soil.

$$\frac{m_t\omega^2}{Gr_0} = \varkappa^2 b\beta$$

Consequently,

$$A_z^* \frac{Gr_0}{P} = \frac{1}{\sqrt{[3.84/(1-\nu) - \varkappa^2 b\beta]^2 + 15.35\varkappa^2 b\xi^2\beta/(1-\nu)}} \qquad \text{(III-1-40)}$$

Taking the derivative of the right-hand side of Eq. (III-1-40) and equating it to zero, we find that resonance corresponds to the following value of the independent variable $\varkappa$:

$$\varkappa = \sqrt{\frac{3.84(1-2\xi^2/\beta)}{b(1-\nu)\beta}}$$

Let us assume that for selected values ν and b, the value of the amplitude of vibration computed from Eq. (III-1-32) will attain its maximum at $\varkappa = \varkappa_0$. Assuming that the maximums of amplitudes computed from Eqs. (III-1-40) and (III-1-32) correspond to the same value of the independent variable $\varkappa$, we obtain

$$\sqrt{\frac{3.84(1-2\xi^2/\beta)}{q(1-\nu)\beta}} = \varkappa_0 \qquad \text{(III-1-41)}$$

Let us assume also that the maximum values of amplitudes of vibrations computed from Eqs. (III-1-40) and (III-1-32) coincide with each other. Then we obtain:

$$\frac{1}{\sqrt{[3.84/(1-\nu) - \varkappa_0^2 b\beta]^2 + 15.35\varkappa_0^2 b\xi^2/(1-\nu)}} = a_r \qquad \text{(III-1-42)}$$

Equations (III-1-40) and (III-1-42) may be transformed as follows:

$$\frac{\xi^2}{\beta} = \frac{1}{2}\left[1 - \sqrt{1 - \left(\frac{1-\nu}{3.84a_r}\right)^2}\right] \qquad \text{(III-1-43)}$$

$$\beta = \frac{3.84(1-2\xi^2/\beta)}{b\varkappa_0^2(1-\nu)} \qquad \text{(III-1-44)}$$

Figure III-8 presents graphs of ξ and β as functions of b for soils with different values of Poisson's ratio ν.

If we take from these graphs values of the coefficients ξ and β corresponding to various values of b and ν, and then, using Eq. (III-1-40), plot resonance curves of forced vibrations, we shall see that these curves coincide fairly well with graphs plotted on the basis of Eq. (III-1-32). Such comparisons were made for $b = 2.5$, 10, and 20 and $\nu = 0$, 0.25, and 0.50. The results of computations from Eq. (III-1-40) using values of coefficients ξ and β taken from Fig. III-8a and b coincide so well with the results of computations from Eq. (III-1-32) that the two curves merge completely.

Thus it is established that if numerical values of the reduced coefficient of damping ξ and the coefficient of increase in mass β are taken from the graphs of Fig. III-8a and b, then the results of computations of amplitudes

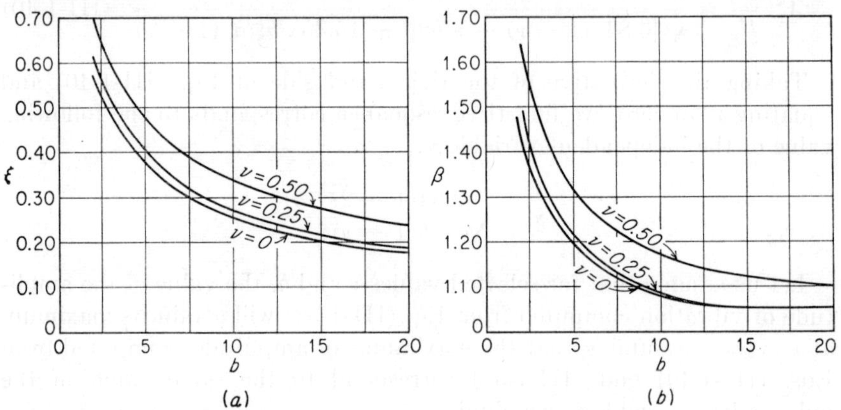

FIG. III-8. Auxiliary diagrams for the determination of the reduced damping coefficient ξ and the coefficient of mass increase β.

of forced vibrations from the approximate formula (III-1-40) (the latter obtained on the basis of an assumption that soil is represented by weightless springs) coincide fairly well with the results of a more precise theory which takes into account inertial properties of soil.

An analysis of Fig. III-8a and b leads to some general conclusions in respect to the effect of soil inertia on foundation vibrations. It follows directly from Fig. III-8b that, for all values of ν, with a decrease of b the coefficient of increase of mass β will increase; hence the effect of soil inertia on forced vertical vibrations of the foundation will grow.

According to Eq. (III-1-35) the value of b is proportional to the pressure p_{st} on the foundation contact area and inversely proportional to the square root of the area. Small values of b correspond to foundations with small height and large contact area. Such foundations are influenced more strongly by the effect of soil inertia than are foundations with considerable height and relatively small contact area. Thus foundations having the shape of thick slabs cause larger masses of soil to vibrate than do foundations having the shape of high blocks.

Numerical values of b for machine foundations attain 7 to 15. For these values, the coefficient of increase of mass β, even for $\nu = 0.5$, does not surpass the value 1.23. Hence, the soil mass participating in foundation vibrations does not exceed 23 per cent of the foundation mass. Since the period of natural vertical vibrations is inversely proportional to the square root of mass, it follows that the results of calculations of the period of natural vibrations of a foundation which include soil inertial effects are not more than 10 per cent smaller than those which neglect soil inertial properties. Thus a correction accounting for the influence of soil inertia will not be significant. Errors in computations of natural frequencies or amplitudes of forced foundation vibrations are usually not less than 10 to 15 per cent; therefore the effect of soil inertia on machinery foundations may be considered to be so slight that it may be neglected in many engineering calculations.

The foregoing discussion on the effect of soil inertia refers to a foundation resting on the soil surface. Where the foundation is embedded and the soil reacts not only on the horizontal foundation base area, but also on the foundation sides, there can be a considerable soil-inertia effect.

III-2. Rocking Vibrations of Foundations

Let us consider vibrations of foundations due to the action of a rocking external moment which changes with time according to the function $M \sin \omega t$ and which lies in one of the principal vertical planes of the foundation (Fig. III-9). It is assumed that the center of inertia of the mass of the foundation and the centroid of its horizontal base area lie on a vertical line located in the plane of the rocking moment.

Let us further assume that the elastic resistance of the soil against sliding of the foundation is so large in comparison to the resistance of the foundation against rocking that it may be considered to be infinitely great.

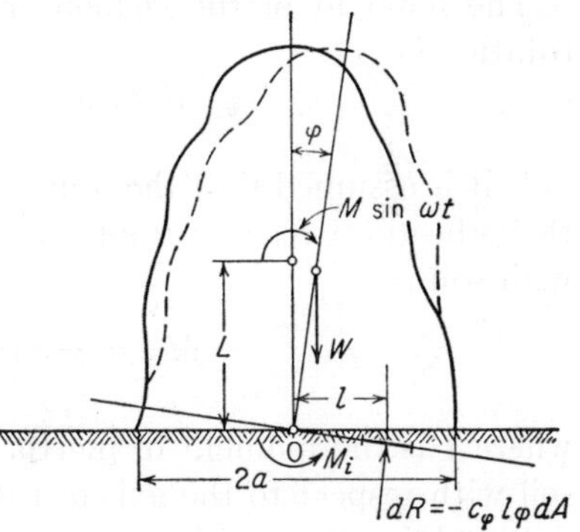

FIG. III-9. Analysis of rocking motion of a foundation.

In this case, the motion induced by an external moment $M \sin \omega t$ will be a rocking around the axis passing through the centroid of the area of foundation in contact with soil, perpendicular to the plane of vibrations. The position of the foundation is determined by one independent variable: the angle of rotation φ of the foundation around the axis passing through the point O (Fig. III-9).

Let us assume that at a certain instant the foundation has rotated a small angle φ around this axis.

The equation of its motion will be

$$-W_0\ddot{\varphi} + \Sigma M_i = 0 \tag{III-2-1}$$

where W_0 = moment of inertia of mass of foundation and machine with respect to axis of rotation

ΣM_i = sum of all external moments with respect to same axis

In this case, the foundation weight and the soil reaction are external forces.

a. Foundation Weight. The moment of this force W in respect to the axis of rotation is

$$LW\varphi$$

where L is the distance between the axis of rotation and the center of gravity of the vibrating mass.

b. Soil Reaction. An element dA of the foundation area in contact with soil, located at a distance l from the axis of rotation, is acted upon by the soil reaction

$$dR = c_\varphi l\varphi \, dA$$

where c_φ is the coefficient of elastic nonuniform soil compression.

The moment of the elementary force dR with respect to the axis of rotation is

$$dM_r = -l \, dR = -c_\varphi l\varphi \, dA$$

If it is assumed that the foundation does not lose contact with the soil, then the total reactive moment against the foundation area in contact with soil is

$$M_r = -c_\varphi \varphi \int_A l^2 \, dA = -c_\varphi I\varphi \tag{III-2-2}$$

where I is the moment of inertia of the foundation area in contact with soil with respect to the axis of rotation of the foundation.

By adding the exciting moment $M \sin \omega t$ to the two other moments, we obtain the equation of forced vibrations of the foundation:

$$-W_0\ddot{\varphi} + WL\varphi - c_\varphi I\varphi + M \sin \omega t = 0$$

or

$$W_0\ddot{\varphi} + (c_\varphi I - WL)\varphi = M \sin \omega t \tag{III-2-3}$$

By equating M to zero, we obtain the equation of free rocking vibrations with respect to the y axis:

$$W_0\ddot{\varphi} + (c_\varphi I - WL)\varphi = 0 \tag{III-2-4}$$

The solution of this equation is

$$\varphi = C \sin (f_{n\varphi}t + \varphi_0) \tag{III-2-5}$$

where

$$f_{n\varphi} = \frac{c_\varphi I - WL}{W_0} \tag{III-2-6}$$

and $f_{n\varphi}$ = natural frequency of rocking vibrations of foundation

c, φ_0 = arbitrary constants determined from initial conditions of motion of foundation

The solution of Eq. (III-2-3) will have a form similar to that for forced vertical vibrations [Eq. (III-1-13)], but instead of P, m, and f_{nz}, the values of M, W_0, and $f_{n\varphi}$ should be inserted.

In the case under consideration, the following expression will be obtained for the amplitude of vibrations:

$$A_\varphi = \frac{M}{W_0(f_{n\varphi}^2 - \omega^2)} \tag{III-2-7}$$

Since the product WL is usually small in comparison with $c_\varphi I$, that term may be neglected in Eq. (III-2-6); we then obtain for the frequency of natural rocking vibrations

$$f_{n\varphi}^2 = \frac{c_\varphi I}{W_0} \tag{III-2-8}$$

If the foundation base area has a rectangular form with sides a and b, and a is the side perpendicular to the axis of rotation, we have

$$I = \frac{ba^3}{12}$$

$$f_{n\varphi}^2 = \frac{ba^3}{12} \frac{c_\varphi}{W_0}$$

It follows from the latter formula that the length of the side of the foundation area in contact with soil and perpendicular to the axis of rotation has considerable effect on the natural frequency of rocking vibrations of the foundation. Depending on the selected length, the natural frequency may change considerably; hence the amplitude of forced vibrations will also change. The length of the other side of the foundation area, i.e., the one parallel to the axis of rotation, does not much influence the values of $f_{n\varphi}$ and A_φ. This length is usually selected on the basis of design considerations.

The amplitude of the vertical component of vibrations of the edge b of the foundation area in contact with soil is

$$A = \frac{a}{2} A_\varphi$$

or

$$A = \frac{Ma}{2W_0(f_{n\varphi}^2 - \omega^2)}$$

Rocking vibrations occur mostly in high foundations under machines having unbalanced horizontal components of exciting forces and exciting moments. For example, such vibrations occur in foundations under sawmill log-sawing frames. These foundations are usually high and project above the first floor of the sawmill. Figure III-10 illustrates the distribution of amplitudes of horizontal forced vibrations along the height of the foundation under a log frame. It is seen from the graph that the centroid of the foundation area in contact with soil is subjected to forced rocking vibrations.

Therefore Eq. (III-2-7) may be used for the computation of the amplitude of forced vibrations of foundations under log frames induced by the horizontal component of exciting forces developed in these frames.

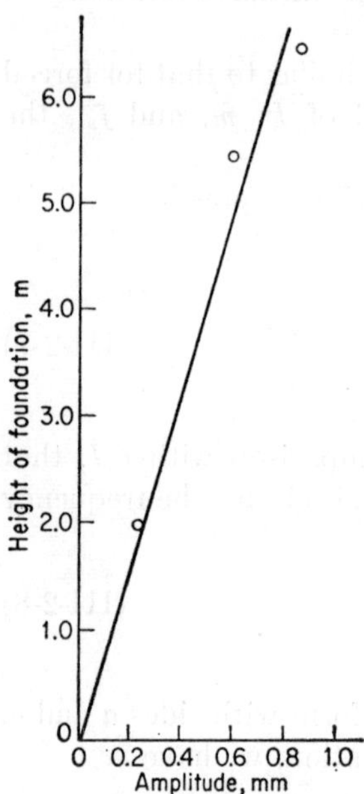

FIG. III-10. Variation of the horizontal amplitude of rocking forced vibrations along the height of a foundation.

III-3. Vibrations of Pure Shear

If the resistance of soil to compression is large in comparison with the resistance to shear, then displacement of the foundation under the action of horizontal forces will occur mainly in the direction of the action of horizontal exciting forces.

Let us assume that a horizontal exciting force $P_T \sin \omega t$ acts on the foundation. The equations of forced and free vibrations will be analogous to the equations of vertical vibrations of a foundation, e.g., Eq. (III-1-11), in which c_τ should be inserted instead of c_u; thus the equation of forced horizontal vibrations will be

$$\ddot{x} = f_{nx}^2 x = p_T \sin \omega t \qquad \text{(III-3-1)}$$

where x is the horizontal displacement of the center of gravity of the foundation and

$$f_{nx}^2 = \frac{c_\tau A}{m} \qquad \text{(III-3-2)}$$

f_{nx} is the natural frequency of vibrations in shear.

The solution of Eq. (III-3-1) is

$$A_x = \frac{P_T}{m(f_{nx}^2 - \omega^2)} \tag{III-3-3}$$

All conclusions and formulas obtained while considering vertical vibrations of foundations apply also to vibrations in shear.

In addition to the foregoing type of vibrations of foundations, characterized by horizontal displacement of the center of gravity of the vibrating mass, vibrations in shear may have a form of rotational vibrations with respect to the vertical axis passing through the center of gravity of the foundation and the centroid of the area of its base. Letting

W_z = moment of inertia of vibrating mass with respect to above axis
J_z = polar moment of foundation base area
ψ = angle of torsion of foundation
$M_z \sin \omega t$ = exciting moment acting in horizontal plane
c_φ = coefficient of elastic nonuniform shear

we obtain the following equation of forced vibrations of a foundation induced by an exciting moment:

$$W_z\ddot{\psi} + c_\psi J_z \psi = M_z \sin \omega t \tag{III-3-4}$$

A particular solution of this equation may be presented in the form

$$\psi = \frac{M_z}{W_z(f_{n\psi}^2 - \omega^2)} \tag{III-3-5}$$

where $f_{n\psi}^2 = c_\psi J_z/W_z$ is the square of the natural frequency of vibrations of a foundation for the vibrations under consideration.

III-4. Vibrations of Foundations Accompanied by Simultaneous Rotation, Sliding, and Vertical Displacement

a. Equations of Vibration. The foregoing discussion concerned cases of vibrations of massive foundations in which the soil was characterized by infinite rigidity with respect either to compression or to shear. Let us consider now the simplest case of vibrations of foundations, where the soil is able to offer elastic resistance both to compression and to shear. As before, it is assumed that the center of gravity of the foundation and machine and the centroid of the foundation base area are located on a vertical line which lies in the main central plane of the foundation. External exciting forces induced by the machine also lie in this plane; if O (Fig. III-11) is the center of gravity of the vibrating mass, then these forces may be reduced to the exciting force $P(t)$ applied at O and a couple with moment $M(t)$. Let us match the origin of a coordinate system with the center of gravity of the foundation and machine mass at an instant when the foundation is motionless; the direction of coordinate axes is shown in Fig. III-11.

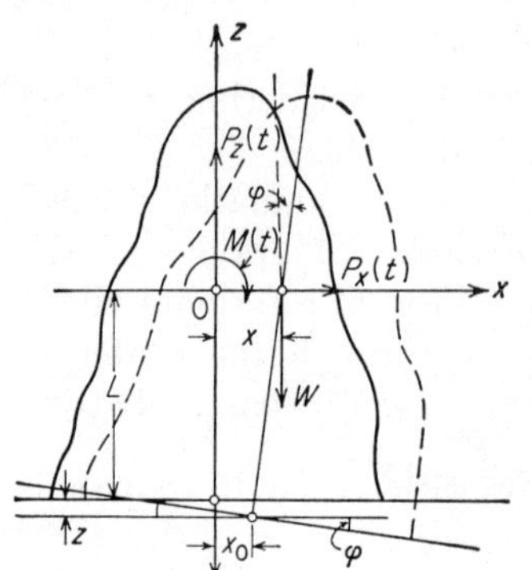

FIG. III-11. Analysis of combined types of foundation motion.

Under the action of loads $P(t)$ and $M(t)$, the foundation will undergo a two-dimensional motion determined by the values of three independent parameters: the projections x and z of displacement of the foundation center of gravity on the coordinate axes and the angle φ of rotation of the foundation with respect to the y axis which passes through the center of gravity of the foundation and machine, perpendicular to the plane of vibrations.

By projecting all forces acting on the foundation at time t on the x and z axes and adding to them the projections on the same axes of the inertia forces, we obtain, according to d'Alembert's principle,

$$\begin{aligned} -m\ddot{x} + \Sigma X_i &= 0 \\ -m\ddot{z} + \Sigma Z_i &= 0 \end{aligned} \qquad \text{(III-4-1)}$$

where m = foundation mass

X_i, Z_i = projections on x, z axes of all external forces acting on foundation

The equation of the moments with respect to the y axis should also be added to Eqs. (III-4-1).

$$-M_m\ddot{\psi} + \Sigma M_i = 0 \qquad \text{(III-4-2)}$$

where M_m is the moment of inertia of the mass with respect to the y axis.

The following forces are acting on the foundation at time t:

1. Weight W of foundation and machine. A projection of the force W on the x axis equals zero, that on the z axis equals

$$Z_1 = -W$$

2. Soil reaction caused by settlement of the foundation under the action of weight. A projection of this force on the x axis also equals zero; the projection of this force on the z axis equals

$$Z_2 = c_u A z_{st}$$

where c_u = coefficient of elastic uniform compression of soil

A = area of foundation in contact with soil

z_{st} = elastic settlement caused by action of weight of foundation and machine

The soil reaction is applied to the centroid of the contact area of foundation and soil. It produces a moment with respect to the y axis:

$$M_1 = WL\varphi$$

where L is the distance from the center of gravity of the mass to the foundation base.

If at a given instant of time t, the foundation has a displacement z measured from the equilibrium position, then the soil reaction induced by this displacement is

$$Z_3 = c_u Az$$

3. Horizontal reaction of elastic resistance of soil. Its projection on the x axis is

$$X_1 = -c_\tau A x_0$$

where c_τ = coefficient of elastic uniform shear of soil
x_0 = displacement centroid of contact area of foundation

$$x_0 = x - L\varphi$$

where x is horizontal displacement of the common center of gravity of foundation and machine.

Substituting this value of x_0 into the expression for X_1, we obtain

$$X_1 = -c_\tau A(x - L\varphi)$$

The moment of this force with respect to the y axis is

$$M_2 = c_\tau AL(x - L\varphi)$$

4. Reactive resistance of soil induced by rotation of foundation base area. In order to compute the moment caused by this resistance, let us single out an infinitely small element dA of the foundation area in contact with soil. The reaction dR of soil on this element is

$$dR = c_\varphi l\varphi\, dA$$

where c_φ = coefficient of elastic nonuniform compression of soil
l = distance between area element dA and axis of rotation

The moment of this elementary reaction with respect to the y axis is

$$dM_3 = -c_\varphi l^2\varphi\, dA$$

By integrating over the whole foundation area in contact with soil, we obtain the total reactive moment of soil developed when the foundation base contact area turns an angle φ. This moment is

$$M_3 = -c_\varphi I\varphi$$

where I is the moment of inertia of foundation contact area with respect to the axis passing through the centroid of this area, perpendicular to the plane of vibrations.

To these four forces there should be added the projections $P_x(t)$ and $P_z(t)$ of the external exciting force $P(t)$ on the coordinate axes and the moment $M(t)$.

Substituting into Eqs. (III-4-1) and (III-4-2) the established values of the projections of forces on the x and z axes, as well as the values of moments with respect to the y axis, we obtain, after some elementary transformations, a system of three differential equations of forced vibrations of a foundation:

$$m\ddot{z} + c_u A z = P_z(t) \qquad \text{(III-4-3)}$$

$$\left.\begin{aligned} m\ddot{x} + c_\tau A x - c_\tau A L \varphi &= P_x(t) \\ M_m \ddot{\varphi} - c_\tau A L x + (c_\varphi I - WL + c_\tau A L^2)\varphi &= M(t) \end{aligned}\right. \qquad \text{(III-4-4)}$$

N. P. Pavliuk[32] was the first to give these equations of foundation vibration.

Equations (III-4-4) are interdependent because each of them includes x and φ. Equation (III-4-3) in no way depends upon Eqs. (III-4-4). Hence it follows that vertical vibrations of foundation occur independently of vibrations associated with the other two coordinates. If a foundation is acted upon by exciting loads having no vertical components, then no vertical vibrations of the foundation develop. In this case the foundation will undergo rotation around the y axis and horizontal displacement in the direction of the x axis. If a foundation is acted upon by an exciting load producing only a vertically centered force, then the foundation will undergo only vertical vibrations.

In the same way, if the equilibrium of a foundation at a certain instant of time is disturbed only by a vertical displacement of its center of gravity, or if at this moment the foundation is given a velocity in the vertical direction, then the foundation will undergo only vertical natural vibrations. If the equilibrium of a foundation is disturbed by a displacement of its center of gravity in the horizontal direction or if it is given a velocity in the horizontal direction, then no vertical vibrations appear. In this case the foundation motion will be characterized by changes in two coordinates: x and φ. The same coordinates characterize the foundation motion if its equilibrium is disturbed by changes in either x or φ.

The fact that vertical vibrations of foundations are independent of vibrations in the directions x and φ gives us a chance to consider each type of vibration separately. An investigation of vertical vibrations has already been made in Art. III-1. Therefore it remains to investigate vibrations corresponding to the system of Eqs. (III-4-4).

b. Free Vibrations. If the equilibrium of a foundation is disturbed by subjecting it at the initial instant of time to certain changes in the coordinates x and φ and the velocities $\dot{x}$ and $\dot{\varphi}$, then during the time which follows the foundation will be subjected to elastic soil reactions and inertial forces and will undergo free vibrations. The equations of these vibrations are as follows:

$$\begin{aligned} m\ddot{x} + c_\tau Ax - c_\tau AL\varphi &= 0 \\ M_m\ddot{\varphi} - c_\tau ALx + (c_\varphi I - WL + c_\tau AL^2)\varphi &= 0 \end{aligned} \qquad \text{(III-4-5)}$$

Particular solutions of these equations may be written in the form

$$x = A_a \sin (f_n t + \alpha) \qquad \varphi = B_a \sin (f_n t + \alpha)$$

where A_a, B_a, and α are arbitrary constants.

Substituting these solutions into (III-4-5) and reducing all terms by eliminating $\sin (f_n t + \alpha)$, we obtain two homogeneous equations:

$$\begin{aligned} (c_\tau A - mf_n^2)A_a - c_\tau ALB_a &= 0 \\ -c_\tau ALA_a + (c_\varphi I - WL + c_\tau AL^2 - M_m f_n^2)B_a &= 0 \end{aligned} \qquad \text{(III-4-6)}$$

The constants A_a, B_a, and f_n should satisfy these equations if the particular solutions are to satisfy the system of differential equations of free vibrations of the foundation.

System (III-4-6) does not permit the determination of values of all three constants A_a, B_a, and f_n. In order to do this, it is necessary to know the initial conditions of the foundation. However, if we consider that in Eqs. (III-4-6) only A_a and B_a (that is, only the amplitudes of vibrations) are unknown, then we obtain from the first equation

$$A_a = \frac{c_\tau AL}{c_\tau A - mf_n^2} B_a$$

Substituting this expression for A_a into the second equation, we obtain

$$B_a[-c_\tau^2A^2L^2 + (c_\varphi I - WL + c_\tau AL^2 - M_m f_n^2)(c_\tau A - mf_n^2)] = 0$$

If B_a does not equal zero, then, in order to satisfy the above equation, it is necessary to assume that the factor in brackets equals zero. Then we obtain the frequency equation

$$\Delta(f_n)^2 = -c_\tau^2A^2L^2 + (c_\varphi I - WL + c_\tau AL^2 - M_m f_n^2)(c_\tau A - mf_m^2) = 0 \qquad \text{(III-4-7)}$$

This equation contains only one unknown constant f_n, the natural frequency of vibrations of the foundation.

Let us transform Eq. (III-4-7) by opening brackets and grouping members containing the same powers of f_n. Then we obtain a second-degree equation for f_n^2.

After dividing all members of the new equation by mM_m, it may be rewritten as follows:

$$f_n^4 - \left(\frac{c_\tau I - WL}{M_m} + \frac{c_\tau A}{m}\frac{L^2m + M_m}{M_m}\right) f_n^2 + \frac{c_\varphi I - WL}{M_m}\frac{c_\tau A}{m} = 0$$

Let us denote by M_{m0} the moment of inertia of the total vibrating mass (the foundation and machine) with respect to the axis passing through the centroid of the base contact area and perpendicular to the plane of vibrations; this moment equals

$$M_{m0} = M_m + mL^2$$

Let

$$\frac{M_m}{M_{m0}} = \gamma$$

where $1 > \gamma > 0$.

Substituting $M_m = \gamma M_{m0}$ into the equation for frequencies, we rewrite it as follows:

$$f_n^4 - \left(\frac{c_\varphi I - WL}{M_{m0}} + \frac{c_\tau A}{m}\right)\frac{f_n^2}{\gamma} + \frac{c_\varphi I - WL}{\gamma M_{m0}}\frac{c_\tau A}{m} = 0$$

But according to (III-2-6) and (III-3-2) the expressions

$$\frac{c_\varphi I - WL}{M_{m0}} = f_{n\varphi}^2 \qquad \text{and} \qquad \frac{c_\tau A}{m} = f_{nx}^2$$

represent limiting frequencies of the foundation when the resistance of soil to shear is very large in comparison to its resistance to rotational vibrations or vice versa.

Using these two expressions we obtain the final equation of frequencies in the following form:

$$\Delta(f_n^2) = f_n^4 - \frac{f_{n\varphi}^2 + f_{nx}^2}{\gamma} f_n^2 + \frac{f_{n\varphi}^2 f_{nx}^2}{\gamma} = 0 \qquad \text{(III-4-8)}$$

This equation will have two positive roots f_{n1} and f_{n2} corresponding to the two principal natural frequencies of the foundation.

It can be proved that the natural frequencies which are the roots of Eq. (III-4-8) have the following interrelationship with the limiting frequencies $f_{n\varphi}$ and f_{nx}: the smaller of the two natural frequencies (for example, f_{n2}) is smaller than the smallest of the two limiting frequencies; the larger natural frequency is always larger than $f_{n\varphi}$ and f_{nx}.

In the case under consideration, involving a foundation with two degrees of freedom, specific forms of vibrations correspond to the frequencies f_{n1} and f_{n2} of the foundation; these vibrations are characterized by a certain interrelationship between the amplitudes A_a and B_a which depends on the foundation size and the soil properties, but does not depend on the initial conditions of foundation motion.

Let us determine from the first equation of system (III-4-6) the ratio A_a/B_a:

$$\rho = \frac{A_a}{B_a} = \frac{f_{nx}^2 L}{f_{nx}^2 - f_n^2} \tag{III-4-9}$$

If the foundation vibrates at the lower frequency f_{n2}, then, according to the above statement,

$$f_{nx}^2 - f_2^2 > 0$$

and ρ also is larger than zero; consequently, the amplitudes A_a and B_a have the same sign. It means that during vibration at frequency f_{n2}, when the center of gravity deviates from the equilibrium position,

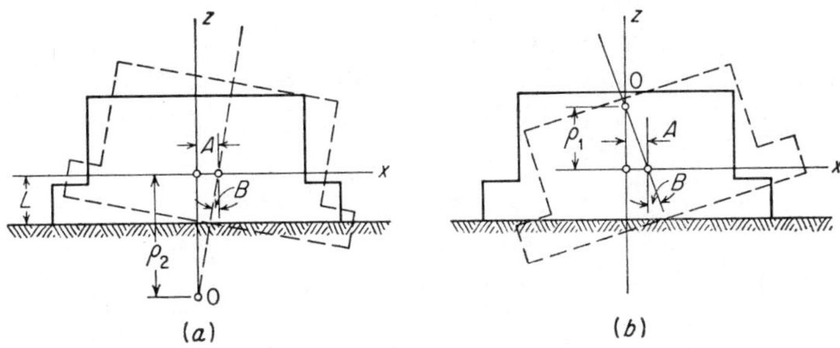

FIG. III-12. Two types of foundation vibrations which correspond to Eq. (III-4-9).

for example, in the positive direction of the x axis, the rotation of the foundation will be also positive, and changes of amplitudes A_a and B_a will be in phase. The form of vibrations in this case will be analogous to that shown in Fig. III-12*a*; i.e., the foundation will undergo rocking vibrations with respect to a point situated at a distance ρ_2 from the center of gravity of the foundation. The value of ρ_2 is determined by the absolute value of expression (III-4-9) if f_{n2} is substituted for f_n. However, if a foundation vibrates at the higher frequency f_{n1}, then, since $f_{nx}^2 - f_{n1}^2 < 0$, ρ will be negative, and A_a and B_a will be 180° out of phase. Figure III-12*b* illustrates the form of vibrations corresponding to this case. Here the foundation vibrates around a point which lies higher than the center of gravity and at a distance ρ_1 determined from expression (III-4-9) if f_{n1} is substituted for f_n.

There is a simple relationship between ρ_2 and ρ_1:

$$\rho_1\rho_2 = i^2$$

where

$$i^2 = \frac{M_m}{m}$$

i is the radius of gyration of the mass of foundation and machine.

If the main dimensions of a foundation which determine its mass, base area, and moments of inertia are selected, then the limiting natural frequencies $f_{n\varphi}$ and f_{nx} will depend only on the coefficients of elastic nonuniform compression and shear c_φ and c_τ of the soil.

Often the exact values of these coefficients are not known and only the range of the most probable values of c_φ and c_τ may be assumed. Then the computation of the natural frequencies f_{n1} and f_{n2} should be performed for the whole range of values of these coefficients.

The natural frequencies f_{n1} and f_{n2} of the foundation are determined as the roots of Eq. (III-4-8):

$$f_{n1,2}{}^2 = \frac{1}{2\gamma}\left[f_{n\varphi}{}^2 + f_{nx}{}^2 \pm \sqrt{(f_{n\varphi}{}^2 + f_{nx}{}^2)^2 - 4\gamma f_{n\varphi}{}^2 f_{nx}{}^2}\right]$$

If it is necessary to compute a range of possible values of frequencies corresponding to the most probable values of elastic coefficients of the soil, then it is more convenient to transform the latter expression as follows:

$$\beta_{1,2} = \frac{f_{n1,2}{}^2}{f_{n\varphi}{}^2} = \frac{1}{2\gamma}\left[1 + \mu \pm \sqrt{(1+\mu)^2 - 4\gamma\mu}\right]$$

where μ is the ratio of the squares of the limiting natural frequencies:

$$\mu = \frac{f_{nx}{}^2}{f_{n\varphi}{}^2}$$

If the dimensions of a foundation are selected, then the value β depends only on the assumed values of the coefficients of elastic compression and shear. After the selection of a range of values for these coefficients, it is easy to calculate all possible values of frequency.

Figures III-13*a* and *b* present graphs of $\beta_{1,2}$ as functions of μ. The values of β_1 and β_2 are plotted along the y axis, the values of μ along the x axis. Curves are plotted for different values of γ, from $\gamma = 0.4$ (high foundations) up to $\gamma = 0.9$ (low foundations).

With these graphs it is easy to determine a possible range of changes in f_{n1} and f_{n2}, using a given range of values of c_φ and c_τ. The frequencies are determined from the formula

$$f_{n1,2}{}^2 = f_{n\varphi}{}^2\beta_{1,2}$$

Since the values of $f_{n1,2}$ depend not only on $\beta_{1,2}$, but also on $f_{n\varphi}$, in order to determine the smallest and the largest value of f_{n1} and f_{n2}, it is necessary to compute the minimum and maximum values of the right-hand part of the above expression.

c. Forced Vibrations. Returning to Eqs. (III-4-4), describing forced vibrations of foundations, let us consider separately several particular cases of the action of exciting loads.

Assume that a horizontal force of magnitude $P \sin \omega t$ is applied at the center of gravity of the foundation and machine. This case is of the greatest interest in engineering practice. Equations (III-4-4) (forced vibrations of a foundation accompanied by changes with time in the

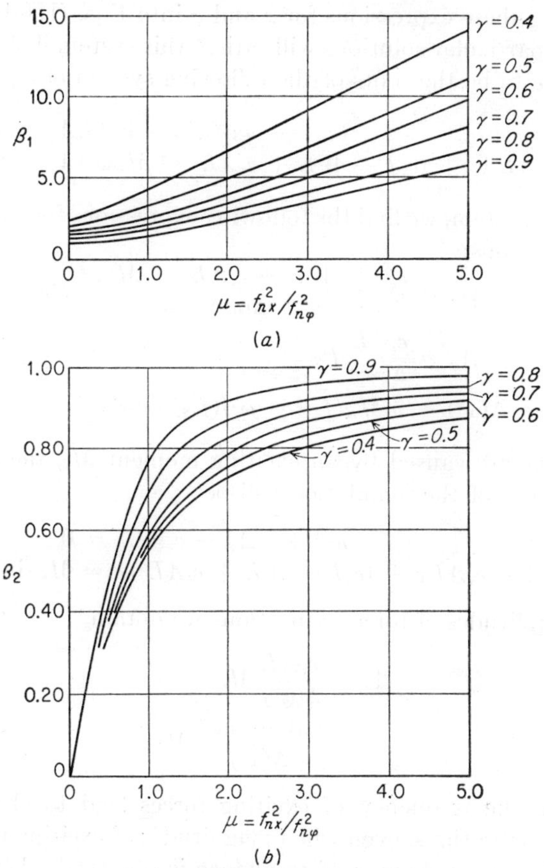

FIG. III-13. Variation of the coefficients β_1 and β_2 (which govern the two main natural frequencies f_{n1} and f_{n2} of the foundation shown in Fig. III-12) with the ratio μ of natural horizontal and rocking frequencies and the ratio γ of moments of inertia which inversely reflects the height of a foundation.

angle of rotation φ and the horizontal component x of the coordinates of the center of gravity) will be rewritten as follows:

$$\begin{aligned} m\ddot{x} + c_\tau A x - c_\tau A L \varphi &= P_T \sin \omega t \\ M_m \ddot{\varphi} - c_\tau A L x + (c_\varphi I - WL + c_\tau A L^2)\varphi &= 0 \end{aligned} \qquad \text{(III-4-10)}$$

We shall seek particular solutions of this system, corresponding only to the forced vibrations of a foundation, in the form

$$x = A_x \sin \omega t$$
$$\varphi = A_\varphi \sin \omega t$$

Substituting these expressions for x and φ into Eqs. (III-4-10), we find that selected particular solutions will satisfy this system if the coefficients A_x and A_φ are to be the roots of the following system of equations:

$$(c_\tau A - m\omega^2)A_x - c_\tau ALA_\varphi = P_T$$
$$-c_\tau ALA_x + (c_\varphi I - WL + c_\tau AL^2 - M_m\omega^2)A_\varphi = 0$$

Solving this system, we find the following expressions for the amplitudes of forced vibrations:

$$A_x = \frac{c_\varphi I - WL + c_\tau AL^2 - M_m\omega^2}{\Delta(\omega^2)} P_T$$
$$A_\varphi = \frac{c_\tau AL}{\Delta(\omega^2)} P_T \tag{III-4-11}$$

where $\Delta(\omega^2) = mM_m(f_{n1}{}^2 - \omega^2)(f_{n2}{}^2 - \omega^2)$

If vibrations are caused by an exciting moment M_i, the equations of forced vibrations of the foundation will be

$$m\ddot{x} + c_\tau A_x - c_\tau AL\varphi = 0$$
$$M_m\ddot{\varphi} - c_\tau ALx + (c_\varphi I - WL + c_\tau AL^2)\varphi = M_i \sin \omega t$$

For the amplitudes of forced vibrations we obtain

$$A_x = \frac{c_\tau AL}{\Delta(\omega^2)} M_i$$
$$A_\varphi = \frac{c_\tau A - m\omega^2}{\Delta(\omega^2)} M_i \tag{III-4-12}$$

Changes in the frequency of exciting forces lead to changes in the amplitudes of vibrations, even when magnitudes of exciting forces remain the same. The phenomenon of resonance is observed when one of the natural frequencies coincides with the frequency of exciting forces. Since in the case under consideration, the foundation has two natural frequencies, two resonances are possible when amplitudes grow rapidly. Figure III-14 illustrates the general character of resonance curves for the forced vibrations of foundations under discussion; here an experimental resonance curve is plotted on the basis of data obtained by the author during investigations of a test foundation with a 4-m^2 base area in contact with soil.

To every frequency of forced vibrations of a foundation there corresponds a particular form of vibrations which is characterized by the magnitude and sign of the radius vector ρ connecting the center of gravity of the foundation and the point O (Fig. III-12a and b), around which the foundation rotates. The magnitude and sign of ρ are determined from the equation

$$\rho = \frac{A_x}{A_\varphi} \qquad \text{(III-4-13)}$$

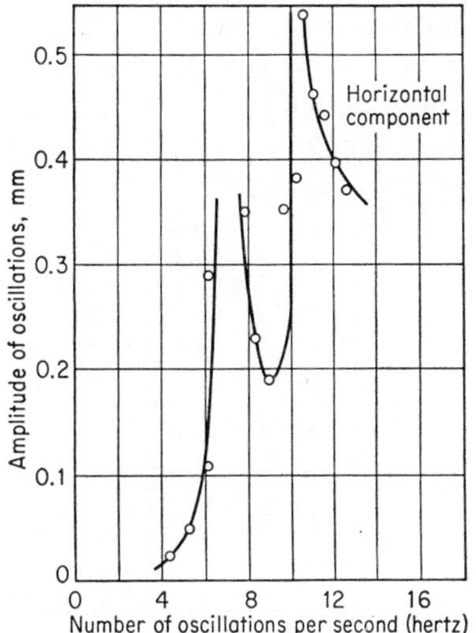

FIG. III-14. General character of resonance curves corresponding to Eq. (III-4-12).

The character of the exciting loads causing forced vibrations of a foundation also has an effect on the dependence of the form of vibrations upon changes in the frequency of the exciting force.

If a foundation is subjected only to the action of the exciting moment M_i, then according to (III-4-12),

$$\rho = \frac{f_{nx}^2}{f_{nx}^2 - \omega^2} L \qquad \text{(III-4-14)}$$

When ω is small in comparison with f_{nx}, ρ does not differ much from L, i.e., at low exciting frequencies, the foundation will vibrate with respect to the axis passing through the center of gravity of the base contact area, perpendicular to the plane of vibrations. With an increase in exciting frequency, the denominator of (III-4-14) will decrease rapidly and ρ will grow, i.e., the foundation vibrations will be accompanied not only by changes in ρ, but also by changes in x_0; in other words, the foundation contact area will undergo sliding. If $\omega = f_{nx}$, then ρ will be indefinitely large. In this case the foundation will undergo only vibrations of shear (sliding) with a certain amplitude. With a further increase in the exciting frequency, ρ changes its sign; with an increase in ω, ρ continuously decreases, approaching zero as ω becomes infinitely large.

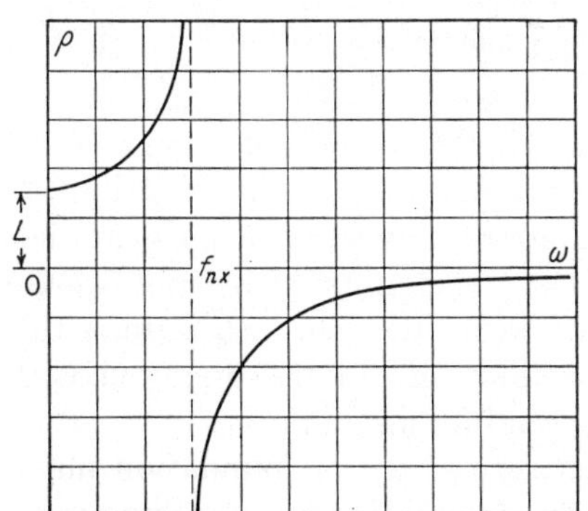

FIG. III-15. General character of changes in the radius vector ρ with the frequency of vibrations ω.

This means that at exciting frequencies considerably larger than the limiting frequency f_{nx}, a foundation will principally undergo rotation around the axis passing through its center of gravity. Figure III-15 illustrates the general character of changes in ρ, depending on ω.

Figure III-16 gives position of the main vertical axis versus frequency ω, plotted for a test foundation and computed from Eq. (III-4-14) on the basis of an experimental value of f_{nx} and a selected value of ω. Circles plotted on the same figure show amplitudes of the horizontal component of vibrations corresponding to some magnitudes of exciting frequency. These amplitudes were measured at various foundation heights. The experimental points agree well with values established on the basis of the foregoing theory.

d. The Effect on the Natural Frequencies of Eccentric Distribution of the Foundation and Machine Mass. An eccentric distribution of the machine

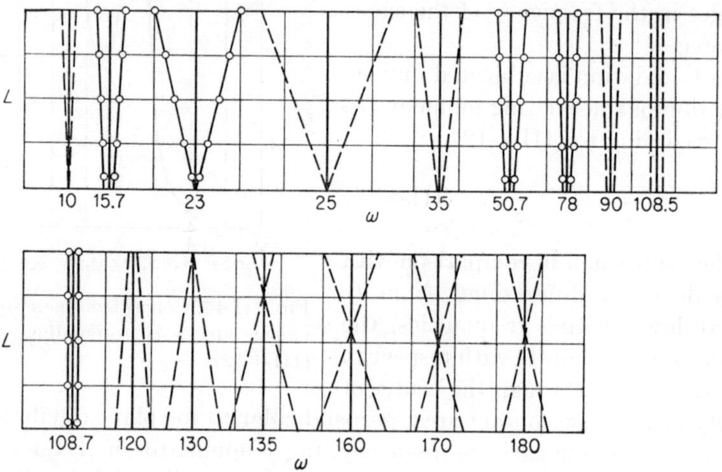

FIG. III-16. Variations in the position of the main vertical axis (Fig. III-12) with changes in the frequency ω.

mass may occur when a machine and a generator or a motor are coupled on the same shaft. Sometimes an eccentricity in the mass distribution is caused by asymmetry of the foundation resulting from various cavities, channels, etc. The asymmetry can often be eliminated by moving the centroid of the foundation area in contact with the soil. Sometimes this cannot be done; then foundation vibrations should be computed with the asymmetric distribution of mass taken into account.

Let us consider the simplest case of asymmetry of foundation mass, that in which the center of gravity of the foundation and machine mass and the centroid of the foundation contact area lie in one of the main foundation planes, but not on the same vertical line. We shall investigate foundation vibrations in the main plane, in which both the centers of gravity lie.

The foundation motion is again determined by three parameters: the projections x and z of the displacement of the foundation center of gravity on the corresponding coordinate axes and angle φ of rotation of the foundation around the y axis, passing through the center of gravity and perpendicular to the plane of vibrations.

Let us assume that foundation motion has been caused by an initial disturbance (for example, an impact) and examine the forces acting on the foundation during its motion. Then we shall set up the differential equations for this motion taking into consideration the soil reactions produced by foundation displacements only.

As before, we match the origin of the coordinate system used for the study of the foundation with the center of gravity, when the foundation is at rest.

We assume that at a certain instant of time, the projections of the displacement of the center of gravity of the foundation will equal x and z, and the projection of the rotation vector on the y axis will equal φ. We measure these values from the equilibrium position of the foundation when it is subjected to the action of weight and to static soil reactions.

The horizontal displacement of the centroid of the foundation area in contact with soil equals

$$x_0 = x - L\varphi$$

where L, as before, is a distance between the center of gravity of the foundation and the foundation base; hence, the soil reaction caused by horizontal displacement of the centroid of the contact area equals

$$R_x = -c_\tau A(x - L\varphi)$$

The vertical component of the displacement of the centroid of the contact area equals

$$z_0 = z - \epsilon\varphi$$

where ϵ is the eccentricity of the foundation and machine mass.

The soil reaction caused by this displacement of the centroid is

$$R_z = -c_\tau A(z - \epsilon\varphi)$$

Let us compute the moments of all forces with respect to the y axis.

The moment of the force of gravity equals zero.

The moment of the soil reaction caused by the static action of weight equals

$$M_1 = WL\varphi$$

where W is the total weight of the foundation and machine.

The horizontal component of the soil reaction produces the moment

$$M_2 = c_\tau A(x - L\varphi)L$$

The moment due to the vertical component of the soil reaction is

$$M_3 = c_u A(z - \epsilon\varphi)\epsilon$$

The reactive moment produced by soil due to rotation of the foundation by the angle φ equals

$$M_4 = -c_\varphi I \varphi$$

where I is the moment of inertia of the foundation contact area with respect to the axis passing through its centroid, perpendicular to the plane of vibrations.

Substituting the above expressions for forces and moments into Eqs. (III-4-1) and (III-4-2), we obtain the following differential equations for the eccentric distribution of masses now under consideration:

$$\begin{aligned} m\ddot{x} + c_\tau A x - c_\tau A L \varphi &= 0 \\ m\ddot{z} + c_u A z - c_u A \epsilon \varphi &= 0 \\ M_m \ddot{\varphi} - c_\tau A L x + (c_\varphi I - WL + c_u A \epsilon^2 + c_\tau A L^2)\varphi - c_u A \epsilon z &= 0 \end{aligned} \tag{III-4-15}$$

Equations (III-4-15) show that—unlike the previous case, in which the center of inertia of mass and the center of gravity of the foundation contact area lay on the same vertical line—here the three differential equations of motion are interrelated. Therefore, if at the initial moment a change took place in only one parameter related to the motion, then as a consequence there would be changes in all three parameters determining the position of the foundation. Thus, if at the initial instant the foundation is subjected to the action of a disturbance inducing only a horizontal displacement of its center of gravity, then it will move not only in this direction, but also in the vertical direction, and will undergo rotational vibrations around the y axis as well.

If the asymmetry in the distribution of masses is very small ($\epsilon \cong 0$), then Eqs. (III-4-15) at this limit value of ϵ are transformed into the systems (III-4-3) and (III-4-4).

We shall proceed as before in order to obtain the frequency equation for the asymmetrical case under consideration.

First of all, let us simplify the differential equations of motion by denoting

$$\begin{aligned} c_\tau A &= c_x \\ c_u A &= c_z \\ c_\varphi I - WL + (c_u \epsilon^2 + c_\tau L^2)A &= c \end{aligned}$$

Inserting these notations into Eqs. (III-4-15), we obtain

$$\begin{aligned} m\ddot{x} + c_x x - c_x L \varphi &= 0 \\ m\ddot{z} + c_z z - c_z \epsilon \varphi &= 0 \\ M_m \ddot{\varphi} - c_x L x + c\varphi - c_z \epsilon z &= 0 \end{aligned} \tag{III-4-16}$$

We will then seek a particular solution of Eqs. (III-4-16) in the form

$$x = A_a \sin f_n t \qquad z = B_a \sin f_n t \qquad \varphi = C_a \sin f_n t$$

where A_a, B_a, and C_a are arbitrary constants.

Substituting these solutions into Eqs. (III-4-16), we obtain three homogeneous equations:

$$
\begin{aligned}
(c_x - mf_n^2)A_a - c_x L C_a &= 0 \\
(c_z - mf_n^2)B_a - c_z \epsilon C_a &= 0 \\
(c - M_m f_n^2)C_a - c_x L A_a - c_z \epsilon B_a &= 0
\end{aligned}
\qquad \text{(III-4-17)}
$$

The constants A_a, B_a, C_a, and f_n should satisfy these equations if our selected particular solutions are to satisfy the system (III-4-16) of differential equations. Equations (III-4-17) are linear and homogeneous. In order for these equations to have solutions other than zero for A_a, B_a, and C_a, their determinant should be identically reduced to zero:

$$
(c_x - mf_n^2)(c_z - mf_n^2)(c - M_m f_n^2) - (c_x - mf_n^2)c_z^2\epsilon^2 - (c_z - mf_n^2)c_x^2L^2 = 0 \qquad \text{(III-4-18)}
$$

If the eccentricity in distribution of mass equals zero, then this equation of frequencies is reduced to the equation

$$
(c_z - mf_n^2)[(c_x - mf_n^2)(c_1 - M_m f_n^2) - c_x^2L^2] = 0
$$

Thus, when $\epsilon \to 0$,

$$
f_n^2 = f_{nz}^2 = \frac{c_z}{m}
$$

$$
\Delta(f_n^2) \equiv (c_x - mf_n^2)(c_1 - M_m f_n^2) - c_x^2L^2 = 0 \qquad \text{(III-4-19)}
$$

where $$c_1 = c_\varphi I - WL + c_\tau L^2 A$$

Equation (III-4-19) is identical with Eq. (III-4-8). Denoting the natural frequencies of vibrations of the foundation which correspond to the limiting case $\epsilon = 0$ by $\bar{f}_{n1}$ and $\bar{f}_{n2}$, and assuming that $\bar{f}_{n1} > \bar{f}_{n2}$, we may rewrite Eq. (III-4-19) in the form

$$
\Delta(f_n^2) = mM_m(f_n^2 - \bar{f}_{n1}^2)(f_n^2 - \bar{f}_{n2}^2) = 0 \qquad \text{(III-4-20)}
$$

On the basis of the general characteristics of the interrelationships between frequencies of systems with a limited number of degrees of freedom, we may state that the following dependence exists between the frequencies f_{n1}, f_{n2} and f_{n3}, corresponding to $\epsilon \neq 0$, and the frequencies $\bar{f}_{n1}$, $\bar{f}_{n2}$ when $\epsilon = 0$.

$$
f_{n3} < \bar{f}_{n2} < f_{n2} < \bar{f}_{n1} < f_{n1}
$$

We rewrite Eq. (III-4-18) in the form

$$
(c_x - mf_n^2)(c_z - mf_n^2)(c_1 + c_z\epsilon^2 - M_m f_n^2) - (c_x - mf_n^2)c_z^2\epsilon^2 - (c_z - mf_n^2)c_x^2L^2 = 0
$$

Dividing by m, we obtain

$$
i(f_{nz}^2 - f_n^2)(\bar{f}_{n1}^2 - f_n^2)(\bar{f}_{n2}^2 - f_n^2) - \epsilon^2(f_{nx}^2 - f_n^2)f_n^2 f_{nz}^2 = 0
$$

from which we find

$$\epsilon^2 f_n^2 = \frac{i(f_{nz}^2 - f_n^2)(\bar{f}_{n1}^2 - f_n^2)(\bar{f}_{n2}^2 - f_n^2)}{(f_{nx}^2 - f_n^2)f_{nz}^2} \quad \text{(III-4-21)}$$

where

$$i = \frac{M_m}{m}$$

The right-hand part of Eq. (III-4-21) does not depend on ϵ. Let us draw a graph of this member, taking f_n^2 as an independent variable (Fig. III-17). This graph will be formed by two separate curves with an asymptote corresponding to the value $f_n^2 = f_{nx}^2$. Branch A meets the x axis at point $f_n^2 = \bar{f}_{n2}^2$. Branch B crosses the same axis at point $f_n^2 = \bar{f}_1^2$ and point $f_n^2 = f_{nz}^2$ (Fig. III-17 is plotted on the assumption that $f_{nz} > \bar{f}_{n1}$). Straight line C corresponds to the left member of Eq. (III-4-21). The abscissas of points at which the curves cross this straight line give the unknown roots f_{n3}^2, f_{n2}^2, and f_{n1}^2 of Eq. (III-4-21). The graph clearly illustrates the effect of eccentricity ϵ on the natural frequencies of foundation vibration. It is seen that in the case of an eccentric distribution of mass, the two smaller frequencies f_{n3} and f_{n2} become somewhat lower, while the largest frequency becomes higher. If the eccentricity ϵ is small, the frequencies f_{n3}, f_{n2}, and f_{n1} do not differ much from $\bar{f}_{n1}, \bar{f}_{n2}$, and $\bar{f}_{nz}$, and at the limit, when $\epsilon = 0$, they become equal.

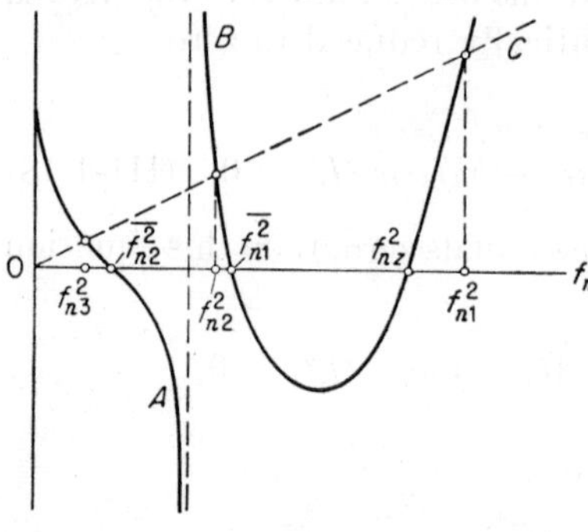

FIG. III-17. Graph illustrating Eq. (III-4-21).

For foundations having a relatively small eccentricity, say 5 per cent of the length of a side of the foundation contact area, its effect may be neglected and computations may be based on formulas derived for $\epsilon = 0$.

III-5. Experimental Investigations of Vibrations of Massive Foundations

a. Verification of the Theory of Vertical Vibrations. The first field investigations of vibrations of foundations were performed by the author together with A. Mikhalchuk[3] on a porous water-saturated silty clay with some sand. On the site of investigations the clay had a thickness of 4.5 m and was underlaid by a sand bed having a thickness of about 4 m. The sand rested on a thick layer of clay. The ground-water level was 20 to 30 cm higher than the base of the test foundations, which were all placed in the same excavation at a depth of about 2.5 m. For dynamic investigations three test foundations were employed with areas in contact with soil of 2.0, 4.0, and 8 m² and weights up to 30 tons.

In addition to dynamic investigations of forced and free vibrations, static investigations were also performed in order to determine the coefficient of elastic uniform compression of soil.

Figure III-18 presents resonance curves of vertical vibrations of a foundation with an area in contact with soil of 8 m² for different eccentricities ϵ of unbalanced mass of a vibromachine. Analogous curves were obtained for other foundations. Table III-1 presents data on computed

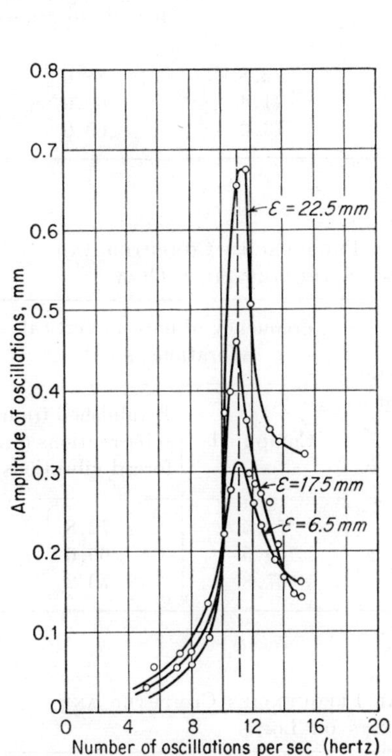

FIG. III-18. Resonance curves of vertical vibrations for three different eccentricities of unbalanced mass of a vibromachine.

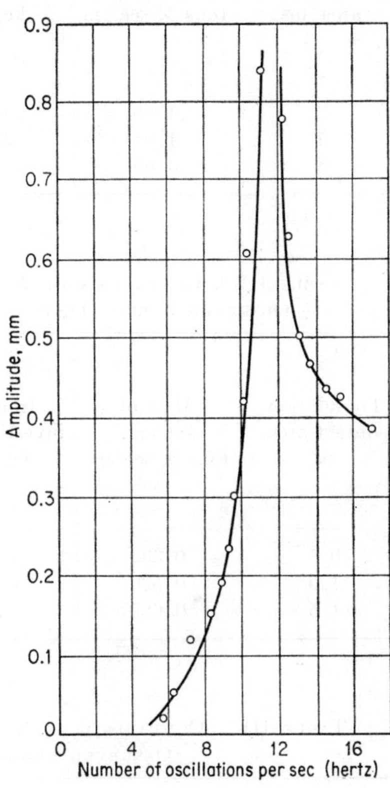

FIG. III-19. Resonance curve of 1.5-m² test of Table III-2.

and experimentally established values of natural frequencies of foundation vibrations. The computation was performed on the basis of values of c_u established by static investigations.

The author and Ya. N. Smolikov performed analogous investigations on a water-saturated soft gray clay with an admixture of organic silt. Investigations were performed on test foundations with base areas of 0.5, 1.0, and 1.5 m².

Figure III-19 presents one of the resonance curves of forced vibrations obtained for a foundation with a contact base area of 1.5 m². In addition

Table III-1. Comparison of Natural Frequencies Computed and Observed during Tests on Water-saturated Clay with a 4-m-thick Sand Layer at 4.5 m Depth

Foundation contact area, m²	Mass of system, tons × sec²/m	c_u from static investigations, kg/cm³	Frequency of natural vertical vibrations, sec⁻¹	
			Computed	Established from observations of forced vibrations
2	1.66	4.40	72.8	88.0
4	1.92	2.45	71.4	60.0
8	3.05	2.05	73.2	69.0

Table III-2. Comparison of Natural Frequencies Computed and Observed during Tests on Soft Saturated Silty Clay

Foundation contact area, m²	Mass of system, tons × sec²/m	c_u from static investigations, kg/cm³	Frequency of natural vertical vibrations, sec⁻¹	
			Computed	Established from observations of forced vibrations
0.5	0.332	3.5	72.5	72.8
1.0	0.520	2.52	69.5	69.0
1.5	0.685	2.1	67.8	70.2

Table III-3. Comparison of Natural Frequencies Computed and Observed during Tests on Loess

Foundation contact area, m²	Mass of system, tons × sec²/m	c_u from static investigations, kg/cm³	Frequency of natural vertical vibrations, sec⁻¹		
			Computed	Established from observations of natural vibrations	Established from observations of forced vibrations
0.81	0.44	14.2	162	158	159
1.40	1.08	10.8	118	113	107
2.00	1.10	10.3	137	117	117
4.00	1.74	8.2	137	118	121

to dynamic investigations, the coefficient c_u was also determined by the static method. Results are presented in Table III-2.

Similar investigations were performed by the author, Ya. N. Smolikov, and P. A. Saichev on loessial loams and on loess.

Table III-3 gives values of c_u secured from static investigations and natural frequencies of vertical vibrations established by two different dynamic methods.

Table III-4 gives results of similar investigations performed on water-saturated gray fine dense sands containing, in places, peat and organic silt. The first two foundations listed in Table III-4 were placed on sand with an admixture of peat and organic silt; the remaining three rested on pure sand.

TABLE III-4. COMPARISON OF NATURAL FREQUENCIES COMPUTED AND OBSERVED DURING TESTS ON FINE SATURATED SANDS

Foundation contact area, m^2	Mass of the system, tons × sec^2/m	c_u from static investigations, kg/cm^3	Frequency of natural vertical vibrations, sec^{-1}		
			Computed	Established from observations of natural vibrations	Established from observations of forced vibrations
1.0	0.38	3.96	102	103	95
4.0	0.84	4.45	145	136	143
4.0	0.84	7.54	189	155	181
8.0	2.76	5.55	126	126	130
15.0	3.60	4.00	127	121	124

In all tests the foundations were placed either directly on the soil surface, or at the bottoms of excavations. Therefore the soil reacted only along the foundation base area. The free vertical vibrations were excited by impacts; the forced vibrations by special vibromachines.

Frequencies of natural vertical vibrations of test foundations were computed from Eq. (III-1-5), which does not take into account the inertial properties of soil.

Errors in evaluation of results of experiments are about 10 per cent. The analysis of Tables III-2 to III-4 leads to the conclusion that there is only a small difference between the values of natural frequencies of vertical vibrations as computed by the two methods, i.e., from the data of static investigations and from those established on the basis of observation of free and forced vibrations. Hence, Eq. (III-1-5) is in rather good agreement with experimental data.

The conclusion is also possible that in all foundations studied, the soil inertia had small effect on the natural frequencies of vertical vibrations of foundations. Apparently this is explained by the fact that for these foundations the value b, found from Eq. (III-1-35), was comparatively large, reaching 20 in some cases. Therefore the effect of soil inertia on the natural frequency of vertical foundation vibrations was comparatively small, being within the range of errors involved in the experiments and the evaluation of results.

b. Experimental Investigations of the Coefficient of Damping. Table III-5 summarizes the results of the determination of the coefficient of damping of vibrations ξ. The values of ξ were determined from the measured amplitudes of forced vibrations at resonance. This was done as follows: at resonance $\omega = f_{nz}$, and Eq. (III-1-21) becomes:

$$A_z^* = A_{res} = \frac{P}{m2cf_{nz}} = \frac{P}{2m\xi f_{nz}{}^2}$$

since from Eq. (III-1-17*a*), $c = \alpha/2m$ and from Eq. (III-1-39),

$$\xi = \frac{\alpha}{(2m_t f_{nz})} = \frac{\alpha}{(2mf_{nz})}$$

if one sets $m_t = m$ (i.e., if one considers only the mass m of the foundation and neglects the mass m_s of the soil). Then:

$$\xi = \frac{P}{A_{res}2mf_{nz}{}^2}$$

The values of b were established from Eq. (III-1-35).

The analysis of Table III-5 leads to the conclusion that the coefficient of damping is much smaller for soft gray silty clays with some sand than

Table III-5. Summary of Test Results for the Determination of the Reduced Coefficient of Damping ξ

Soil description	Foundation contact area, m²	Weight of system, tons	b	ξ
Water-saturated brown silty clay with some sand	2	16.3	16.5	0.145
	4	18.8	6.7	0.133
	8	30.0	4.0	0.181
Water-saturated soft gray clay with sand and organic silt	0.5	3.25	26.0	0.071
	1.0	5.10	14.5	0.058
	1.5	6.72	9.0	0.051
Water-saturated fine dense gray sand	1	6.76	19.0	0.132
	1	6.76	20	0.190
	1	6.76	20	0.175
	1	16.4	55	0.083

for brown silty clay with some sand and for fine gray sands. In foundations placed on sands, much higher values of b were observed than in foundations placed on brown silty clay with some sand. Hence it follows that for similar values of b, the coefficient of damping will be larger in sands than in brown clays.

The value of the coefficient of damping is strongly influenced by several factors very difficult to take into account (for example, the backfilling of foundation excavations). Figure III-20 presents two resonance curves of vertical forced vibrations of a foundation with a contact area of 1.0 m^2; curve 1 corresponds to the situation in which the foundation is exposed along all its height; curve 2 characterizes the same foundation, but back-filled. The depth of backfilling was around 2 m. The foundation was placed on silty clays interbedded with sands; the groundwater level was considerably below the foundation base.

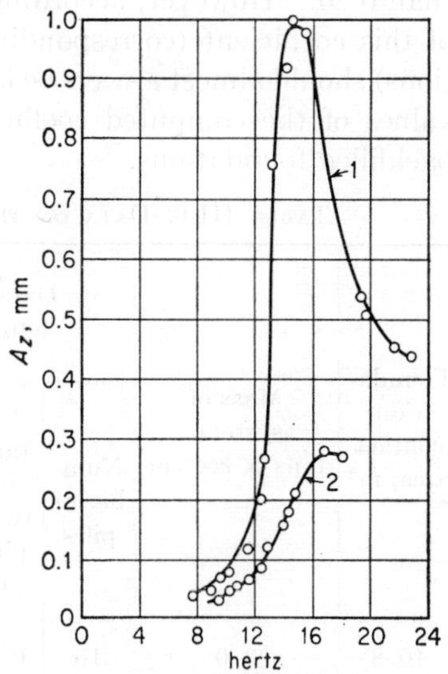

FIG. III-20. Resonance curves of a 1.0-m^2 test foundation: curve 1, sides of footing free; curve 2, sides of foundation backfilled.

It follows from the comparison of graphs 1 and 2 of Fig. III-20 that amplitudes of vibrations of a backfilled foundation at resonance are about 3.5 times smaller than those of an exposed foundation. Since the coefficient of damping is inversely proportional to the amplitude of vibrations at resonance, it follows that in the case under consideration, the value of the coefficient of damping ξ for a backfilled foundation will be approximately 3.5 times larger than that for an exposed foundation.

A considerable effect of backfilling on the value of ξ was also observed in investigations of foundations placed on gray sands. For example, ξ increased from 0.19 to 0.32 when a foundation was backfilled to the height of 1 m.

Even when foundation sides are not subjected to the influence of soil reactions, but are flooded by ground water, the value ξ grows. For example, investigations on the same fine gray sands established that submerging of the foundation to a height of 1 to 1.5 m is accompanied by an increase in the coefficient of damping by 1.5 to 2 times its value.

The increase in the coefficient of damping when foundation sides are not free is explained by the increase in the total foundation surface dissipating energy into the soil. There is an increase in the dissipation of energy of foundation vibrations, and therefore an increase in the coefficient of damping. Besides, the value of ξ is affected by the forces of friction, whose magnitude increases with an increase of the area of backfilling of the foundation.

Experimentally established values of ξ were in most cases smaller than computed values taken from graphs of Fig. III-8*a*, corresponding to selected values of the coefficient b. For foundations characterized by data presented in Table III-5, no experimental value of ξ was greater than 0.20. However, according to the graphs of Fig. III-8*a*, the values of this coefficient (corresponding to the value of b for the test foundations) should almost always be larger than 0.20. Apparently the absolute values of the computed coefficients are close to the values typical for backfilled foundations.

TABLE III-6. DATA ON THE VIBRATION OF PILE FOUNDATIONS

Foundation contact area, m^2	Mass of system, tons $\times$ sec^2/m	Data on pile foundation			K_z from static investigations, kg/cm	Frequency of natural forced vibrations, sec^{-1}	
		Number of piles	Distance between piles, m	Length of piles, m		Computed	Established from observation of forced vibrations
10.8	3.0	16	0.81	5.4	153×10^4	227	201
8.6	2.3	12	0.81	5.6	104×10^4	215	186
8.3	1.7	12	0.81	5.4	105×10^4	247	235
6.5	2.0	9	0.81	5.6	55×10^4	166	138

If one plots a resonance curve of forced vertical vibrations of a foundation on the basis of computed values of the coefficient ξ at resonance, it will turn out that the computed amplitudes of forced vibrations are not in full agreement with experimentally established amplitudes. This divergence is partly explained by errors involved in the operation of frequency-measuring devices. In addition, values of ξ apparently depend on the frequency of vibrations and, possibly, on the amplitude; therefore the values of ξ are different for different sections of the resonance curve.

Consequently, amplitudes computed on the basis of an assumption that ξ remains constant only approximately correspond to the true values.

Available experimental data permit the assertion that if the coefficients of elasticity of the soil are correctly selected, then the divergence between computed and test values of amplitude outside the resonance zone will not exceed 10 to 20 per cent.

c. Investigations of Vibrations of Pile Foundations. Investigations of vertical vibrations were performed on test pile foundations. Results are presented in Table III-6.

Piles were driven into water-saturated fine dense sands. The natural frequency of vertical vibrations of the foundation was computed from Eq. (III-1-5); the foundation mass was considered to be the only vibrating mass. It is natural, therefore, that the natural frequencies of vertical vibrations established by static investigations turned out to be somewhat higher than the frequencies obtained from investigations of forced vibrations. For all foundations investigated, a magnitude of vibrating mass was computed on the basis of values of the coefficient of elastic uniform compression, established by static investigations, and on the basis of the resonance frequency of vertical forced vibrations. This vibrating mass was, on the average, 30 per cent larger than the mass restricted to the foundation only.

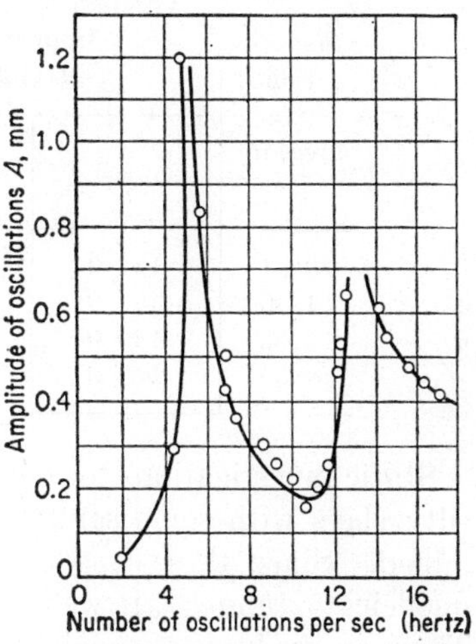

FIG. III-21. Resonance curves of a 1.5-m² test foundation of silty clay with sand subjected to forced horizontal vibrations.

d. Experimental Investigations of Complicated Forms of Foundation Vibrations. Other forms of vibrations of foundations were also investigated. For example, forced vibrations in one of the principal planes, induced by a horizontal force, were studied. Figure III-14 gives a resonance curve of forced horizontal vibrations of a foundation with a contact area of 4 m^2 placed on brown silty clay with sand; Fig. III-21 shows a similar curve for a foundation with a contact area of 1.5 m^2 placed on water-saturated soft silty clays with sand. Similar investigations were performed on loess and on gray sands. In the first two cases, for all foundations investigated, two frequencies f_{n1} and f_{n2} were obtained which correspond to a sharp rise in the amplitudes of vibrations.

Thus, there is experimental corroboration of a theoretical conclusion concerning two maximums in the resonance curves of forced vibrations of foundations, corresponding to the two frequencies f_{n1} and f_{n2}. No static tests for determining c_φ were performed during the investigations of foundations on water-saturated silty clays with some sand. Therefore, in this case there is no way to verify directly how far the frequencies f_{n1} and f_{n2}, computed from values of c_φ and c_τ (established by means of static investigations), coincide with those which were established experimentally. Static tests for determining c_φ and c_τ were performed on foundations resting on loess. Using the established values, it is possible to compute the frequencies f_{n1} and f_{n2} and to verify how close the computed values of these frequencies are to those established experimentally. Results of this analysis are presented in Table III-7.

TABLE III-7. DATA ON THE TWO FUNDAMENTAL FREQUENCIES OF A FOUNDATION SUBJECTED TO FORCED HORIZONTAL VIBRATION

Foundation contact area, m^2	Frequencies computed from static investigations		Experimentally established values of f_{n2}	
	f_{n1}, sec^{-1}	f_{n2}, sec^{-1}	From free vibrations	From forced vibrations
0.81	174.0	58.4	65.3	48.3
1.40	181.0	73.5	73.5	64.0
2.60	140.0	65.2	69.1	50.2
4.00	167.0	89.2	77.8	54.0

Static investigations for determining c_u and c_τ were performed on gray silty clays with some sand; also f_{n1} and f_{n2} were experimentally determined. Since these frequencies are connected with the limiting frequencies $f_{n\varphi}$ and f_{nx}, it was possible to establish the latter analytically. Then, on the basis of the values of $f_{n\varphi}$ and f_{nx}, the values of c_τ were determined. Results of this processing of data from dynamic investigations on gray silty clay with some sand are presented in Table III-8.

TABLE III-8. DATA ON THE COEFFICIENT c_τ OF ELASTIC UNIFORM SHEAR OF SOIL

Foundation contact area, m^2	c_τ, kg/cm^2	
	From experimental frequencies f_{n1} and f_{n2}	From static investigations
0.5	1.88	1.90
1.0	1.64	1.58
1.5	1.27	1.40

It is seen from the tables that the values of c_τ computed on the basis of static investigations and those obtained as a result of the investigations of vibrations coincide with a satisfactory degree of accuracy.

Available data, including those cited above, lead to the conclusion that the theory of vibrations of foundations for several-degrees-of-freedom systems, as presented in this chapter, is supported by experiments.

IV

FOUNDATIONS UNDER RECIPROCATING ENGINES

IV-1. General Directives for the Design of Foundations

a. Design Values of Permissible Amplitudes of Foundation Vibrations. Many types of reciprocating engines belong to the group of unbalanced machines which are dangerous in respect to vibrations. The fact that these engines usually operate at comparatively low speed increases the probability that vibrations may develop in adjoining buildings or structures. Therefore a thorough analysis of vibrations in such foundations is of the utmost importance.

The greater the amplitude of vibrations of the foundation, the more danger there is to adjoining structures. In addition, if the amplitude of foundation vibrations is large, the foundation may lose its stability and undergo a nonuniform settlement endangering the normal work of the engine. Finally, vibrations of large amplitude may lead to the destruction of the foundation and to damage of the engine; for example, it has been observed that sometimes crankshafts of log-sawing frames fail as a result of vibrations of their foundations and frames, to which the crankshafts are rigidly tied. Similarly, foundation vibrations often cause dangerous vibrations of machine connections.

It is extremely difficult to establish a limit for the permissible value of amplitude of foundation vibrations on the basis of general principles. There are some cases in which vibrations with an amplitude up to 0.4 to 0.5 mm did not have any harmful effects. However, many cases have been observed in which foundations under engines with low-frequency vibrations underwent vibrations at smaller amplitudes than those cited above, but induced strong vibrations of structures located at a distance of several tens of meters. It may happen that even when the amplitude of vibration of a machine foundation is smaller than an accepted permissible limit, the adjoining structure will vibrate due to resonance.

On the strength of data gained by experience it is possible to state that if no resonance is to occur in adjoining buildings and structures, then the amplitude of vibrations of a foundation should not exceed 0.20 to 0.25 mm. This range of amplitude values may serve as a basis for evaluation of the adequacy of foundation-design computations.

b. Machine Data Required for Foundation Design. Data supplied by the maker of the engine or motor is the basic information for a foundation designer, together with data on soil conditions and on the exciting loads imposed by the engine. The following information should be given:

1. The normal speed and power of the engine
2. The character, magnitude, and point of application of dynamic loads which will develop in the process of operation of the engine. If this data cannot be supplied, the designer of the machine foundation should be given all the data needed for the computation of exciting forces
3. The distribution of static loads imposed by the engine over the foundation surface
4. The size and shape of the engine supporting plate
5. The location of openings and grooves in the foundation provided for anchor bolts, pipe lines, the flywheel, etc.

c. Foundation Material. The following loads are imposed on a machine foundation:

1. The weight of the machine and equipment
2. The dynamic loads which develop in the process of machine operation

For diesel engines, the total load is such that the reduced pressure on the upper surface of the foundation usually does not exceed 3 to 5 kg/cm^2.

For horizontal piston engines this value is still smaller. In any case, the permissible bearing values of concrete and masonry are considerably higher. It is natural that since pressures imposed on foundations are small, the stress analysis should be performed only for cross sections weakened by large openings or grooves.

Thus the question of selection of a material for the foundation—concrete or masonry—is first of all a question of cost and of availability of material on the site. Concrete type 100† is usually employed for foundations under reciprocating machinery.

† Translation Editor's Note: The figure 100 indicates the 28-day compressive cube strength in kilograms per square centimeter of the concrete mixture used (100 kg/cm^2 equals 1,420 psi).

d. Comments in Regard to Design. Foundations under reciprocating engines are usually built as massive blocks provided with grooves and channels for machine details and openings for anchor bolts.

Due to the massive shapes of such foundations, when studying their vibrations it is possible to consider them as absolutely rigid bodies and to use in the computations of frequencies and amplitudes the theory of vibrations of a solid resting on an elastic base, as presented in Chap. III.

The main condition to be observed when designing a machine foundation is as follows: the minimum dimensions of the foundation should be selected in such a way that the amplitudes of its forced vibrations will not exceed the permissible value.

If a foundation is erected on a natural soil base, its depth should not be less than that of frost penetration. There is an established opinion among practicing engineers that in order to decrease the transmission of vibrations the depth of a machine foundation should be no less than the depth of the footings of adjoining walls and columns. Theoretical and experimental data on wave propagation in soils, presented in Arts. VII-2 and VII-5, lead to the conclusion that provision for machines of foundations deeper than footings for walls has no effect on the transmission of vibrations to walls. Therefore the depth of a machine foundation may be selected without taking into account the transmission of vibrations.

In order to obtain uniform settlement of the foundation, t is recommended to place the common center of gravity of the system (i.e., of the foundation and machine) on the same vertical line with the centroid of the foundation area in contact with the soil. In any case the eccentricity in the distribution of masses should not exceed 5 per cent of the length of the side of the contact area. Satisfying this condition makes it possible to simplify the computation of foundation vibrations. When the common center of gravity does not lie on the same vertical line as the centroid of the foundation contact area, it is necessary, as stated above, to solve at least three interrelated differential equations of vibrations. In order to simplify the computations, an attempt should be made to achieve a condition such that the plane of action of the exciting forces imposed by the machine coincides with one of the principal planes of inertia of the foundation.

In order to decrease the transmission of vibrations to adjoining parts of buildings, it is necessary to leave a gap between the foundation of an unbalanced machine and the adjoining structures (footings, walls, floors, and so on). As a rule, the machine foundation is not allowed to serve as a support for other parts of the building or for mechanisms not related to a given machine. If it is not possible to avoid placing unimportant parts of a building on the machine foundation, measures should be taken to soften the connection by providing gaskets made of rubber, cork, felt, or other insulating materials.

If several machines are to be installed in the same shop, and if the distances between these machines are comparable to the foundation dimensions, then, in soft soils, it is recommended to place the foundations under similar machines on one common mat of sufficient thickness. The rigidity of this mat should be selected so that its possible deformations remain small in comparison with the amplitudes of vibrations. Only then may a group of machine foundations installed on the same mat be regarded as an absolutely solid block resting on an elastic base. The computation of vibrations of such groups of foundations is very difficult. If an asymmetry and an alternating phase shift are observed, then computations are practically impossible.

Therefore if several foundations are erected on the same mat, it is conditionally broken up into sections corresponding to separate foundations; the computations of vibrations proceed as if each foundation were installed separately. Then the design value for the permissible amplitude of vibrations may be increased somewhat (by 25 to 30 per cent).

To avoid a distorted tilt of the master shaft of the machine, its external bearing should be placed on the same machine foundation. This directive refers also to the installation of motors coupled directly to reciprocating engines such as electromotors and generators.

In any case, the foundation area in contact with soil should be selected in such a way that pressure on the soil does not exceed permissible values.

The larger the foundation contact area, the smaller the reduced pressure on the soil and the higher the natural frequencies of the foundation. This is of considerable importance for low-frequency machines, including most of the reciprocating engines. The easiest way to change the natural frequencies of a foundation is to increase or decrease the dimensions of the foundation contact area and change its configuration in plan. Therefore a final selection of the contact area should be based on requirements obtained as a result of design computations regarding vibrational loads on the foundation.

Foundations under low-frequency machines should be designed so that their natural frequencies are much higher than the operational frequencies of the machines.

The natural frequencies of foundations are affected by the absolute value of the foundation mass and by its distribution in space. The designer should try to distribute the mass so that the smallest possible value of its moment of inertia is obtained with respect to the principal axis passing through the centroid of the foundation contact area. To meet this requirement, the minimum foundation height should be selected.

IV-2. Unbalanced Inertial Forces in Reciprocating Engines

Forced vibrations in foundations under piston engines are largely caused by the unbalanced inertia forces in the moving parts of crank mechanisms.

a. Single-line Machines. Figure IV-1 illustrates the main features of a reciprocating mechanism. The piston A and the piston rod B execute an alternating motion; the connection rod C executes a complicated periodic motion; all points of the crank D execute a rotational motion around the main axis O. Any of these parts may have unbalanced inertial forces which independently may cause foundation vibrations.

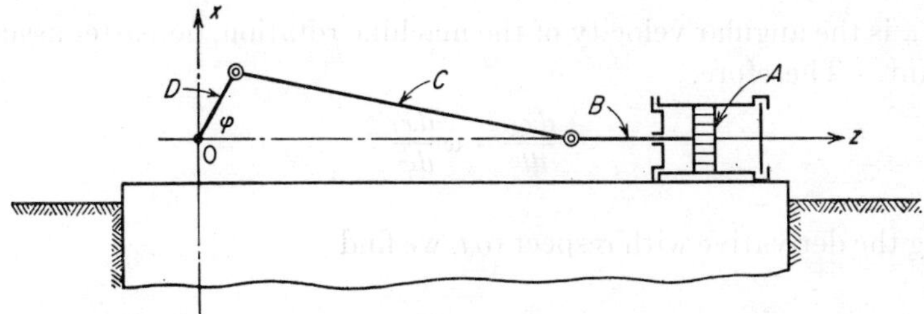

Fig. IV-1. The main parts of a reciprocating engine.

According to the laws of statics, we may replace all these forces by forces acting at point O and place at this point the origin of a coordinate system x, y, z.

Let us place the z axis in the direction of piston movement, the x axis perpendicular to this direction, and the y axis perpendicular to the plane of the drawing. Since all points of the crank mechanism move in the plane xz, the y ordinates will remain constant for all these points. As a result of replacing all unbalanced inertia forces by forces acting at point O, we obtain one force and one couple. Resolving these into their components along the coordinate axes, we obtain the force components P_x and P_z and the moment M_y.

The position of any point of the crank mechanism is determined by one independent variable: the angle of rotation φ of the crankshaft. Therefore the coordinates x_i, y_i, and z_i of any point i of the moving machine parts are functions of the angle φ, the latter being a function of time.

Let us denote by $\ddot{x}_i$ and $\ddot{z}_i$ the projections of the acceleration of an element i of the crank mechanism on the x and z axes; then the projections of the inertial forces acting on this element will be as follows:

$$P_{xi} = m_i\ddot{x}_i \qquad P_{zi} = m_i\ddot{z}_i$$

The projections of the inertial forces acting on all elements of the crank mechanism will be

$$P_x = \Sigma m_i \ddot{x}_i \qquad P_z = \Sigma m_i \ddot{z}_i \tag{IV-2-1}$$

Considering the x_i and z_i coordinates as functions of t, we obtain

$$\frac{dx_i}{dt} = \frac{dx_i}{d\varphi}\frac{d\varphi}{dt}$$

but

$$\frac{d\varphi}{dt} = \omega$$

where ω is the angular velocity of the machine rotation, hereafter assumed constant. Therefore,

$$\frac{dx_i}{dt} = \omega \frac{dx_i}{d\varphi}$$

Taking the derivative with respect to t, we find

$$\ddot{x}_i = \omega^2 \frac{d^2x_i}{d\varphi^2}$$

similarly,

$$\ddot{z}_i = \omega^2 \frac{d^2z_i}{d\varphi^2}$$

Substituting the values for $\ddot{x}_i$ and $\ddot{z}_i$ into Eqs. (IV-2-1), we obtain

$$\begin{aligned} P_x &= \omega^2 \sum m_i \frac{d^2x_i}{d\varphi^2} \\ P_z &= \omega^2 \sum m_i \frac{d^2z_i}{d\varphi^2} \end{aligned} \tag{IV-2-2}$$

The total resultant inertial force of the crank mechanism evidently will equal the sum of the inertial forces of its moving parts: the crank, the piston, and the connecting rod. Consequently the component of the resultant of the inertial forces acting in the direction of the piston motion can be described by the formula

$$P_z = P_{z1} + P_{z2} + P_{z3}$$

where P_{z1} = projection of inertial forces of crank on the z axis
P_{z2} = projection of inertial forces of rod, crosshead, and piston
P_{z3} = projection of inertial forces of connecting rod

Further,

$$P_{z1} = \omega^2 \sum m_{1i} \frac{d^2 z_{1i}}{d\varphi^2}$$

$$P_{z2} = \omega^2 \sum m_{2i} \frac{d^2 z_{2i}}{d\varphi^2}$$

$$P_{z3} = \omega^2 \sum m_{3i} \frac{d^2 z_{3i}}{d\varphi^2}$$

The first moment, for example, of the mass of the crank with respect to the rotation axis will equal $\Sigma m_{1i} z_{1i}$. On the other hand, if M_1 is the mass of the crank and x_1 and z_1 are the coordinates of its center of gravity, then

$$\Sigma m_{1i} z_{1i} = M_1 z_1$$

Differentiating this equation twice with respect to φ, we obtain

$$\sum m_{1i} \frac{d^2 z_1}{d\varphi^2} = M_1 \frac{d^2 z_1}{d\varphi^2}$$

Using this relationship, we find

$$P_{z1} = M_1 \omega^2 \frac{d^2 z_1}{d\varphi^2}$$

$$P_{z2} = M_2 \omega^2 \frac{d^2 z_2}{d\varphi^2}$$

$$P_{z3} = M_3 \omega^2 \frac{d^2 z_3}{d\varphi^2}$$

similarly,

$$P_{x1} = M_1 \omega^2 \frac{d^2 x_1}{d\varphi^2}$$

$$P_{x2} = M_2 \omega^2 \frac{d^2 x_2}{d\varphi^2}$$

$$P_{x3} = M_3 \omega^2 \frac{d^2 x_3}{d\varphi^2}$$

The expressions for the projections of the resultant inertial force will be

$$\begin{aligned} P_x &= \omega^2 \left(M_1 \frac{d^2 x_1}{d\varphi^2} + M_2 \frac{d^2 x_2}{d\varphi^2} + M_3 \frac{d^2 x_3}{d\varphi^2} \right) \\ P_z &= \omega^2 \left(M_1 \frac{d^2 z_1}{d\varphi^2} + M_2 \frac{d^2 z_2}{d\varphi^2} + M_3 \frac{d^2 z_3}{d\varphi^2} \right) \end{aligned} \qquad \text{(IV-2-3)}$$

Without limiting the general validity of the solution, and without involving any significant error, it is possible to concentrate the entire mass of the crank mechanism not at three points, as has been done above, but at two points. This will simplify the expressions obtained for P_x and P_z.

It has been assumed that the crank executes its motion at a uniform rate. Therefore the magnitude of its inertial force will be the magnitude of the centrifugal force; i.e.,

$$P_c = R_1 M_1 \omega^2$$

where R_1 is the distance between the center of gravity of the crank and the axis of rotation.

This force will be directed along the radius of rotation. Let us consider that it is applied not to the center of gravity of the crank, but to point a, i.e., to the crankpin (Fig. IV-2). In order to obtain a force equal to P_c and applied at point a, it is necessary to assume that mass M_{11} is concentrated at this point and is smaller than M_1 in the same proportion as R_1 is smaller than R; thus we should set

$$M_{11} = \frac{R_1}{R} M_1$$

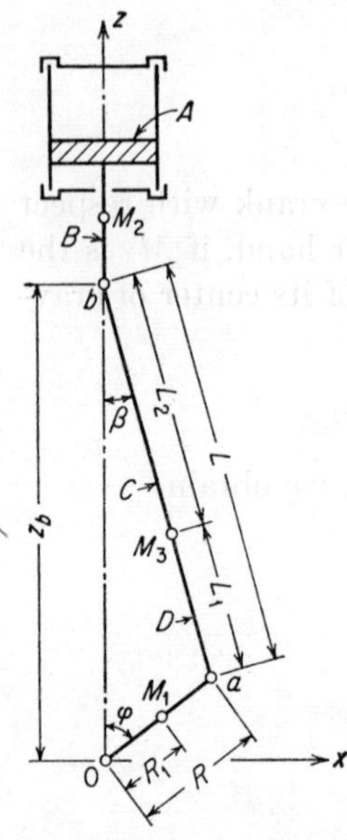

FIG. IV-2. Reciprocating engine, illustrating the derivation of Eq. (IV-2-4).

Since point b (the crosshead) executes a reciprocating motion which does not differ from the motion of the center of gravity of mass M_2, we may consider that mass M_2 is concentrated at point b; this will not change the magnitude of its induced inertial force.

Finally the mass M_3 of the connecting rod may be replaced by masses M_{31} and M_{32} concentrated at points a and b. However, this distribution of mass M_3 should be made so as not to change the magnitude of the inertial force of the connecting rod computed on the assumption that its mass is concentrated at the center of gravity (x_3,z_3).

Denoting the coordinates of points a and b respectively by (x_a,z_a) and (x_b,z_b), we obtain

$$P_{z3} = \omega^2 \left(M_{31} \frac{d^2 z_a}{d\varphi^2} + M_{32} \frac{d^2 z_b}{d\varphi^2} \right)$$

On the other hand,

$$P_{z3} = \omega^2 M_3 \frac{d^2 z_3}{d\varphi^2}$$

Equating the right-hand parts of these expressions, we obtain

$$M_{31} z_a + M_{32} z_b = M_3 z_3$$

Similarly, deriving the expression for the projection of the inertial force of the connecting rod on the x axis, we obtain the second equation:

$$M_{31}x_a + M_{32}x_b = M_3x_3$$

It follows from the equations obtained for M_{31} and M_{32} that the values of these masses should be selected so that their center of gravity lies at point (x_3,z_3), i.e., the masses should be distributed in an inverse relationship to their distances from mass M_3. Denoting these distances by L_1 and L_2, we obtain

$$\frac{M_{31}}{M_{32}} = \frac{L_2}{L_1}$$

On the other hand,

$$M_{31} + M_{32} = M_3$$

Solving these equations for M_{31} and M_{32}, we find

$$M_{31} = \frac{L_2}{L} M_3 \qquad M_{32} = \frac{L_1}{L} M_3 \qquad L = L + L_2$$

where L is the length of the crank.

Thus the three masses, concentrated at the centers of gravity corresponding to the parts of the crank mechanism, may be replaced, without changing the magnitudes of the inertial forces of the mechanism, by two masses:

1. Mass M_a, concentrated at the crankpin:

$$M_a = M_{11} + M_{31} = \frac{R_1}{R} M_1 + \frac{L_2}{L} M_3 \tag{IV-2-4}$$

2. Mass M_b, concentrated at the crosshead:

$$M_b = M_2 + M_{32} = M_2 + \frac{L_1}{L} M_3 \tag{IV-2-5}$$

The projections of the inertial force of mass M_a on the coordinate axes will be

$$P_{xa} + M_a\omega^2R \sin \varphi \qquad P_{za} = M_a\omega^2R \cos \varphi$$

The inertial force of mass M_b will only have a projection on the z axis; this will be equal to

$$P_{zb} = -M_b\omega^2\ddot{z}_b$$

The projections of the resultant inertial force of the whole mechanism will equal

$$\begin{aligned} P_x &= M_a\omega^2R \sin \varphi \\ P_z &= M_a\omega^2R \cos \varphi - M_b\omega^2\ddot{z}_b \end{aligned} \tag{IV-2-6}$$

It follows directly from Fig. IV-2 that

$$z_b = R \cos \varphi + L \cos \varphi$$

We have, from the triangle Oab,

$$\sin \beta = \frac{R}{L} \sin \varphi = \alpha \sin \varphi$$

from which

$$\cos \beta = \sqrt{1 - \alpha^2 \sin^2 \varphi}$$

Expanding $\cos \beta$ into a series according to Newton's binomial theorem, we obtain

$$\cos \beta = (1 - \alpha^2 \sin^2 \varphi)^{1/2} = 1 - \tfrac{1}{2}\alpha^2 \sin^2 \varphi - \tfrac{1}{8}\alpha^4 \sin^4 \varphi - \tfrac{1}{16}\alpha^6 \sin^6 \varphi - \cdots$$

Using the formula converting an even exponential trigonometric function into a linear one,

$$2^{n-1}(-1)^{n/2} \sin^{n\varphi} = \cos n\varphi - n \cos (n-2)\varphi + \frac{n(n-1)}{1.2} \cos (n-4)\varphi - \cdots + (-1) \frac{(n/2)n(n-1) \cdots (n/2+1)}{2 \cdot 1 \cdot 2 \cdots n/2}$$

we replace powers of sines by the cosines of multiples of 2φ. Then

$$\cos \beta = A_0 + A_2 \cos 2\varphi + A_4 \cos 4\varphi + \cdots$$

where $A_0, A_2, A_4, \ldots$ are constants depending only on the characteristic number α of the crank mechanism:

$$A_0 = 1 - \tfrac{1}{4}\alpha^2 - \tfrac{3}{64}\alpha^4 - \tfrac{5}{256}\alpha^6 - \cdots$$
$$A_2 = \tfrac{1}{4}\alpha^2 + \tfrac{1}{16}\alpha^4 + \tfrac{15}{512}\alpha^6 + \cdots$$
$$A_4 = -(\tfrac{1}{64}\alpha^4 + \tfrac{3}{256}\alpha^6 + \cdots)$$

Substituting the expression established for $\cos \beta$ into Eq. (IV-2-6), we obtain

$$z_b = R \left[\cos \varphi + \frac{1}{\alpha} (A_0 + A_2 \cos 2\varphi + A_4 \cos 4\varphi + \cdots) \right]$$
$$\frac{d^2 z_b}{d\varphi^2} = -R(\cos \varphi + B_2 \cos 2\varphi + B_4 \cos 4\varphi + \cdots)$$

where $B_2 = \dfrac{4A_2}{\alpha} \qquad B_4 = \dfrac{16A_4}{\alpha}$

Substituting into these formulas the expressions for A_2 and A_4, and, in view of the small value of α, disregarding all terms containing its fourth or higher powers, we obtain

$$B_2 = \alpha\left(1 + \frac{\alpha^2}{4}\right) \qquad B_4 = -\frac{\alpha^3}{4}$$

Therefore,

$$\frac{d^2z_b}{d\varphi^2} = -R\left[\cos\varphi + \alpha\left(1 + \frac{\alpha^2}{4}\right)\cos 2\varphi - \frac{\alpha^3}{4}\cos 4\varphi\right]$$

Substituting the expression established for $d^2z_b/d\varphi^2$ into the formula for P_{zb}, we find that

$$P_{zb} = M_bR\omega^2\left[\cos\varphi + \alpha\left(1 + \frac{\alpha^2}{4}\right)\cos 2\varphi - \frac{\alpha^3}{4}\cos 4\varphi\right]$$

Substituting this expression for P_{zb} into Eq. (IV-2-6) describing the projection on the z axis of the resultant inertial force of the machine and replacing φ by ωt, where ω is the angular velocity of machine rotation, we finally obtain

$$\begin{aligned} P_x &= R\omega^2 M_a \sin\omega t \\ P_z &= R\omega^2\left[(M_a + M_b)\cos\omega t + \alpha M_b\left(1 + \frac{\alpha^2}{4}\right)\cos 2\omega t \right. \\ &\qquad \left. - \frac{M_b\alpha^3}{4}\cos 4\omega t\right] \end{aligned} \qquad \text{(IV-2-7)}$$

Thus the formula describing the exciting loads causing the forced vibrations of a foundation contains terms depending not only on the frequency ω of machine rotation, but also on the double, quadruple, etc. of this frequency. However, the coefficients preceding cos $2\omega t$ and cos $4\omega t$ decrease very quickly, and these terms may be disregarded in engineering calculations.

The terms containing cos ωt are called primary inertial forces (the first harmonics); those containing cos $2\omega t$, secondary forces (secondary harmonics); and so on.

The foregoing discussion leads to the conclusion that rotating machinery masses produce primary inertial forces; reciprocating masses produce both primary inertial forces and forces of even higher orders.

By installing counterweights of mass M'_a on a shaft, it is possible to balance inertial forces induced by mass M_a. If M'_a is fixed on the shaft so that the angle between the radius vectors of masses M_a and M'_a equals π, then, in order to have the inertial forces of the rotating parts balanced, one of the two following conditions should be satisfied: Either

$$M_a - \frac{l}{R}M'_a = 0$$

or

$$\frac{R'}{R}M_1 + \frac{L_2}{L}M_3 - \frac{l}{R}M'_a = 0$$

In order to balance fully the projections of primary inertial forces in the direction of piston motion it is necessary to select mass M'_a and distance l (between its center of gravity and the axis of rotation) so that either

$$M_a + M_b - \frac{l}{R} M'_a = 0$$

or

$$\frac{R_1}{R} M_1 + M_2 + M_3 - \frac{l}{R} M'_a = 0 \qquad \text{(IV-2-8)}$$

If the selected values of l and M'_a satisfy one of the above equations, then in the expression for P_z there will remain terms depending only on $\cos 2\omega t$ and $\cos 4\omega t$, while the expression for P_x will be as follows:

$$P_x = R\omega^2 M_b \sin \omega t$$

Usually M_b is larger than M_a; therefore the selection of a counterweight mass satisfying Eq. (IV-2-8) leads to an enlargement of inertial forces in the direction perpendicular to the sliding of the piston.

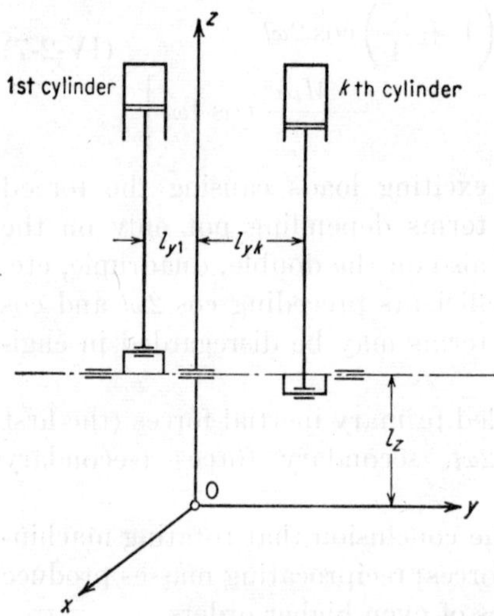

FIG. IV-3. Multicylinder engine, illustrating the derivation of Eq. (IV-2-9).

b. Multicylinder Engines. The method of determination of exciting loads in multicylinder engines is in principle the same as in single-cylinder engines.

Consider a vertical engine in which the cylinders are situated in the same plane, parallel to each other (the so-called linear arrangement of cylinders). Usually the number n of cylinders does not exceed 10. Unbalanced inertial forces are calculated similarly for vertical and horizontal reciprocating engines.

Let us direct the y axis (Fig. IV-3) along the crankshaft of the engine, the x axis perpendicular to the shaft and horizontal, the z axis upward, along the axis of sliding of the pistons. Let us place the origin at the mass center of the foundation and engine and let us assume that the yz plane passes through the principal axis of engine rotation.

We confine ourselves to the case in which the engine has only main cylinders (no auxiliaries such as compressor and exhaust cylinder). We denote by β_k the angle between the crank of the kth cylinder and the first crank (the wedging angle). By the reasoning of Art. IV-2-a we obtain the following expressions for the component exciting force along the x and z axes for the kth cylinder:

$$P_{xk} = R_k\omega^2 M_{ak} \sin (\omega t + \beta_k)$$
$$P_{zk} = R_k\omega^2[(M_{ak} + M_{bk}) \cos (\omega t + \beta_k) + M_{bk}\alpha_k \cos (\omega t + \beta_k)]$$

The terms $$M_{bk} \frac{\alpha k^3}{4} \cos 2(\omega t + \beta_k)$$

and $$\frac{M_{bk}\alpha k^3}{4} \cos 4(\omega t + \beta_k)$$

have been neglected.

In order to obtain the resultant exciting force transmitted to the foundation from all engine cylinders, it suffices to sum the above expressions for all n cylinders. Then we have:

$$P_x = \omega^2 \sum_{k=1}^{n} R_k M_{ak} \sin (\omega t + \beta_k)$$
$$P_z = \omega^2 \sum_{k=1}^{n} R_k[(M_{ak} + M_{bk}) \cos (\omega t + \beta_k) + M_{bk}\alpha_k \cos 2(\omega t + \beta_k)] \qquad \text{(IV-2-9)}$$

In addition to exciting forces, there are exciting moments; their magnitudes equal

$$M_x = \sum_{k=1}^{n} P_{zk}l_{yk} \qquad M_y = \sum_{k=1}^{n} P_{xk}l_{zk} \qquad M_z = \sum_{k=1}^{n} P_{xk}l_{yk}$$

If the crank mechanisms are identical in all cylinders, then the equations for the exciting force will be simplified:

$$P_x = R\omega^2 M_a \sum_{k=1}^{n} \sin (\omega t + \beta_k)$$
$$P_z = R\omega^2 \left[(M_a + M_b) \sum_{k=1}^{n} \cos (\omega t + \beta_k) + M_b\alpha \sum_{k=1}^{n} \cos 2(\omega t + \beta_k) \right]$$

Hence it follows that to balance the first harmonics of the exciting forces, the following equations should be satisfied:

$$\sum_{k=1}^{n} \cos (\omega t + \beta_k) = 0 \qquad \sum_{k=1}^{n} \sin (\omega t + \beta_k) = 0$$

The second harmonics will be satisfied if

$$\sum_{k=1}^{n} \cos 2(\omega t + \beta_k) = 0$$

To balance the exciting moments of the first harmonics, the following equations should be satisfied:

$$\sum_{k=1}^{n} l_{xk} \cos (\omega t + \beta_k) = 0$$
$$\sum_{k=1}^{n} l_{zk} \sin (\omega t + \beta_k) = 0$$
$$\sum_{k=1}^{n} l_{yk} \sin (\omega t + \beta_k) = 0$$

Similar conditions hold for the second harmonics.

Let us consider several particular computations of exciting loads imposed by multicylinder engines, assuming all cylinders are identical and neglecting all higher harmonics of exciting loads.

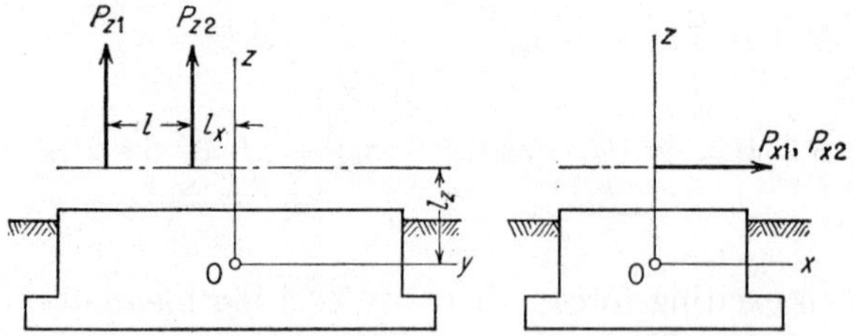

FIG. IV-4. Derivation of equations for a vertical two-cylinder engine.

c. Vertical Two-cylinder Engines. Let us assume that the engine is mounted asymmetrically on the foundation (Fig. IV-4). Both cylinders are identical.

CASE 1. CRANKS IN SAME DIRECTION. Here,

$$\beta_1 = 0 \qquad \beta_2 = 2\pi$$

Assuming in Eq. (IV-2-9) that $k = 1.2$, we obtain

$$P_{x1} = P_{x2} = R\omega^2 M_a \sin \omega t$$
$$P_{z1} = P_{z2} = R\omega^2 (M_a + M_b) \cos \omega t$$

The resultant components of the exciting forces will be

$$P_x = 2R\omega^2 M_a \sin \omega t$$
$$P_z = 2R\omega^2 (M_a + M_b) \cos \omega t$$

The components of the exciting moment equal

$$M_x = P_{z1}(l + 2l_x)$$
$$M_y = 2P_{x1} l_z$$
$$M_z = P_{x1}(l + 2l_x)$$

The values of l, l_x and l_z are shown in Fig. IV-4.

The engine under consideration belongs to the class of highly unbalanced engines, dangerous with respect to vibrations.

CASE 2. TWO-CYLINDER ENGINE WITH 90° CRANK ANGLE. On the basis of Eqs. (IV-2-9), we have

$$P_{x1} = R\omega^2 M_a \sin \omega t$$
$$P_{x2} = R\omega^2 M_a \sin\left(\omega t + \frac{\pi}{2}\right) = R\omega^2 M_a \cos \omega t$$
$$P_{z1} = R\omega^2 (M_a + M_b) \cos \omega t$$
$$P_{z2} = R\omega^2 (M_a + M_b) \cos\left(\omega t + \frac{\pi}{2}\right) = -R\omega^2 (M_a + M_b) \sin \omega t$$

The resultant components of exciting forces are

$$P_x = R\omega^2 M_a(\sin \omega t + \cos \omega t) = \sqrt{2}\, R\omega^2 M_a \sin\left(\omega t + \frac{\pi}{4}\right)$$
$$P_z = R\omega^2 (M_a + M_b)(\cos \omega t - \sin \omega t)$$
$$= \sqrt{2}\, R\omega^2 (M_a + M_b) \cos\left(\omega t + \frac{\pi}{4}\right)$$

Hence it follows that the resultant components of exciting forces are 1.41 times the resultant forces in each cylinder.

Let us determine the components of the exciting moment:

$$M_x = P_{z1}(l + l_y) + P_{z2} l_y$$

Analogously,

$$M_y = (P_{x1} + P_{x2}) l_z$$
$$M_z = P_{x1}(l + l_y) + P_{x2} l_y$$

This case, the 90° crank angle, is the most characteristic for two-cylinder engines.

CASE 3. TWO-CYLINDER ENGINE WITH 180° CRANK ANGLE. Here $\beta_1 = 0$, $\beta_2 = \pi$. According to Eq. (IV-2-9), the resultant components of the exciting forces will equal

$$P_x = 0 \qquad P_z = 0$$

The components of the exciting moment equal

$$M_x = P_{z1} l \qquad M_y = 0 \qquad M_z = P_{x1} l$$

d. Reciprocating Horizontal Compressors. These engines usually have two cylinders with 90° crank angles. The expressions for the exciting forces here are the same as for a vertical engine; the only difference is that in the equations x should be changed to z and vice versa.

The exciting moments will equal (Fig. IV-5)

$$M_x = P_{z1}l_{y1} - P_{z2}l_{y2}$$
$$M_y = P_z l_x + P_x l_z$$
$$M_z = P_{x1}l_{y1} - P_{x2}l_{y2}$$

e. Vertical Three-cylinder Engine. These engines usually have 120° crank angles; i.e., $\beta_1 = 0$, $\beta_2 = 120°$, and $\beta_3 = 240°$. Since

$$\cos 0 + \cos 120° + \cos 240° = 0$$
$$\sin 0 + \sin 120° + \sin 240° = 0$$

the first harmonics of the exciting forces are balanced:

$$P_x = 0 \qquad P_z = 0$$

If all three cylinders are spaced alike, then the exciting moments of the engine are

$$M_x = P_{z1}(2l + l_y) + P_{z2}(l + l_y) + P_{z3}l_y$$
$$M_y = 0$$
$$M_z = P_{x1}(2l + l_y) + P_{x2}(l + l_y) + P_{x3}l_y$$

f. Vertical Four-cylinder Engine. This engine is so designed that $\beta_1 = 0°$, $\beta_2 = 180°$, $\beta_3 = 180°$, and $\beta_4 = 360°$. All components of exciting forces and moments are balanced as a result of this arrangement.

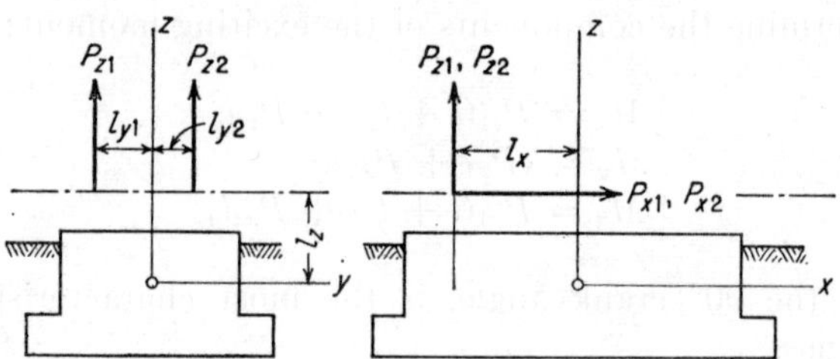

FIG. IV-5. Derivation of equations for horizontal piston compressors.

g. Vertical Six-cylinder Engine. In this engine crank angles are usually as follows:

$$\beta_1 = 0 \qquad \beta_2 = \frac{2\pi}{3} \qquad \beta_3 = \frac{4\pi}{3} \qquad \beta_4 = \frac{4\pi}{3} \qquad \beta_5 = 2\pi \qquad \beta_6 = \frac{8}{3}\pi$$

For such crank positions, the first and second harmonics of disturbing forces are balanced. The exciting moments equal

$$M_x = \sqrt{3}\,P_z l \qquad M_y = 0 \qquad M_z = \sqrt{3}\,P_x l$$

Their absolute values are comparatively small and they cannot cause vibrations with an amplitude exceeding the permissible value. Therefore in design computations of foundations under six-cylinder engines there is no need to compute forced vibrations.

If the engines have auxiliary cylinders (a compressor and an exhaust cylinder in addition to the main cylinders), then in the computations of exciting forces a load imposed by the auxiliaries should be added to the loads produced by the main cylinders. However, the exciting loads caused by auxiliary cylinders are small in comparison to loads caused by the main cylinders, and they may often be neglected in computations of foundation vibrations.

IV-3. Stresses Imposed by Belt Pull

In many cases reciprocating engines set into rotary motion some types of operating machines, usually electric generators, by means of a belt

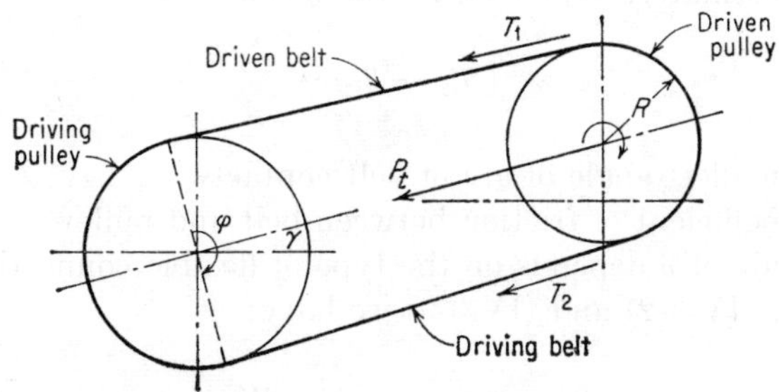

FIG. IV-6. Diagram of forces transmitted to pulleys by belt.

drive. On the other hand, some reciprocating engines are given rotary motion by means of a belt drive from an electromotor. Saw frames and compressors belong to this group.

When a belt drive is in operation, the force of the belt pull acts on the engine bearings and consequently on the foundation.

Let us consider a reciprocating engine set in rotation by means of a belt drive and examine the stresses transmitted to the foundation.

If T_1 is the magnitude of belt tension in the slack side of a belt (Fig. IV-6) and T_2 is tension in the driving side, the resultant force of pull transmitted to the bearings of the engine, and consequently to the foundation, equals

$$T_1 + T_2 = P_t \tag{IV-3-1}$$

The peripheral tension transmitted by the belt to the driven pulley is the difference between tensions in the driving and driven belts:

$$P_r = T_1 - T_2 \tag{IV-3-2}$$

If W is the engine power and v is the peripheral speed of the belt, then it is known that

$$P_r = \frac{W}{v}$$

Since
$$v = R\omega = \frac{2\pi}{60} NR$$

where N = speed of engine, rpm
R = radius of driven pulley

then
$$P_r = \frac{60}{2\pi} \frac{W}{NR} \cong 9.55 \frac{W}{NR} \qquad \text{(IV-3-3)}$$

The interrelationship between the pull values in the driven and driving belts is approximately expressed by the formula

$$T_2 = T_1 e^{\varphi\mu} \qquad \text{(IV-3-4)}$$

where φ = smallest angle of arc of belt contact
μ = coefficient of friction between belt and pulley
The magnitude of μ depends on the type of flexible connection used.

From Eqs. (IV-3-2) and (IV-3-3) we have:

$$T_1 - T_2 = 9.55 \frac{W}{RN}$$

Substituting here the expression for T_2 from Eq. (IV-3-4), we obtain

$$T_1 = 9.55 \frac{W}{RN} \frac{1}{1 - e^{\varphi\mu}}$$

and consequently,

$$T_2 = 9.55 \frac{W}{RN} \frac{e^{\varphi\mu}}{1 - e^{\varphi\mu}} \qquad \text{(IV-3-5)}$$

Substituting these expressions for T_1 and T_2 into the right-hand part of Eq. (IV-3-1), we obtain the following expression for the force transmitted to the foundation by the belt pull:

$$P_r = 9.55 \frac{W}{RN} \frac{1 + e^{\varphi\mu}}{1 - e^{\varphi\mu}} \qquad \text{(IV-3-6)}$$

The direction of this force depends on the respective locations of the axes of the driving and driven pulleys.

If the straight line passing through the axes of rotation of the driving and driven pulleys forms an angle γ with the horizontal (Fig. IV-6), then the horizontal component of belt pull equals

$$P_{tx} = P_t \cos \gamma$$

The vertical component of pull tension usually may be neglected since it is small in comparison with the engine weight. In cases in which the engine is driving, the expression for P_t remains the same but the direction sign changes.

Therefore if the driven and driving pulleys are mounted on the same foundation, the forces imposed by a driving gear represent internal forces and do not influence the displacement of the foundation.

IV-4. Examples of Dynamic Analyses of Foundations for Reciprocating Engines

Example 1. *Dynamic computations of foundation for a vertical compressor coupled on a shaft with an electromotor*

1. Design Data. A two-cylinder compressor has the following characteristics: crank angles, $\beta_1 = 0$, $\beta_2 = \pi/2$; compressor weight, 12 tons; electromotor weight, 4 tons; operational speed, 480 rpm.

The first harmonics of the exciting forces equal (in tons): in the direction of sliding of the piston,

$$P_{z1} = 3.0 \cos \omega t \qquad P_{z2} = -3.0 \sin \omega t$$

and in the horizontal direction perpendicular to the shaft axis,

$$P_{x1} = 0.4 \sin \omega t \qquad P_{x2} = 0.4 \cos \omega t$$

where ω is the angular velocity of rotation of the compressor, equaling

$$\omega = 0.104 \times 480 = 50 \text{ sec}^{-1}$$

The base of the foundation consists of silty clays with some sand characterized by the following design elastic coefficients:

$$c_u = 5.0 \times 10^3 \text{ tons/m}^3 \qquad c_\varphi = 10.0 \times 10^3 \text{ tons/m}^3 \qquad c_\tau = 2.5 \times 10^3 \text{ tons/m}^3$$

2. Design Diagram of Foundation. To simplify computations, it is advisable to shape the foundation in plan as simply as possible, avoiding all small grooves, projections, asymmetry, and so on. Figure IV-7 gives a design diagram for the foundation under consideration, selected on the basis of the foregoing reasoning. Somewhat larger dimensions were selected for the projection of the foundation slab on the left-hand side, due to the eccentric distribution of the equipment on the foundation.

3. Centering of the Foundation Area in Contact with Soil and Determination of Static Pressure on Soil. Let us determine the coordinates x_0, y_0, and z_0 of

the common center of gravity of the system (the foundation and compressor with the electromotor) with respect to the axes shown in Fig. IV-7:

$$x_0 = \frac{\Sigma m_i x_i}{m} \qquad y_0 = \frac{\Sigma m_i y_i}{m} \qquad z_0 = \frac{\Sigma m_i z_i}{m}$$

where m_i = masses of single elements of system
x_i, y_i, z_i = coordinates of centers of gravity of single elements with respect to axes
m = mass of system

We will consider that the masses of the compressor and electromotor are concentrated at the height of the level of the master-shaft axis (at a distance of 0.8 m from the foundation surface).

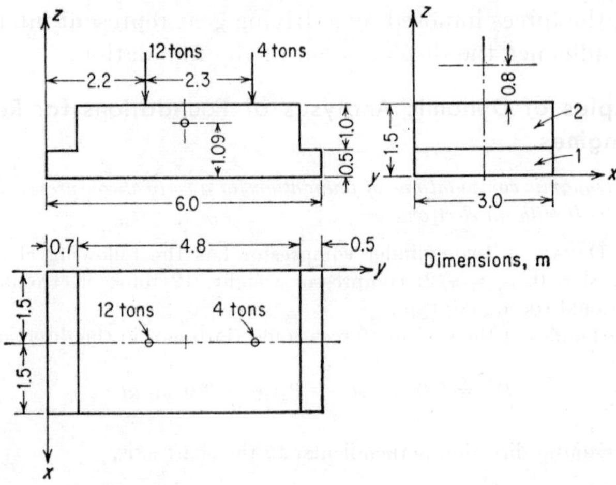

FIG. IV-7. Design diagram of foundation, example 1: (1) foundation slab; (2) upper part of foundation.

The results of computations of static moments of single elements of the system are given in Table IV-1. Using those data, we obtain the coordinates of the common center of gravity of the system:

$$x_0 = \frac{10.35}{6.91} = 1.5 \text{ m} \qquad y_0 = \frac{20.55}{6.91} = 2.98 \text{ m} \qquad z_0 = \frac{7.52}{6.91} = 1.09 \text{ m}$$

The relative values of the eccentricity in the directions of the x and y axes equal, in per cent:

$$\epsilon_x = 0 \qquad \epsilon_y = \frac{3.0 - 2.98}{3.0} \times 100 = 0.7$$

These values of eccentricity in the distribution of the masses are so small that they may be neglected in further computations of the foundation. Thus we obtain: the weight of the whole system,

$$W = mg = 6.91 \times 9.81 = 67.5 \text{ tons}$$

the foundation area in contact with soil,

$$A = 6.0 \times 3.0 = 18 \text{ m}^2$$

and the static pressure on soil,

$$p_{st} = \frac{W}{A} = \frac{67.5}{18} = 3.8 \text{ tons/m}^2 = 0.38 \text{ kg/cm}^2$$

4. Possible Forms of Foundation Vibrations and Design Values of Exciting Loads. The foregoing data lead to the conclusion that horizontal components of the

Table IV-1. Summary of Data for the Solution of Example IV-4-1.

Elements of system	Dimensions of elements, m			Mass of element, tons × sec²/m	Coordinates of center of gravity of element, m			Static moments of mass of elements, tons × sec²		
	a_x	a_y	a_z		x_i	y_i	z_i	m_ix_i	m_iy_i	m_iz_i
Compressor.....	...	...	...	1.23	1.5	2.2	2.3	1.85	2.70	2.82
Electromotor...	...	...	...	0.41	1.5	4.5	2.3	0.62	1.84	0.90
Foundation slab.........	3	6	0.5	2.02	1.5	3.0	0.25	3.03	6.06	0.55
Upper part of foundation...	3	4.8	1.0	3.25	1.5	3.0	1.0	4.85	9.75	3.25
Total..........				6.91				10.35	20.55	7.52

disturbing forces of the compressor are small in comparison with vertical components. Therefore the dynamic analysis of the foundation may be confined merely to determining the amplitudes of forced vibrations caused by vertical components of the exciting forces and of their moments.

The resultant vertical component of the disturbing forces equals (see Art. IV-2)

$$\begin{aligned} P_z &= P_{z1} \cos \omega t - P_{z2} \sin \omega t = 3.0(\cos \omega t - \sin \omega t) \\ &= 4.2 \cos \left(\omega t + \frac{\pi}{4} \right) \end{aligned}$$

The design value of the vertical component of the exciting loads will be

$$P_z = 4.2 \text{ tons}$$

This load will induce vertical forced vibrations of the foundation.

Due to the asymmetric position of the compressor, the foundation will be subjected to the action of the disturbing moment M_x with respect to the x axis. The magnitude of this moment is

$$M_x = P_{z1}(l + l_y) + P_{z2}l_y$$

where l = distance between cylinder axes; in the case under consideration $l = 1.3$ m
l_y = distance between second cylinder and vertical axis passing through center of gravity of complete system; in the case under consideration $l_y = 0.2$ m

Thus
$$M_x = 3.0(1.3 + 0.2) \cos \omega t - 3.0 \times 0.2 \sin \omega t$$
$$= 4.6 \cos \left(\omega t + \frac{\pi}{4} \right)$$

The design value of the disturbing moment should equal its greatest magnitude:

$$M_x = 4.6 \text{ tons} \times \text{m}$$

Under the action of this moment, vibrations will develop in the plane parallel to yz; they will be accompanied by a simultaneous sliding of the foundation in the direction of the y axis and a rotation of the foundation with respect to an axis parallel to the x axis and passing through the common center of gravity.

5. Computations of the Amplitude of Forced Vertical Vibrations of the Foundation. From Eq. (III-1-5) we determine the frequency of vertical natural vibrations of the foundation:

$$f_{nz}^2 = \frac{5.0 \times 10^3 \times 18}{6.91} = 13.0 \times 10^3 \text{ sec}^{-2}$$

The amplitude of forced vertical vibrations is found from Eq. (III-1-13):

$$A_z = \frac{4.2}{6.91(13.0 - 2.5) \times 10^3} = 0.058 \times 10^{-3} \cong 0.06 \text{ mm}$$

Hence it follows that the amplitude of vertical vibrations of the foundation will be much smaller than permissible (0.15 mm).

6. Determination of the Moments of Inertia of the Foundation Area in Contact with Soil and of the Mass of the Whole System. The moment of inertia I of the foundation contact area with respect to the axis passing through its center of gravity perpendicular to the plane of vibrations is

$$I_a = \frac{3 \times 6^3}{12} = 54.0 \text{ m}^4$$

The moments of inertia of the masses of separate elements of the system with respect to the same axis are: for the compressor, whose mass is considered to be concentrated at the height of the shaft axis,

$$I_{01} = m_1(0.8^2 + 2.3^2) = 1.23 \times 5.94 = 7.3 \text{ tons} \times \text{m} \times \text{sec}^2$$

for the electromotor,

$$I_{02} = m_2(1.5^2 + 2.3^2) = 0.41 \times 7.55 = 3.1 \text{ tons} \times \text{m} \times \text{sec}^2$$

for the foundation slab,

$$I_{03} = \frac{m_3}{12} (a_{3y}^2 + a_{3z}^2) + m_3 h_3^2 = \frac{2.02}{12} (6.0^2 + 0.5^2) + 2.02 \times 0.25^2 = 6.1 \text{ tons} \times \text{m} \times \text{sec}^2$$

(h_3 is the distance between the center of gravity of the mat and the foundation contact area).

For the upper part of the foundation, located above the mat, the moment of inertia, from an analogous formula, is

$$I_{04} = \frac{3.25}{12}(4.8^2 + 1.0^2) + 3.25 \times 1.0^2 = 9.8 \text{ tons} \times \text{m} \times \text{sec}^2$$

The total moment of inertia of the mass of the whole system with respect to this axis is

$$W_0 = \sum_{i=1}^{4} I_{0i} = 7.3 + 3.1 + 6.1 + 9.8 = 26.3 \text{ tons} \times \text{m} \times \text{sec}^2$$

The moment of inertia of the whole system with respect to the axis passing through the center of gravity of the whole system perpendicular to the plane of vibrations is

$$I = I_0 - mh^2 = 26.3 - 6.91 \times 1.09^2 = 18.2 \text{ tons} \times \text{m} \times \text{sec}^2$$

since $h = z_0 = 1.09$ m.

The ratio between the moments of inertia is

$$\gamma = 18.2/26.3 = 0.69$$

7. Computation of Amplitudes of Forced Vibrations of a Foundation Accompanied Simultaneously by Sliding and Rocking. The limiting natural frequency of rocking vibrations of the foundation, according to Eq. (III-2-6), is

$$f_{n\varphi}^2 = \frac{10 \times 10^3 \times 54 - 67.5 \times 1.09}{26.3} = 20.5 \times 10^3 \text{ sec}^{-2}$$

The limiting frequency of vibrations in shear, from Eq. (III-3-2), is

$$f_{nx}^2 = \frac{2.5 \times 10^3 \times 18}{6.91} = 6.5 \times 10^3 \text{ sec}^{-2}$$

The frequency equation for the foundation [Eq. (III-4-8)] is

$$f_n^4 - \frac{(20.5 + 6.5) \times 10^3}{0.69} f_n^2 + \frac{20.5 \times 6.5}{0.69} 10^6 = 0$$

or

$$f_n^4 - 39.2 \times 10^3 f_n^2 + 193.0 \times 10^6 = 0$$

By solving this equation we find the natural frequencies of vibrations of the system:

$$f_{n1}^2 = 33.4 \times 10^3 \text{ sec}^{-2} \qquad f_{n2}^2 = 5.8 \times 10^3 \text{ sec}^{-2}$$

We compute the coefficient $\Delta(\omega^2)$:

$$\begin{aligned}\Delta(\omega^2) &= mI(f_{n1}^2 - \omega^2)(f_{n2}^2 - \omega^2)\\ &= 6.91 \times 18.2(33.4 - 2.5)(5.8 - 2.5) \times 10^6\\ &= 13.8 \times 10^9\end{aligned}$$

From Eqs. (III-4-12) we determine the amplitudes of sliding shear and rotation of the foundation. The amplitude of sliding shear of the center of gravity of the whole system is

$$A_y = \frac{2.5 \times 10^3 \times 18 \times 1.09}{13.8 \times 10^9} 4.6 = 0.016 \times 10^{-3} \text{ m} = 0.016 \text{ mm}$$

The amplitude of rotation is

$$A_\varphi = \frac{2.5 \times 10^3 \times 18 - 6.91 \times 2.5 \times 10^3}{13.8 \times 10^9} 4.6 = 0.009 \times 10^{-3} \text{ radians}$$

The maximum horizontal displacement of the foundation surface in the plane yz is

$$\begin{aligned} A &= A_y + h_1 A_\varphi = (0.016 + 0.41 \times 0.009) \times 10^{-3} \\ &= 0.020 \times 10^{-3} \text{ m} \cong 0.02 \text{ mm} \end{aligned}$$

The foregoing computations show that the amplitude of horizontal vibrations, as well as the amplitude of vertical vibrations, lies within the range of permissible values. Hence the conclusion is possible that the dimensions of the foundation for the machine under consideration were selected properly.

It is clear that in the case under review an increase in foundation height would lead to greater amplitudes of vibrations; hence an increase in height would not only raise the cost of the construction, but would also have a negative effect on the dynamic condition of the foundation.

This conclusion holds for all cases in which the natural frequencies of a foundation supported on soil are higher than the operational frequency of the engine mounted on the foundation. This occurs in the overwhelming majority of reciprocating engines.

Example 2. *Dynamic analysis of a foundation for a reciprocating horizontal compressor*

1. Design Data. The reciprocating horizontal compressor has two cylinders. The distance between the axis of the engine master shaft and the foundation surface is 0.9 m. The operational speed is 167 rpm.

Maximum values of unbalanced inertia forces of the engine are: horizontal component in the direction of piston motion P_x = 12.8 tons; vertical component P_z = 0.73 tons.

The foundation rests on a soil of medium strength having a permissible bearing value of 2 kg/cm². The design values of the coefficients of elasticity of the soil may be selected according to Table I-8 as follows:

Coefficient of elastic uniform compression:

$$c_u = 4.0 \times 10^3 \text{ tons/m}^3$$

Coefficient of elastic nonuniform compression:

$$c_\varphi = 8.0 \times 10^3 \text{ tons/m}^3$$

Coefficient of elastic uniform shear:

$$c_\tau = 2.0 \times 10^3 \text{ tons/m}^3$$

The dimensions of the foundation are not limited by structures, communication lines, or plant equipment.

2. Selection of a Design Diagram for the Foundation. The dimensions of the foundation are to be selected according to design considerations based on the requirements of the plant management. Figure IV-8 gives the design diagram selected on the basis of these considerations. Concrete type 100† is to be employed for the foundation.

3. Centering of the Foundation Area in Contact with Soil and Determination of Pressure on the Soil. Let us determine the coordinates x_0, y_0, and z_0 of

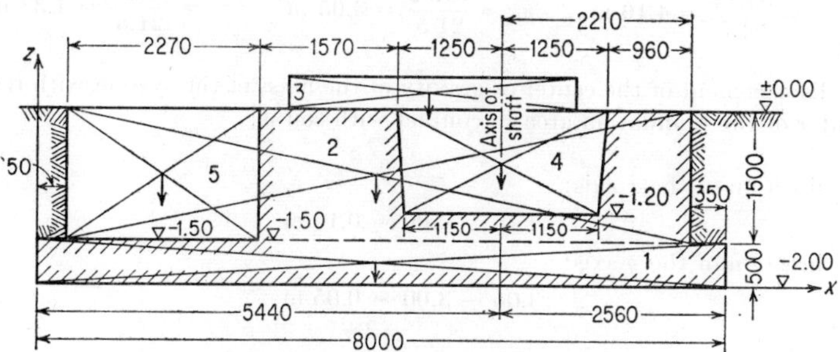

Section along longitudinal axis of the foundation

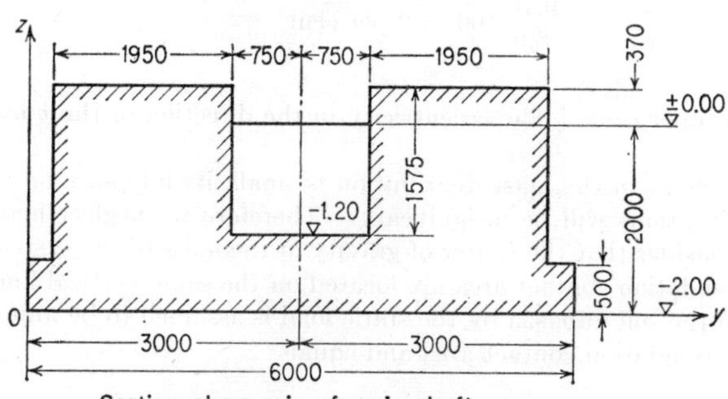

Section along axis of main shaft

Fig. IV-8. Design diagram of foundation, example 2.

the center of gravity of the whole system (the foundation and engine) with respect to the axes shown in Fig. IV-8:

$$x_0 = \frac{\Sigma m_i x_i}{m} \qquad y_0 = \frac{\Sigma m_i y_i}{m} \qquad z_0 = \frac{\Sigma m_i z_i}{m}$$

where m_i = masses of separate elements of system

x_i, y_i, z_i = coordinates of centers of gravity of these elements with respect to x, y, z axes

m = mass of complete system

Separate elements of the foundation are marked in Fig. IV-8 by the numbers 1, 2, and so on.

The foundation should be divided into elements of such shape that the data on magnitudes of masses and coordinates of centers of gravity of separate elements may

† See footnote in Art. IV-1-c, p. 132.

be used later, when the moment of inertia of the mass of the whole system will be computed.

The data for the computation of coordinates of the center of gravity of the system are given in Table IV-2. Masses corresponding to cavities in the foundation are shown with minus signs. From the data of Table IV-2 we find the coordinates of the center of gravity of the system:

$$x_0 = \frac{89.14}{21.5} = 4.16 \text{ m} \qquad y_0 = \frac{64.52}{21.5} = 3.05 \text{ m} \qquad z_0 = \frac{28.55}{21.5} = 1.33 \text{ m}$$

The displacement of the center of gravity of the mass of the system with respect to the center of the foundation area in contact with soil is:

In the direction of the x axis:

$$4.16 - 4.00 = 0.16 \text{ m}$$

In the direction of the y axis:

$$3.05 - 3.00 = 0.05 \text{ m}$$

The relative magnitude of the eccentricity in the direction of the x axis is

$$\frac{0.16}{8.0}\,100 = 2 \text{ per cent}$$

which is less than 5 per cent. The eccentricity in the direction of the y axis is even smaller.

Since the eccentricity in the mass distribution is small, its influence on the amplitudes of forced vibrations will be insignificant. Therefore we neglect hereafter the eccentricity and consider that the center of gravity of the mass of the system and the centroid of the foundation contact area are located on the same vertical line.

The pressure on the soil imposed by the static load is assumed to be uniformly distributed over the foundation contact area and equals

$$p_{st} = \frac{mg}{A} = \frac{21.5 \times 9.81}{48} = 4.4 \text{ tons/m}^2 = 0.44 \text{ kg/cm}^2$$

Thus the static pressure on the soil is considerably smaller than the permissible pressure.

4. Computation of Amplitudes of Forced Vibrations of the Foundation. Since the horizontal component of the unbalanced inertial forces of the engine in the direction perpendicular to the motion of the piston is zero, and since the vertical component of the above forces is insignificant, we compute the amplitudes of forced vibrations only for foundation vibrations caused by the horizontal component of unbalanced inertial forces in the direction of piston motion (the system will be subjected to vibrations in the xz plane). We also neglect the action of exciting moments tending to produce rocking vibrations of the foundation.

We begin by establishing the data needed for the computation of amplitudes of foundation vibrations. The frequency of machine rotation equals

$$\omega = 0.105N = 0.105 \times 167 = 17.3 \text{ sec}^{-1}$$
$$\omega^2 = 300 \text{ sec}^{-2}$$

Table IV-2. Summary of Data for the Solution of Example IV-4-2

Elements of system (engine and foundation)	Dimensions of elements, m			Mass, tons × sec²/m	Coordinates of center of gravity of element with respect to x, y, z axes, m			Static moment of mass of element with respect to x, y, z axes, tons × sec²			Moment of inertia of mass of element with respect to axes passing through center of gravity of element, tons × sec² × m	Distance between center of gravity of element and common center of gravity, m		r
	a_{xi}	a_{yi}	a_{zi}	m_i	x_i	y_i	z_i	m_ix_i	m_iy_i	m_iz_i	$\frac{m_i}{12}(a_{xi}^2 + a_{zi}^2)$	x_{0i}	z_{0i}	$m_i(x_{0i}^2 + z_{0i}^2)$
Compressor	....	....	....	2.23	3.61	2.84	2.90	8.05	6.35	6.47	......	0.39	1.57	5.98
Motor.....	....	....	....	1.47	5.44	3.18	2.90	8.00	4.67	4.26	......	1.44	1.57	6.76
1..........	8.00	6.00	0.50	5.38	4.00	3.00	0.25	21.52	16.14	1.34	28.80	0	0.81	6.03
2..........	7.30	5.40	1.50	13.27	4.00	3.00	1.25	53.08	39.81	16.59	61.00	0	0.08	0.04
3(2).......	3.36	1.95	0.37	1.10	4.66	3.00	2.19	5.13	3.30	2.41	1.05	0.66	0.86	1.33
−4........	2.40	1.50	1.20	−0.97	5.44	3.00	1.40	−5.27	−2.91	−1.35	−0.58	1.44	0.07	−2.02
−5........	2.27	1.25	1.50	−0.95	1.44	3.07	1.25	−1.37	−2.92	−1.19	−0.58	2.56	0.08	−6.03
Total......				21.5				89.14	64.52	28.55	89.69			12.09

The distance from the axis of the master shaft of the engine to the common center of gravity of the mass of the system is $h_1 = 1.53$ m. The exciting moment of the engine is then

$$M = P_x h_1 = 12.8 \times 1.53 = 19.6 \text{ tons} \times \text{m}$$

The moment of inertia of the foundation contact area with respect to the axis passing through its center of gravity perpendicular to the plane of vibrations equals

$$I = \frac{6.00 \times 8.00^3}{12} = 256 \text{ m}^4$$

The weight of the whole system is

$$W = mg = 21.5 \times 9.81 = 211 \text{ tons}$$

The moment of inertia of the mass of the whole system with respect to the axis passing through the common center of gravity perpendicular to the plane of vibrations equals

$$\begin{aligned} W_m &= 1/12 \Sigma m_i(a_{xi}^2 + a_{zi}^2) + m\Sigma(x_{0i}^2 + z_{0i}^2) \\ &= 89.69 + 12.09 = 101.78 \cong 102 \text{ tons} \times \text{m} \times \text{sec}^2 \end{aligned}$$

The moment of inertia of the mass of the whole system with respect to the axis passing through the centroid of the foundation contact area perpendicular to the plane of vibrations equals

$$W_0 + I_m = mh^2 = 102 + 21.5 \times 1.33^2 = 140 \text{ tons} \times \text{m} \times \text{sec}^2$$

The ratio between the moments of inertia of the masses is

$$\gamma = 102/140 = 0.73$$

The limit value of natural frequency of rocking vibrations of the foundation is determined from Eq. (III-2-6):

$$f_{n\varphi}^2 = \frac{8 \times 10^3 \times 256 - 1.33 \times 211}{140} = 14.6 \times 10^3 \text{ sec}^{-2}$$

The limit value of the natural frequency of sliding shear vibrations, from Eq. (III-3-2), is

$$f_{nx}^2 = \frac{2 \times 10^3 \times 48}{21.5} = 4.46 \times 10^3 \text{ sec}^{-2}$$

We set up the frequency equation of the foundation according to Eq. (III-4-8):

$$f_n^4 - \frac{14.6 \times 10^3 + 4.46 \times 10^3}{0.73} f_n^2 + \frac{14.6 \times 10^3 \times 4.46 \times 10^3}{0.73} = 0$$

$$f_n^4 - 26.0 \times 10^3 f_n^2 + 89.0 \times 10^6 = 0$$

Solving this equation,

$$f_{n1,2}^2 = [13.0 \pm \sqrt{(13.0^2 - 89.0)}]10^3$$

Hence

$$f_{n1,2}^2 = (13.0 \pm 8.9)10^3$$

Thus the natural frequencies of the foundation will be

$$f_{n1} = 21.9 \times 10^3 \text{ sec}^{-2} \qquad f_{n2}^2 = 4.1 \times 10^3 \text{ sec}^{-2}$$

We compute the coefficient:

$$\begin{aligned}\Delta(\omega^2) &= mW_m(f_{n1}^2 - \omega^2)(f_{n2}^2 - \omega^2)\\ &= 21.5 \times 102(21.9 - 0.30)(4.1 - 0.30) \times 10^6 = 18.0 \times 10^{10}\end{aligned}$$

We then compute the amplitudes of forced vibrations induced by the horizontal force P_x and by the moment $M = P_x h_1$; according to Eqs. (III-4-11) and (III-4-12) the horizontal displacement of the common center of gravity of the foundation and the engine is

$$A_x = \frac{(8 \times 10^3 \times 256 - 211 \times 1.33 + 2 \times 10^3 \times 48 \times 1.33 - 102 \times 17.3^2)12.8}{18.0 \times 10^{10}} + \frac{2 \times 10^3 \times 48 \times 1.33 \times 19.6}{18.0 \times 10^{10}} = 0.17 \times 10^{-3}\ \text{m} = 0.17\ \text{mm}$$

From the same equations we find the amplitude of rocking vibrations of the foundation about the horizontal axis passing through the center of gravity of the foundation perpendicular to the plane of vibrations:

$$\begin{aligned}A_\varphi &= \frac{(2 \times 10^3 \times 48 \times 1.33 \times 12.8 + (2 \times 10^3 \times 48 - 21.5 \times 17.3^2)19.6}{18.0 \times 10^{10}}\\ &= 0.019 \times 10^{-3}\ \text{radians}\end{aligned}$$

Thus the amplitude of forced vibrations of the upper edge of the foundation equals

$$A = 0.17 \times 10^{-3} + 1.04 \times 0.019 \times 10^{-3} = 0.19\ 10^{-3}\ \text{m} < 0.2\ \text{mm}$$

The design value of the amplitude of vibrations does not exceed the permissible value; hence the dimensions of the foundation are selected correctly.

The foregoing computations show that vibrations of the foundation are produced mainly by its horizontal displacement in the direction of the action of the horizontal component of the disturbing force of the compressor. This is explained by the fact that the dimensions of the foundation in the direction of the action of this force is large in comparison with the height of the foundation. Therefore rocking results only in small dynamic displacements.

Hence the rocking vibrations of a foundation may be neglected when computing the amplitude of forced vibrations if the foundation is elongated in the direction of action of the horizontal exciting force; in this case the vibrations of the foundation may be considered to be vibrations of sliding shear. This assumption greatly simplifies dynamic computations.

However, this simplification of computations should be very cautiously applied. For example, if such a simplification were made in the case of the foundation under consideration, then from Eq. (III-3-3) we would obtain for the amplitude of horizontal displacements of the foundation

$$A_x = \frac{12.8}{21.5(4.46 - 0.30) \times 10^3} = 0.14 \times 10^{-3}\ \text{m} = 0.14\ \text{mm}$$

The computed amplitude is 26 per cent smaller than that obtained by means of the foregoing computations (0.19 mm). This cannot be admitted as a good approximation.

The results of computations will be more accurate if we add to the amplitude of vibrations of sliding shear a displacement produced by rocking vibrations of the foundation, computed from the formula

$$A_{x\varphi} = A_\varphi h$$

where A_φ is the amplitude of rocking vibrations when no shear is present, determined from Eq. (III-2-7), and h is the full height of the foundation.

Assuming that the horizontal exciting force acts at height h_1 from the base of the foundation, and assuming that, in Eq. (III-2-7), $M = P_x h_1$, we obtain

$$A_x = \frac{P_x h h_1}{W_0(f_n^2 - \omega^2)}$$

For the foundation under consideration, $h = 2.0$ m and $h_1 = 2.9$ m; consequently,

$$A_{x\varphi} = \frac{12.8 \times 2.0 \times 2.9}{140(14.6 - 0.30) \times 10^3} = 0.038 \times 10^{-3} \text{ m} = 0.04 \text{ mm}$$

Thus the total amplitude of the horizontal displacement of the foundation will equal

$$A = A_x + A_{x\varphi} = 0.14 + 0.04 = 0.18 \text{ mm}$$

An amplitude computed by means of the above approximate method will not differ much from the value obtained as a result of computations taking into account vibrations of the foundation accompanied by simultaneous sliding shear and rocking.

IV-5. Methods for Decreasing Vibrations of Existing Foundations

a. Counterbalancing of Exciting Loads Imposed by Engines. As stated in Art. IV-2, there are different methods of balancing primary inertial forces by means of counterweights.

It is possible to counterbalance completely a component in the direction perpendicular to piston motion and partly a component in the direction of piston motion. Or, the dimensions of counterweights and their distances from axes of rotation may be selected to counterbalance completely the first harmonic of the component exciting forces in the direction of piston motion. Then the component in the perpendicular direction will increase.

Usually the first method is employed for the counterbalancing of engines because stresses in the engine itself are smaller than those occurring when the other method is used. Another advantage of the first method is that it requires a smaller counterweight mass.

The efficiency of a certain method of counterbalancing the exciting forces induced by an engine for the purpose of decreasing foundation vibrations depends on the type of engine and on special features of the foundation.

For a horizontal reciprocating engine, the most dangerous foundation vibrations are those which are accompanied simultaneously by rocking and sliding. In this case, a decrease in the vibrations of the foundation may be achieved by counterbalancing the inertial forces of the engine by the second method, even if this leads to some increase in vertical vibrations. Therefore, if such an engine was counterbalanced by means of the first method but impermissible horizontal vibrations were observed after the construction of the foundation, then counterbalancing by means of the second method (i.e., by changing the character of counterbalancing) may be recommended as one of the simplest measures to decrease these vibrations.

In cases in which vertical vibrations of an impermissible amplitude are present in systems with horizontal motors, the second method is unsuitable, and the first method should be applied.

Similarly, for a vertical motor, the method of counterbalancing selected depends on the type of foundation vibrations—vertical, horizontal, or rocking.

The installation of counterweights for balancing a motor does not require dismantling or prolonged interruption of operation. The interruption is only for the time needed to attach the counterweight to the sides of the crank.

b. Chemical Stabilization of Soils. If a foundation rests on sandy soil, then, in order to decrease vibrations, chemical or cement stabilization of the soil under the foundation may be used. Such soil stabilization will result in an increase in the rigidity of the base and consequently in an increase in the natural frequencies of the foundation. Therefore this method is very effective when natural frequencies of the foundation on a nonstabilized soil are higher than the operational frequency of the engine—which is usually the case. An increase in rigidity will increase still further the difference between the frequency of natural vibrations and the frequency of the engine; consequently the amplitudes of foundation vibrations will decrease. When a foundation resting on a natural soil has natural frequencies smaller than the operational frequency of the engine, then soil stabilization may cause an increase in the amplitudes of vibrations. This may be undesirable if a soil is stabilized to such a degree that frequencies of natural vibrations of the foundation merely approach the operational frequency. But if a soil is thoroughly stabilized and natural frequencies of the foundation became much higher than the operational frequency of the engine, then such soil stabilization may result in a considerable decrease in amplitudes of vibrations.

Chemical and cement stabilization of soils is economically advantageous, since its costs are low in comparison, for example, with structural measures. The principal advantage of this method lies in the fact that it can be applied without a prolonged interruption in the work of the engine. The interruption is only for the period of direct work connected with soil stabilization and then for 2 to 3 days more. Thus the over-all result is that the engine will be inactive only for a few days.

The limits of the stabilized zones of soil and their shape are determined by the character of the vibrations. If, for example, a foundation is subjected mainly to rocking vibrations about an axis passing through the centroid of the base contact area, then it suffices to stabilize the soil near the foundation edges, perpendicular to the plane of vibrations, and it is not necessary to stabilize the soil under the entire foundation. The depth of the stabilized zone should be no less than 1 to 2 m.

This method of decreasing vibrations was applied at one of the Soviet plants when it became necessary to decrease the amplitudes of vibrations of an operating horizontal compressor without a long interruption in its work. Soil was stabilized to a depth of about 1.0 m; the zone extended horizontally 30 cm beyond the foundation edges. The results of foundation vibration measurements before and after stabilization showed that the amplitudes of vibrations, on the average, decreased by 50 per cent. The work of the compressor was stopped only for the period of injection of the silicates; the engine was set in motion immediately after silicatization was completed. It can therefore be assumed that when the compressor renewed its motion, the stabilized soil had not as yet formed a sufficiently rigid base, and it is possible that foundation vibrations acted unfavorably on the stabilized zone of soil, which had not fully hardened.

c. Structural Measures. The use of structural measures for decreasing foundation vibrations often requires a long interruption in the engine's operation and considerable expense of funds and materials. Therefore the use of this method may be suggested only in cases in which for some reason no other methods may be applied. At the same time, it should be noted that the correct change in foundation design may prove very effective in decreasing the amplitude of vibrations.

Structural measures are applied with the purpose of changing the natural frequencies of a foundation in such a way as to achieve the largest possible difference between them and the operational frequency of the engine. The choice of structural measure depends on the nature of the vibrations and the interrelationships between the frequencies of natural and forced vibrations. The operational frequencies of reciprocating engines are usually lower than the fundamental frequencies of foundations; therefore most of the structural measures are directed towards increasing still further the natural frequencies of the foundation. This is achieved by increasing the foundation contact area and its moments of inertia, as well as by increasing the rigidity of its base by means of piles.

In addition, it is possible to increase the foundation mass without inducing changes in the frequency of foundation vibrations. This results in a decrease in the amplitudes of vertical vibrations.

When check calculations of the natural frequencies of a vibrating foundation show that they are lower than the operational frequencies of the engine, an enlargement of the foundation contact area or an increase in the soil rigidity not only may not decrease the amplitudes of vibrations, but may even increase them. In this case, it is better to decrease still more the natural frequency of the foundation. This may be achieved by enlarging the foundation mass without an increase in its area in contact with soil.

The selection of particular structural measures depends on local conditions. For example, if a vibrating foundation lies close to another foundation, it can be attached to the latter. As an illustration we will

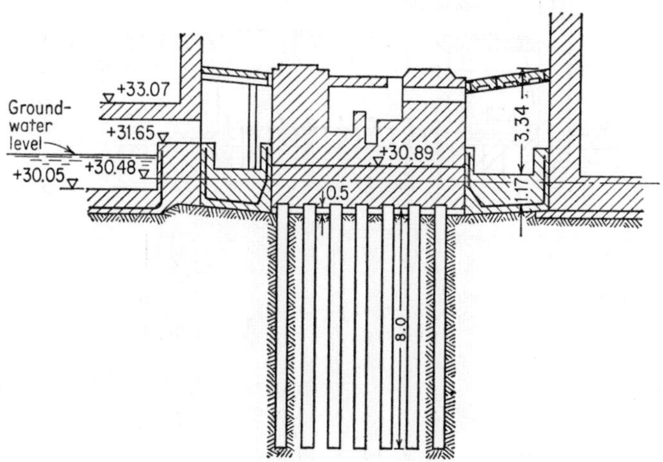

FIG. IV-9. Machine foundation which developed horizontal vibrations of high amplitude.

describe here the structural measures which were applied to a foundation under a horizontal compressor in order to decrease its vibrations.

The horizontal component of the exciting forces induced by the whole system was 30 tons. The foundation rested on a medium-grained sand with clay laminae.

The foundation consisted of a block about 4.6 m high, with a base area 7 by 8 m^2, placed on 55 situ-cast piles. The length of the piles was about 8 m. Figure IV-9 shows a cross section of the foundation.

Horizontal vibrations of extremely large amplitude (around 0.9 mm) were observed while the engine was in operation. At the same time, there occurred settlement of the basement of an adjoining structure, under which no piles were provided. On the side nearest the foundation under discussion, settlement of the basement reached 70 mm. It appears that this considerable settlement of the basement was caused by vibrations transmitted from the foundation. These vibrations furthered the loosening of the soil and the carrying away of soil particles from beneath the foundation and basement by ground water. When reinforcement of the foundation was started, it was found that soil under it was washed out or had subsided to a depth of about 0.5 m. No damage was found in the foundation block.

The ground-water level was approximately 1.5 m above the level of the foundation base.

Reinforcement of the foundation was undertaken with the purpose of decreasing its vibrations. After the ground-water level had been artificially lowered, soil was removed from beneath the foundation to a depth of 0.75 m. Thus an excavation of a total depth of about 1.25 m was formed. This excavation was filled with concrete. The foundation area in contact with the soil was extended on all sides and 33 new situ-cast piles were installed. Figure IV-10 illustrates the measures recommended for reinforcement. Reinforcement of the new part of the foundation provided a good connection with the old part. To avoid settlement of the footings under the building walls due to excavation of the soil, sheetpiling and chemical stabilization of soil were used beneath the footings under the walls.

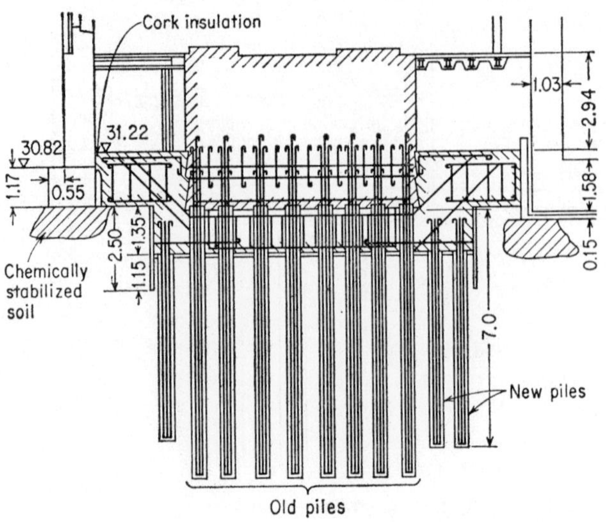

FIG. IV-10. Reinforcement of foundation shown in Fig. IV-9.

This reinforcement of the foundation was very effective. The amplitudes of foundation vibrations decreased from 0.9 to 0.05 mm, i.e., 18 times.

This decrease in amplitudes was caused by considerable increase in the natural frequencies of foundation vibrations due to an increase in the foundation area in contact with the soil, as well as to an increase in the moment of inertia of the contact area. In addition, a considerable effect was produced by the extra foundation mass and the increase in the rigidity of the base due to the installation of supplementary piles.

To facilitate the application of various structural measures for decreasing foundation vibrations, it is recommended in doubtful cases to leave projecting reinforcement which may be used, if necessary, for the attachment of an additional mass to the foundation or for the extension of its area in contact with soil. These measures, of course, should be applied only after recognizing the fact that the foundation is undergoing vibrations of an impermissible magnitude.

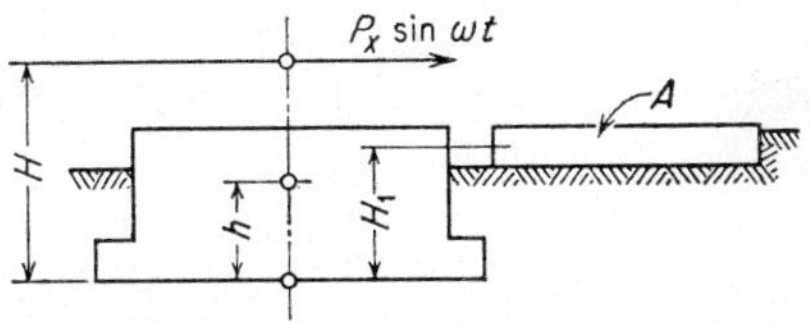

Fig. IV-11. Analysis of the effect of attaching a slab A to an engine foundation.

The use of special slabs, first proposed by Professor N. P. Pavliuk and Engineer A. D. Kondin,[33] may be considered as a structural measure to decrease foundation vibrations. By means of these slabs, it is possible in some cases to decrease the amplitudes of rocking and horizontal vibrations of foundations.

Let us assume (Fig. IV-11) that slab A, resting on soil, is attached to a foundation undergoing rocking vibrations around the axis passing through the centroid of the foundation area in contact with soil. Let us set up the equation of forced rocking vibrations of the foundation. The following symbols will be used:

W_0 = moment of inertia of foundation mass and of mass of engine with respect to axis of vibrations

I = moment of inertia of foundation area in contact with soil, with respect to same axis

$P_T \sin \omega t$ = magnitude of horizontal exciting force induced by engine and transmitted to foundation, where ω = frequency of engine rotation

H = distance between line of action of exciting force and foundation contact area

h = distance between center of mass of foundation and engine, and foundation contact area

m_1 = mass of attached slab

A_1 = contact area of attached slab

H_1 = distance between place of connection of foundation with attached slab, and foundation contact area

W = foundation weight

c_φ, c_τ = coefficients of elastic nonuniform compression, shear of soil.

The differential equation of forced vibrations of the foundation together with the attached slab will be as follows:

$$(W_0 + m_1H_1^2)\ddot{\varphi} + (c_\psi I - Wh + H_1^2c_\tau A_1)\varphi = P_TH \sin \omega t \quad \text{(IV-5-1)}$$

From this we obtain in the usual way the expression for the natural frequency of rocking vibrations of the foundation with attached slab:

$$f_{n\varphi 1}^2 = \frac{c_\varphi I - Wh + H_1^2c_\tau A_1}{W_0 + m_1H_1^2} \quad \text{(IV-5-2)}$$

The amplitudes of rocking vibrations of the foundation will be found from the equation

$$A_{\varphi 1} = \frac{P_TH}{(W_0 + m_1H_1^2)(f_{n\varphi 1}^2 - \omega^2)} \quad \text{(IV-5-3)}$$

Equation (IV-5-2) shows that under certain conditions an attached slab may have no effect on the natural frequency $f_{n\varphi}$ of rocking vibrations of the foundation. To determine these conditions, let us use the following expression:

$$f_{n\varphi 1}^2 \geq f_{n\varphi}^2 \quad \text{(IV-5-4)}$$

or since

$$f_{n\varphi}^2 = \frac{c_\varphi I - Wh}{W_0}$$

we substitute into the left-hand part of (IV-5-4) the expression for $f_{n\varphi 1}^2$ from Eq. (IV-5-2); then, neglecting the term containing Wh because of its smallness, we obtain

$$\frac{c_\varphi I + H_1^2c_\tau A_1}{W_0 + m_1H_1^2} \geq \frac{c_\varphi I}{W_0} \quad \text{(IV-5-5)}$$

Hence

$$\frac{c_\tau A_1}{m_1} \geq \frac{c_\varphi I}{W_0}$$

or

$$f_{nx1}^2 \geq f_{n\varphi}^2 \quad \text{(IV-5-6)}$$

Thus if one selects the attached slab so that the frequency of its natural vibrations of pure shear equals the frequency of rocking vibrations of the foundation, then the attached slab will have no effect on the magnitude of the frequency of natural vibrations of the foundation. Besides, the amplitude of forced vibrations of the foundation will decrease according to the ratio

$$a = \frac{A_{\varphi 1}}{A_\varphi} = \frac{1}{1 + \delta^2} \quad \text{(IV-5-7)}$$

where

$$\delta = \frac{m_1H_1^2}{W_0}$$

Usually in foundations under engines, $f_{n\varphi 1}$ is considerably larger than ω; therefore, approximately,

$$A_{\varphi 1} \approx \frac{P_T H}{c_\varphi I + H_1^2 c_\tau A_1}$$

To determine the amplitude of foundation vibrations before the slab is attached, if $f_{n\varphi} \gg \omega$ and the value of WH in comparison with $c_\varphi I$ is small, we have from Eq. (IV-5-3):

$$A_\varphi = \frac{P_T H}{c_\varphi I}$$

Hence

$$a = \frac{A_{\varphi 1}}{A_\varphi} = \frac{c_\varphi I}{c_\varphi I + H_1^2 c_\tau A_1}$$

or

$$a = \frac{1}{1 + H_1^2 c_\tau A_1 / c_\varphi I} \tag{IV-5-8}$$

It follows from Eq. (IV-5-8) that the effect of the attached slab on the decrease in foundation vibrations will be proportional to the frequency f_{nx} of natural vibrations of shear of the slab and proportional to the height H_1 of the slab above the foundation base. The foundation area in contact with soil should always be as large as possible; the contact area of the slab A_1 is limited by local conditions and economic considerations. Since the value of f_{nx1} depends also on the coefficient of the elastic shear of soil c_τ, a pile foundation may be installed under the slab to increase f_{nx1} as much as possible. Short frictional reinforced-concrete piles should be used.

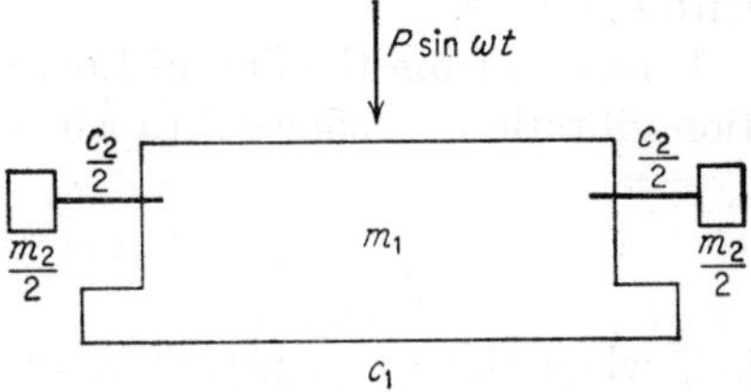

FIG. IV-12. Analysis of the effect of attaching dampers $m_2/2$ to an engine foundation m_1.

Professor N. P. Pavliuk and A. D. Kondin describe a case in which a reinforced-concrete slab was used to decrease the vibrations of a foundation under a compressor. By installation of the slab, foundation vibrations were practically reduced to zero.

d. Dynamic Vibration Dampers. Consider a foundation with mass m_1 (Fig. IV-12) subjected to the action of external exciting forces induced by an engine and producing only vertical vibrations. Let us assume that two masses, each equaling $m_2/2$, are attached to this foundation by means of elastic ties (elastic rods, springs, etc.). Then the foundation with attached masses will have not one but two degrees of freedom. Formulas from the theory of vibrations of a system with two degrees of freedom may be directly applied when considering the forced vibrations of the system.

For example, for the amplitudes A_1 and A_2 of forced vibrations of the foundation and the attached masses, respectively, we have:

$$A_1 = \frac{f_{22}^2 - \omega^2}{\Delta(\omega^2)} p$$

$$A_2 = \frac{f_{22}^2}{\Delta(\omega^2)} p \qquad \text{(IV-5-9)}$$

Also,

$$f_{na}^2 = \frac{c_2}{m_2}$$

where f_{na} = natural frequency of vibrations of attached masses
c_2 = coefficient of elastic rigidity of tie between these masses and foundation
p = magnitude of exciting force per unit of foundation mass

$$\Delta(\omega)^2 = (f_{n1}^2 - \omega^2)(f_{n2}^2 - \omega^2) \qquad \text{(IV-5-10)}$$

where f_{n1} and f_{n2} are natural fundamental frequencies of the foundation with dampers.

It follows from the first of Eqs. (IV-5-9) that the amplitude of foundation vibrations becomes zero when

$$\omega^2 = f_{na}^2 \qquad \text{(IV-5-11)}$$

i.e., when the frequency of natural vibrations of the attached masses equals the frequency of the exciting force. In order to determine the amplitude of vibrations of the damper masses, we substitute f_{22} into the right-hand part of Eq. (IV-5-10) in place of ω; then

$$\Delta(\omega^2) = \frac{c_2}{m_1}$$

Substituting this expression for $\Delta(\omega^2)$ into the second of Eqs. (IV-5-9), we obtain the following expression for the amplitude of vibrations of the damper:

$$A_2 = \frac{m_1}{c_2} p$$

Since

$$p = \frac{P(t)}{m_1}$$

it follows that

$$A_2 = \frac{P(t)}{c_2} \qquad \text{(IV-5-12)}$$

Thus, the amplitude of vibrations of the damper equals its static deflection as produced by a force of magnitude equal to the maximum value of the exciting force $P(t)$.

Equations (IV-5-11) and (IV-5-12) determine the selection of the damper. It should be noted that neither the frequencies of the damper nor the amplitudes of its vibrations depend on the properties of the soil base or the mass of the foundation.

It follows from Eq. (IV-5-9) that theoretically it is possible to damp vibrations of infinitely large foundations subjected to the action of periodic exciting forces by attaching dampers to these foundations, even dampers with small masses. However, the smaller the mass of the damper, the smaller should be the rigidity c_2 of its elastic tie with the foundation, and consequently the larger will be the amplitude of its vibrations.

At values of c_2 smaller than a certain limit, amplitudes of vibrations of the damper may attain magnitudes endangering its strength. Therefore the minimum value of the damper mass is limited by permissible values of stresses in the elastic tie between the damper and the foundation.

It has already been mentioned that when dampers are used, the foundation has not one, but two natural frequencies of vibrations, determined as roots of the equation

$$f_n^4 - [f_{nz}^2 + (1 + \mu)f_{na}^2]f_n^2 + f_{nz}^2 f_{na}^2 \qquad \text{(IV-5-13)}$$

where $$f_{nz}^2 = \frac{c_1}{m_1} \qquad c_1 = c_u A$$

The roots of Eq. (IV-5-13) are:

$$f_{n1,2}^2 = \tfrac{1}{2}(f_{nz}^2 + f_{na}^2(1 + \mu) \pm \sqrt{[f_{nz}^2 + f_{na}^2(1 + \mu)^2]^2 - 4f_{nz}^2 f_{na}^2})$$

where f_{nz} = frequency of natural vertical vibrations of foundation
$f_{na} = \omega$ = average operational machine rotation
$\mu = m_2/m_1$ = ratio between masses of dampers and foundation mass

When a damper is installed, f_{n1} will be larger than both f_{nz} and ω, and f_{n2} will be smaller than these frequencies. Besides, either f_{n1} or f_{n2} will lie close to ω, and the other will be close to f_{nz}.

Let us assume that $f_{nz} > \omega$. Then the lower fundamental frequency f_{n2} will be close to ω; the higher one, f_{n1}, will be close to f_{nz}. If the engine has varying angular frequency, then, with the installation of the damper, the danger arises that one of the values of ω will coincide with f_{n2}, i.e., that resonance will occur with the lower frequency of the system "foundation and damper." In this case, the amplitudes of foundation vibrations will be sharply increased and the masses attached to the foundation will work not as dampers, but as intensifiers of the vibrations.

In order to avoid such an intensification of vibrations, m_2 should be selected so that the maximum decrease in operational frequency of the engine, as compared with the average value of this frequency, is smaller than the difference between the average operational frequency of the engine ω_{av} (equaling the natural frequency f_{22} of the damper) and the lower frequency f_{n2} of the foundation. Hence, the following condition should be satisfied when the damper sizes are selected:

$$f_{n2}^2 < \omega_{min}^2$$

or $\frac{1}{2}(f_{nz}^2 + f_{na}^2(1 + \mu) - \sqrt{[f_{nz}^2 + f_{na}^2(1 + \mu)]^2 - 4f_{nz}^2 f_{na}^2}) < \omega_{min}^2$

Solving this inequality for μ and noting that

$$f_{na}^2 = \omega_{av}^2$$

we obtain

$$\mu > \frac{f_{nz}^2(\omega_{av}^2 - \omega_{min}^2) + \omega_{min}^4}{\omega_{min}^2 \omega_{av}^2} - 1$$

Let us assume that the nonuniformity in engine speed is as follows:

$$\epsilon = 1 - \frac{\omega_{min}}{\omega_{av}}$$

Hence

$$\omega_{min} = \omega_{av}(1 - \epsilon)$$

Substituting this expression for ω_{min} into the right-hand part of the inequality obtained for μ,

$$\mu > \frac{\beta^2 - (1 - \epsilon)^2}{(1 - \epsilon)^2} \epsilon(2 - \epsilon)$$

where

$$\beta = \frac{f_{nz}}{\omega_{av}} > 1$$

If $0 < \beta < 1$, then $\omega_{max} = \omega_{av}(1 + \epsilon)$ and we obtain:

$$\mu > \frac{(1 + \epsilon)^2 - \beta^2}{(1 + \epsilon)^2} \epsilon(2 + \epsilon) \qquad \text{(IV-5-14)}$$

The inequalities obtained show that the selection of a proper interrelationship between the damper mass and the foundation mass depends not only on the values of the irregularity in the engine speed, but also on the interrelationship between the natural frequency of the foundation and the average operational speed of the engine.

If the frequency of natural vibrations is higher than the operational frequency of the engine (i.e., $\beta > 1$), then the damper mass should be selected in proportion to the value of β. If $\beta < 1$, the value of μ decreases with an increase in β. If $\beta = 1$, μ should have its lowest value.

The nonuniformity in the speed of reciprocating engines lies in the range from 0.01 to 0.10. The most uniform speed is observed in multi-cylinder diesels with flywheel, where ϵ is about 0.01 to 0.02; in saw frames ϵ is 0.05 to 0.10.

The frequencies of natural vertical vibrations of foundations are usually higher than the operational frequencies of low-speed reciprocating engines, i.e., usually $\beta > 1$. With variations in the angular speed of the engine, the foundation may develop resonance with the lower frequency of the "foundation-damper" system. If one assumes that the average smallest irregularity in engine speed is around 0.075, then in order to avoid resonance, it is necessary that the lower frequency of the system differ by at least 3 per cent from the average operational speed of the engine. Thus, in calculations of the smallest value of μ, the design value of ϵ should be taken not less than 0.03. For this value of ϵ, and with $\beta = 1.3$, the value $\mu = 0.05$; if $\beta = 1.6$, then the damper mass should be about one-tenth the foundation mass. If the natural frequency of vertical vibrations is two times larger than the operational frequency of the engine, the damper mass should not be less than 20 per cent of the foundation mass.

The weights of foundations under reciprocating engines may reach several hundred tons. For the previously mentioned values of ϵ and β the damper weight will equal several tens of tons. In practice it is difficult to attach a mass of this size elastically to the foundation so that the frequency of natural vibrations of this mass corresponds exactly to the average operational speed of the engine. For example, for a foundation weighing 300 tons, for $\epsilon = 0.03$ and $\beta = 1.5$, the damper should weigh not less than 27 tons; for $\epsilon = 0.05$ it should weigh 45 tons. The difficulty in attaching such blocks to the foundation limits the use of dynamic dampers even for machines with uniform speed. It is out of the question for such machines as saw frames, in which the irregularity in speed attains 0.1.

For high-frequency engines, such as turbogenerators and electromotors with small irregularity in speed, the employment of dynamic dampers may be effective because of the low value of ϵ and because usually $\beta < 1$ for their foundations.

By introducing damping into the system of the damper, it is possible to increase the difference $\omega_{av} - f_{n2}$ and decrease the amplitude of foundation vibrations when ω approaches f_{n2}. However, these effects will take place only for some optimum values of damping. In order to achieve in practice these optimum values, the damper should be thoroughly tuned up. The maintenance of constant damping is especially difficult under working conditions. Temperature, moisture, and contamination may affect a damper's natural period and its damping, thus upsetting its tuning.

When foundations undergo vibrations close to those of pure sliding shear, all the above interrelationships will be valid, except that f_{nx} should be inserted everywhere in the equations instead of f_{nz}. They are valid also when foundation vibrations are close to being pure rotational vibrations around an axis passing through the centroid of the foundation area in contact with soil. In this case, μ designates the ratio between the

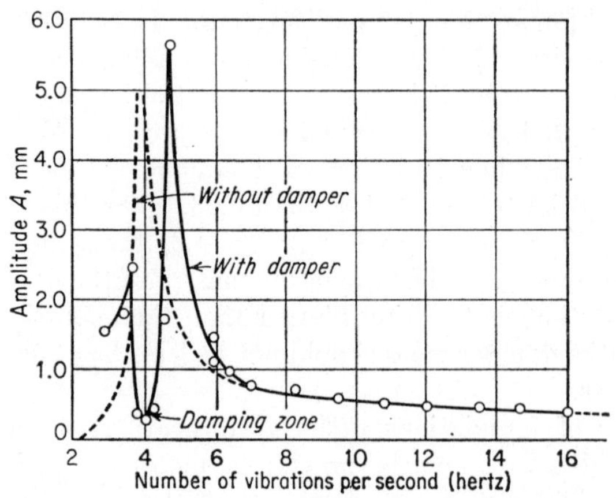

FIG. IV-13. Experimentally determined shift of resonance peak as a result of the use of a vibration damper.

moment of inertia of the damper masses with respect to the axis passing through the center of gravity of the foundation perpendicular to the plane of vibrations and the moment of inertia of both the foundation and engine mass with respect to the same axis.

Let us note in conclusion that the author and his associates investigated experimentally the effects of dampers on model foundations. These experiments confirmed the fundamental theoretical conclusions. For example, the experiments verified that after the damper is installed, one of the resonances of the newly formed "foundation-damper" system (Fig. IV-13) appears close to the operational frequency of the engine. This resonance is dangerous even at negligible changes in frequency of engine rotation.

IV-6. Analysis and Design of Foundations with Vibration Absorbers

In some cases much lower than usual permissible amplitude values of machine foundation vibrations are necessary. It is very difficult to decrease these amplitudes by means of proper selection of the mass or the foundation contact area or by increasing the rigidity of the base.

However, the amplitudes of foundation vibrations under reciprocating engines may be considerably decreased by means of special spring absorbers.

These absorbers are comparatively inexpensive, reliable in operation, and effective in decreasing the amplitudes of forced vibrations of foundations. Absorbers considerably decrease vibrations produced not only by the main (first) harmonics, but also by higher harmonics of exciting loads, as well as by loads developing as a result of various factors not taken into account by design computations. Therefore spring absorbers are sometimes used to decrease vibrations of machine foundations having unbalanced second harmonics. This is done in order to eliminate completely the transmission of the inevitable vibrations to adjacent structures and especially to equipment and precision measuring instruments. Human beings feel vibrations of even very small amplitudes. Sometimes small vibrations interfere with the work of precise devices or are the reason for undesirable distortions in various technological processes (for example, in the operation of precision devices, in molding, etc.).

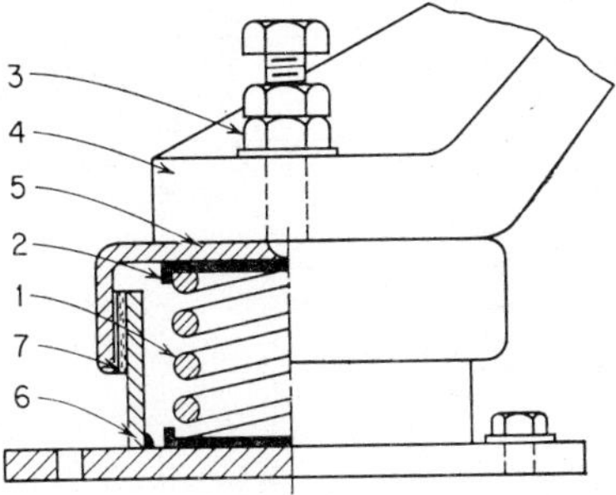

FIG. IV-14. Small one-spring vibration absorber.

There are various types of absorbers employed, depending on the type of machine to which they will be attached, on the static load transmitted to the absorber, and on special requirements in regard to assembling and adjusting.

Figure IV-14 shows a sketch of a small one-spring absorber used for small engines producing no considerable unbalanced exciting loads. This absorber consists of a coil spring 1 which fits into the adjusting slab 2. The regulating bolt 3 rests on this slab. The frame 4 is placed on the lid 5 of the absorber. The position of the frame is adjusted by turning the regulating bolt 3. To eliminate harmful external effects on the spring, the latter is enclosed in the housing 6, having insulating pads 7 (made of rubber or cork) which protect the spring from water and dirt.

Such light absorbers are used for vibroisolation of small diesel engines, ventilation units, presses, pumps, and other machines.

For the vibroisolation of reciprocating engines of medium and high capacities, absorbers containing several springs are used. An absorber of this type with four springs is shown schematically in Fig. IV-15. The housing and parts of the absorber are made of steel plates and several other metals.

The main characteristics of the springs (i.e., the diameters and number of coils) are selected according to the results of dynamic computations.

In addition to spring absorbers, rubber absorbers may also be employed for vibroisolation of light engines and devices. In comparison with spring absorbers, rubber absorbers are simpler and less expensive. Besides, they are characterized by a larger coefficient of resistance to vibrations, useful when they are used for vibroisolating machines of

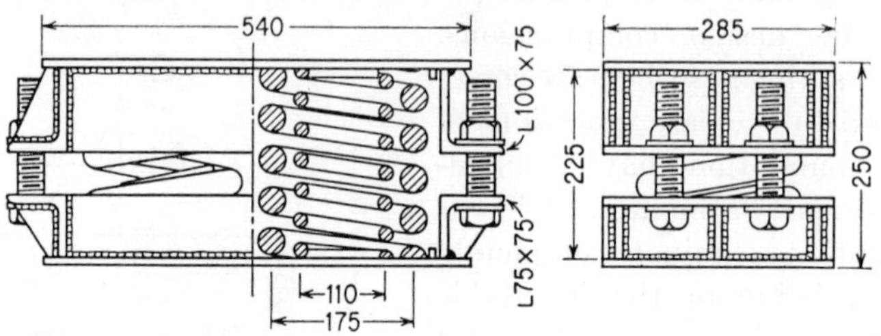

Fig. IV-15. Four-spring vibration absorber.

irregular performance. A disadvantage of rubber absorbers is the variation in their modulus of elasticity, which depends on the load. Computations related to vibroisolation always involve relatively large errors if rubber is used.

Depending on the balance of the engine and its operational speed, different arrangements may be used for the vibroisolation of foundations by spring absorbers. Fundamentally these arrangements can be reduced to two types: supporting and suspension springs.

When designing the vibroisolation of a foundation for a high-speed engine (more than 300 rpm) which is relatively well balanced, i.e., no first harmonics of exciting loads are present, it is not necessary to provide a heavy foundation above the springs. It may be designed as a reinforced-concrete slab of comparatively small thickness. In this case, a "supported" type of vibroisolation is employed, in which the absorbers are installed directly under the mass above the springs. Figures IV-16 and IV-17 illustrate vibroisolations of this type. Figure IV-16 shows the arrangement of absorbers employed for the vibroisolation of a six-cylinder diesel engine operating on one shaft with a generator. Here the absorbers are installed directly under a metal frame made of rolled steel shapes, used instead of the cast-iron frame of the diesel and generator. The mass above the springs consists here of the mass of the motor with the generator and the supporting frame; no portion of the foundation lies above the springs. This arrangement of absorbers can be used only for high-speed engines. The supporting frame should be very rigid to avoid the harmful effects of its deformations on the connector of the shaft.

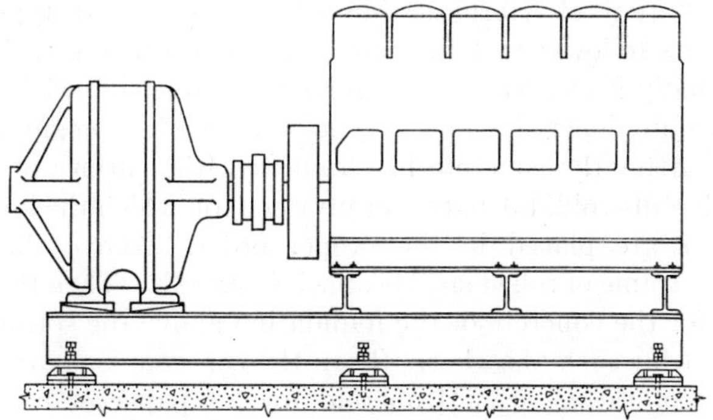

FIG. IV-16. Supporting-spring type of vibroisolation of a six-cylinder diesel with generator on the same shaft.

Figure IV-17 shows the "supported" vibroisolation of a high-speed two-cylinder diesel engine having unbalanced first harmonics of exciting loads.

In this case, the mass above the springs was increased by means of a special thick reinforced-concrete slab under which the absorbers were placed. They rest on supporting slabs which also support rolled steel beams embedded in the lower part of the foundation above the springs.

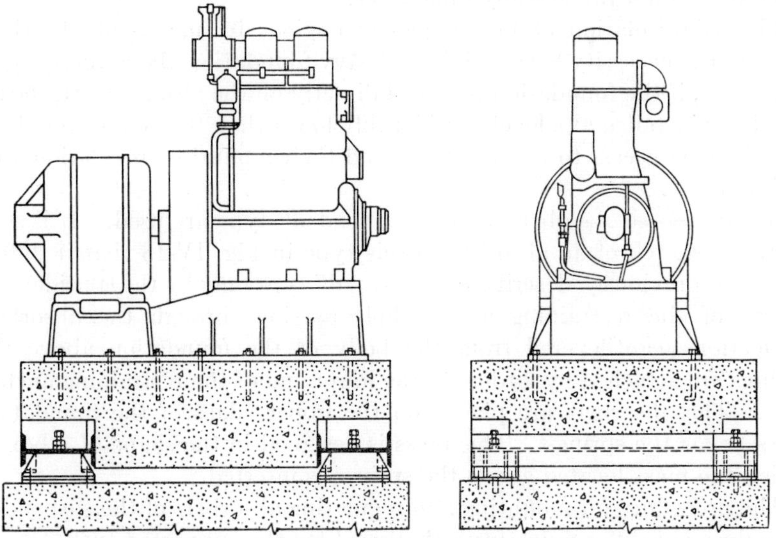

FIG. IV-17. Vibroisolation of high-speed two-cylinder diesel with generator on the same shaft. System has unbalanced first harmonics of exciting loads.

A rigid coupling between the absorbers and the beams is formed by bolting a beam to the cover of each absorber. Special cavities are left in the concrete above the springs to permit access to the absorbers.

The installation of vibroisolation of the supported type proceeds successively as follows: first the foundation beneath the springs is concreted. Usually it consists of a reinforced-concrete slab with a thickness of 0.20 to 1.00 m, depending on the type and size of the engine and on soil properties. After the concrete has hardened, the surface of the slab is covered with Ruberoid, tar paper, or plywood, on which the lower slabs of the absorbers are placed in the proper order. Above these slabs a prefabricated frame of rolled steel beams is installed. Then the formwork is prepared for the concrete of the foundation above the springs; cavities should be left for each absorber. Then the concrete is poured. Due to the presence of the layer of Ruberoid (or tar paper or plywood), the concrete of the upper part of the foundation will not bind to the concrete beneath the springs.

After the concrete of the upper part of the foundation has hardened, the absorbers are mounted. The springs are placed on the lower slabs of the absorbers and are covered by the upper supporting slabs, which are bolted to girders. Finally, the restraining anchor bolt is installed to permit lifting of the mass above the springs. The lifting is carefully regulated by means of a level. If the absorbers are installed correctly, the lifting and regulation of the mass above the springs do not take much time and do not involve any difficulties.

The vibroisolation of low-frequency engines by means of absorbers leads to the necessity for providing a heavy foundation above the springs. Therefore, if the foundation is placed directly on the absorbers, the latter should be installed at a level considerably lower than the floor of the shop. This hinders access to the absorbers and their mounting, regulation, and maintenance.

In such cases, absorbers of the "suspended" type are used. It is seen from the sketch of an absorber of this type in Fig. IV-18 that it differs from the previously described "supported" type only by the considerable length of the restraining anchor bolt passing through the absorber. Projections cantilevered from the body of the foundation above the springs are attached by girders to the lower end of the restraining anchor bolt. The absorbers are placed on the upper edges of the foundation mass below the springs. This mass is designed in the shape of a box in which the mass located above the springs is inserted.

The procedure for mounting and regulating absorbers of the suspended type does not differ much from that used for the supported type.

Generally the dynamic computations of a foundation with absorbers are reduced to an investigation of vibrations of a system having up to

12 degrees of freedom. However, since absorbers are mostly used for the vibroisolation of engines with vertical cylinders, the analysis may in many cases be limited to an investigation of vertical vibrations only. Then the problem of foundation vibrations is reduced to an investigation of a system with 2 degrees of freedom.

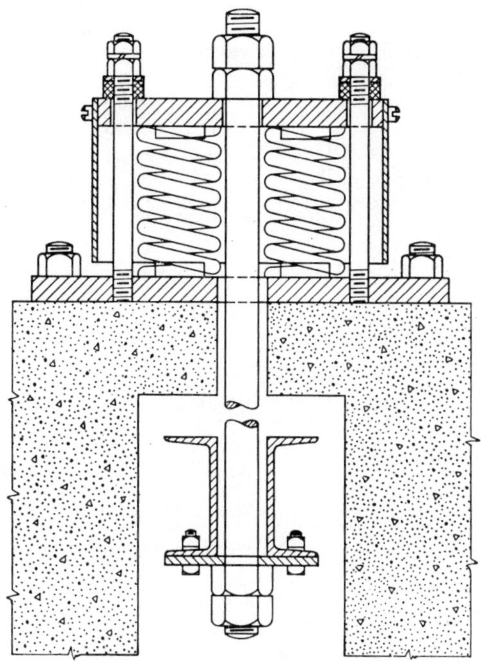

FIG. IV-18. Suspended-type absorber.

Let us assume that the masses of the foundation above and beneath the springs are concentrated in their centers of gravity, located on the same vertical line. Let us further assume that an exciting force $P(t) \sin \omega t$ acts on the mass m_2 above the springs (Fig. IV-19). The differential equations of forced vertical vibrations of the system under consideration will be as follows:

$$\begin{aligned} m_1\ddot{z}_1 + c_1z_1 - c_2(z_2 - z_1) &= 0 \\ m_2\ddot{z}_2 + c_2(z_2 - z_1) &= P(t) \sin \omega t \end{aligned} \qquad \text{(IV-6-1)}$$

where $P(t)$ = magnitude of exciting force

ω = frequency of exciting force

z_1, z_2 = vertical displacements of centers of gravity of masses below and above springs m_1, m_2

c_1 = coefficient of elastic rigidity of base under foundation beneath springs

$$c_1 = c_uA \qquad \text{(IV-6-2)}$$

c_u = coefficient of elastic uniform compression of base
A = area of foundation beneath springs, in contact with soil
c_2 = total coefficient of rigidity of all springs

$$c_2 = \frac{n_1 n_2 d^4}{8nD^3} G \qquad \text{(IV-6-3)}$$

n_1 = number of springs in each absorber
n_2 = number of absorbers
n = number of coils in each spring
d = diameter of spring
D = diameter of coil
G = modulus of elasticity of material of springs

Each spring should be designed so that stresses developed therein under the action of static and dynamic loads will not exceed a permissible value.

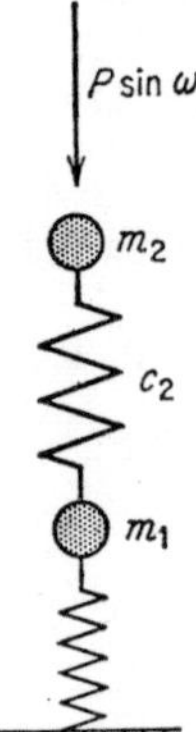

FIG. IV-19. Derivation of Eq. (IV-6-1) concerning the vibration of foundation masses above and below absorber springs.

Limiting our discussion to forced vibrations only, we take the solution of the system (IV-6-1) in the form

$$z_1 = A_1 \sin \omega t \qquad z_2 = A_2 \sin \omega t$$

where the amplitudes A_1 and A_2 of forced vibrations of the foundation beneath and above the springs are

$$A_1 = \frac{f_{nl}{}^2}{m_1 \Delta(\omega^2)} P(t) \qquad \text{(IV-6-4)}$$

$$A_2 = \frac{(1 + \mu) f_{nlz}{}^2 + \mu f_{nl}{}^2 - \omega^2}{m_2 \Delta(\omega^2)} P(t) \qquad \text{(IV-6-5)}$$

where f_{nl} is the limiting frequency of natural vibrations of the foundation above the springs, computed on the basis of the assumption that the foundation beneath the spring is infinitely large; the value of f_{nl} is determined by the equation

$$f_{nl}{}^2 = \frac{c_2}{m_2} \qquad \text{(IV-6-6)}$$

f_{nlz} is the limiting frequency of natural vibrations of the complete system when it is assumed that no absorbers are used:

$$f_{nlz}{}^2 = \frac{c_1}{m_1 + m_2} \qquad \text{(IV-6-7)}$$

Finally,

$$\mu = \frac{m_2}{m_1}$$

The coefficient $\Delta(\omega^2)$ is determined by the expression

$$\Delta(\omega^2) = \omega^4 - (1 + \mu)(f_{nl}^2 + f_{nlz}^2)\omega^2 + (1 + \mu)f_{nl}^2 f_{nlz}^2 \quad \text{(IV-6-8)}$$

Returning to Eq. (IV-6-4), let us investigate the dependence of the amplitude of forced vibrations of the foundation beneath the springs on f_{nl}, which is proportional to the rigidity c_2 of the absorbers. The value of the exciting force is proportional to the square of the frequency of engine rotation; therefore

$$P(t) = \gamma\omega^2$$

where γ is a coefficient which depends on parameters of the engine.

Substituting this expression for P into the right-hand part of Eq. (IV-6-4) and dividing the numerator and denominator by ω^2, we obtain the following expression for the vibration amplitude of the foundation beneath the springs:

$$A_1 = \frac{\gamma}{m_1} \frac{\xi_l^2}{1 - (1 + \mu)(\xi_l^2 + \xi_{lz}^2 - \xi_l^2\xi_{lz}^2)} \quad \text{(IV-6-9)}$$

where $\xi_l = \dfrac{f_{nl}}{\omega} \qquad \xi_{lz} = \dfrac{f_{nlz}}{\omega}$

If no absorbers are used and the upper and lower parts of the foundation are rigidly connected, then according to Eq. (III-1-13) the amplitude of vertical forced vibrations will equal

$$A_z = \frac{\gamma}{m_1} \frac{1}{(1 + \mu)(\xi_{lz}^2 - 1)} \quad \text{(IV-6-10)}$$

The degree of absorption of vibrations will be

$$\eta = \frac{A_z}{A_1} = \frac{1 - (1 + \mu)(\xi_l^2 + \xi_{lz}^2 - \xi_l^2\xi_{lz}^2)}{(1 + \mu)(\xi_{lz}^2 - 1)\xi_l^2} \quad \text{(IV-6-11)}$$

Let us investigate the effect of changes in the value of ξ_l on the value of η. We assume that f_{nl} is very small in comparison with ω; i.e., the value ξ_l is also very small. It follows directly from Eq. (IV-6-11) that

$$\text{If } \xi_l \to 0, \text{ then } \eta \to \infty$$

hence it follows that if the natural frequency of foundation vibrations above the springs is small in comparison with the frequency of engine rotation, then the amplitude A_1 of the foundation with absorbers is small in comparison with the amplitude of vibrations of the same foundation without absorbers.

Suppose $\xi_l \to \infty$; this corresponds to a very large value of f_{nl}. One can see from Eq. (IV-6-11) that in this case $\eta \to 1$; i.e., absorbers will not have any influence on the amplitudes of foundation vibrations.

Figure IV-20 gives a graph of changes in η depending on changes in ξ_l. It is evident that the absorbers cause a decrease in the amplitudes of vibrations only when $\eta > 1$. It is seen from the graph that the zone of usefulness of absorbers is limited to a very narrow range of values of ξ_l lying between

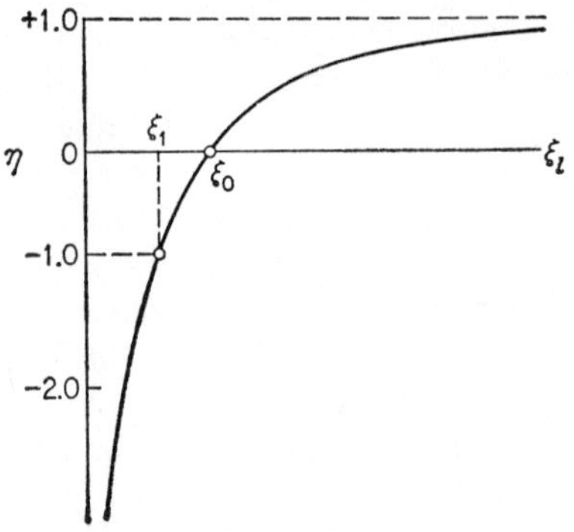

FIG. IV-20. Diagram illustrating Eq. (IV-6-12) and the limits of usefulness of vibration absorbers.

$$\xi_l = 0$$

$$\text{and} \quad \xi_l = \xi_1 = \sqrt{\frac{(1+\mu)\xi_{lz}^2 - 1}{2(1+\mu)(\xi_{lz}^2 - 1)}} \tag{IV-6-12}$$

When $\xi_l > \xi_1$, the following inequality exists:

$$|\eta| < 1$$

Consequently, the use of absorbers does not bring any advantage, but on the contrary is harmful, because when $|\eta| < 1$ the amplitude of vibrations of the foundation with absorbers is larger than that of the same foundation without absorbers.

The foregoing theory of vibrations of foundations with absorbers leads to the conclusion that in order for absorbers to have a favorable effect on the amplitudes of foundation vibrations, the following condition is necessary: the frequency of natural vibrations of the mass above the springs should be as small as possible in comparison with the frequency of engine rotation. A decrease in the frequency of natural vibrations of the foundation above the springs may be achieved by the use of absorbers of a suitable stiffness and by an increase in the mass above the springs. For high-frequency engines the required relationship between ω and f_{nl} can be easily achieved without a considerable increase of the weight of the foundation above the springs. For low-frequency engines the relationship is usually difficult to achieve by just decreasing the rigidity of the absorber because, due to strength requirements, this decrease cannot extend below a certain limit determined by the strength of the springs. In such cases a decrease in f_{nl} is achieved by providing a massive foundation above the springs.

If the degree of absorption of foundation vibrations η is specified, then from Eq. (IV-6-11) we obtain for ξ_l:

$$\xi_l^2 = \frac{1 - (1+\mu)\xi_{lz}^2}{(1+\mu)(\eta - 1)(\xi_{lz}^2 - 1)} \tag{IV-6-13}$$

Example. Design computations for a foundation with absorbers under a vertical compressor

1. DATA. A foundation for a 120-kw vertical compressor is to be designed. Foundation vibrations are objectionable, since a precision apparatus is adjacent to the foundation. The base of the foundation is formed by a soil characterized by a coefficient of elastic uniform compression c_u of 2×10^3 tons/m³.

In order to avoid harmful influences of the foundation on the apparatus, the amplitude of foundation vibrations should not exceed 0.03 mm.

The main exciting force imposed by the compressor is the vertical exciting force $P_z = 2.6$ tons; the compressor speed is 480 rpm.

2. COMPUTATIONS. In order to have an amplitude of foundation vibrations under the compressor smaller than 0.03 mm, the foundation should be very heavy and should have a large area in contact with the soil.

For example, assuming that the ratio between the frequency of natural vertical vibrations of the foundation and the engine frequency equals 2, for $A_z = 0.03 \times 10^{-3}$ m and $P_z = 2.6$ tons, we obtain from Eq. (III-1-13) the following foundation weight:

$$W = \frac{2.6 \times 9.81}{0.03 \times 10^{-3} \times 3 \times 2.5 \times 10^3} = 115 \text{ tons}$$

Then the foundation area in contact with soil should equal

$$A = \frac{f_{nz}^2 W}{c_u g} = \frac{4 \times 2.5 \times 10^3 \times 115}{2 \times 10^3 \times 9.81} = 57.5 \text{ m}^2$$

For such a low-power engine as the compressor under consideration it would be unreasonable to construct a foundation with weight and dimensions as large as those obtained above. In order to meet the requirements in regard to the amplitude of foundation vibrations, it is better to use spring absorbers. The selected dimensions of the foundation with absorbers are shown in Fig. IV-21.

The data for the computations are as follows: the foundation area in contact with soil $A = 12.5$ m²; the weight of the foundation beneath the springs is 21.0 tons; the weight of the foundation above the springs (together with the engine) is 35.0 tons. The coefficient of rigidity of the base is

$$c_1 = c_u A = 2 \times 10^3 \times 12.5 = 25.0 \times 10^3 \text{ tons/m}$$

The mass of the foundation beneath the springs is

$$m_1 = 21.0/9.81 = 2.15 \text{ tons} \times \text{sec}^2/\text{m}$$

The mass of the foundation above the springs is

$$m_2 = 35.0/9.81 = 3.56 \text{ tons} \times \text{sec}^2/\text{m}$$

The limit natural frequency of vertical vibrations of the whole system (assuming that no absorbers are used) equals

$$f_{nz}^2 = \frac{c_1}{m_1 + m_2} = \frac{25.0 \times 10^3}{2.15 + 3.56} = 4.38 \times 10^3 \text{ sec}^{-2}$$

The coefficient ξ_{lz} is computed to be

$$\xi_{lz} = \frac{f_{nz}^2}{2} = \frac{4.38 \times 10^3}{2.5 \times 10^3} = 1.75$$

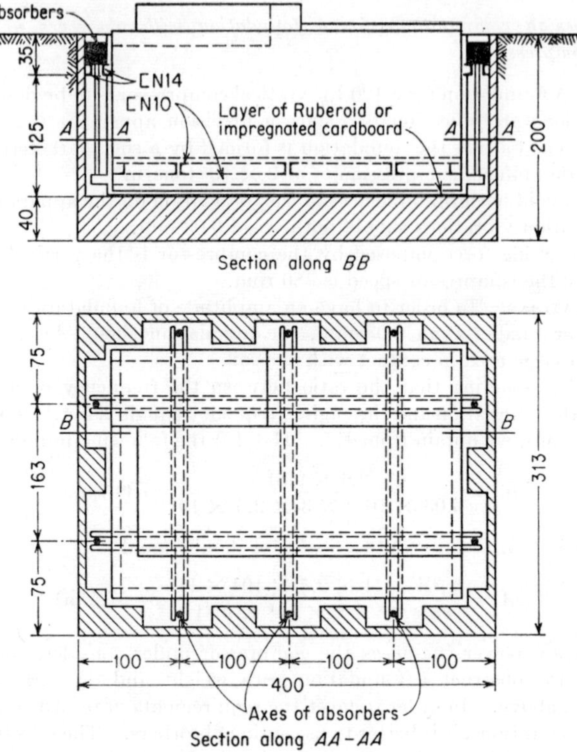

FIG. IV-21. Example of design computations for foundation with absorbers.

The ratio between values of the masses is

$$\mu = \frac{m_2}{m_1} = \frac{3.56}{2.15} = 1.65$$

If no absorbers were present, then for the selected dimensions of the foundation the amplitude of vertical vibrations would equal

$$A_z = \frac{2.6}{(2.15 + 3.56)(4.38 - 2.5)10^3} = 0.25 \times 10^{-3} = 0.25 \text{ mm}$$

In order that the permissible amplitude of vibrations of the foundation with absorbers is not exceeded, the degree of absorption of vibrations should equal

$$\eta = \frac{A_z}{A_1} = \frac{0.25}{0.03} = 8.4$$

Let us assume that the design value of η equals -10. From Eq. (IV-6-13) we determine the required value of the coefficient ξ_l:

$$\xi_l^2 = \frac{1 - (1 + 1.65)1.75}{(1 + 1.65)(-10 - 1)(1.75 - 1)} = 0.165$$

We determine the required value of the limiting frequency f_{nlz} of natural vertical vibrations of the mass of the foundation above the springs:

$$f_{nlz}^2 = \xi_l^2\omega^2 = 0.165 \times 2.5 \times 10^3 = 414 \text{ sec}^{-2}$$

The required total rigidity of all the absorbers will be

$$c_2 = f_{nlz}^2 m_2 = 414 \times 3.56 = 1{,}480 \text{ tons/m}$$

If c_{sp} is the rigidity of one spring, then

$$c_{sp} = \frac{c_2}{n_1 n_2}$$

where n_1 = number of springs in each absorber
n_2 = number of absorbers

Assume $n_1 = 2$ and $n_2 = 8$; then the required rigidity of one spring will be

$$c_{sp} = \frac{1{,}480}{2 \times 8} = 92.0 \text{ tons/m}$$

On the other hand, using Eq. (IV-6-3), we obtain

$$c_{sp} = \frac{G}{8}\frac{d^4}{D^3}\frac{1}{n} \tag{IV-6-14}$$

where G is the modulus of elasticity in shear of the spring material; its value may be assumed to be 7.5×10^6 tons/m².

We assume there are five coils in a spring. Substituting values of n and G into formula (IV-6-14) we obtain

$$92.0 = \frac{7.5 \times 10^6}{8}\frac{d^4}{D^3}\frac{1}{5} = 1.88 \times 10^5 \frac{d^4}{D^3}$$

Hence

$$\frac{d^4}{D^3} = \frac{92.0}{1.88 \times 10^5} = 4.9 \times 10^{-4}$$

Let us assume the diameter of the spring $d = 2.5 \times 10^{-2}$ m; then

$$D^3 = \frac{d^4}{4.9 \times 10^{-4}} = \frac{39.0 \times 10^{-8}}{4.9 \times 10^{-4}} = 8.0 \times 10^{-4} \text{ m}^3$$

$$D = 9.3 \text{ cm}$$

3. Stress Analysis of the Spring. The permissible load on the spring equals

$$P_p = \frac{\pi d^3 R_p}{8D} \tag{IV-6-15}$$

where R_p is the permissible torsional stress for the spring material; we assume its value is 40×10^3 tons/m².

Assuming in accordance with the foregoing computations that $d = 2.5 \times 10^{-2}$ m and $D = 9.3 \times 10^{-2}$ m, we obtain for the permissible load on the spring

$$P_p = \frac{3.14 \times 15.6 \times 10^{-6} \times 40 \times 10^3}{8 \times 9.3 \times 10^{-2}} = 2.64 \text{ tons}$$

In order to find the actual load on each spring, it is necessary to determine the amplitude of forced vibrations of the foundation above the spring. From Eq. (IV-6-8), we determine the value of $\Delta(\omega^2)$:

$$\Delta(\omega^2) = 6.25 \times 10^6 - (1 + 1.65)(0.414 \times 10^3 + 4.38 \times 10^3) \times 2.5 \times 10^3 + (1 + 1.65) \times 0.414 \times 10^3 \times 4.38 \times 10^3 = 21.0 \times 10^6$$

According to (IV-6-5),

$$A_2 = \frac{(1 + 1.65) \times 4.38 \times 10^3 + 1.65 \times 0.414 \times 10^3 - 2.5 \times 10^3}{3.56 \times 21.0 \times 10^6} 2.64 = 0.34 \times 10^{-3} \text{ m}$$

The dynamic force imposed on the springs as a result of vibrations is

$$W_2 = 0.34 \times 10^{-3} \times 1.48 \times 10^3 \times 3.56 = 0.5 \text{ tons}$$

Thus, the actual load on each spring equals

$$P_{act} = \frac{0.5 + 35.0}{16} = 2.22 \text{ tons}$$

which is smaller than the permissible load.

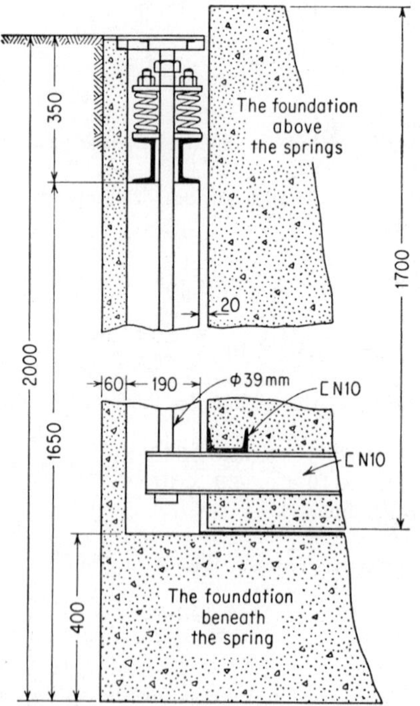

FIG. IV-22. Detail of absorbers with suspension system.

A schematic diagram of the main part of the arrangement of the foundation with suspended absorbers is shown in Fig. IV-22.

The construction of a foundation with absorbers proceeds analogously to the procedure described in Art. V-7.

V

FOUNDATIONS FOR MACHINES PRODUCING IMPACT LOADS

V-1. General Directives for the Design of Foundations for Forge Hammers

a. Classification of Forge Hammers. Forge hammers are divided into two groups: drop hammers for die stamping and forge hammers proper.

The side frame of a drop hammer is mounted on an anvil (Fig. V-1), thus giving rigidity to the system. The side frame, together with guides for the ram, contributes to the precision of blows required in forging. This peculiarity in the design of drop hammers predetermines to a certain degree the design of their foundations, since the foundation block under the anvil serves as a support for the whole hammer.

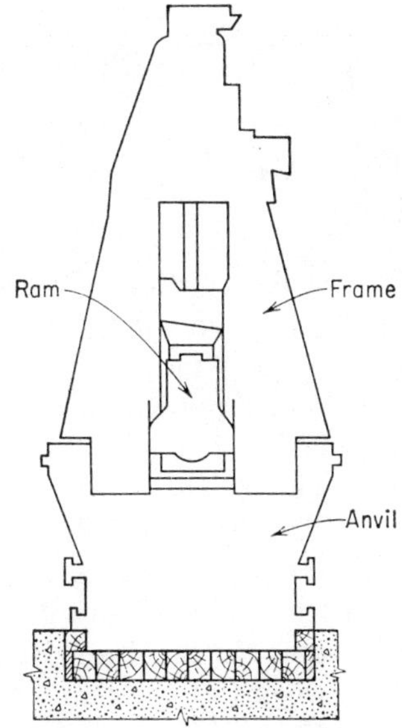

FIG. V-1. Drop hammer with frame mounted on anvil.

Free forging operations are usually performed by forge hammers proper. The anvil and the side frame, as a rule, are mounted separately. Forge hammers are built as single-support frames (Fig. V-2) and as double-support frames (Fig. V-3). The latter can be of the arch or bridge types. Pneumatic hammers with air compressors are single-frame hammers.

b. Design Data for a Hammer Foundation. The following technological data are required for the design of a hammer foundation:

1. The nominal weight of dropping parts, which usually characterizes the power of the hammer. In drop hammers the real weight of dropping parts, in addition to the weight of the ram, piston, and rod, includes also the upper half of the die. Therefore in these hammers the real weight of dropping parts is greater than the nominal weight or that shown in catalogues.

The design of a hammer foundation is made on the basis of the real weight of dropping parts. The total average height of the upper and

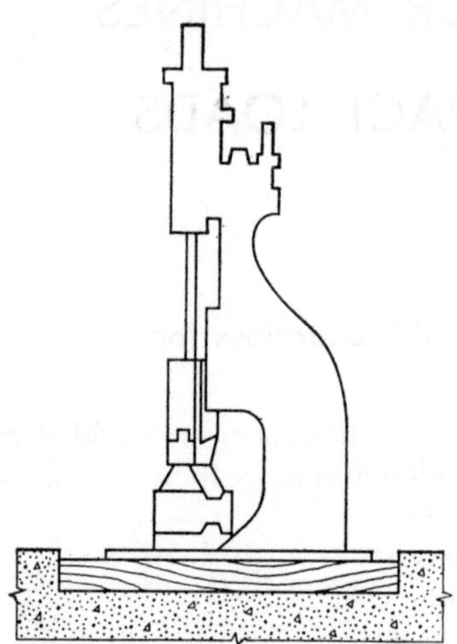

FIG. V-2. Hammer with single support frame.

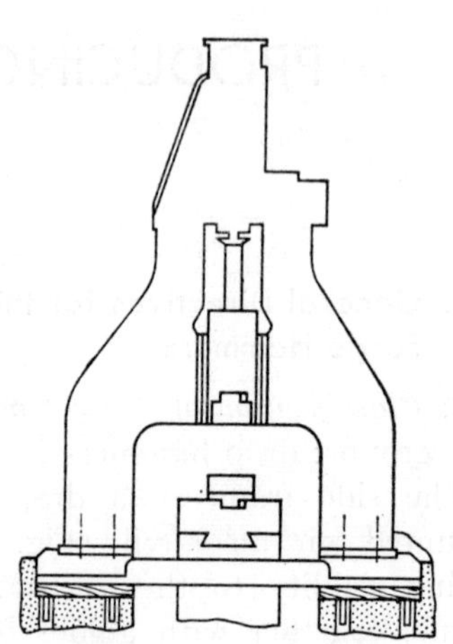

FIG. V-3. Hammer with double support frame.

lower halves of the die lies within the range 250 to 500 mm. It should be noted that some dies for long units (axes, shafts, etc.) are provided with heavy upper parts, whose weight reaches 100 per cent of the nominal weight of dropping parts. Such cases should be mentioned in the technological assignment.

2. The weight of the anvil and the dimensions of its base area.

3. The weight of the hammer, including side frames, ram cylinders with the anchor plate, etc., but excluding the anvil weight. When no data are available on the weight of the anvil and frames, it is permissible to assume the anvil weight to be 15 to 20 times the weight of the ram; the total weight of the hammer and anvil is taken to be 25 to 30 times the weight of the ram.

4. The maximum height of the ram drop (or the maximum piston stroke).

5. The upper piston area.

6. The average working pressure on the piston.

For the design and analysis of the foundation, either the machine assembly drawing or the following data should also be made available:

7. Dimensions in plan, the thickness and elevation of the top of the pad under the frame and anvil of the forge hammer.

8. The position, diameter, and length of anchor bolts.

9. The elevation of the anvil base with respect to the floor level of the shop.

10. Dimensions in plan and the thickness of the wooden pad under the anvil.

11. The location of the hammer in the shop with respect to adjacent foundations under engines, machinery, and supporting structures of the building; the dimensions, elevations, and depths of these foundations.

c. Material of the Foundation and Pad under the Anvil. Foundations under hammers with a weight of dropping parts in excess of 1 ton are made of concrete type 150,† with coarse aggregate of hard rocks with a compressive strength not less than 250 kg/cm^2. Normal portland cement, of a type not below 300, is used for concrete. The latter is reinforced according to design data or according to instructions given on the job.

The pad under the anvil is usually made of oak. Experience in the operation of hammers under war conditions showed that for hammers with a weight of dropping parts up to 2 tons, pine and larch timber may be used as material for under-anvil pads. Timbers of the best quality, having a moisture content below 15 to 18 per cent, should be used.

d. Directives for Design. Until recently there was a tendency to design foundations under hammers as massive blocks embedded to a considerable depth in the soil.[44] The purpose was to provide such dimensions of the foundation that its static elastic settlement would be larger than the amplitude of its vertical vibrations. Since design values of amplitudes of foundation vibrations under hammers were selected within the range 2.0 to 2.5 mm, the height of the foundation had to be increased considerably to obtain the desired static settlement. Figure V-4*a* shows a typical design of a foundation under a hammer, as used until recently. Here for each 1 ton of dropping parts weight, there correspond 80 to 100 and often 120 tons of foundation weight.

The discussion which follows will show that the use of such heavy foundations is not necessary, especially since they involve a considerable increase in the cost of construction.

Foundations under hammers should be designed as blocks or slabs loaded from above by backfill. Figure V-4*b* is a typical design of such a block foundation. The ratio between the weight of the foundation and that of the dropping parts is about 40.

† See footnote, Art. IV-1-c, p. 132.

In block-type foundations the thickness of the part below the anvil should be selected as follows: for a ram weight up to 0.75 tons, the thickness should be not less than 0.75 m; for a weight of 0.75 to 2.5 tons, the thickness should be 1.5 m; for a weight of 2.5 tons and more, the thickness must be 1.25 to 2.50 m, depending on the power of the hammer.

Previously, foundations under forge hammers were built under the anvils separated from the footings under the frames. This decreased the stresses in the hammer frames during eccentric forging. However, the separation of the foundation elements results in their tilting with respect to each other and in considerable nonuniform settlement of the foundation under the anvil. In recent years foundations under forge hammers have been designed by the method shown in Fig. V-4*c*; i.e., the footings

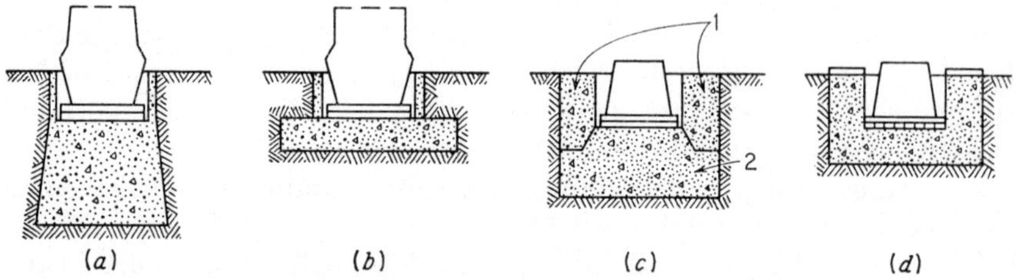

FIG. V-4. Types of hammer foundations: (*a*) deep block foundation; (*b*) slab block foundation weighted by backfill; (*c*) anvil block foundation 2 supports frame 1; (*d*) single block foundation for anvil and frame.

under the frame were not placed directly on the soil, but on the block under the anvil, and boards 2 to 3 cm thick (or several layers of Ruberoid) were placed between these two footings. Thus tilting of the anvil with respect to the frame was prevented. However, due to insufficient elasticity of the pad, this foundation design was not very effective in decreasing the stresses which developed as a result of nonuniform forging. These stresses may be decreased much more efficiently by means of spring washers in foundation bolts and by oak timbers installed under the anvil and under the frame of the hammer. In such cases the foundation under the forge hammer should be designed as a single block (Fig. V-4*d*).

Reinforcement is placed as directed on the job. The reinforcement used for the foundation under the anvil consists of 2 to 4 horizontal steel grillages formed by 8- to 12-mm bars spaced at 10 to 20 cm; the upper grillage is placed at a distance of 2 to 3 cm from the foundation surface.

Near the foundation surface in contact with soil, the reinforcement consists of 1 or 2 horizontal grillages formed by 12- to 20-mm bars and spaced 15 to 30 cm apart. Distances between the grillages are 10 to 15 cm in the part of the foundation under the anvil and 15 to 30 cm near the foundation contact surface.

The number of grillages is determined by the power of the hammer. It should be kept in mind that double-acting hammers belong to the group of heaviest hammers with respect to their impact effect on the foundation.

Pads under the anvil are made of square timbers from 10 by 10 cm to 20 by 20 cm in cross section. Timbers are laid flat in one or several rows, one over the other. Each row is braced by transverse bolts every 0.5 to 1.0 m and forms a separate mat.

If several rows of timbers are employed, then in order to decrease wear and tear and make the pad more rigid, the timbers are placed in the form of grillages. The upper row of timbers is laid along the short side of the anvil base. The mats must be strictly horizontal, smoothly planed, and easily fitted into the excavation. They should be checked by means of a water level. To prevent decay resulting from moisture, it is advisable to impregnate timbers with wood-preserving solutions.

TABLE V-1. THICKNESS OF TIMBER PADS UNDER THE ANVIL

No.	Type of hammer	Thickness of pad, m, if weight of dropping parts of hammer is:		
		Up to 1 ton	1–3 tons	Over 3 tons
1	Double-action drop hammer	Up to 0.20	0.20–0.60	0.60–1.20
2	Single-action drop hammer	Up to 0.10	0.10–0.40	0.40–0.90
3	Forge hammer	Up to 0.20	0.20–0.60	0.60–1.00

The space between the pad and the excavation walls may be filled with petroleum asphalt. In order to prevent a horizontal displacement of the anvil along the pad, four timbers are placed around it near the base.

The pad thickness is shown on the assembly drawings of the hammer or indicated in the technical assignment. Tentative values of pad thickness under the anvil are given in Table V-1.

The thickness of the pad should be selected so that the stresses therein do not exceed permissible values, which are as follows:

Oak: 300 to 350 kg/cm^2
Pine: 200 to 250 kg/cm^2
Larch: 150 to 200 kg/cm^2

e. Remarks on Construction Procedures. Concrete for the footings should be placed using vibrators. In the presence of ground water containing chemicals which may produce deterioration of concrete, pozzolan cement should be used or special measures should be provided to protect concrete from the action of water; the velocity of water flow and possible fluctuations of ground-water level should be taken into account.

In the process of foundation construction, special care should be taken to provide accurate location of holes for anchor bolts (if the latter are foreseen by the design) and the excavation for the anvil or frame.

The lower part of the excavation for the anvil should be strictly horizontal; no additional corrective pouring of concrete is permitted. If supplementary cement grout has to be placed under the frame of the forge hammer, then the foundation surface in contact with the cement grout should be roughened, cleaned, and washed. The underlying soil should be compacted by tamping in of broken stone. In moist soils a working mat of concrete type 50† is placed under the foundation.

Concrete should be placed in horizontal layers and, as a rule, without interruption in the work. In case of an emergency interruption, the following measures should be taken to secure the monolithic character of the foundation:

1. Dowels of 12 to 16 mm diameter should be embedded on both sides of the joint to a depth of 30 cm at a distance of 60 cm from each other.
2. Prior to placing a new layer of concrete, the previously laid surface should be roughened, thoroughly cleaned, washed by a jet of water, and covered by a layer of a rich 1:2 cement grout, 20 mm thick. The placement of concrete should be started not later than 2 hr after this mixture is laid on the surface.

The anvil may be mounted on the foundation only after hardening of the concrete, i.e., not less than a week after its placement. The foundation may be put in operation as soon as the concrete attains the design strength value.

Vibrations of the forge-hammer anvil may result in some soil falling into the space between the anvil base and the upper row of the timber pad. This may lead to tilting of the anvil or damage to the pad. To avoid this, after the anvil is built a protective wall is usually installed around it, extending from the top of the foundation to floor level. Such a protective box permits an easy and rapid inspection of the anvil and pad; it also simplifies the mounting and dismounting of the anvil.

V-2. Initial Conditions of Foundation Motion under Impact Action

a. The Velocity of Dropping Parts at the Beginning of Impact. Large hammers may be divided into two groups: those with an unrestricted drop of the ram, and those with a restricted ram movement.

The first group includes hammers with frictional hoisting of the ram and hammers in which the ram, rigidly tied to the piston, is lifted by

† See footnote, Art. IV-1-c, p. 132.

steam pressure from underneath. In frictional hammers, the ram is connected to a plate which moves between two frictional disks pressed against this plate. When the ram is lifted to the height desired, one of the disks is moved away from the plate and the ram falls, moving along its guides.

In single-acting hammers, the ram, which is rigidly tied to the piston by means of a rod, is lifted by the pressure of steam released through a valve located under the piston and opened when the latter is in its extreme low position. After the piston is raised to the height desired, the access of steam into the cylinder under the piston is stopped, the valve opens, and the steam or compressed air escapes. The piston, together with the ram, drops at increasing speed.

After the access of steam is discontinued and the exhaust valve opens, steam cannot escape at once from the space in the cylinder under the piston. Therefore a counterpressure against the ram drop is created, resulting in a loss both in the ram's velocity and in the kinetic energy of its drop.

The velocity v of the ram drop under the condition of unrestricted motion equals

$$v = \eta \sqrt{2gh} \qquad \text{(V-2-1)}$$

where g = acceleration of gravity
h = height of ram drop
η = coefficient which takes into account counterpressure and frictional forces

The numerical value of η depends on the design of the hammer, its working order, the regulation of valves, etc.

Modern forging practice mostly employs the large double-acting hammers. In these hammers, steam or compressed air acts on the ram not only while it is being lifted, but also during its drop; therefore the velocity and kinetic energy are considerably larger at the moment of impact of the ram against the workpiece.

Changes in steam pressure during the drop of the ram, both under the piston (counterpressure) and above the piston (active pressure), depend on many varying factors such as the proper operation of the valves, the tightness of the piston in the cylinder, and the working order of the stuffing box. It is impossible to take into account all these factors with sufficient accuracy. Therefore in computations of ram velocity, one usually works with the average pressure of steam and air in the feed pipe. Then the velocity of the forced motion of the ram under the action of its own weight and the steady pressure will equal

$$v = \eta \sqrt{\frac{2g(W + Ap)h}{W}} \qquad \text{(V-2-2)}$$

where A = piston area
p = total pressure on piston
W = total weight of dropping parts
η = correction coefficient

b. Experimental Determination of the Correction Coefficient η. The values of the correction coefficient in Eqs. (V-2-1) and (V-2-2) may be determined only empirically by comparing the velocities obtained from the equations with the corresponding values obtained experimentally.

TABLE V-2. RESULTS OF HAMMER EFFICIENCY MEASUREMENTS

Type of hammer	Nominal weight of dropping parts, tons	Velocity at beginning of impact, m/sec		Ratio η between measured and computed velocities
		Measured	Computed from Eq. (V-2-2) or (V-2-1) for $\eta = 1$	
Double-acting hammer	5.4	6.2	9.0	0.69
	3.6	6.0	8.4	0.71
	2.25	5.4	8.6	0.63
	1.8	4.5	8.1	0.56
	1.125	6.2	8.6	0.72
	1.0	6.8	8.5	0.80
	1.0	5.8	9.8	0.59
	0.635	5.5	9.0	0.61
Hammer with unrestricted action	0.54	3.3	3.56	0.96
	1.125	3.5	3.93	0.89

The author performed such measurements for ten drop hammers of different powers and makes. The measurements were made under working conditions in shops, without any special adjustment. Therefore the results of these measurements may be considered to be characteristic for working conditions.

The results of these measurements are presented in Table V-2, which also gives velocities computed from Eqs. (V-2-1) and (V-2-2) for $\eta = 1$.

This table shows that the measured velocity of dropping parts at the moment of impact is much lower in double-acting hammers than the values computed from Eq. (V-2-2). For these hammers, the ratio between the values of measured and computed velocities lies within the range 0.45 to 0.80; the average value of η in Eq. (V-2-2) may be taken to equal 0.65.

In addition, it follows from Table V-2 that the absolute velocity does not depend much on the power of the hammer. This is explained in part by the fact that usually the height of drop of the ram and the steam or air pressure vary for different hammers only within comparatively narrow ranges. Therefore in many cases the design velocity value may be taken to equal approximately the average value of velocities measured in hammers with different powers. This value for double-acting die-stamping hammers equals 6.0 to 6.5 m/sec.

In hammers with unrestricted drop, especially frictional hammers, no counterpressure of steam is encountered, and a decrease in the velocity of dropping parts is mostly caused by friction in the guides. When a hammer is properly adjusted, the effect of friction will be negligible; therefore the correction coefficient in Eq. (V-2-1) will be close to unity. This conclusion is confirmed by data for hammers with unrestricted drop, presented in Table V-2.

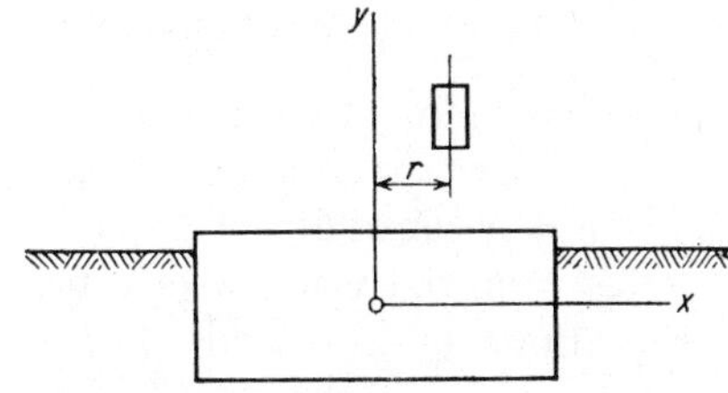

FIG. V-5. Derivation of Eq. (V-2-4).

c. Initial Velocities of Foundation Motion. Let us investigate the values of velocity characteristic for the foundation at the end of a vertical eccentric impact. Such an impact occurs in drop hammers when edge grooves are stamped.

We assume that no pad is present under the anvil and that the anvil and foundation form one single body whose elasticity may be neglected in comparison with the elasticity of the soil. Let us also assume that the foundation can be represented by the body shown in Fig. V-5, that the eccentric impact produced by the falling mass occurs in a plane which we shall consider to be one of the principal planes of the foundation, and that the center of gravity of the foundation and the centroid of its contact area with the soil lie on the same vertical line. Due to the impact, the foundation undergoes vibrations which occur in the aforementioned principal plane.

If the ram and foundation are considered as one closed system, it may be assumed that linear momentum is conserved during the impact. The foundation is motionless before the impact; at that time linear momentum equals the momentum of the ram, i.e., m_0v, where v is the velocity of the falling ram of mass m_0 at the moment it touches the foundation (the beginning of impact).

After impact, i.e., during the period following the instant when the ram detaches itself from the foundation, the momentum of the ram and foundation is

$$m_0v_1 + mv_0$$

where v_1 = velocity at which ram rebounds from foundation
m = foundation mass
v_0 = initial velocity of forward motion of center of mass of foundation

The momentum of the system before the impact equals the momentum after the impact; therefore

$$m_0 v = m_0 v_1 + m v_0 \tag{V-2-3}$$

In addition to progressive downward motion under the action of an eccentric impact, the foundation undergoes a rotational movement around the axis passing through its center of gravity, perpendicular to the plane in which impact occurs. The moment of momentum will be

$$m_0 v r = m_0 v_1 r + I \varphi_0 \tag{V-2-4}$$

where I = moment of inertia of mass of foundation and hammer in regard to axis of rotation
r = eccentricity of impact
φ_0 = initial velocity of rotation of foundation

Equations (V-2-3) and (V-2-4) include three unknown values. In order to derive a third equation, let us use Newton's hypothesis concerning the restitution of impact. According to this hypothesis, if there occurs an impact between two bodies moving in relation to each other, the relative velocity after the impact is proportional to the relative velocity before the impact. The ratio between these two depends only on the material of the bodies which underwent the impact. The foundation was motionless before impact; therefore the relative velocity of the ram equals v. After impact, the absolute velocity of ram motion equals v_1, but the point of the foundation which was subjected to impact acquired a velocity whose vertical component equals $v_0 + r\varphi_0$; it follows that the relative velocity of the ram after the impact equals $v_0 + r\varphi_0 - v_1$. According to Newton's hypothesis,

$$e = \frac{v_0 + r\varphi_0 - v_1}{v} \tag{V-2-5}$$

where e is the coefficient of restitution.

From Eqs. (V-2-3) to (V-2-5) we obtain expressions for initial velocities of the foundation motion:

$$v_0 = (1 + e) \frac{\rho}{(1 + \mu)(r^2 + \rho^2) - r^2} v \tag{V-2-6}$$

$$\varphi_0 = \frac{\mu r (1 + e)}{(1 + \mu)(r^2 + \rho^2) - r^2} v \tag{V-2-7}$$

where $$\rho^2 = \frac{I}{m_0} \qquad \mu = \frac{m}{m_0}$$

If the impact was at the center of the foundation, then $r = 0$ and

$$v_0 = \frac{1 + e}{1 + \mu} v \qquad \varphi_0 = 0 \tag{V-2-8}$$

When the elasticity of the pad cannot be neglected, it should be considered that the initial velocity of motion is acquired only by the anvil (for forge hammers) or by the anvil and frame (for drop hammers). Then Eqs. (V-2-6) and (V-2-7) remain the same, but the symbols m and I denote the mass and the moment of inertia either of the anvil or of both the anvil and the frame, without taking into account the foundation mass.

d. Coefficient of Restitution e. It follows from the foregoing formulas that the initial velocities of motion of the foundation or anvil depend considerably on the coefficient of restitution e. If the impact was perfectly elastic, then $v = v_1$, and consequently $e = 1$. For the impact of a rigid body against a plastic one, $v_1 = 0$, and consequently $e = 0$. For real bodies, the numerical values of e lie within the range $0 < e < 1$.

In forge hammers, e depends on many factors, the most important of which are: the temperature of a forged piece, the dimensions and forms of grooves (in stamping hammers), and the elastic properties of materials of the ram, head, and anvil.

Since the design values of the amplitude of hammer foundation vibrations depend on the selected values of the coefficient of restitution e, the designer of a foundation naturally has a practical interest in knowing its real values. However, the answer to this question is poorly elucidated in special publications on heat treatment of metals. In this connection the author carried out special measurements to determine numerical values of the coefficient of restitution of hammers.[11] Measurements were performed under working conditions with both single- and double-acting hammers. The computation of e was made from measured values of the heights of fall and rebound of the ram after impact or from the interval of time between two rebounds of the ram.

The results showed that the values of the coefficient of restitution depend to a great extent on the state of a forged piece. Figure V-6 presents a graph of changes in e as a function of the number of blows on a forged piece under a hammer having a weight of dropping parts equaling 5.3 tons. Analogous graphs were obtained for other hammers. It follows from these plots that during the first blows against the forged piece, when its temperature is high and it is in a plastic state, the coefficient of impact velocity restitution is very small, equaling approximately 0.10. As the number of blows increases, the temperature of the forged piece decreases, the impact rigidity increases, and consequently the value of e increases. For the last blows, when a comparatively cooler piece is being forged, the coefficient of restitution approaches 0.5. Measurements of this coefficient during idle blows and under conditions of cold forging showed that its value does not exceed 0.5.

Since computations of hammer foundation vibrations should be performed for the most unfavorable conditions of operation, the design value of the coefficient of restitution for hammers forging steel parts should be taken as 0.5.

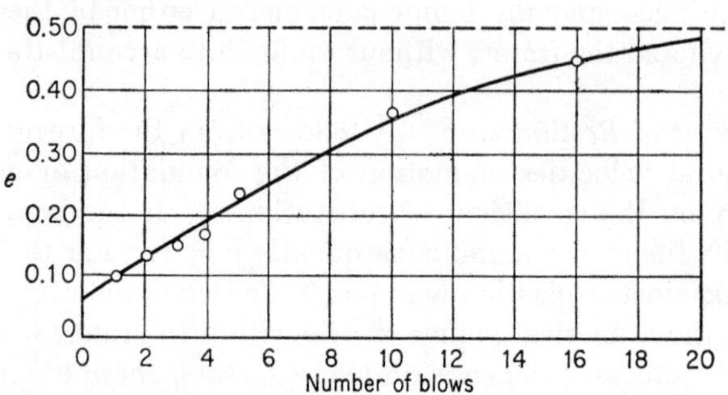

FIG. V-6. Variation of the coefficient of impact velocity restitution e with the number of hammer blows on the forge piece.

Values of e for forge hammers proper are much smaller than those for stamping hammers, and corresponding design values may be taken to equal 0.25.

Finally, for hammers forging nonferrous metals, this coefficient is considerably smaller than for hammers working on steel parts and may be considered to equal zero.

V-3. Natural Vibrations of a Hammer and Its Foundation as a Result of a Centered Impact

a. The Main Assumptions Involved in Design Computations. The foundation and hammer present a system which includes at least seven bodies: the frame, the dropping parts, the forged piece, the anvil, the elastic pad under the anvil, the foundation block, and, finally, the soil. From the point of view of mechanics, the phenomena which develop as a result of the impact of the ram against a forged piece lying on the anvil are extremely complicated and may be analyzed only with a high degree of approximation.

The main problems in computations for a hammer foundation are to determine the amplitude of foundation vibrations and to establish the values of stresses in the pad under the anvil.

The solution of these problems is usually based on the assumption that the hammer frame, the forged piece, the anvil, the elastic pad under the anvil, and the foundation block form one solid body. Such an assumption in regard to the foundation, the anvil, and the frame is justified by the fact that the deformation (due to impact) of each of these bodies is small in comparison with soil settlement under the foundation and therefore may be neglected.

However, deformation of the pad under the anvil may be much larger than soil settlement. Therefore the assumption that the pad has an infinitely large rigidity may lead in some cases to large errors in computation. This assumption is permissible only when the masses of both the anvil and the frame, if the latter is placed directly on the anvil, are comparatively small in relation to the foundation mass. Only in this case will the pad have no considerable effect on the amplitude of foundation vibrations. Otherwise, the elasticity of the pad cannot be neglected. In the case under consideration, the computation setup will be reduced to a system of three bodies: the ram, which is the striking body; the anvil, which is separated from the foundation by an elastic connection; and the foundation on an elastic base. The anvil and the foundation are the impact-receiving bodies.

In determining the amplitudes of foundation vibrations, it is possible to assume that the time of actual impact is small in comparison with the period of natural vibrations of the system; therefore, during the impact, there is no time for the foundation and anvil to undergo displacements comparable to their displacements during the vibrations which follow the impact. Since the reactions of the pad and the soil depend only on the displacements of the anvil and the foundation (we neglect damping reactions), it is possible to assume that during impact no additional reactions occur from the pad and soil. Thus only static reactions develop, imposed by the weight of the foundation, hammer, and anvil. These reactions existed before the impact and balanced the weight of the installation.

Therefore during impact, the foundation (with anvil and frame) and the dropping ram, in the first approximation, may be considered to be free bodies. Then an analysis of the impact of the system may be reduced to the analysis of a free impact of two or more absolutely solid bodies moving with given initial velocities.

The striking body (the ram) in all computations is assumed to be absolutely rigid.

b. Equations of the Vibrations of Foundation and Anvil. We begin by considering the simplest conditions: those in which pad elasticity may be neglected and the vibrations of the foundation, anvil, and frame occur as vibrations of a body with only one degree of freedom.

In this case the equation of vertical free vibrations of the foundation will be (Art. III-1)

$$\ddot{z} + f_{nz}^2 z = 0 \tag{V-3-1}$$

where z = vertical displacement of center of mass of foundation and anvil, measured from equilibrium position

f_{nz}^2 = square of frequency of natural vibrations of foundation:

$$f_{nz}^2 = \frac{c_u A}{m}$$

A = foundation area in contact with soil

m = total vibrating mass

c_u = coefficient of elastic uniform compression of soil

Equation (V-3-1) is the equation of free vibrations of the foundation without damping. The general solution of this equation is

$$z = A \sin f_{nz}t + B \cos f_{nz}t \tag{V-3-2}$$

The constants A and B, as usual, are determined from the initial conditions of motion. Taking as the beginning of readings the instant when the impact of the ram against the anvil ends, we obtain, for $t = 0$,

$$z = 0 \qquad \dot{z} = v_0$$

Using these initial conditions, we obtain

$$A = \frac{v_0}{f_{nz}} \qquad B = 0$$

Equation (V-3-2) will take the form

$$z = \frac{v_0}{f_{nz}} \sin f_{nz}t \tag{V-3-3}$$

The maximum deflection of the foundation will occur after time t_1:

$$t_1 = \frac{\pi}{2f_{nz}}$$

Its value will be

$$A_z = \frac{v_0}{f_{nz}} \tag{V-3-4}$$

If one is to take into account the soil reactions which are proportional to the velocity of foundation displacement, then the amplitude of real vibrations will be smaller than the one computed without considering the damping forces. However, it is a difficult task to evaluate the influence of damping forces by means of computations. As stated in Chap. III, these forces depend on many factors (for example, the foundation area in contact with soil, the foundation mass and the period of its free vibrations, and the foundation depth).

The pad under the anvil is fairly elastic in comparison with the anvil and foundation; therefore the anvil and frame (if the latter rests on the anvil) will not only participate in vibrations of the foundation on soil, but will undergo some vibration with respect to the foundation.

In order to evaluate the amplitude of vibrations of the anvil in relation to the foundation, it is necessary to consider vibrations of a system with two degrees of freedom. Free vibrations of such a system are determined by the following differential equations:

$$\begin{aligned} m_1\ddot{z}_1 + c_1z_1 - c_2(z_2 - z_1) &= 0 \\ m_2\ddot{z}_2 + c_2(z_2 - z_1) &= 0 \end{aligned} \qquad \text{(V-3-5)}$$

where m_1, m_2 = masses of foundation, anvil (with frame, if latter is mounted on anvil)

$c_1 = c_uA$ = coefficient of rigidity of soil base under foundation

$c_2 = (E/b)A_2$ = coefficient of rigidity of pad under anvil

A_2 = base area of pad

b = thickness of pad

E = Young's modulus of material of pad

z_1, z_2 = displacements of foundation, anvil measured from equilibrium position

We denote by f_{n1} and f_{n2} the natural frequencies of the system whose motion is determined by Eqs. (V-3-5); by

$$f_{na}^2 = \frac{c_2}{m_2}$$

we denote the frequency of natural vibrations of the anvil with the frame (or for forge hammers proper that of the anvil on a motionless foundation); then we obtain a general solution of the system of Eqs. (V-3-5):

$$\begin{aligned} z_1 &= C_1(f_{na}^2 - f_{n1}^2) \sin (f_{n1}t + \alpha_1) + C_2(f_{na}^2 - f_{n2}^2) \sin (f_{n2}t + \alpha_2) \\ z_2 &= C_1f_{na}^2 \sin (f_{n1}t + \alpha_1) + C_2f_{na}^2 \sin (f_{n2} + \alpha_2) \end{aligned} \qquad \text{(V-3-6)}$$

Setting

$$C^{(1)} = C_1 \cos \alpha_1 \qquad C^{(2)} = C_1 \sin \alpha_1$$
$$C^{(3)} = C_2 \cos \alpha_2 \qquad C^{(4)} = C_2 \sin \alpha_2$$

we obtain

$$\begin{aligned} z_1 = {} & C^{(1)}(f_{na}^2 - f_{n1}^2) \sin f_{n1}t + C^{(2)}(f_{na}^2 - f_{n1}^2) \cos f_{n1}t \\ & + C^{(3)}(f_{na}^2 - f_{n2}^2) \sin f_{n2}t - C^{(4)}(f_{na}^2 - f_{n2}^2) \cos f_{n2}t \\ z_2 = {} & C^{(1)}f_{na}^2 \sin f_{n1}t + C^{(2)}f_{na}^2 \cos f_{n1}t + C^{(3)}f_{na}^2 \sin f_{n2}t \\ & + C^{(4)}f_{na}^2 \cos f_{n2}t \end{aligned} \qquad \text{(V-3-7)}$$

The natural frequencies f_{n1} and f_{n2} are determined as roots of the equation

$$f_n^4 - (f_l^2 + f_{na}^2)(1 + \mu_1)f_n^2 + (1 + \mu)f_l^2 f_{na}^2 = 0 \qquad \text{(V-3-8)}$$

where

$$\mu = \frac{m_2}{m_1} \qquad f_l^2 = \frac{c_1}{m_1 + m_2}$$

f_l^2 is the limiting frequency of the foundation together with the hammer placed on soil (for the condition that the pad is infinitely rigid).

The initial conditions of motion in the case under consideration (at $t = 0$) are as follows:

$$z_1 = z_2 = 0 \qquad \dot{z}_1 = 0 \qquad \dot{z}_2 = v_a$$

where v_a is the initial velocity of motion of the anvil,

$$v_a = \frac{1 + e}{1 + \mu_a} v$$

and

$$\mu_a = \frac{m_2}{m_0}$$

Particular solutions of system (V-3-5) which correspond to these initial conditions are as follows:

$$\begin{aligned} z_1 &= \frac{(f_{na}^2 - f_{n2}^2)(f_{na}^2 - f_{n1}^2)}{f_{na}^2(f_{n1}^2 - f_{n2}^2)} v_a \left(\frac{\sin f_{n1}t}{f_{n1}} - \frac{\sin f_{n2}t}{f_{n2}} \right) \\ z_2 &= \frac{v_a}{f_{n1}^2 - f_{n2}^2} \left(\frac{f_{na}^2 - f_{n2}^2}{f_{n1}} \sin f_{n1}t - \frac{f_{na}^2 - f_{n1}^2}{f_{n2}} \sin f_{n2}t \right) \end{aligned} \qquad \text{(V-3-9)}$$

With these expressions it is possible to compute stresses which develop in the pad as a result of combined vibrations of the anvil and foundation.

The maximum stress σ in the pad evidently will equal

$$\sigma = \frac{c_2}{A_2}(z_2 - z_1) \qquad \text{(V-3-10)}$$

V-4. Experimental Studies of Vibrations of Foundations under Forge Hammers

a. Introduction. The theory of vertical vibrations of hammer foundations, presented in Art. V-3, is based on some assumptions which may be verified only by comparing the results of computations with experimental data. This refers primarily to the negligibility of the mass and damping properties of soil. As stated in Chap. IV, due to the fact that forced vibrations of foundations under reciprocating engines are usually characterized by a frequency different from the natural frequency of foundation vibrations, the influence of damping soil reactions in such cases is small and may be ignored.

Under an impact, the foundation below a hammer undergoes free vertical vibrations. Therefore the damping soil reactions will have considerable influence on the amplitudes of foundation vibrations. The introduction of damping-reaction values into the computations of these vibrations will somewhat complicate the formula used, but the calculations will still be practicable from a mathematical point of view. However, in order that this complication, caused by the introduction of damping reactions, should be of some practical value, it is necessary to know the constants characterizing the dissipative properties of the soil. Great difficulties are involved in establishing these constants, because their values depend not only on the soil, but also on the design of the foundation (in particular, on the depth of the foundation, the ratio between the length and width of the foundation, the foundation height, and the material and density of the backfill). It is very difficult to take into account the influence of all these factors on the value of the damping constant of a soil.

The inertial properties of the soil, which were not considered by the theory of vertical vibrations presented in Art. V-3, also may have great effect. In addition, the results of computations may be influenced by values of the coefficient of elastic rigidity c_2 of the pad under the anvil. This coefficient depends not only on the properties of the material of the pad under the anvil, but also on its design and on other special features which cannot always be taken into account by computations.

The pad under the anvil in hammer foundations of conventional design usually consists of several shields made of timber beams bolted together. The horizontal surfaces of these shields, just as the base of the anvil and the surface of the foundation under the anvil, are not ideally smooth surfaces and consequently do not come into contact with each other at all points. Because of this, some sections of the surfaces of the pad, the anvil, and the foundation are subjected to considerable stresses while others are not loaded at all. As a result, the elastic properties of the whole pad depend not only on its material, but also on the conditions of its contact with the surfaces of the foundation and anvil.

Only by means of measurements of vibrations occurring in a sufficiently large number of operating hammer foundations is it possible to elucidate the influence of all the above factors on vertical foundation vibrations. It is obvious that measurements do not give us an opportunity to establish separately the influence of each of these factors; for example, that of damping and inertial properties of soil. However, the measurement data do make it possible to introduce corrective coefficients into the formulas of Art. V-3, which then permit the adjustment of the results of computations performed on the basis of these formulas to the results of vibration measurements.

As early as 1939, the author carried out a large-scale investigation of foundation vibrations.[8] He studied 47 foundations under hammers located at six different plants.

That study had a threefold purpose: the verification of formulas for computations of hammer foundation vibrations, the collection of data on the design of normally operating foundations, and the determination of values of vibration amplitudes which could be accepted by designers as permissible and on which the main dimensions of the foundation would depend.

Table V-3. Data on Plant Sites where Hammer-foundation Measurements Were Made

Plant no.	*Geological and hydrogeological description of site*
1	A brown sandy clay with yellow inclusions comes to the surface everywhere on the site of the plant. This clay has a thickness of 0.1–2.5 m and is underlaid by a fine quartz sand alternating with lenses of clayey sand and of clay with some sand and silt. The thickness of these layers is not uniform over the area of the plant, varying from 0.3–6 m. With increasing depths, sands free of clay admixture predominate. Ground-water level is at a depth of 7 m, i.e., below all the hammer foundations.
2	Yellow medium-grained dense sands at a natural moisture content
3	Yellow medium-grained dense sands at a natural moisture content
4	Fine dense sand, ground-water level at a depth of 2.2–2.8 m, i.e., above the base of the hammer foundations
5	Medium-grained sands of medium density reaching to a depth of 9.4 m
6	Heavy brown clays with some sand and silt

Foundations of various designs were studied. Slab-shaped foundations predominated at one plant only. At other plants, only deeply embedded block-type foundations were present. This design of foundations was very popular at the time of the investigation (1939); slab-shaped foundations embedded to a small depth were seldom used then.

b. Description of Bases and Foundations. The greatest part of the forge hammers mounted on the foundations studied (35 out of 47) were double-acting steam or air stamping hammers. Only 6 foundations were under drop hammers of unrestricted action. The remaining 6 foundations were under forge hammers proper. The foundations investigated were located at six different plants. The geological conditions for each plant are given in Table V-3.

c. Results of Measurements of Foundation and Anvil Vibrations. Preliminary measurements of foundation vibrations showed that, in addition to vertical vibrations, hammer foundations also undergo rocking vibrations. However, the latter are less important because their amplitudes are much smaller than those of vertical vibrations.

The vibration amplitudes of the foundation, the anvil, and the frame are strongly affected by the state of the forged piece. For example, during the first impacts of the hammer against the piece, the energy of impact is largely consumed in plastic deformation of the metal, and the coefficient of restitution is small, as are the vibration amplitudes. The amplitudes of vibrations of the anvil and foundation grow with each subsequent impact. The largest amplitudes come with the last few impacts, when the forged piece is already deformed to such a degree that the greater part of the impact is taken not by the forged piece, but by the lower die, which transfers the impact energy to the anvil and foundation. Since the last few impacts induce the most unfavorable dynamic conditions for the foundation and anvil, their vibrations were measured during these impacts.

Figure V-7 shows samples of vibrograms obtained for some of the hammers investigated. It is seen from these vibrograms that vibrations of the hammer foundation and anvil, in most cases, differ considerably from damped sinusoids, which could be assumed on the basis of theoretical considerations. This shows that the foundation together with the anvil presents a much more complicated vibrating system than was assumed in Art. V-3, wherein vibration equations were derived.

In addition, as was to be expected, the vibrograms reveal considerable influence of the damping reactions. In some cases this influence is so large that the motion is almost aperiodic.

Finally it was found that identical foundations built under the same geologic conditions and subjected to the action of identical impacts underwent vibrations of varying amplitudes, sometimes sharply differing from one another. For example, two identical foundations under 3.6-ton hammers were investigated; one of them had 0.48 mm amplitude of vibrations, the other 0.78 mm. In the same way, two identical foundations under 2.25-ton hammers had 0.80- and 1.80-mm amplitudes of vibrations.

These data on vibrations of existing foundations permit the assumption that they are greatly affected by factors not considered by theory. Thus the observed differences in the amplitudes of foundation vibrations under hammers operating under the same conditions are apparently explained by the influence of the following factors: (1) the state of the timber pad under the anvil; (2) the contacts between this pad and both the anvil and the foundation; (3) the backfill of the foundation; there may also be influences of other factors which are difficult to include in design computations.

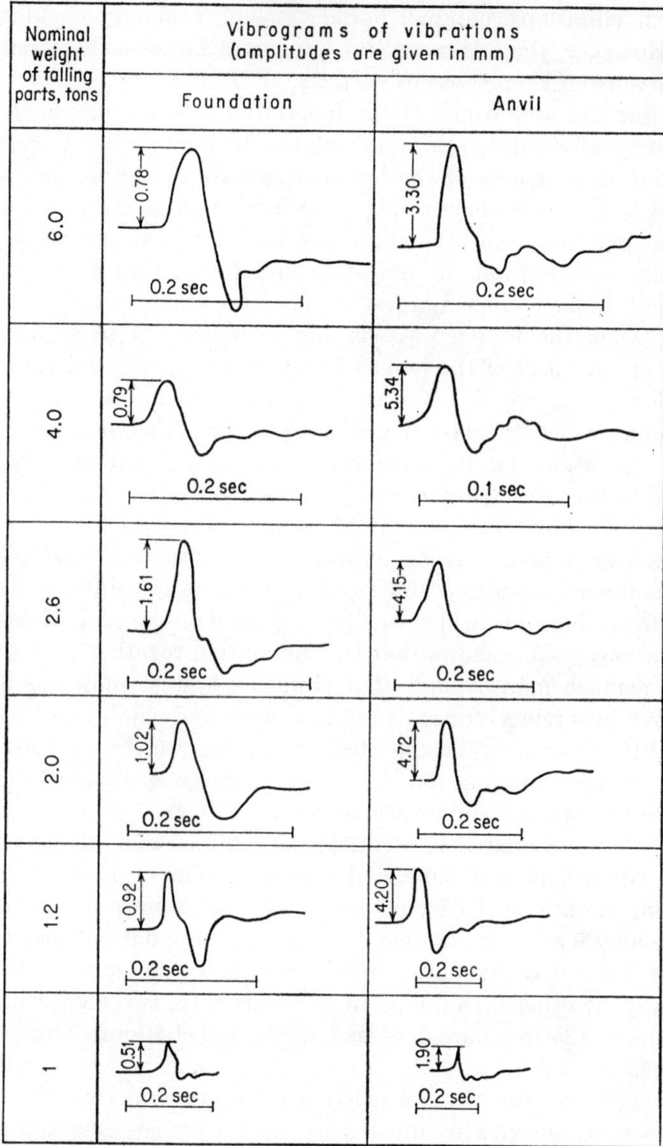

FIG. V-7. Typical vibrograms of operating hammer foundations.

Therefore the results of computations carried out on the basis of the formulas of Art. V-3 should be considered as tentative values only, showing the order of magnitude of vibration amplitudes, but not their absolute values.

A comparison of computed and measured vibration amplitudes of hammer foundations leads to the conclusion that if data required for calculation are selected correctly, then an average error in computation will be around ±30 per cent.

d. Amplitudes of Foundation Vibrations. On the basis of investigations of 47 foundations, points are plotted in Fig. V-8 giving measured vibration amplitudes versus actual weights of dropping parts of hammers. It is seen that the amplitudes of hammer vibrations never attain the value of 2.0 mm, i.e., the value which in the past was taken as permissible in computations.[44] With a decrease in power of the hammer, a

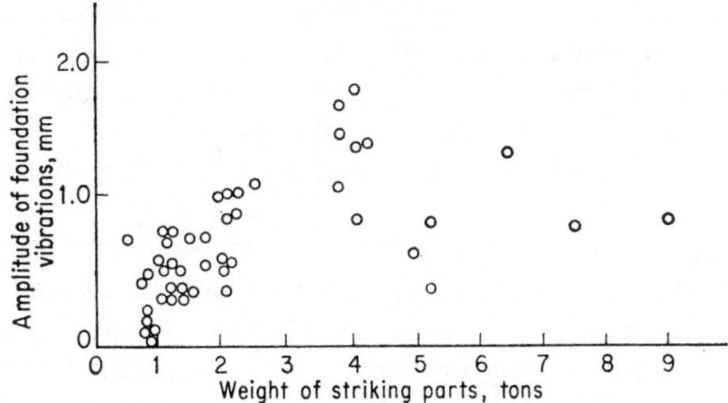

FIG. V-8. Measured vibration amplitudes of 47 hammer foundations plotted against the weight of the striking parts.

decrease in the amplitudes of foundation vibrations is observed. The overwhelming majority of foundations studied had vibration amplitudes of about 1.0 to 1.3 mm or less. Foundation vibrations characterized by these amplitudes did not exercise any noticeable harmful influence on the structures of forge shops. Similarly, no considerable settlements of foundations were observed where amplitudes of vibrations were of the order of 1.0 to 1.3 mm.

However, foundations having vibration amplitudes greatly exceeding 1.0 mm underwent considerable settlements. For example, a foundation with an amplitude of around 1.8 mm underwent a settlement reaching 0.3 m.

The above discussion leads to the conclusion that a design value of vertical vibrations of hammer foundations may be taken in the range 1 to 1.2 mm.

e. Amplitudes of Anvil Vibrations. Amplitudes of vibrations of anvils and frames of stamping hammers are much larger than amplitudes of foundation vibrations. For powerful hammers having thick pads under the anvil, absolute values of anvil vibrations reach 5 mm, although most of the hammers studied had amplitudes in the range 2 to 4 mm. With an increase in power of the hammer, the amplitude of anvil vibrations increases. For hammers with a weight of dropping parts up to 1 ton, a typical amplitude reaches 1 mm; for 2-ton hammers this amplitude reaches 2 mm; for hammers in which the weight of dropping parts exceeds 3 tons, the amplitude of anvil vibrations is usually 3 to 4 mm. These values of amplitudes of anvil and hammer vibrations may be taken as permissible.

The above values show that when stamping hammers are in operation, the anvil rebounds on the pad. Shocks of the anvil and frame (if the latter is attached to an anvil whose amplitude of vibrations reaches considerable values) have a harmful effect on the condition of the hammer. In addition, the larger the amplitude of vibrations of the anvil, the more kinetic energy of impact is consumed by these vibrations and consequently the smaller the hammer's efficiency. Loss of impact energy due to vibrations reaches 10 per cent of the work of the hammer's dropping parts.

The large vibration amplitudes of anvils are explained by insufficient rigidity of the pad under the anvil, which in some hammers had a thickness of 1.5 m. There is no reason to use such thick pads either from the point of view of the forging process or from a structural point of view. The pad thickness is usually assigned by the hammer supplier on the basis of traditional recommendations of the manufacturer and is not substantiated by any design data. Therefore different plants producing hammers of the same power recommend pads of different thicknesses.

The thickness of the pad should be selected so that the vibration amplitudes of the anvil do not exceed a particular value; in addition, stresses in the pad should not be greater than is permissible. Table V-1 was compiled on the basis of these considerations. Thicknesses of pads, as recommended in that table, are somewhat smaller than those which usually have been employed up to the present time. The decrease in pad thickness as compared with usually accepted thicknesses is based on considerations concerning the harmful effects of anvil vibrations of large amplitude.

f. The Determination of Elastic Constants of the "Anvil-Foundation" System. If one is to consider a foundation together with the anvil mounted thereon as a system with two degrees of freedom, as was done in Art. V-3, then vibrograms of the anvil and foundation vibrations will show two sinusoids of different periods superimposed on each other. It follows from theory that the amplitudes of these sinusoids will be inversely proportional to their frequencies.

As stated before, the shapes of the measured foundation vibrograms in many cases approach aperiodic curves. In no case was it possible to determine from vibrograms both natural frequencies of the combined vibrations of the anvil and foundation. Vibrograms usually reveal only the vibrations at the lower principal frequency. Therefore, it is possible to consider (with a precision sufficient for practical purposes) that in Eqs. (V-3-9) the amplitude of vibrations for sin $f_{n1}t$ (where $f_{n1} > f_{n2}$) equals zero. Then approximate expressions for dynamic displacement of the foundation and anvil will be as follows:

$$z_1 = -\frac{(f_{na}^2 - f_{n2}^2)(f_{na}^2 - f_{n1}^2)}{f_{na}^2(f_{n1}^2 - f_{n2}^2)f_{n2}^2} v_a \sin f_{n2}t$$
$$z_2 = -\frac{f_{na}^2 - f_{n1}^2}{(f_{n1}^2 - f_{n2}^2)f_{n2}} v_a \sin f_{n2}t \tag{V-4-1}$$

Hence

$$\beta = \left|\frac{z_2}{z_1}\right| = \frac{1}{1 - \gamma^2} \tag{V-4-2}$$

where

$$\gamma^2 = \frac{f_{n2}^2}{f_{na}^2} \tag{V-4-3}$$

Thus, having found, from vibrograms obtained for the anvil and foundation, the value β and the lower natural frequency of vibrations, one can establish from Eq. (V-4-3) the limiting frequency f_{na} of vibrations of the anvil on the pad. From the formula

$$f_{na}^2 = \frac{EA}{bm_2}$$

one can establish the value of the modulus of elasticity E of the pad under the anvil.

Then no difficulties are involved in establishing the value of the second higher frequency f_{n1}, as well as the limiting frequency f_l of the natural vertical vibrations of the entire installation on the soil. After some transformations (not shown here), we obtain

$$f_{n1}^2 = \frac{1 + \mu - \gamma^2}{(1 - \gamma^2)\gamma^2} f_{n2}^2$$
$$f_l^2 = \frac{1 + \mu - \gamma^2}{(1 + \mu)(1 - \gamma^2)} f_{n2}^2 \tag{V-4-4}$$

Here, as before, μ is the ratio between the anvil mass (in drop hammers the frame mass is also included) and the foundation mass. Knowing f_l^2 from the formula

$$f_l^2 = \frac{c_u A}{m_1 + m_2}$$

one can establish the real value of the coefficient of elastic uniform compression c_u of the base under the hammer foundation.

The results of computations of moduli of elasticity for pads and coefficients of uniform compression c_u for bases, performed in accordance with the above methods for the several foundations studied, lead to the following conclusions:

As was to be expected, the computed moduli of elasticity E for pads under the anvils of different hammers vary within a comparatively wide range of values. The probable reasons for this have already been discussed. The moisture content and working life of the pad should also be mentioned, as their influence may be very noticeable.

The average value of E was established from computations to equal 4.7×10^4 tons/m^2, i.e., approximately two times smaller than the value customarily used in stress analyses of oak beams when the latter are compressed across their fibers. On the basis of these results, it is recommended that a design value of 5×10^4 tons/m^2 be used for the modulus of elasticity of timber pads.

In most hammers the dynamic stress in the pad under the anvil does not exceed 200 tons/m^2, i.e., it is much lower than the permissible value of about 300 to 350 tons/m^2 for oak timbers compressed across their fibers. This attests to the fact that pads which were employed up to this time have had a considerable safety factor. As stated before, this is because the thickness of the pad has usually been taken much larger than was necessary from the point of view of dynamic computations.

In low-power hammers, dynamic stresses in pads do not exceed 100 tons/m^2. Therefore it is possible to employ in these hammers pads made of pine or larch instead of oak.

Special investigations showed that the design value of the coefficient c_u of elastic uniform compression of the soil base of hammer foundations was about 4.0 kg/cm^3. However, the average value of this coefficient obtained from measurements of hammer vibrations was around 25 kg/cm^3, i.e., approximately six times larger. Such a large divergence between these values attests to the fact that the amplitudes of foundation vibrations under hammers are greatly affected by factors not considered by the theory presented in Art. V-3. In particular, this theory, as stated above, does not consider the influence of the damping and inertial properties of the soil, but this influence may be considerable, just as in the case of natural foundation vibrations. In addition, the value of c_u may be affected by the backfill of the foundation.

Investigations of a test foundation showed that with backfilling the value of the coefficient c_u of elastic uniform compression increases approximately two times as compared with the value established from tests on an exposed foundation. Consequently, in computations of natural vertical vibrations of foundations under hammers, the value of the coefficient of elastic uniform compression c_u should not be taken equal to that used in computations of vibrations of other machines whose foundations do not undergo natural vertical vibrations. According to the above data, in design computations of hammer foundations the coefficient c'_u should be used instead, where

$$c'_u = kc_u \tag{V-4-5}$$

With allowance for some safety reserve for dynamic stability of the foundation, the value of the correction coefficient k may be taken as equal to 3.

g. Comparison of Different Formulas for the Computation of Amplitudes. To simplify the practical computations of vibrations of the foundation and anvil, the vibrations are considered to be independent of each other. This is equivalent to the assumptions that the presence of the pad under the anvil has no influence on the amplitudes of vibrations of the foundation and that the vibrations of the anvil are not affected by elastic properties of the soil base or the mass of the foundation. This assumption leads to considerable simplification of formulas for the computation of vibration amplitudes of the foundation and anvil. According to Eqs. (V-3-4), the amplitude of vibrations of the foundation can be established from

$$A_z = \frac{(1+e)W_0 v}{(W_1 + W_2)f_{nz}} \tag{V-4-6}$$

and the amplitude of vibrations of the anvil from

$$A_a = \frac{(1+e)W_0 v}{W_2 f_{na}} \tag{V-4-7}$$

It is interesting to establish and compare the computational errors involved in these equations and the more accurate Eqs. (V-3-9). Taking into account the fact that in Eqs. (V-3-9) the amplitudes for $\sin f_{n1}t$ are much smaller than the amplitudes for $\sin f_{n2}t$, we may neglect the terms containing $\sin f_{n1}t$ in these formulas. Then the vibration amplitudes of the anvil and foundation will be determined by Eqs. (V-4-1).

Table V-4 gives the results of computations of vibration amplitudes of the foundation and anvil, performed for several foundations studied; both methods of computation were employed. The same values of the modulus of elasticity of the pad under the anvil and the coefficient of elastic uniform compression of the base under the foundation were taken for all hammers (respectively, 5×10^4 tons/m^2 and 20×10^3 tons/m^3). The same velocity of dropping parts (6.5 m/sec) was used for all hammers.

It is seen from Table V-4 that there is considerable difference between the amplitude values of foundation vibrations as computed by Eqs. (V-4-1) and (V-4-6). Hence it follows that a control computation of

TABLE V-4. COMPARISON OF VIBRATION AMPLITUDES COMPUTED BY DIFFERENT PROCEDURES

W, tons[a]	m_1, tons × sec²/m	m_2, tons × sec²/m	c_1, tons/m	c_2, tons/m	A_1, mm[b]	A_2, mm[c]	A_z, mm[d]	A_a, mm[e]
7.50	51.2	19.8	136.2×10^4	42.2×10^4	0.915	4.13	0.74	4.10
4.10	17.2	9.17	56.8×10^4	39.6×10^4	1.32	2.81	0.85	2.80
5.30	29.4	9.54	62.4×10^4	23.1×10^4	0.55	4.5	0.55	4.34
3.25	17.4	4.92	60.0×10^4	21.9×10^4	1.61	6.7	1.25	7.36
2.10	9.37	3.44	40.0×10^4	22.0×10^4	1.74	4.46	1.04	4.90
1.20	5.05	2.85	29.4×10^4	28.6×10^4	0.77	2.00	0.58	2.28

[a] These are the actual weights of the dropping parts; they differ from the nominal values.
[b] $A_1 = z_1$ = amplitude of foundation vibration computed from Eqs. (V-4-1).
[c] $A_2 = z_2$ = amplitude of anvil vibration computed from Eqs. (V-4-1).
[d] A_z = amplitude of foundation vibration computed from Eq. (V-4-6).
[e] A_a = amplitude of anvil vibration computed from Eq. (V-4-7).

vibrations of a foundation under a hammer should be performed using Eqs. (V-4-1), considering the effect of the pad under the anvil on these vibrations.

However, Table V-4 shows that there is only a small difference between computations of anvil vibration amplitudes by Eq. (V-4-7) and by the more precise Eqs. (V-4-1). Therefore it is permissible to use the simplified formulas in computations of anvil vibration amplitudes and in stress analysis of the pad under the anvil.

V-5. Selection of the Weight and Base Area of a Hammer Foundation

It was formerly held that the weight of the foundation for a hammer and the size of its area in contact with the soil should be selected in such a way as to meet the following requirements: the total pressure on the soil should not exceed the bearing capacity of this soil; and the foundation should not bounce on the soil. These conditions may be written as follows:

$$p_{st} + p_{dy} \leq \alpha p_0 \tag{V-5-1}$$

$$p_{st} > p_{dy} \tag{V-5-2}$$

where p_{st}, p_{dy} = static, dynamic pressures on soil

p_0 = permissible bearing value under condition that only static load is acting

α = coefficient of required reduction

The condition expressed by Eq. (V-5-1) is based on an assumption that either the static and dynamic pressures are equivalent or that the "coefficient of required reduction" is not a constant for a given soil and foundation. As stated in Chap. II, dynamic pressure transmitted to soils (especially to granular soils) may induce settlements and deformations which are tens and hundreds of times larger than those caused by static pressure of the same magnitude. Therefore if one of the items of the left-hand part of expression (V-5-1) changes, but the sum of these items remains constant, the total settlements and deformations will change. Hence it follows that under the assumption that α has a constant value for a given soil and foundation, the condition expressed by Eq. (V-5-1) cannot be accepted, because it is contrary to the physical nature of the phenomenon.

The condition expressed by Eq. (V-5-2) has no practical significance, because the bouncing of a foundation is of no essential importance and cannot be observed under working conditions. However, an observance of this condition led to carrying foundations down to a considerable depth. Since the design value of vibration amplitude was taken to equal 2 mm and more, in order to obtain a static settlement of this value it was necessary to increase considerably the heights of foundations. The result was that all hammer foundations erected with designs complying with the conditions of Eqs. (V-5-1) and (V-5-2) represented massive blocks carried down to a considerable depth. Figure V-4*a* shows a typical foundation of this kind.

Limitation of the vibration amplitude of a hammer foundation is the most important condition which must be satisfied by the design of such a foundation. The smaller the vibration amplitude, the smaller the influence of vibrations on adjacent structures and buildings, on the foundation, on the soil, and on the hammer. Another significant condition is the limiting of the value of static pressure on soil; as shown in Art. II-3, the smaller the static pressure, the smaller the settlement of the foundation (other conditions remaining equal).

Thus instead of the conditions of Eqs. (V-5-1) and (V-5-2), the foundation design should satisfy the following two conditions:

$$A_z < A_0 \tag{V-5-3}$$

$$p_{st} \leq \alpha p_0 \tag{V-5-4}$$

As indicated in Art. V-4, an average value of vibration amplitudes of hammer foundations, obtained from the results of numerous investigations of operating hammers, is approximately 1 mm. This may be taken as a design value for the permissible amplitude. Therefore the condition of Eq. (V-5-3) may be rewritten as follows:

$$A_z < 10^{-3} \text{ m} \tag{V-5-5}$$

In the simplest case, under the assumption that the foundation together with the anvil presents a system with one degree of freedom, the value of the vibration amplitude is determined by Eq. (V-4-6), and the condition of Eq. (V-5-5) may be written in the form

$$\frac{(1+e)W_0 v}{\sqrt{kc_u W A g}} < 10^{-3} \tag{V-5-6}$$

where all dimensions are in tons, meters, and seconds.

From Eqs. (V-5-4) and (V-5-6) values of foundation contact area and weight can be found for which the amplitude of foundation vibrations will not exceed 1 mm and the static pressure on the soil will not exceed the value αp_0; thus we obtain

$$A \gg \frac{(1+e)W_0 v}{\sqrt{kc_u \alpha p_0 g}} 10^3 \qquad \text{meters} \tag{V-5-7}$$

$$W_f = \frac{(1+e)W_0 v \sqrt{\alpha p_0}}{\sqrt{kc_u g}} 10^{-3} - W_a \qquad \text{tons} \tag{V-5-8}$$

where W_f = weight of foundation together with backfill, if present
W_a = weight of anvil and frame

Dividing both parts of Eq. (V-5-8) by W_0, we obtain a formula for determining the reduced foundation weight corresponding to a unit of actual weight of dropping parts of the hammer:

$$n_f = \frac{(1 + e)\sqrt{\alpha p_0}}{\sqrt{kc_u g}} v \times 10^3 - n_a \qquad \text{(V-5-9)}$$

where $$n_f = \frac{W_f}{W_0} \qquad n_a = \frac{W_a}{W_0}$$

In accordance with the values of p_0 and c_u for different soils given in Art. I-2, the following approximate relationship can be used:

$$\frac{p_0}{c_u} = 0.07$$

The corrective coefficient k, in accordance with data presented in Art. V-4, we set equal to 3.0; the coefficient of reduction of bearing capacity α, we set equal to 0.4.

TABLE V-5. VALUES OF SOME HAMMER COEFFICIENTS

Type of hammer	v, m/sec	e	n_a	n_f
Stamping hammers:				
Double-acting hammers (stamping of steel pieces)	6.5	0.5	30	48
Unrestricted hammers:				
Stamping of steel pieces	4.5	0.5	20	34
Stamping of nonferrous metals	4.5	0.0	...	16
Forge hammers proper:				
Double-acting	6.5	0.25	30	35
Unrestricted	4.5	0.25	20	25

Substituting the values of these coefficients into (V-5-9), we obtain a simple formula for the tentative determination of the foundation weight depending on the velocity of dropping parts of the hammer and the coefficient of restitution:

$$n_f = 8.0(1 + e)v - n_a \qquad \text{(V-5-10)}$$

According to data of machine-building plants, one can take approximately:

For double-acting hammers: $n_f = 30$
For unrestricted hammers: $n_a = 20$

Numerical values of n_f for different hammers are given in Table V-5.

In order to compare the computed values of n_f with data secured from experience, Table V-6 shows values of n_f for some of the drop-hammer slab foundations investigated at one of the plants.

The average experimental value of n_f for the hammers of Table V-6 is 46. It should be mentioned, however, that the actual weight of dropping parts of the hammers studied is much larger than their nominal weight. In most other cases, the operating dies are lighter and consequently the difference between the actual and the nominal weights of the dropping parts is smaller. Therefore such hammers have a larger value of n_f, reaching 50 to 55 for the same foundation weights. For massive foundations, which until recently were accepted by all design organizations, this ratio is much larger, lying in the range 70 to 80 and often reaching 100 to 120.

TABLE V-6. COMPARISON OF ACTUAL AND COMPUTED REDUCED FOUNDATION WEIGHTS n_f [EQ. (V-5-10)] AND REDUCED FOUNDATION AREAS a_f [EQ. (V-5-11)] OF SEVERAL HAMMER SLAB FOUNDATIONS

Nominal weight of dropping parts, tons	W_0, tons	W_f, tons	A, m²	Actual value of n_f	Actual value of a_f	Computed value of n_f	$\frac{W_b}{W_0}$†	Computed value of a_f
1.0	1.2	49.7	14.7	41.1	12.2	48.0	32.6	12.2
1.25	2.10	92.0	20.0	43.8	9.6	48.0	31.3	9.5
5.4	7.5	502	65.6	67.0	8.8	48.0	30.8	8.8
2.25	4.10	217	38.2	53.0	9.4	48.0	32.4	9.3
1.35	2.0	73.5	16.5	36.8	8.2	48.0	36.8	8.2
0.54	0.75	28.4	9.9	38.0	13.0	48.0	31.0	13.2
2.25	4.10	168	40	41.1	10.0	48.0	34.0	9.8

† W_b = weight of concrete foundation without consideration of the backfill.

If in slab foundations one takes into account only the weight of concrete, neglecting the weight of backfill above the slab, then the value of n_f will be around 30 to 35. In massive foundations no such backfill is present; therefore the value of n_f of 70 to 120 represents the ratio between the weight of the foundation and the actual weight of dropping parts. Thus the expenditure of material for slab foundations is two times smaller than for massive foundations. And, as stated in Art. V-4, the results of instrumental investigations of hammer slab foundations show that the amplitudes of their vibrations lie within the range of permissible values (around 1 mm and less).

Equation (V-5-7) for the selection of the foundation area in contact with soil may be simplified on the basis of the following considerations:

According to available data on the values of c_u and p_0 for different soils, it can be stated that

$$\frac{p_0}{c_u} \cong 0.5 \times 10^{-3} \text{ m}$$

Hence

$$c_u = 2 \times 10^{-2} p_0$$

where all dimensions are tons and meters.

Dividing both parts of Eq. (V-5-7) by W_0, we obtain an expression for the reduced contact area of foundation per unit of the actual weight of dropping parts:

$$a_f = \frac{A}{W_0} = \frac{(1+e)v \times 10^3}{p_0 \sqrt{\alpha k g} \times 2 \times 10^{-2}}$$

Setting as before $k = 3.0$ and $\alpha = 0.4$, we obtain a simple formula for the tentative determination of a_f and consequently of the entire contact area of the foundation:

$$a_f = \frac{20(1+e)v}{p_0} \qquad \text{(V-5-11)}$$

Equation (V-5-11) establishes the dependence of the foundation contact area not only on the hammer characteristics, but also on soil properties; the required dimensions of the foundation contact area increase in an inverse proportion to the bearing capacity of soil. Table V-7 presents values of a_f computed for different types of soils.

TABLE V-7. VALUES OF REDUCED FOUNDATION CONTACT AREAS a_f REQUIRED FOR DIFFERENT SOILS

Type of hammer	Values of a_f for following groups of soil:		
	Weak soils, $p_0 \leq$ 1.5 kg/cm²	Soils of medium strength, $p_0 =$ 1.5–3.5 kg/cm²	Soils of high strength, $p_0 =$ 3.5–6 kg/cm²
Stamping hammers:			
Double-acting hammers (stamping of steel pieces)	13	13–5.5	5.5–3.3
Unrestricted hammers			
Stamping of steel pieces	9	9–4	4–2.5
Stamping of nonferrous metals	6	6–2.5	2.5–1.5
Forge hammers Proper			
Double-acting	11	11–5	5–3
Unrestricted	7.5	7.5–3	3–2

For example, it is seen from this table that double-acting drop hammers used for stamping steel pieces, i.e., the type most frequently employed in forge shops, are characterized by the following ratio: when the soil is of medium strength, on the average about 9 m² of foundation contact area will be required per unit of actual weight of dropping parts.

Table V-6 gave values of a_f which were used for hammer slab foundations when $p_0 = 2.5$ kg/cm². It follows from the data of this table that values of a_f established from Eq. (V-5-11) are close to those accepted for the design of hammer slab foundations. Massive hammer foundations designed as blocks are characterized by values of a_f of 7 to 8, i.e., by somewhat smaller values than those used for slab foundations.

V-6. Design of a Hammer Foundation

Example. Dynamic computations for the design of a hammer foundation

1. Design Data. A double-acting stamping hammer has the following specifications:

Nominal weight of dropping parts:

$$3.0 \text{ tons}$$

Actual weight of dropping parts:

$$W_0 = 3.5 \text{ tons}$$

Height of drop:

$$h = 1.0 \text{ m}$$

Piston area from above:

$$A = 0.15 \text{ m}^2$$

Steam pressure:

$$p = 8 \text{ atm}$$

Weight of the anvil and frame:

$$W_2 = 90 \text{ tons}$$

Base area of the anvil:

$$A_2 = 4.75 \text{ m}^2$$

Thickness of pad under anvil:

$$b = 0.60 \text{ m}$$

Soils on the site of the foundation consist of brown clays with some sand and silt, with a permissible pressure $p_0 = 2$ kg/cm² if only static pressure is acting.

2. Velocity of Dropping Parts at the Beginning of Impact. From Eq. (V-2-2),

$$v = 0.65\sqrt{\frac{2 \times 9.81 \times 1.0(3.5 + 80 \times 0.15)}{3.5}} = 6.1 \text{ m/sec}$$

3. Preliminary Computation of the Required Values of Foundation Weight and Soil Contact Area. Design computations for determining the required weight of the foundation corresponding to a unit weight of dropping parts are made from Eq. (V-5-10). The coefficient of restitution is taken as $e = 0.5$. The weight of the anvil and frame corresponding to a unit weight of dropping parts will be:

$$n_a = \frac{90}{3.5} = 25.7$$

According to Eq. (V-5-10), the weight of the foundation corresponding to a unit weight of dropping parts equals

$$n_f = 8.0(1 + 0.5)6.1 - 25.7 = 47.3$$

The required weight of the foundation (together with backfill) equals

$$W = 3.5 \times 47.3 = 166 \text{ tons}$$

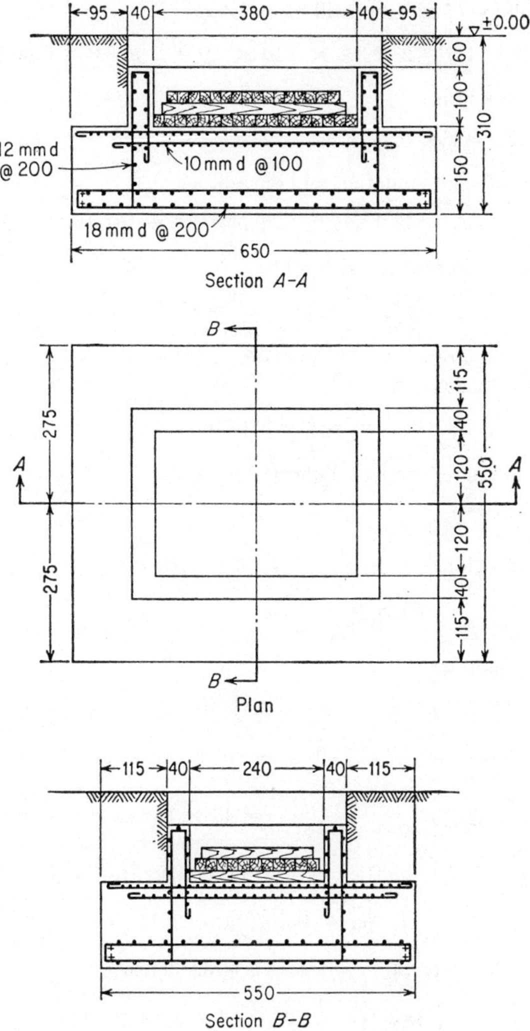

FIG. V-9. Design of foundation for example of Art. V-6.

The foundation contact area corresponding to a unit weight of dropping parts is determined from Eq. (V-5-11):

$$a_f = \frac{20(1 + 0.5)6.1}{20} = 9.2 \text{ cm}^2$$

The required foundation area in contact with soil equals

$$A = 9.2 \times 3.5 = 32.2 \text{ m}^2$$

4. DESIGN OF FOUNDATION. On the basis of the values determined above for the required foundation weight and area in contact with soil, we design the foundation in the form of a slab. The outline of the foundation design is shown in Fig. V-9.

The actual volume of concrete will be as follows:

$$V_c = 6.5 \times 5.5 \times 1.5 + 2 \times 3.2 \times 0.40 \times 1.0 + 2 \times 3.8 \times 0.40 \times 1 = 58.0 \text{ m}^3$$

The volume of backfill is

$$V_b = 2 \times 0.95 \times 5.5 + 2 \times 4.60 \times 1.15 = 20.9 \text{ m}^3$$

The total weight of the foundation and backfill is

$$W_1 = 58.0 \times 2.2 + 20.9 \times 1.6 = 161.4 \text{ tons}$$

The foundation area in contact with soil is

$$A = 6.50 \times 5.5 = 35.7 \text{ m}^2$$

5. Amplitude of Foundation Vibrations. We take the modulus of elasticity of the pad under the anvil to equal

$$E_2 = 50 \times 10^3 \text{ tons/m}^2$$

We take the thickness of the pad under the anvil from the design data:

$$b = 0.60 \text{ m}$$

The coefficient of rigidity of the pad under the anvil will equal

$$c_2 = \frac{50 \times 10^3 \times 4.75}{0.60} = 39.5 \times 10^4 \text{ tons/m}$$

The mass of the hammer is

$$m_2 = 90/9.81 = 9.18 \text{ tons} \times \text{sec}^2/\text{m}$$

The limiting frequency of natural vibrations of the anvil on the oak timber pad is

$$f_{na} = \frac{39.5 \times 10^4}{9.18} = 43 \times 10^3 \text{ sec}^{-2}$$

We set the coefficient of elastic uniform compression to equal

$$c_u = 4 \times 10^3 \text{ tons/m}^3$$

and the value of the correction coefficient $k = 3$. Then

$$c_u' = 3 \times 4 \times 10^3 = 12 \times 10^3 \text{ tons/m}^3$$

The coefficient of rigidity of the base under the foundation equals

$$c_1 = 12 \times 10^3 \times 35.7 = 43.8 \times 10^4 \text{ tons/m}$$

The mass of the foundation together with the backfill is

$$m_1 = \frac{161.4}{9.81} = 16.5 \text{ tons} \times \text{sec}^2/\text{m}$$

The square of the limiting frequency of natural vibrations of the whole system is then

$$f_l^2 = \frac{42.8 \times 10^4}{16.5 + 9.18} = 16.7 \times 10^3 \text{ sec}^{-2}$$

The ratio between the mass of the hammer and the mass of the foundation together with the backfill is

$$\mu_1 = \frac{9.18}{16.5} = 0.557$$

Using Eq. (V-3-8), we set up the equation for determining the frequencies of natural vibrations of the foundation-hammer system:

$$f_n^4 - (1 + 0.557)(43.0 \times 10^3 + 16.7 \times 10^3)f_n^2 + (1 + 0.0557) \times 43.0 \times 10^3 \times 16.7 \times 10^3 = 0$$

Or $$f_n^4 - 92.5 \times 10^3 f_n^2 + 1115 \times 10^6 = 0$$

Solving this equation, we obtain

$$f_{n1,2}^2 = [46.25 \pm \sqrt{(46.25)^2 - 1115}]10^3 = (46.25 \pm 32.2)10^3$$

Hence we have the frequencies:

$$f_{n1}^2 = 78.5 \times 10^3 \text{ sec}^{-2} \qquad f_{n1} = 288 \text{ sec}^{-1}$$
$$f_{n2}^2 = 14.1 \times 10^3 \text{ sec}^{-2} \qquad f_n^2 = 199 \text{ sec}^{-1}$$

We determine the initial velocity of the motion of the anvil together with the frame:

$$v_a = \frac{(1 + 0.5)3.5 \times 6.1}{3.5 + 90.0} = 0.342 \text{ m/sec}$$

From Eqs. (V-4-1) we establish the amplitudes of vibration of the foundation and anvil. The amplitude of vibration of the foundation is

$$A_z = -\frac{(43.0 \times 10^3 - 14.1 \times 10^3)(43.0 \times 10^3 - 78.5 \times 10^3)}{43.0 \times 10^3(78.5 \times 10^3 - 14.1 \times 10^3)119} 0.342$$
$$= 1.07 \text{ mm}$$

The amplitude of vibrations of the anvil together with the frame is

$$A_a = -\frac{(43.0 \times 10^3 - 78.5 \times 10^3)\sqrt{342}}{(78.5 \times 10^3 - 14.1 \times 10^3)119} = 1.6 \times 10^{-3} \text{ m} = 1.6 \text{ mm}$$

Thus the results of computations show that the amplitude of vibrations of the foundation will not exceed the permissible value of 1.0 to 1.2 mm.

The dynamic stresses in the pad under the anvil approximately equal

$$\sigma = \frac{c_2(A_a - A_z)}{A_2} = \frac{39.5 \times 10^4(1.6 \times 10^{-3} + 1.07 \times 10^{-3})}{4.75} = 222 \text{ tons/m}^2$$

which is much smaller than the permissible value of 300 to 350 tons/m².

6. REINFORCEMENT OF THE FOUNDATION. The foundation is reinforced as shown in Fig. V-9 according to practical requirements pointed out in Art. V-1. Concrete type 150† is used for the foundation.

Standard Illustrative Designs of Hammer Foundations. Computation and design of foundations for stamping hammers of different powers, as

† See footnote, Art. IV-I-c, p. 132.

well as foundation design for forge hammers proper, may be performed in a manner similar to the preceding numerical example.

Figures V-10 to V-17 show several standard illustrative foundation designs for stamping hammers and forge hammers of several systems. In the preparation of these examples, actual data on hammers were borrowed from instructive design manuals which had been compiled with the author's participation as a consultant.[18]

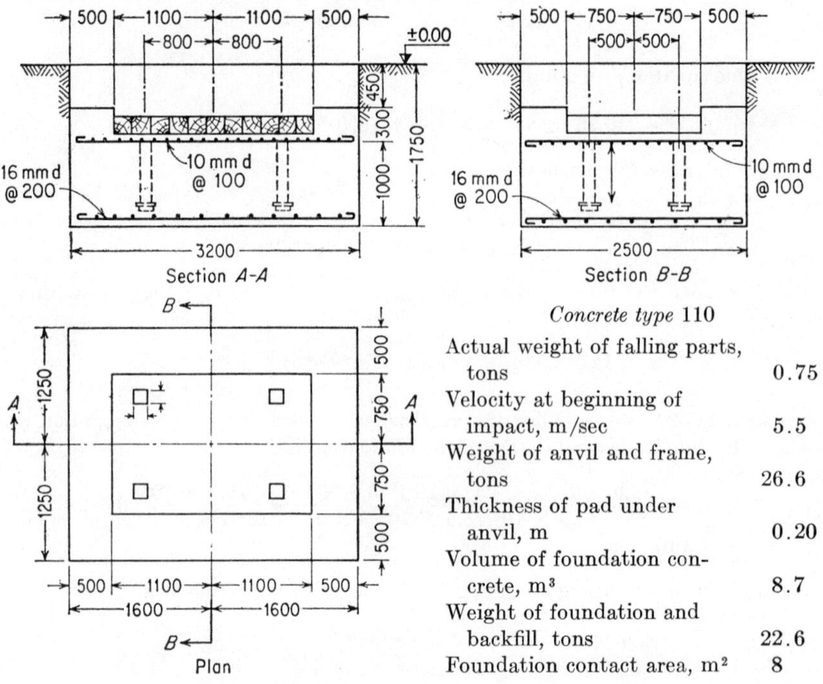

FIG. V-10. Foundation for 750-kg frictional hammer of "KLF" plant.

Design computations for these foundations were performed for a soil of medium strength with the coefficient of elastic uniform compression of soil c_u equal to approximately 4 kg/cm^3.

V-7. Computation and Design of Hammer Foundations with Vibration Absorbers

a. General Directives on Computation and Design. Sometimes the decrease in vibration amplitudes of hammer foundations is of great practical importance.

It follows from the approximate Eqs. (V-4-6) and (V-4-7) that vibration amplitudes of the foundation and anvil are inversely proportional to the square roots of the products of the base rigidity and mass. Conse-

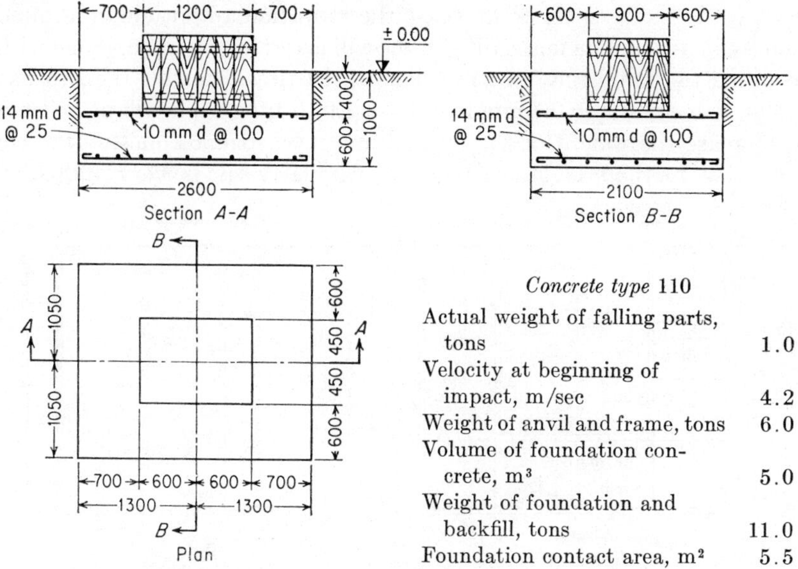

Concrete type 110

Actual weight of falling parts, tons	1.0
Velocity at beginning of impact, m/sec	4.2
Weight of anvil and frame, tons	6.0
Volume of foundation concrete, m^3	5.0
Weight of foundation and backfill, tons	11.0
Foundation contact area, m^2	5.5

FIG. V-11. Foundation for 1.0-ton free-falling cable hammer.

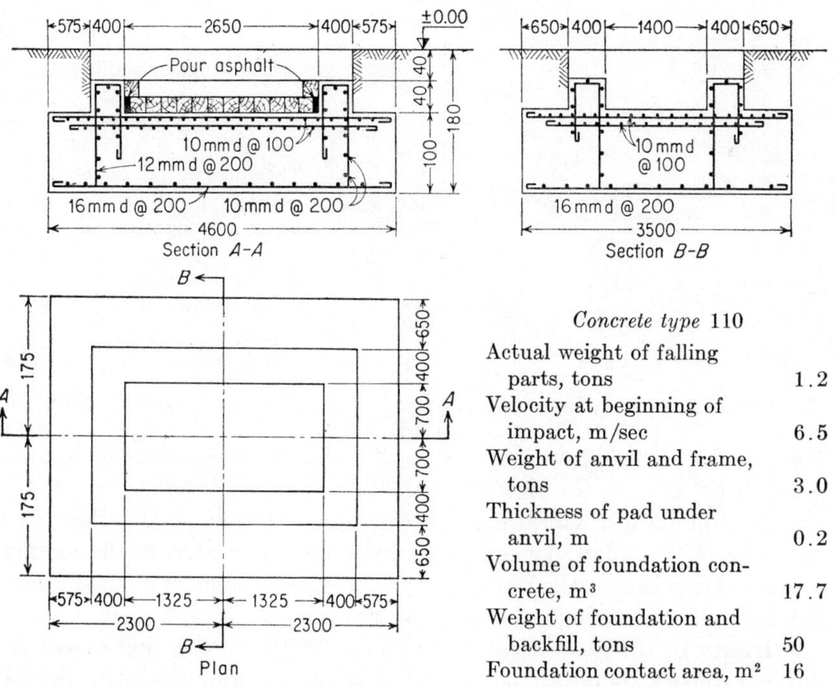

Concrete type 110

Actual weight of falling parts, tons	1.2
Velocity at beginning of impact, m/sec	6.5
Weight of anvil and frame, tons	3.0
Thickness of pad under anvil, m	0.2
Volume of foundation concrete, m^3	17.7
Weight of foundation and backfill, tons	50
Foundation contact area, m^2	16

FIG. V-12. Foundation for 1.2-ton steam or compressed-air stamping hammer.

quently, if one is to try to decrease the vibration amplitude of a foundation by increasing its mass, difficulties will arise; for example, if one wishes to make the amplitude of vibrations three times smaller (i.e., to use a design value of 0.3 to 0.4 mm instead of 1.0 to 1.2 mm) it will be necessary to increase the weight of the foundation at least nine times. It is clear that this method is impracticable. Similarly, it is very difficult to

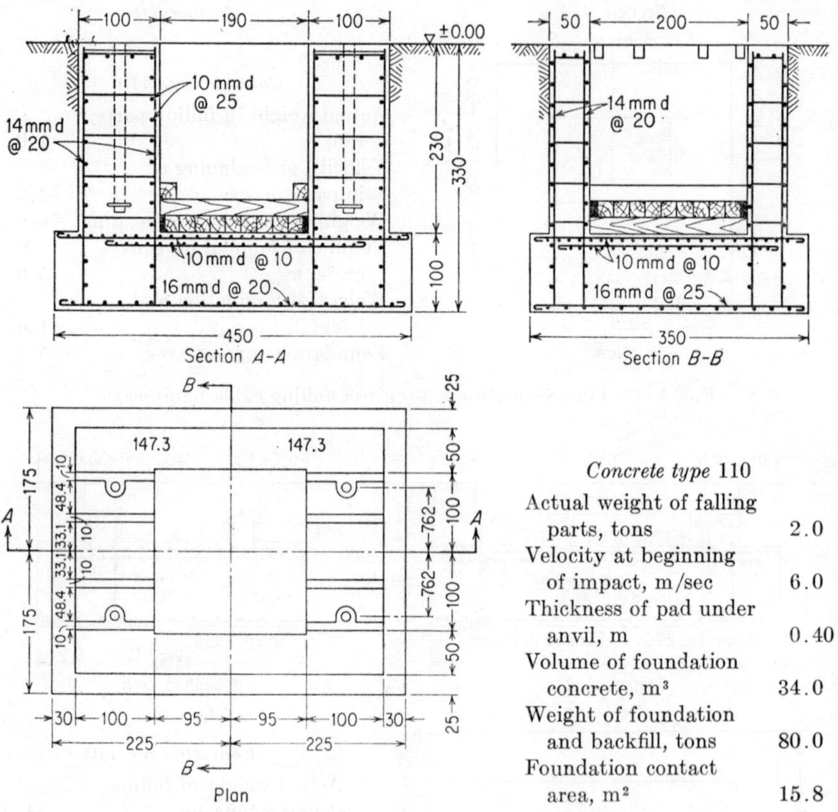

FIG. V-13. Foundation for 2-ton steam or compressed-air forging hammer.

decrease the amplitude of foundation vibrations by increasing the foundation contact area or the rigidity of the base.

Since foundation vibrations, if they are considered together with vibrations of the anvil, depend not only on the parameters of the foundation c and m_1 (i.e., on the rigidity of the base and the mass of the foundation beneath the springs), but also on the parameters c_2 and m_2 (i.e., on the rigidity of the absorbers and the mass of the foundation above the springs), theoretically it is possible to decrease the amplitudes of foundation vibrations by selecting suitable values of c_2 and m_2. These parameters should be selected so as not to increase sharply the vibration amplitude of the anvil in comparison with values customary in forging practice.

Thus the problem is reduced to the following: values of c_2 and m_2 should be found for which the corresponding vibration amplitudes of the foundation and anvil do not exceed selected values. Hammer characteristics (the weight of dropping parts, their velocity at the beginning of

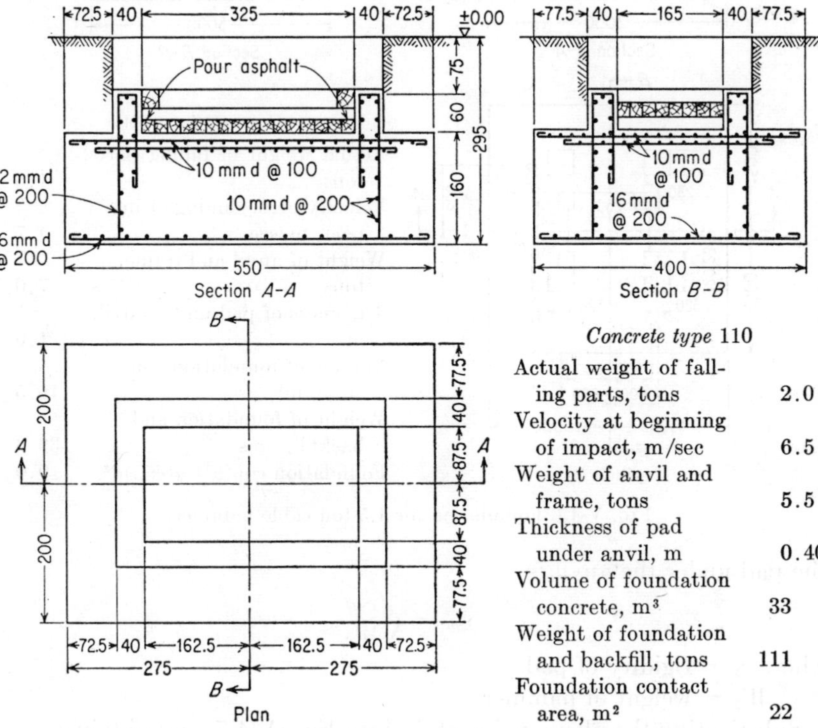

FIG. V-14. Foundation for 2-ton steam or compressed-air stamping hammer.

impact, and the coefficient of restitution) are considered to be assigned and fixed.

Let us consider vibrations of the foundation above the springs (the anvil), as a first approximation, to be a system with one degree of freedom. Then, according to Eq. (V-4-7), the amplitude of vibrations of this portion of the foundation equals

$$A_a = \frac{\alpha}{\sqrt{c_2 W_2}} \tag{V-7-1}$$

Equation (V-7-1) was derived from Eq. (V-4-7) as follows, assuming the foundation does not move: the natural frequency f_{na} of the hammer on

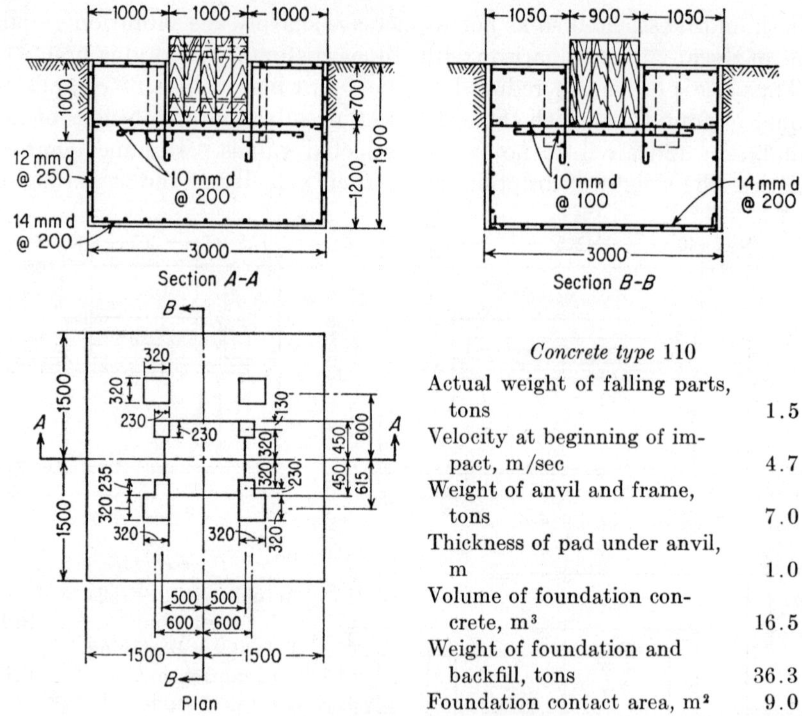

Concrete type 110

Actual weight of falling parts, tons	1.5
Velocity at beginning of impact, m/sec	4.7
Weight of anvil and frame, tons	7.0
Thickness of pad under anvil, m	1.0
Volume of foundation concrete, m³	16.5
Weight of foundation and backfill, tons	36.3
Foundation contact area, m²	9.0

FIG. V-15. Foundation for 1.5-ton cable hammer.

the pad under the anvil is

$$f_{na} = \sqrt{\frac{c_2 g}{W_2}}$$

where c_2 = rigidity of pad

W_2 = weight of hammer

Substituting the above value of f_{na} into Eq. (V-4-7), we obtain

$$A_a = \frac{(1+e)W_0 v}{W_2\sqrt{c_2 g/W_2}} = \frac{(1+e)W_0 v}{\sqrt{g}\sqrt{c_2 W_2}} = \frac{\alpha}{\sqrt{c_2 W_2}}$$

where α is a coefficient depending only on the characteristics of the hammer and equaling

$$\alpha = \frac{(1+e)W_0 v}{\sqrt{g}} \qquad \text{(V-7-2)}$$

We denote by z_2 the static settlement of the foundation above the springs on absorbers; then

$$z_2 = \frac{W_2}{c_2} \qquad \text{(V-7-3)}$$

We consider z_2 to be fixed.

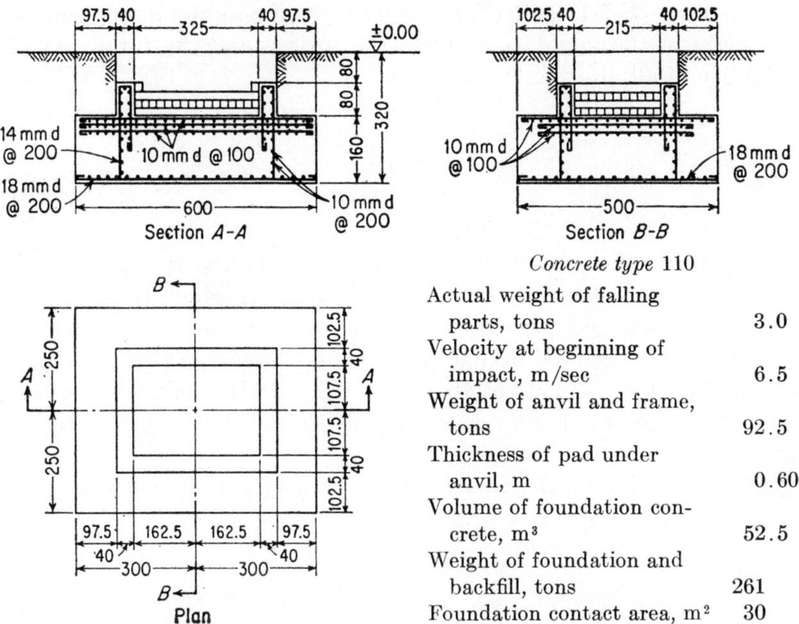

Concrete type 110

Actual weight of falling parts, tons	3.0
Velocity at beginning of impact, m/sec	6.5
Weight of anvil and frame, tons	92.5
Thickness of pad under anvil, m	0.60
Volume of foundation concrete, m^3	52.5
Weight of foundation and backfill, tons	261
Foundation contact area, m^2	30

FIG. V-16. Foundation for 3-ton steam or compressed-air stamping hammer.

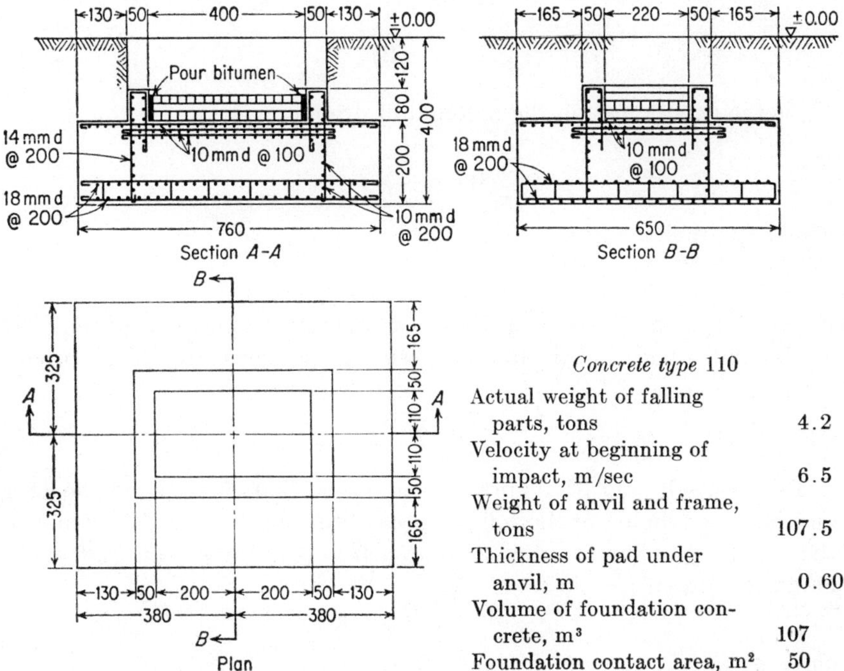

Concrete type 110

Actual weight of falling parts, tons	4.2
Velocity at beginning of impact, m/sec	6.5
Weight of anvil and frame, tons	107.5
Thickness of pad under anvil, m	0.60
Volume of foundation concrete, m^3	107
Foundation contact area, m^2	50

FIG. V-17. Foundation for 4.2-ton steam or compressed-air stamping hammer.

From Eqs. (V-7-1) and (V-7-3) we determine approximate values for the weight of the foundation above the springs and the total coefficient of rigidity of all absorbers:

$$W_2 = \frac{\alpha}{A_a} \sqrt{z_2} \tag{V-7-4}$$

$$c_2 = \frac{W_2}{z_2} \tag{V-7-5}$$

The mass and the area of the foundation under the springs in contact with the soil are selected on the basis of design considerations.

From the selected parameters of the system, the amplitudes of its vibrations are computed taking into account the fact that the system has not one but two degrees of freedom. The design values of vibration amplitudes computed from Eqs. (V-3-9) should not exceed permissible values.

The installation of absorbers should not decrease the efficiency of the hammer, which is

$$\eta = 1 - \frac{W_1 + K_{in}}{W}$$

where W is the work done by the dropping parts of the hammer, equaling

$$W = \frac{W_0 v^2}{2g}$$

W_1 is the energy lost on the rebound of dropping parts:

$$W_1 = \frac{W_0 v_1^2}{2g}$$

where, approximately,

$$v_1 = ev$$

K_{in} is the maximum value of the kinetic energy of vibrations of the anvil on the timber pad or of the foundation above the springs on the absorbers:

$$K_{in} = \frac{A_a^2 c_2}{2}$$

The values W and W_1 do not depend on the characteristics of the anvil and its base; therefore the following condition must be satisfied so that the efficiency of the hammer with absorbers is not less than that of the hammer without absorbers:

$$A_{a0}^2 c_{20} > A_a^2 c_2$$

The subscript 0 refers to the design without absorbers.

Since the amplitudes of foundation vibrations above the springs are about the same whether or not absorbers are used, the last condition may be rewritten as follows:

$$c_2 < c_{20}$$

Example. Computations for a stamping-hammer foundation with absorbers

1. DATA. The following specifications are given:
Weight of dropping parts of the hammer:

$$W_0 = 2.0 \text{ tons}$$

Weight of anvil together with frame:

$$W_a = 33.7 \text{ tons}$$

Coefficient of restitution:

$$e = 0.5$$

Velocity of dropping parts:

$$v = 6.0 \text{ m/sec}$$

Design values of permissible amplitudes are as follows:

For the anvil: $A_a = 3$ mm
For the foundation: $A_z = 0.2$ mm

2. COMPUTATIONS. The soil is of medium strength, with a coefficient of elastic uniform compression c_u equaling 3.3 kg/cm³. According to data of Art. V-4, the value of the coefficient of rigidity of the base under the hammer foundation will be

$$c_u' = kc_u = 3 \times 3.3 = 10 \text{ kg/cm}^3$$

Let us assume that the static settlement of the mass above the springs equals 0.01 m. We determine the hammer coefficient from Eq. (V-7-2):

$$\alpha = \frac{(1 + 0.5)2 \times 6.0}{\sqrt{9.81}} = 5.75$$

From Eq. (V-7-4) we determine the tentative value of the weight of the mass above the springs:

$$W_2 = \frac{5.75}{3 \times 10^{-3}} \sqrt{10^{-2}} = 191 \text{ tons}$$

The weight of the concrete block of the foundation above the springs, which is added to the weight of the hammer, is:

$$W_{f2} = 191 - 33.7 = 157.3 \text{ tons}$$

From Eq. (V-7-5) we determine the required rigidity of the absorbers:

$$c_2 = 191/10^{-2} = 19{,}100 \text{ tons/m}$$

The foundation above the springs is designed as a block of height 2.3 m and 6.0 by 5.0 m² in plan. It has a depression for the anvil, which is placed not on timber beams but on a pad made from steel wool.

The foundation under the springs is designed in the shape of a box. The thickness of the protecting walls will be 0.3 m. The cross-sectional dimensions of the columns which support the absorbers will be 0.45 m; the thickness of the supporting slab will be 1.0 m. Thus the area of the foundation beneath the springs in contact with soil equals

$$A = 7.76 \times 6.80 = 52.7 \text{ m}^2$$

Figure V-18 gives a sketch of the foundation with absorbers. The weight of the mass under the springs equals $W_1 = 180$ tons.

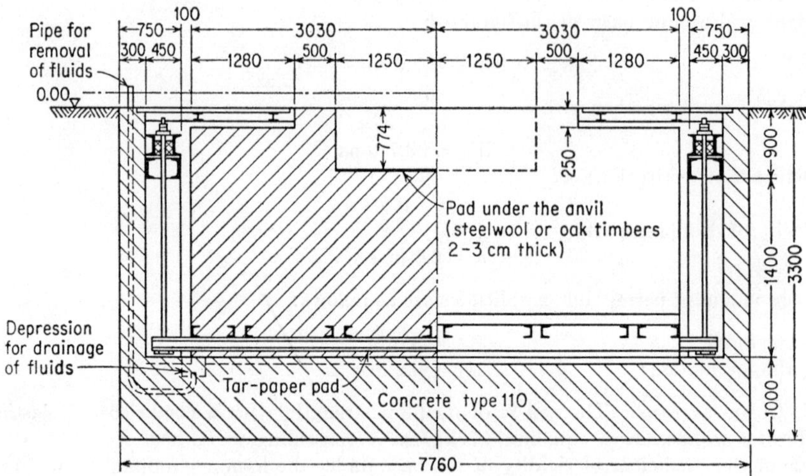

FIG. V-18. Design of foundation for example of Art. V-7.

The foundation both above and beneath the springs is built of properly reinforced concrete type 110.†

Vibrations are computed from Eqs. (V-3-9). The coefficient of rigidity of absorbers is

$$c_2 = 1.9 \times 10^4 \text{ tons/m}$$

The mass above the springs is

$$m_2 = 191/9.81 = 19.5 \text{ tons} \times \text{sec}^2/\text{m}$$

The coefficient of rigidity of the base under the foundation equals

$$c_1 = c'_u A = 10 \times 10^3 \times 52.7 = 52.7 \times 10^4 \text{ tons/m}$$

The mass of the foundation beneath the springs is

$$m_1 = 189/9.81 = 18.4 \text{ tons} \times \text{sec}^2/\text{m}$$

The square of the frequency of natural vertical vibrations of the foundation above the springs is

$$f_a{}^2 = \frac{c_2}{m_2} = \frac{1.9 \times 10^4}{19.5} = 0.975 \times 10^3 \text{ sec}^{-2}$$

† See footnote, Art. IV-1-c, p. 132.

The square of the frequency of vibrations of the foundation beneath the springs is

$$f_b^2 = \frac{c_1}{m_1 + m_2} = \frac{52.7 \times 10^4}{18.4 + 19.5} = 13.4 \times 10^3 \text{ sec}^{-2}$$

The ratio between masses is

$$\mu_1 = \frac{19.5}{18.4} = 1.06$$

We then set up the frequency Eq. (V-3-8):

$$f_n^4 - (0.975 \times 10^3 + 13.4 \times 10^3)(1 + 1.06)f_n^2 + (1 + 1.06) \times 13.4 \times 10^3 \times 0.975 \times 10^3 = 0$$

or

$$f_n^4 - 29.7 \times 10^3 f_n^2 + 27.0 \times 10^6 = 0$$

Solving this equation, we find the natural frequencies of the system:

$$f_{n1}^2 = 29.7 \times 10^3 \text{ sec}^{-2}$$
$$f_{n2}^2 = 0.946 \times 10^3 \text{ sec}^{-2}$$

We then determine the initial velocity of motion of the foundation above the springs:

$$v_a = \frac{(1+e)W_0 v}{W_0 + W_2} = \frac{(1 + 0.5) \times 2.0 \times 6.0}{(2.0 + 191)} = 0.093 \text{ m/sec}$$

The displacements of separate parts of the foundation are found from Eqs. (V-3-9). The displacement of the foundation beneath the springs is

$$z_1 = \frac{(0.975 \times 10^3 - 0.946 \times 10^3)(0.975 \times 10^3 - 29.7 \times 10^3)}{0.975 \times 10^3(29.7 \times 10^3 - 0.942 \times 10^3)} \times 0.093 \left(\frac{\sin f_{n1}t}{172} - \frac{\sin f_{n2}t}{30.8}\right) = -0.0199 \sin f_{n1}t + 0.109 \sin f_{n2}t \quad \text{mm}$$

The displacement of the foundation above the springs is

$$z_2 = \frac{0.093}{29.7 \times 10^3 - 0.946 \times 10^3}\left(\frac{0.975 \times 10^3 - 0.946 \times 10^3}{172} \sin f_{n1}t - \frac{0.975 \times 10^3 - 29.7 \times 10^3}{30.8} - \sin f_{n2}t\right) = 0.0007 \sin f_{n1}t + 3.03 \sin f_{n2}t \quad \text{mm}$$

Neglecting terms containing $\sin f_{n1}t$, we obtain for the amplitudes of vibrations

$$A_1 = 0.101 \text{ mm} \qquad A_2 = 3.03 \text{ mm}$$

Thus the selected dimensions of the foundation mass above the springs and the selected value of the coefficient of rigidity of the absorbers lead to an amplitude of vibrations of the foundation above the springs approximately one-half the design value (0.2 mm). The foundation will be practically motionless and no harmful influence will be exercised by the vibrating foundation on structures or on technological processes.

Absorbers are made of cylindrical standard springs used in railway rolling stock. The dimensions of the springs are chosen as follows:

Diameter of the coil D:	80 mm
Diameter of the spring d:	30 mm
Number of coils n:	5.5

If the number of absorbers is n_1 and the number of springs in each absorber is n_2, then the required rigidity of each spring will be

$$c_{sp} = \frac{c_2}{n_1 n_2}$$

On the other hand, as mentioned in Art. IV-6,

$$c_{sp} = 940 \frac{d^4}{D^3 n} 10^3 \qquad \text{tons/m}$$

Equating the left-hand parts of the two latter expressions, we obtain

$$n_1 n_2 = \frac{c_2 D^3 n}{940 d^4} 10^{-3}$$

Substituting here the corresponding numerical values as given above,

$$n_1 n_2 = \frac{1.91 \times 10^4 \times 8^3 \times 10^{-6} \times 5.5}{940 \times 3^4 \times 10^{-8}} \times 10^{-3} = 70$$

If each absorber is made of four springs, then the required number of absorbers will be

$$n_1 = \frac{70}{n_2} = \frac{70}{4} = 18$$

Thus the actual number of springs will be 72.

The permissible torsional stress for the spring is

$$T_0 = 40 \times 10^3 \text{ tons/m}^2$$

Then the permissible load on each spring will be

$$P_p = \frac{\pi d^3 T_0}{8D} = \frac{3.14 \times 3^3 \times 10^{-6} \times 40 \times 10^3}{8.8 \times 8 \times 10^{-2}} = 5.3 \text{ tons}$$

The rigidity of one spring equals

$$c_{sp} = \frac{1.91 \times 10^4}{72} = 264 \text{ tons/m}$$

The permissible deflection of one spring is

$$z_0 = \frac{P_0}{c_{sp}} = \frac{5.3}{264} = 0.020 \text{ m} = 20 \text{ mm}$$

The actual deflection will be smaller, namely,

$$z = z_2 + A_2 = 10 + 3.03 = 13.03 \text{ mm}$$

We design the absorber to be of the suspension type (see Art. IV-6). Each absorber, consisting of four springs, is placed in a case made of steel channels welded together. A general view of the absorber as designed is shown in Fig. V-19. The inside dimensions of the case 5, within which the springs are placed, are 248 by 248 mm. The case is fastened by bolts 14 to the lower supporting plate 2. The lower guide disks 4, for springs 10, are fastened by screws 11 to the same plate. The upper pressure plate 1 is placed on the springs and is also provided with guide disks for the springs. The mass below the springs may be lifted by tightening the regulating bolt 6 by means of nuts 9; the anchor cap of this bolt fits between two edges of girders built into the lower part of the foundation above the springs.

The walls of case 5 containing the absorbers are fastened on shelves formed by girder pieces embedded into projections of the walls of the foundation below the springs.

The cantilevers of girders embedded in the lower part of the foundation above the springs should be designed in such a way that bending stresses produced in them by the action of the weight and inertia forces of the foundation above the springs do not exceed a maximum value.

b. Construction Procedure for a Foundation with Absorbers. A foundation with absorbers should be constructed as follows:

1. Place the concrete of the foundation under the springs, walls, and projections.

2. Place two or three layers of Ruberoid or tar paper on the surface of the foundation slab.

3. Install girders on the Ruberoid or tar paper, thus forming a lower frame; projecting sills of this frame serve as a support for the anchor plates of regulating bolts of the absorbers. Thoroughly check all required dimensions and positions of girders, then weld the frame. Install the frame in a position corresponding to the design location of absorbers.

4. Install absorbers without tightening the springs; insert caps of regulating bolts between each pair of girders forming the frame.

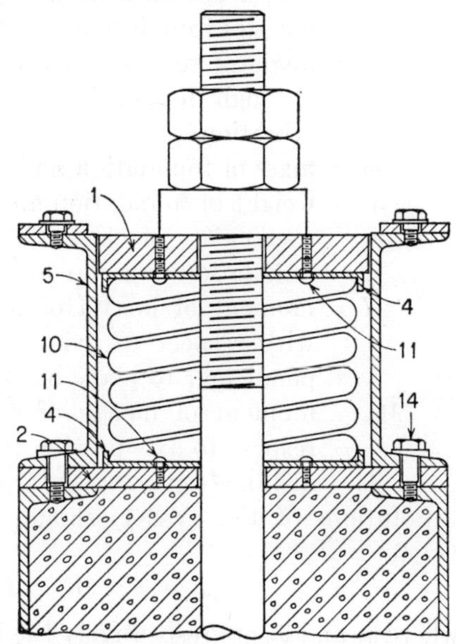

Fig. V-19. Design of vibration absorber for example of Art. V-7.

5. Place the concrete of the foundation above the springs.

6. After the required period of time has elapsed (not less than 10 days), erect the hammer.

7. When the hammer is mounted, lift the foundation above the springs. This is done by gradually tightening the regulating bolts so that the mass above the springs is lifted 1 to 1.5 cm without tilting. A level is used to check that no tilting has occurred.

8. Cover the absorbers and the foundation above the springs with a demountable metal plate.

When the foregoing procedure has been completed, the hammer is ready for operation.

V-8. Pressures in the Base under a Foundation Subjected to Horizontal Impacts

If horizontal impacts are transferred from an operating engine to a massive foundation, natural vibrations of this foundation will develop. When a horizontal impact occurs in one of the vertical principal planes of inertia of the foundation, the equations of this vibration do not differ from Eqs. (III-4-5):

$$\begin{aligned} m\ddot{x} + c_\tau A(x - L\varphi) &= 0 \\ M_m\ddot{\varphi} - c_\tau ALx + (c_\varphi I - WL + c_\tau AL^2)\varphi &= 0 \end{aligned} \tag{V-8-1}$$

where x = projection on a horizontal axis of displacement of center of mass of foundation

φ = angle of rotation of foundation with respect to axis passing through foundation mass center perpendicular to axis of vibrations

m = mass of foundation and engine

W = weight of foundation and engine

c_τ, c_φ = coefficients of elastic shear, elastic nonuniform compression

A = foundation area in contact with soil

I = moment of inertia of foundation area in contact with soil, with respect to axis passing through its centroid and perpendicular to plane of vibrations

M_m = moment of inertia of mass of foundation and engine with respect to axis passing through center of mass

L = distance between center of mass and foundation base

Solutions of Eq. (V-8-1) should satisfy the initial conditions; when $t = 0$,

$$x = \varphi = 0 \qquad \dot{x} = \dot{x}_0 \qquad \dot{\varphi} = \dot{\varphi}_0$$

where $\dot{x}_0$ and $\dot{\varphi}_0$ are respectively the initial velocities of forward motion in the horizontal direction and of rotation around a horizontal axis passing through the center of mass of the system. They are established from Eqs. (V-2-6) and (V-2-7).

Solutions of Eq. (V-8-1) which correspond to these initial conditions are as follows:

$$\begin{aligned} x &= \frac{1}{f_{n1}^2 - f_{n2}^2}\left(\frac{f_{nx}^2\dot{x}_{0c} - f_{n2}^2\dot{x}_0}{f_{n1}}\sin f_{n1}t - \frac{f_{nx}^2\dot{x}_{0c} - f_{n1}^2\dot{x}_0}{f_{n2}}\sin f_{n2}t\right) \\ \varphi &= \frac{1}{f_{nx}^2(f_{n1}^2 - f_{n2}^2)L}\left[\frac{(f_{nx}^2 - f_{n1}^2)(f_{nx}^2\dot{x}_{0c} - f_{n2}^2\dot{x}_0)}{f_{n1}}\sin f_{n1}t - \frac{(f_{nx}^2 - f_{n2}^2)(f_{nx}^2\dot{x}_{0c} - f_{n1}^2\dot{x}_0)}{f_{n2}}\sin f_{n2}t\right] \end{aligned} \tag{V-8-2}$$

where f_{nx} = frequency of natural vibrations in shear, accompanied by sliding of foundation

f_{n1}, f_{n2} = frequencies of foundation determined from solution of Eq. (III-4-8)

$\dot{x}_{0c}$ = initial velocity of centroid of foundation area in contact with soil

$$\dot{x}_{0c} = \dot{x}_0 - L\varphi$$

From these solutions for x and φ it is possible to find the dynamic stresses which develop in the base of a foundation as a result of an impact. Dynamic compressive stress near the edge of the foundation contact area will be

$$\sigma_\varphi = -c_\varphi b \varphi \qquad \text{(V-8-3)}$$

where $2b$ is the length of the foundation contact area.

V-9. Foundations (Bases) for Drop Hammers to Break Scrap Iron

a. Location of Drop Hammer within the Steelworks. Special drop hammers are installed at metallurgical works for breaking up scrap iron, pigs, and large blocks. These hammers are distinguished by the great kinetic energy of ram impact required to break the scrap. While in double-acting 5-ton forge hammers the kinetic energy at the moment of impact against the forged piece does not exceed 10 to 12 tons × m, in modern powerful drop hammers used in scrap yards the kinetic energy of the dropping ram attains 150 tons × m. Therefore these hammers may become a powerful source of elastic waves spreading through soil; sometimes they may also have a harmful influence on various technological processes.

Because of this, scrapyards with drop hammers should be located as far as possible from other structures. It will be shown in Chap. VIII that the propagation of waves through soil is greatly affected by soil properties; therefore the minimum permissible distance from a shop with drop hammers will depend on soil conditions. In addition, it is clear that the greater the kinetic energy of the hammer, the greater the energy of the waves propagated through soil, and consequently, other conditions being equal, the larger should be the values of minimum permissible distances between the drop hammer and other structures.

The location of a scrapyard with drop hammers also depends on the character of technological operations in certain structures and on the vibration amplitudes which are permissible in connection with these operations. It is clear that the distance between a scrapyard with drop hammers and a warehouse can be much smaller than that between a scrapyard and a laboratory with precision instruments or a shop where precision machines are operating.

Generally it is not possible to establish by means of computations the dependence of the minimum safe distance from a scrapyard with drop hammers on the three factors indicated above. In each case this problem should be solved on the basis of the following data: (1) results of experimental investigations of wave propagation at the construction site under study; (2) values of permissible vibration amplitudes for local technological processes; (3) data on construction characteristics of structures.

Table V-8 gives data on tentative values of minimum distances depending on power of drop hammers and soil conditions.

TABLE V-8. DATA ON MINIMUM DISTANCES BETWEEN DROP-HAMMER INSTALLATIONS USED TO BREAK UP SCRAP IRON AND OTHER STRUCTURES

Soil conditions	Minimum distances to the drop hammers, m, for ram weights of:		
	Up to 3 tons	3–7 tons	Over 7 tons
Plastic clays, clays with some sand and silt, moist sands	30	50	Over 70
Sands, clays, clays with some sand and silt below ground-water level	30	50	Over 70
Swamp soils	50	80	Over 100
Dry sandy soils, hard clays, clays with sand and silt, loess, loessial soils	30	40	Over 60
Rocks	20	30	Over 50

b. Design of Crushing Platforms under Drop Hammers. Up to the present time, bases under crushing platforms have been designed according to methods which have much in common with methods of design and computation of forge-hammer foundations. However, the energy of the dropping part (ram) of a crushing drop hammer is many times greater than the energy of extremely powerful forge hammers; therefore if forge-hammer foundation requirements are applied to foundations for crushing platforms, the latter turn out to be extremely heavy blocks sometimes weighing more than a thousand tons.

Figure V-20 shows a massive foundation designed for a breaking hammer with a 10-ton ram weight and a drop height of 30 m. The foundation for the crushing platform was designed similarly to foundations and anvils under forge hammers; the only difference is that sand and crushed rock were used as a pad under the metal anvil instead of the oak timber girders generally used in hammer foundations. To reduce the cost of construction, the lower part was made of cyclopean concrete; the upper part is of heavily reinforced concrete type 130.† The total

† See footnote, Art. IV-1-c, p. 132.

weight of the whole structure reaches 900 tons, the depth of the foundation 9 m, the foundation area in contact with soil 85 m^2.

Technically and economically, such massive foundations cannot be considered rational. The ram velocity of the hammer just described at the moment of impact is 24.2 m/sec, and the kinetic energy is 295 ton $\times$ m. If one considers that the impact occurs not against the scrap lying on the anvil, but directly against the anvil, so that the coefficient of impact velocity restitution is of the order of 0.5, then the foundation should undergo vibrations of an amplitude within the range 5 to 15 mm,

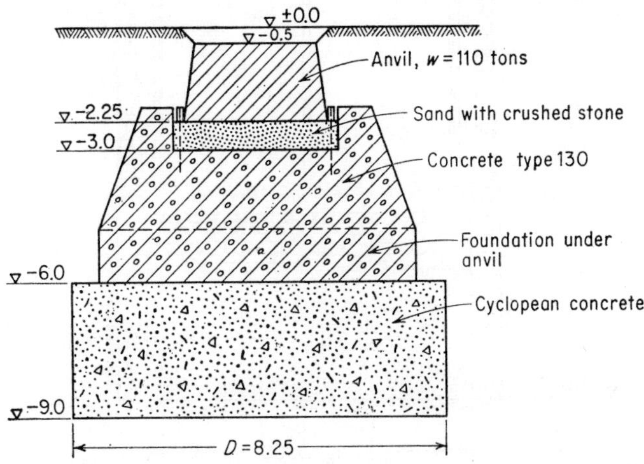

FIG. V-20. A heavy foundation for a scrap-crushing hammer installation.

depending on soil conditions. For a soil with a coefficient of elastic uniform compression equaling 5 kg/cm^3, the amplitude of foundation vibrations will be 7.5 mm, and the dynamic pressure on soil will be of the order of 4 kg/cm^2. The impact of a ram weighing several tons dropping from a height of 20 to 39 m will induce large stresses in the anvil and foundation. Therefore the foundation under the anvil should be made of concrete of better quality and should be thoroughly reinforced. In spite of this, cases have been recorded in which the anvil and the portion of the foundation under the anvil were destroyed in the operation of drop hammers breaking up scrap.

Foundations for crushing platforms can also be designed as hollow cylinders made of reinforced concrete and filled with sand and small scrap. Figure V-21 shows a sketch of this type of base. In order to increase the efficiency of the whole installation, the largest possible degree of compaction should be achieved in filling up the cylinder. Sand may be used for filling; compaction can be accomplished by means of vibrations applied to successive layers, each layer about 0.5 m thick. The sand is covered by a layer of broken scrap to a thickness of 1.5 to 2 m and mixed with the sand which has been subjected to vibration. To protect against flying chips, joists are suspended on hinges from a metallic ring installed above the cylinder and are tied to each other by a rope.

If the walls of the cylinder are sufficiently high, they may screen waves propagated inside of the cylinder and thus may prevent the propagation

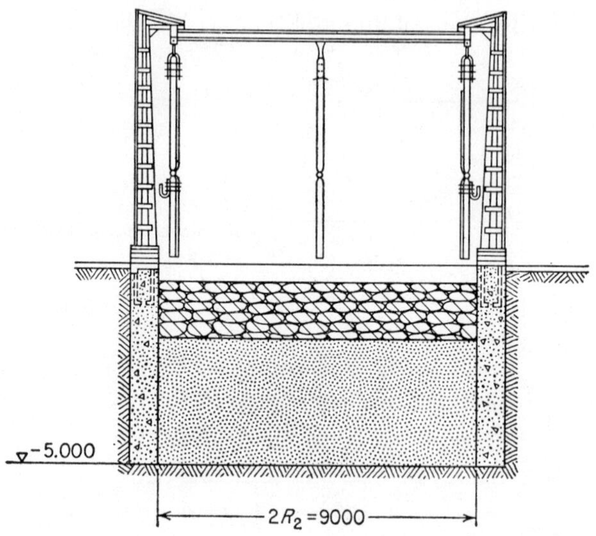

FIG. V-21. Design of foundation for hammer to break up pig-iron scrap.

of waves in the soil beyond it. Therefore a cylindrical foundation may be especially useful if the soils are dangerous in regard to the spreading of vibrations and settlement under the action of vibratory loads.

The larger the depth of the cylinder, the larger its effect on the screening of waves. The frequency of waves induced by the ram impact is smaller in loose soils than in dense soils; therefore, other conditions being equal, cylinder depth should be larger in poor soils than in strong soils. Waves propagated in soil under the action of an impact may be classified as of high frequency, since the number of such waves is of the order of 1,000 min^{-1} and more. As we shall see in Chap. VIII, the dimensions of a screening device should be selected according to the frequency of the propagating waves. For waves of 1,000 cycles/min and more, the depth of the screen in soils of medium strength should not be less than 5 to 6 m.

There is no known accurate stress analysis of a hollow cylinder, filled with a material whose strength properties are other than its own, subjected to the action of elastic waves propagated inside of the cylinder.

Therefore the stress analysis of the cylinder is necessarily limited by a very rough approximation of the real distribution of stresses in its walls.

Let us determine approximately the amplitude of vibrations of the ram which is dropped on the broken scrap; the equation of its vibrations will be

$$\ddot{z} + 2c\dot{z} + f_{nz}^2 z = 0 \tag{V-9-1}$$

where c = damping constant; its value for the case under consideration may be approximately taken to equal 0.5 to 0.7 f_{nz}

f_{nz} = frequency of natural vertical vibrations of ram, equaling

$$f_{nz} = \frac{c_u A g}{W}$$

c_u = coefficient of elastic uniform compression of base subjected to impacts; its value may be approximately taken to be of the order of 3 to 5 $\times$ 10^3 kg/cm^3

A = base area of ram

W = weight of ram

After impact, for a certain time (equal to one-fourth the period of its natural vibrations) the ram will be pressed into the scrap. Its velocity at the beginning of impact will be

$$v = \sqrt{2gh} \tag{V-9-2}$$

where h is drop height.

Let us take as the start of readings the instant at which the ram touches the scrap. The solution of Eq. (V-9-1) for the time $0 < t < T/4$ (where T is the period of natural vibrations of the ram on the scrap) will be

$$z = A_z \sin f_{nz1} t \tag{V-9-3}$$

where A_z is the maximum penetration of the ram into the scrap.

$$A_z = \frac{v}{f_{nz}} \exp\left(\frac{-\pi c}{2f_{nz}}\right) \tag{V-9-4}$$

$$f_{nz1} = \sqrt{f_{nz}^2 - c^2} \cong 0.7 f_{nz}$$

Impact of the ram against the scrap will induce an elastic wave spreading from the point of impact over the volume contained by the cylinder. In the first approximation this wave may be considered to be a spherical three-dimensional wave. After the wave reaches the cylinder walls, it will exert a pressure on them, inducing stresses therein.

Neglecting the absorption of wave energy by the medium filling the cylinder, it can be approximately estimated that the wave amplitudes decrease in an inverse proportion to distance from their source.

If A_z is the amplitude of the source acting on the surface in an area of radius R_0, then the amplitude A_r of soil at a distance r from the source may be approximately taken as

$$A_r = A_z \frac{R_0}{r} \tag{V-9-5}$$

The amplitude of the wave component perpendicular to the cylinder wall is

$$A = A_r \frac{R_2}{\sqrt{R_2^2 + \xi^2}} \tag{V-9-6}$$

where R_2 = inside radius of hollow cylinder

ξ = depth of element of cylinder wall under consideration

Using Eqs. (V-9-4) and (V-9-5) and substituting $\sqrt{R_2^2 + \xi^2}$ for r, we obtain

$$A = \frac{v_0}{f_{nz}} \frac{R_0 R_2}{R_2^2 + \xi^2} \exp\left(-\frac{\pi c}{2f_{nz}}\right) \tag{V-9-7}$$

When computing A we neglected the absorption of vibrations by the mass inside of the cylinder. Besides, it was assumed that all the impact energy is consumed only by the formation of elastic waves. As a matter of fact, a considerable part of the energy is spent in breaking up the iron blocks and on vibrations of the cylinder, together with the mass it contains, acting as a solid body on an elastic base. Therefore the assumption is possible that the values of A established by Eq. (V-9-7) are larger than actual. However, taking into account that the dynamic wave propagating in the mass contained by the cylinder exerts a dynamic pressure on the cylinder walls, it is possible, in static computations of strength, to evaluate stresses with sufficient accuracy by using the amplitudes established from Eq. (V-9-7).

It follows from the condition stipulating continuous contact between the mass included in the cylinder and the cylinder walls that the amplitudes A computed for the soil may be taken as equaling the amplitudes of elastic expansion of the cylinder walls. Stresses in the material of the cylinder should be established from these amplitudes.

The elastic impact wave is not propagated instantaneously in the mass inside the cylinder, but with a certain finite velocity of the order of 2,000 to 2,500 m/sec. First it exerts a pressure on the elements of the wall situated at the level of the impact. As the wave travels in a downward direction, it exerts a pressure on the lower elements of the cylinder. Consequently, under the action of the moving wave, the cylinder is subjected to a nonuniform pressure along its height which results in a bending of the cylinder walls as shown by the dashed line in Fig. V-22. It is very difficult to take into account stresses which develop in the cylinder walls as a result of this bending, but it is necessary to provide longitudinal reinforcement along the inner and outer faces of the cylinder (for example, rods of 16 mm diameter spaced every 0.30 to 0.40 m).

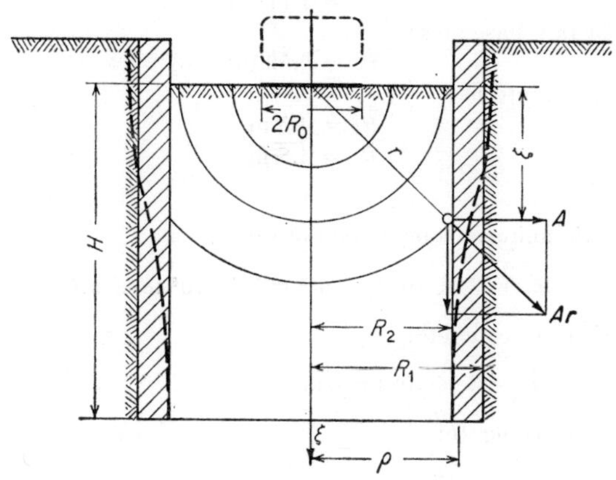

FIG. V-22. Estimation of stresses in cylindrical walls of foundation in Fig. V-21.

For computation of radial and tangential stresses in the cylinder walls, formulas for thick-walled cylinders may be used.† For radial stresses:

$$\sigma_r = \frac{R_2{}^2}{R_1{}^2 - R_2{}^2}\left(1 - \frac{R_1{}^2}{\rho^2}\right)q \tag{V-9-8}$$

For tangential stresses:

$$\sigma_\varphi = -\frac{R_2{}^2}{R_1{}^2 - R_2{}^2}\left(1 + \frac{R_1{}^2}{\rho^2}\right)q \tag{V-9-9}$$

where R_1 is the outside radius of the cylinder and

$$R_2 < \rho < R_1$$

The magnitude of internal pressure acting on a cylinder ring at a depth ξ below the level of impact is determined by the formula

$$q = -\frac{EA}{R_2[(R_1{}^2 + R_2{}^2)/(R_1{}^2 - R_2{}^2) + \nu]} \tag{V-9-10}$$

where ν and E are the Poisson ratio and the modulus of elasticity.

† Cf., for example, S. P. Timoshenko and J. M. Lessels, *Applied Elasticity*, Westinghouse Technical Night School Press, East Pittsburgh, Pa., 1923. (Translated in German and Russian.)

Example. Dynamic computations for a cylindrical base under a crushing platform

1. GIVEN DATA. A scrap hammer has the following dimensions:
Weight of ram breaking up scrap:

$$W = 5 \text{ tons}$$

Area of the ram base:

$$A = 1 \text{ m}^2$$

Reduced radius of ram base area:

$$R_0 = 0.56 \text{ m}$$

Radii of the cylinder: internal:

$$R_2 = 4.5 \text{ m}$$

external: $R_1 = 4.8 \text{ m}$

2. ASSUMED DATA

Coefficient of elastic uniform compression of scrap:

$$c_u = 3 \times 10^3 \text{ kg/cm}^3 = 3 \times 10^6 \text{ tons/m}^3$$

Modulus of elasticity of concrete:

$$E = 3 \times 10^6 \text{ tons/m}^2$$

The Poisson ratio for concrete:

$$\nu = 0.35$$

Damping constant for vibrations of the ram on scrap:

$$c = 0.7 f_{nz}$$

3. CALCULATIONS. We determine the frequency of natural vibrations of the ram:

$$f_{nz} = \sqrt{\frac{3 \times 10^6 \times 1 \times 9.81}{5}} = 2.44 \times 10^3 \text{ sec}^{-1}$$

From Eq. (V-9-2) we determine the velocity of ram motion at the instant of impact against scrap:

$$v_0 = 2 \times 9.81 \times 15 = 17.1 \text{ m/sec}$$

From Eq (V-9-7) we find the amplitude of the normal component of displacement caused by the wave propagating in scrap as a result of impact. The computation is performed for the highest stressed upper zone of the cylinder, i.e., where $\xi = 0$.

$$A = \frac{17.1}{2.44 \times 10^3} \frac{0.56}{4.5} \exp\left(-\frac{0.7\pi}{2}\right) = 0.29 \times 10^{-3} = 0.29 \text{ mm}$$

From Eq. (V-9-10) we find the value of dynamic pressure of the wave on the upper zone of the cylinder:

$$q = \frac{-3 \times 10^6 \times 0.29 \times 10^{-3}}{4.5(4.8^2 + 4.5^2)/(4.8^2 - 4.5^2) + 0.35} = -12.9 \text{ tons/m}^2 = -1.3 \text{ kg/cm}^2$$

Assuming in Eq. (V-9-9) that $\rho = R_2$, we find the maximum value of the tangential stresses in the cylinder wall:

$$\sigma_\varphi = \frac{4.5^2 \times 2}{4.8^2 - 4.5^2} 12.9 = 172 \text{ tons/m}^2 = 17.2 \text{ kg/cm}^2$$

An adequate reinforcement should be installed to resist these tensile stresses. Since the latter decrease along the depth of the cylinder, the lower zones must be reinforced less intensively than the upper zones. From computations performed on the basis of the above formulas, it is easy to see that in a cylinder having, for example, a depth of 5 m, the lower zone will be under the action of tensile stresses which are approximately two times smaller than those acting on the upper zone.

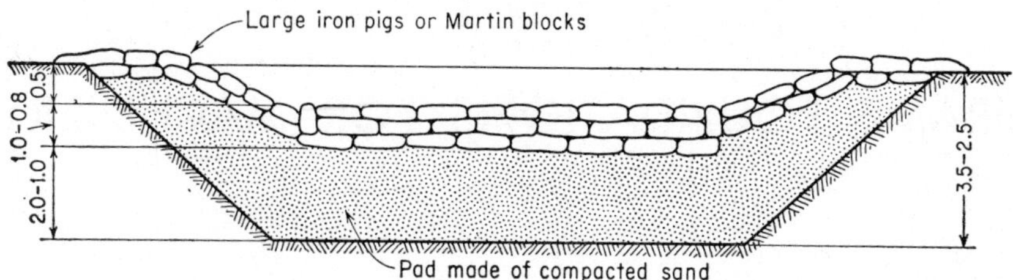

FIG. V-23. Crushing platform on good soil at a considerable distance from building.

If soils are relatively strong and no buildings or shops with technological processes which may be affected by vibrations are located nearby, the base of the crushing platform may be made "without a foundation." It may be formed by iron blocks and scrap placed directly on soil or on a layer of a compacted sand, as shown in Fig. V-23.

VI

FRAME FOUNDATIONS FOR MACHINERY

VI-1. Instructions for the Design and Construction of Frame Foundations

a. Field of Application of Frame Foundations. Frame foundations do not limit a designer in the location of the engine and its auxiliary equipment as do massive foundations. For example, condensers, pipelines, air vents, and electric wiring for turbodynamos and electromotors can be arranged much more conveniently if the machines are mounted on frame foundations.

The use of frame foundations facilitates considerably the inspection of and access to all parts of the machine. Therefore frame foundations are often employed for turbodynamos (turboblowers, turbocompressors, and turbogenerators) of varying power. In the course of recent years a tendency has appeared, in the practice of foundation design for these engines, to limit the use of frame foundations to low-power turbodynamos only (up to 10 to 12,000 kw), and to use massive foundations for turbodynamos of higher power. However, this tendency is not at all justified, since observations of frame foundations under high-power dynamos (up to 100,000 kw) show that these foundations are in many cases more economical than massive foundations and, as has been indicated, they are advantageous in many respects in regard to the mounting and maintenance of the engine. In addition, investigations established that very often cracks are formed in massive foundations under turbodynamos due to the stresses induced by settlement or by temperature changes, while no cracks due to these causes are observed in frame foundations.

Frame foundations can also be successfully used for various electrical machines, such as motor generators, synchronous compensators, high-power dynamos, and electromotors, in which no sudden changes in load occur.

Lately there have been cases in industrial design practice where frame foundations were used for reciprocating engines, in particular for compressors. This use of frame foundations is most rational in cases where for some reason a foundation should have a considerable height; this may happen, for example, if it cuts through a basement.

We shall not consider frame foundations under reciprocating engines, because these foundations are of such high rigidity that they should be computed as rigid bodies resting on elastic bases.

b. Design Assignment. In addition to data on soil conditions, the following information is required for the design of a foundation:

1. Foundation diagrams showing dimensions, distribution and required sizes of pipelines, tunnels, channels, grooves, and openings in the foundation, and distribution and sizes of foundation bolts and pads under bolts
2. A Design Assignment for the installation of the condensation floor within the limits of the edge of the lower slab of the foundation
3. A Design Assignment for the installation of a platform around the turbosystem at floor level of the machine room
4. Data concerning the layout of auxiliary equipment, in particular chambers of the air-cooling apparatus and the generator outlets
5. A diagram of static loads acting on the foundation, imposed by both stationary and rotating parts (the magnitudes of loads and the points of their application should be indicated)
6. Power of the engine in kilowatts and speed
7. The distribution of hot pipelines and the temperatures at the outer insulation surfaces

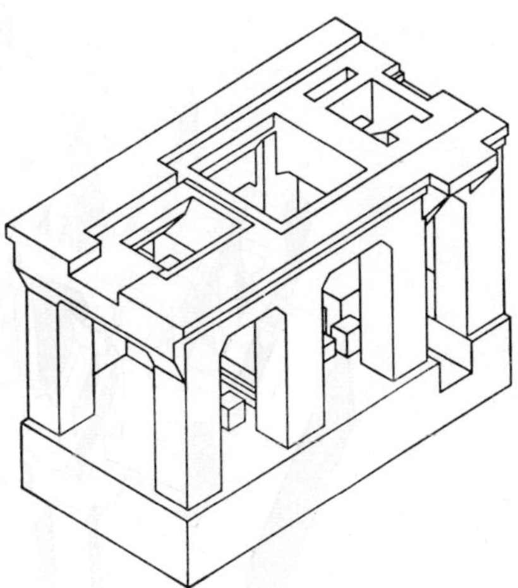

Fig. VI-1. Frame foundation.

c. Instructions for the Design. A frame foundation (Fig. VI-1) is usually designed to be built of three or more transverse frames embedded in a sufficiently thick foundation slab. At the top these frames are tied together by longitudinal girders and an upper (erection) platform having openings necessary for stationary machine parts. Often a layout of the frame foundation is more complicated. Transverse walls are inserted between columns of the transverse frames, or two-story frames are used. Sometimes the rigidity of transverse frames is increased by structural measures to such a degree that the foundation cannot be considered an elastic frame system, but should be treated as an absolutely rigid body.

Figure VI-2 shows an isometric projection of a frame foundation with transverse walls, designed for a 100,000-kw turbogenerator. The computations for such a foundation, particularly dynamic computations, are very complicated. Therefore the foundation should be designed so that the diagram of stresses transferred from the machine to the base is as simple as possible and secures the most efficient distribution of internal stresses in the foundation, as well as the simplest forms of foundation

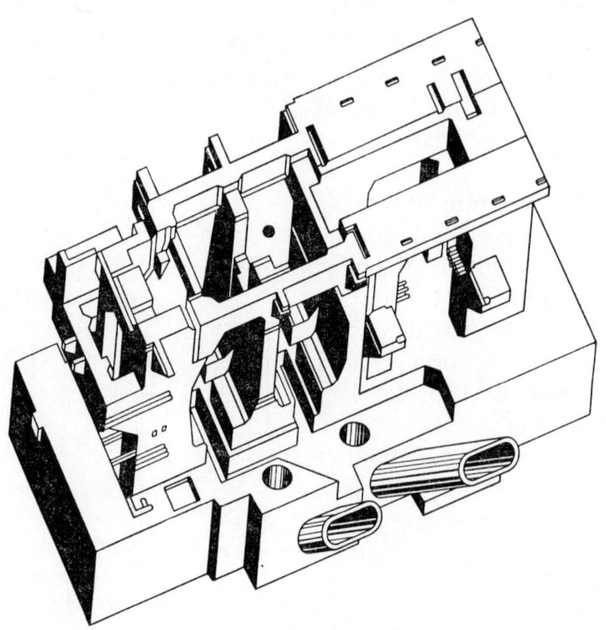

FIG. VI-2. Isometric view of a frame foundation for a 100,000-kw turbogenerator.

vibrations. In this respect, the foundation design should satisfy the following conditions:

The geometric layout of the foundation, the shapes of girder cross sections, and their reinforcement should be basically symmetric with respect to a vertical plane passing through the rotation axis of the engine. The frame beams should be placed directly under bearings, so that centrifugal forces which develop during engine operation are transmitted directly to the transverse frames. Axes of columns and transverse frame beams should lie in the same vertical plane perpendicular to the rotation axis of the motor. To prevent the appearance of torsional stress in transverse girders, eccentric loading of the latter should be avoided as much as possible. The direction of the load should, if possible, pass through the center of gravity of the beam cross section. The beams and girders should be designed of rectangular or T-shaped cross sections. In accordance with the official *Technical Rules and Construction Code*, the minimum cross-sectional dimensions of unloaded elements should be 15 cm for slabs and 25 cm for girders.

The upper erection platform of the foundation should be as rigid as possible in its plane. One method of achieving this is to extend the longitudinal and edge transverse beams towards the outer faces of the foundation. If one attempts to increase the rigidity of the upper platform by extending the dimensions of the horizontal elements of the foundation in the direction of surfaces which limit the space assigned for the installation of machine parts, this change in dimensions should be coordinated with the machine manufacturer.

In order to increase the general rigidity of the frame foundation, haunches should be provided at the intersections of beams and columns.

Turbodynamos and electrical machinery are relatively safe in regard to the transmission of vibrations to buildings. No cases are on record of vibrations of entire buildings induced by these machines. However, occasionally it happens that turbodynamos cause objectionable local vibrations in columns, isolated wall sections, and especially floors and other building elements. An extensive instrumental investigation of foundations under turbogenerators was conducted by the author.[4] In the course of this investigation, considerable vertical floor vibrations were found in places where the foundation was rigidly connected with the floor of the machine room. These vibrations, especially when caused by high-frequency machines with speeds of, for example, 3,000 rpm, produce a very adverse effect on people standing on the vibrating sections, as they cause an unpleasant feeling in the soles of the feet. The vibrations also result in the displacement of pieces of equipment not tied to the floor. These phenomena are observed during floor vibrations with an amplitude of 0.02 mm. For this amplitude and a frequency of 3,000 oscillations per minute the vibration acceleration is about 0.2g.

In order to decrease the transfer of vibrations from the upper erection platform of the foundation under the turbogenerator to the building, and particularly to the floor of the machine room, it is recommended that a gap be provided around the entire contour of the upper foundation platform. The floor beams should be placed on separate columns supported by footings independent of the machine foundation.

Foundations under low-frequency electrical machines cannot produce the floor vibrations described above, since the frequency of natural vibrations of the foundations is considerably higher than the operational frequencies of the machines. Therefore in the design of these foundations there is no necessity to provide a gap between the foundation and the floor of the machine room. Bearing floor elements may be supported directly by the frame beams and the columns of the foundation under low-frequency electrical machines. In some cases such a support may be effective in decreasing the amplitudes of machine foundation vibrations.

Maintenance records of foundations under high-frequency turbo-systems indicate cases of relatively large vibrations of cantilevered parts of the erection platform of the foundation.

Figure VI-3 gives graphs of the distribution of amplitudes of vertical vibrations of cantilevered elements along one of the foundations investigated. These graphs show that in some places the amplitudes of vibrations reached 0.06 mm, which corresponds to an acceleration of vibrations equaling some $0.5g$. Vibrations with such high acceleration resulted in the formation of cracks in the erection platform. Figure VI-3 indicates

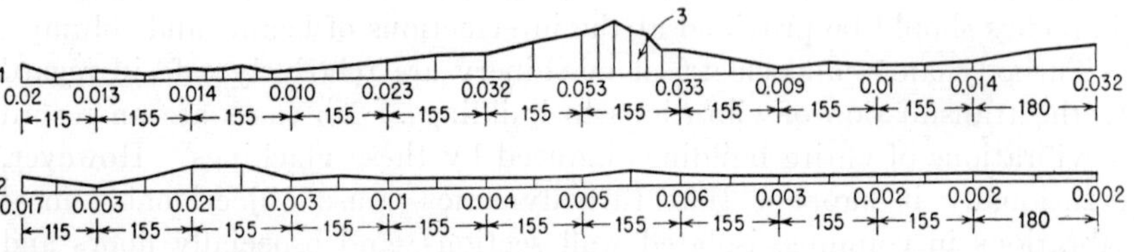

Fig. VI-3. Recorded vibrations (in millimeters) of a floor platform cantilevered around a machine foundation.

(1) a crack in the cantilevered slab and in the edge girder; (2) the zone of assumed deformation in the platform; (3) a crack in the edge girder.

Such vibrations occur only when the frequency of natural vibrations of the erection platform, acting as a cantilever of variable cross section, is close to the frequency of machine rotation. Therefore the cantilever elements of the foundation erection platform should be designed to be much more rigid than is required by static computations; their frequencies of natural vibrations should be much higher than the frequency of machine rotation.

The cantilevered elements of the erection platform usually are T beams of variable cross sections; therefore the computation of the frequencies of natural vibrations of these elements involves some difficulties and is extremely laborious. The design of the erection platform should ensure sufficient rigidity of such cantilever elements. This may be achieved by the installation of a rigid circumferential edge beam resting directly on the cantilevers; another method consists in the installation of special rigid stiffeners. The cross-sectional height of the cantilever at the embedment point should be no less than 60 to 75 per cent of its span.

Turbodynamo and electrical-machine bearings should be thoroughly adjusted, and the shafts should be in strictly horizontal position. Therefore designs of foundations under these machines should ensure proper centering of the masses. For many machine foundations an eccentricity in mass distribution is permissible up to 5 per cent of the side of the foundation area in contact with soil, in the direction in which displacement of the center of gravity occurs. The eccentricity in turbodynamos and electrical machines should, if possible, come close to zero; in any case, its value should not exceed 1 to 2 per cent.

The author studied 36 foundations under turbogenerators and found that only in 2 foundations the amplitudes of vertical vibrations of the lower slabs were 0.002 to 0.003 mm; in 3 foundations the amplitudes of vertical vibrations of the slabs were on the order of 0.001 mm; the amplitudes of vibrations of the lower slabs of the remaining foundations were smaller than 1 micron (0.001 mm). The vibration amplitudes of the foundation slabs were much smaller than the amplitudes of vertical vibrations of the upper parts of the foundations.

The results of instrumental measurements of foundation vibrations lead to the conclusion that in practice the lower slabs of foundations under turbogenerators are not subjected to vibrations and consequently do not transmit any dynamic pressure to the base. Therefore the pressure on the soil under turbogenerator foundations is determined only by static loads, i.e., by the weight of the foundation and equipment thereon. Hence it is clear that it is not necessary to follow the traditions of recent practice in assigning design pressures under turbogenerators not to exceed 0.5 to 0.6 of the permissible pressure on soils determined with respect to static loading only.

The introduction of a coefficient equaling 0.5 to 0.6 and the reduction of permissible pressure on the soil led to the necessity for employing piles, and consequently to considerable rise in construction cost. It should be noted that the above-mentioned extensive investigation of machine foundations established that the use of pile foundations did not safeguard against considerable settlements and tilting of foundations under turbogenerators.

The lower foundation slabs under turbogenerators practically do not vibrate at all; therefore the coefficient of reduction of permissible pressure on soil may be taken to equal 0.8 to 1.0.

The depth of foundation under turbodynamos and electrical machines has no effect on the transmission of vibrations to adjacent structures. Therefore, when necessary because of design considerations or other reasons, the depth of foundation under the machine may be made even smaller than the depth of footings under walls or columns. If a foundation under a turbodynamo is to be erected close to footings under walls, columns, and other machines, then special care should be taken to protect it from nonuniform stresses imposed by adjacent footings. Hence foundations under turbodynamos and electrical machines should be placed at such distances from adjacent foundations that the pearshaped lines of stresses (i.e., the "pressure bulb") in the soil imposed by the latter do not distort significantly the symmetry of the lines of stresses under the machine foundations in question. For the same reason, in some cases it may be useful to increase somewhat the depth of machine foundations with respect to the depth of adjacent footings under walls or columns.

The lower foundation slab should be sufficiently rigid to secure proper embedment of the foundation columns and prevent their nonuniform settlement. In addition, the presence of a lower foundation slab having considerable thickness decreases the height of the common center of gravity of the machine and foundation. Therefore the thickness of the lower foundation slab is usually taken larger than required by static computations. Tentative values of the height of the foundation slab, depending on the power of the machine, are taken as follows:

For machines with power up to 6,000 kw:	0.8 to 1.2 m
For machines with power of 6 to 12,000 kw:	1 to 1.6 m
For machines with power of 12 to 25,000 kw:	1.6 to 2 m
For machines with greater power:	2 to 4 m

Modern turbodynamos use steam of high temperature; consequently proper thermic insulation of steam pipes and air lines conducting hot air should be provided. The pipes should be insulated at least until they leave the foundation. The temperature at the outside surface of the insulation should not exceed 40 to 50°C; otherwise considerable local temperature stresses may develop in the foundation. Therefore the installation of steam and air pipes directly inside the foundation is objectionable.

Frame columns and beams are either reinforced according to design computations or the reinforcement is fitted to field conditions.

In foundation slabs having a thickness of 1.0 m, the vertical reinforcing rods should reach the area in contact with soil. In higher slabs, it is permissible to cut 50 per cent of the reinforcing rods at the half height of the foundation slab. Relevant chapters of the official *Technical Rules and Construction Code for Design of Reinforced-Concrete Structures* should be used in the design of foundation units, and, in addition, the following directions should be taken into account: All units of the foundation should be provided with double reinforcement. A symmetric reinforcement should in all cases be employed in the columns. Reinforcing rods should also be installed along the other two sides of cross sections of beams and columns, even if they are not required by design computations. The amount of reinforcement in separate foundation units should be no less than 30 kg/m³ of concrete. The distance between stirrups in beams should not exceed 25 cm, and in columns 35 cm. To resist stresses induced by settlement, reinforcing rods of 8 to 10 mm diameter are to be installed along three mutually perpendicular directions in massive units of the foundation and are to be spaced 50 to 60 cm apart. The upper and lower reinforcements of the lower foundation slab should be tied together by stirrups (dowels) and spaced 50 to 50 cm apart in a checkerboard pattern. Hooks are to be provided at the ends of steel rods, subjected both to tensile and to compressive stresses. When the reinforcement for the columns is designed, it should be kept in mind that the total steel area of vertical reinforcing rods in a column should be smaller than the total cross-sectional area of the anchor foundation bolts. Additional reinforcing rods should be placed in sections where the foundation is weakened by openings, ducts, etc.

Concrete type 110† is employed for the upper parts of frame foundations, and concrete type 90† is used for the lower foundation slabs.

d. Instructions for Construction Operations. The construction of foundations under turbodynamos and electrical machines should proceed in accordance with all requirements of the applicable official *Technical Rules and Construction Code.*

It should be noted that large cracks observed in foundations under operating turbodynamos and electrical machines are in most cases caused by careless construction work. The construction of foundations under these machines should be carried out with particular care, since the performance of these machines affects the normal work of many plants. Special care should be taken in meeting the following requirements in regard to construction procedures:

Concrete employed for the erection of the foundation should be of plastic consistency, without excessive water; a slump test should show that the cone slump is around 10 to 12 cm (4 to 5 in.). The same concrete mix should be used throughout the construction of the whole upper part of the foundation. The forms for the upper part of the foundation should be fitted with grooves and planed on their inner surface.

Concrete should be poured continuously in horizontal layers. In an emergency, an interruption may be permitted at the level of the upper edge of the lower slab, or at the level of one-third of the column height, where the bending moment has a minimum value. If an interruption in the work occurs, the following measures should be taken to secure the monolithic character of the foundation:

1. Along the cross section of the foundation, where the pouring of concrete was interrupted, 16-mm reinforcing rods should be added to those installed according to the design. Short dowels should be

† See footnote in Art. IV-1-c, p. 132.

embedded to a depth of not less than 0.5 m on both sides of the joint, and their spacing should not exceed 0.2 m.

2. The surface of the joint should be rough. Prior to placing a new layer of concrete, the previously laid surface should be thoroughly cleaned, washed by water, and covered with a rich cement mixture.

As a rule, the placing of the foundation concrete should be mechanized, and a uniform distribution of the concrete aggregates should be assured. A segregation of concrete aggregates into layers usually occurs in places where it is delivered from considerable height. If anchor bolts are embedded into the foundation to considerable depth, it is recommended that pipes of corresponding cross sections be inserted for these bolts; these pipes remain in the concrete permanently. During the concreting of the foundation, the quality control of concrete and of its aggregates is essential, and sample cubes of concrete are to be taken for investigation of its strength properties in accordance with special instructions. All essential points in the process of foundation construction should be recorded in special documents. In any case, the following documents should be compiled: (1) a record of the nature of the soil in the excavation made for the foundation; (2) a record of changes in the type of concrete used for the foundation; it should be noted at what elevation such changes took place; (3) a record concerning the interruption in concreting, if such an interruption occurred; the place where this interruption took place should be noted with a description of measures taken to secure a proper joint; (4) a record of the condition of the concrete after the forms were removed; the length of time the concrete remained in the forms should be noted.

In the process of machine assembly, prior to pouring cement under the machine bedplate, the adjoining foundation surface should be cleaned thoroughly. This surface (erection platform) should be rough to secure the best possible binding of the additionally poured cement to the foundation.

The location of all openings, recesses, etc., should be carefully checked against design drawings.

VI-2. Computations of Forced Vibrations of Frame Foundations

a. Exciting Loads Imposed by Turbodynamos and Electrical Machines. The exciting loads imposed by turbodynamos and electrical machines, unlike those of reciprocating engines and impact mechanisms, cannot be established by computations.

The main moving units of these machines are rotors which execute simple rotating movements. Theoretically the center of gravity of the rotor coincides with the axis of rotation, and consequently the theoretically established values of unbalanced inertia forces equal zero.

However, actual conditions are different. In any engine containing rotating parts, even if this engine is well balanced, there remains a certain unbalanced state caused by the fact that the center of gravity of the rotating parts does not exactly coincide with the axis of rotation. This residual unbalanced state cannot be completely eliminated, and in the course of the machine operation there appear unbalanced inertial forces which induce foundation vibrations.

The magnitude of these exciting loads is proportional to the eccentricity of the rotating parts, the magnitudes of their masses, and the square of the frequency of machine rotation. Rotors of high-power turbodynamos and electrical machines weigh tens of tons, and their speeds can be very large—up to 10,000 rpm. Therefore even for minute eccentricities of rotating masses the magnitudes of the exciting loads may be very large. Consequently, their influence should be taken into account in the design of foundations. For a long time the magnitudes of exciting loads imposed by turbodynamos were unknown; therefore in computations of foundations for turbodynamos some "temporary" loads were taken into account. The static action of these loads was assumed to be equivalent to the dynamic action of actual exciting loads caused by the unbalanced state of the engine.

Many suggestions were offered concerning the selection of these equivalent loads. However, all these suggestions were equally ungrounded, and design computations of foundations were reduced to static stress analyses of the action of arbitrarily selected loads.

However, in the course of recent years, voluminous material has been collected in the U.S.S.R. concerning the balancing of turbodynamos and electromotors, as well as measurements of vibrations of these machines. This material makes it possible to establish design values of exciting forces caused by these machines with a degree of accuracy sufficient for practical purposes. Thus it is no longer necessary to introduce into computations the previously mentioned static equivalents of loads. For the same reason, the method of foundation design changes: instead of computations taking into account static-equivalent loads, computations are performed of forced vibrations of foundations produced by exciting forces and moments. Consequently the foundation may be so designed that forced vibration amplitudes do not exceed permissible values.

This method of design of foundations under turbodynamos and electrical machines does not differ in principle from methods of design accepted, for instance, for foundations under reciprocating engines.

Let us assume that the exciting loads developed by the machine under consideration can be reduced to one unbalanced centrifugal force F, whose plane of action coincides with the plane of symmetry of the machine rotor. This unbalanced state is generally called the static unbalanced state. The unbalanced state of the rotor may be caused by the fact that in addition to the exciting force there exists also an exciting moment (the dynamic unbalanced state). Rotors of electromotor generators, as well as rotors of steam turbines, are short; if the machine under consideration has one rotor whose plane of symmetry coincides with a vertical plane passing through the center of gravity of the foundation, then the exciting moment will be small and may be neglected.

The determination of unbalanced loads is much more difficult for machines with several rotors. If a machine has two rotors (which occurs in the majority of cases), then the force F acting in the vertical transverse plane of the whole installation (the foundation and machine) can be considered in computations as a design exciting load. Then the exciting moment equals

$$M = Fl_s \tag{VI-2-1}$$

where l_s is the distance along the axis of the main shaft between the resultant of exciting forces and the center of mass of the whole installation.

The exciting force of the rotor, being the unbalanced centrifugal inertial force, will rotate with the same frequency as the machine. Therefore the vertical and horizontal components of the exciting force will equal

$$\begin{aligned} F_z &= r_0 m_0 \omega^2 \sin \omega t \\ F_x &= r_0 m_0 \omega^2 \cos \omega t \end{aligned} \tag{VI-2-2}$$

where r_0 = eccentricity of machine rotor
m_0 = mass of rotor
ω = rotation frequency
F_z, F_x = vertical, horizontal components of exciting force (which act in a plane perpendicular to machine shaft)

The exciting moment can be resolved in the same manner into its vertical and horizontal components. Under the action of the vertical component of this moment, the foundation will undergo forced vibrations in the plane parallel to the main shaft of the machine. Measurements show that foundation vibrations often occur in this plane. However, the amplitudes of these vibrations usually are small in comparison with the amplitudes of vertical and horizontal vibrations in a direction perpendicular to the shaft of the machine. Therefore dynamic computation of the foundations under turbodynamos and electrical machines may be limited to computation of the amplitudes of vibrations induced by the exciting force and the horizontal component of the exciting moment, which equals

$$M_z = F_x l_s \tag{VI-2-3}$$

The frequency of rotation and the mass of rotating parts of a turbodynamo or an electrical machine are known; consequently, the eccentricity (unbalance) should be known in order to determine the exciting loads acting on the foundation. The eccentricity can be tentatively determined only from the results of balancing of machines and from measurements of vibrations before and after balancing.

Let us assume that the rotor of the machine under consideration has a static unbalanced state, defined by the force F_0, which causes an amplitude of forced vibrations equal to A_0. Let us further assume that in the process of balancing an additional mass was attached to the rotor at some distance from the axis of rotation. This mass produced a centrifugal force F. Then we assume that as a result of the balancing, the amplitude of forced vibrations decreased to the value A.

There exists a simple proportional relationship between the magnitude of the exciting force and the amplitude of forced vibrations it produces; therefore

$$F_0 - F = \frac{A}{A_0} F_0$$

Hence,

$$F_0 = \frac{A_0}{A_0 - A} F$$

With this relationship it is possible to determine the value of the initial unbalanced exciting force from the results of balancing and measurements of vibrations before and after balancing. From the value of the initial unbalanced exciting force, the mass of the rotor, and the frequency of its rotation, it is easy to determine the eccentricity:

$$r_0 = \frac{F_0}{m_0 \omega^2}$$

Table VI-1 presents data on the balancing of several turbogenerators of various types and powers. From these data we computed values of F_0 and r_0.

All the data presented in Table VI-1 refer to balancing carried out under operating conditions, at times when, in the opinion of workers, machines vibrated with increased amplitudes. Therefore the computed values of exciting forces and eccentricities lie within the range of maximum permissible values.

The amplitudes of vibrations were measured, not on the foundations, but on the bearings, at the same place and in the same directions, both before and after balancing. Vibrations were measured by means of a Geiger vibrograph, which in some cases yields considerably exaggerated readings. In spite of this, the presented values of F_0 and r_0 permit some conclusions which are of interest for dynamic computations of foundations under machines of the type being considered.

Table VI-1 shows that the eccentricity r_0 depends on both the power of the machine and its speed. For machines characterized by 1,500 rpm

TABLE VI-1. RESULTS OF BALANCING MACHINE ROTORS

No.	Power, kw	Weight of rotor, tons	Amplitude of vibrations, mm		F, tons	F_0, tons	r_0, mm
			Prior to balancing A_0	After balancing A			
			Machines with 1,500 rpm				
1	5,000	12.0	0.160	0.035	1.86	2.38	0.086
2	3,000	5.0	0.075	0.042	0.43	0.97	0.085
3	3,000	5.0	0.125	0.023	0.92	1.12	0.097
4	3,000	5.0	0.125	0.025	0.76	0.95	0.082
5	3,000	5.0	0.092	0.027	0.30	0.43	0.038†
6	3,000	5.0	0.600†	0.058	1.23	1.36	0.118
7	3,000	6.0	0.062	0.026	0.52	0.80	0.070
8	50,000	70.0	0.260	0.017	11.41	12.2	0.076
9	50,000	70.0	0.350†	0.030	23.5	25.8	0.160
10	50,000	70.0	0.43†	0.035	25.4	27.3	0.170
11	50,000	37.5	0.150	0.058	3.97	6.5	0.075
			Machines with 3,000 rpm				
12	17,500	18.3	0.087	0.066	1.04	4.3	0.025
13	16,000	18.0	0.157	0.060	1.77	2.87	0.017
14	16,000	7.0	0.133	0.056	1.64	2.84	0.045
15	16,000	18.0	0.170	0.042	2.36	3.10	0.019
16	16,000	18.0	0.140	0.025	8.25	10.0	0.060†
17	25,000	20.0	0.170	0.058	5.10	7.75	0.042
18	25,000	17.5	0.120	0.040	3.02	4.53	0.029
19	6,000	7.2	0.180	0.030	2.18	2.62	0.040
20	6,000	7.2	0.125	0.021	1.00	1.20	0.018

† This figure is not reliable.

and 3,000 kw power, the value of eccentricity lies in the relatively narrow range from 0.070 to 0.118 mm. In only one case the eccentricity was much smaller (0.038 mm). For machines with the same speed but 50,000 kw power, the value of eccentricity varies within a somewhat wider range: from 0.075 to 0.170 mm. However, the order of eccentricity values for these machines remains similar to that for low-power engines. Consequently, the increase in the magnitude of the exciting force observed with an increase in power is mainly explained by the increase in the mass of rotating parts.

The data of Table VI-1 which refer to machines characterized by 3,000 rpm show that the eccentricity for these machines lies in the range 0.017 to 0.045 mm. For these machines, as for machines with speeds of 1,500 rpm, the value of the exciting force grows with an increase in power, and consequently with the weight of the rotor.

Thus there is a difference between the orders of eccentricity for machines with speeds of 1,500 rpm and 3,000 rpm. The maximum value of eccentricity for the low-speed machines may be taken as 0.20 mm, but for machines running at 3,000 rpm the maximum eccentricity does not exceed 0.05 mm. Consequently, it can be held that the eccentricities of the rotating masses of turbodynamos are approximately inversely proportional to the squares of their speeds. As the number of revolutions increases, the weight of rotating machine parts (provided the power is the same) decreases; therefore high-speed turbodynamos are better balanced.

Generalizing the above relationship between the eccentricity and the number of revolutions, and selecting 0.20×10^{-3} m as the design value of eccentricity for machines having a speed of 1,500 rpm, we obtain the following expression for a machine running at N rpm:

$$r_N = \frac{500}{N^2} \qquad \text{meters} \tag{VI-2-4}$$

This relationship may serve as a basis for the selection of tentative design values of eccentricity for rotating machine masses characterized by different speeds.

b. Modulus of Elasticity of Reinforced Concrete. In the design of reinforced-concrete structures subjected only to the action of static loads, the computations mainly determine maximum stresses and deformations appearing under the action of primary loading. In this connection, it is interesting to analyze the behavior of reinforced concrete subjected to primary loading, or loading of the same sign.

The official *Technical Rules and Construction Code* gives values of the modulus of elasticity of concrete established as a result of tests performed on concrete samples under increasing loading. These tests established a mean value of the modulus of elasticity of concrete $E_c = 210{,}000$ kg/cm^2 to be used in the design of reinforced-concrete structures.

When concrete is subjected to primary loading, the relationship between load and deformation is nonlinear; therefore the modulus of elasticity depends on the magnitude of the load and on its sign.

An experimental study of the behavior of reinforced concrete under imposed loads shows that even small stresses result in a simultaneous appearance of elastic and residual deformations. The relative value of residual deformations grows with increase in load; it grows faster than the load. Therefore a "loading-strain" curve tends to turn in the direction of deformations, and the stress-strain relationship in concrete, above a certain value of stresses, has a nonlinear character; therefore the modulus of elasticity depends on the magnitude of stress.

Consequently, the modulus of elasticity established as a result of investigations in which the irreversible part of deformation was not separated from the total deformation is not the actual modulus of elasticity of the material, just as the coefficient of subgrade reaction of soil is not the coefficient of elastic uniform compression of soil. The modulus of elasticity of concrete, which is usually employed in design computations for stresses smaller than the proportionality limit, represents its modulus of linear deformability. The modulus of elasticity may be established after determining the relationship between stresses and the elastic part of deformation. Corresponding static investigations should be carried out by means of repeated loading and unloading of samples.

The amplitude of vibrations and natural frequencies of reinforced-concrete structures depend on the elastic properties of the material, but not on its characteristics corresponding to residual deformations; therefore the modulus of elasticity of concrete may be determined in the simplest way from natural or forced vibrations.

As a result of measurements of vibrations of one frame foundation under a pump and two frame foundations under turbogenerators, the following values of the modulus of elasticity of reinforced concrete were found:

For the foundation under the pump:

$$3 \times 10^6 \text{ tons/m}^2$$

For the first foundation under the turbogenerator:

$$4.2 \times 10^6 \text{ tons/m}^2$$

For the second foundation under the turbogenerator:

$$5.78 \times 10^6 \text{ tons/m}^2$$

These values of E_c are much higher than those usually used for static computations.

N. P. Pavliuk and O. A. Savinov investigated a two-column frame made of concrete type 160† and found the modulus of elasticity to be some

† See footnote in Art. IV-1-c, p. 132.

4×10^6 tons/m². As a result of the investigation of a four-column frame of concrete of about the same type, it was established that the value of E_c lay within the range 4.62 to 3.5×10^6 tons/m².

From results of static investigations of reinforced-concrete beams made of the same concrete, the modulus of elasticity was established to be within the range 3.77 to 3.17×10^6 tons/m².

The modulus of elasticity of concrete may be established by the electro-acoustic method, in which a sound generator excites in the sample

TABLE VI-2. RESULTS OF ACOUSTIC DETERMINATION OF YOUNG'S MODULUS OF CONCRETE

Composition of concrete	Age of concrete, days	Young's modulus, tons/m²
1:2.55:2.55	7 28	3.6×10^6 3.81×10^6
1:3.0:3.0	7 28	3.02×10^6 3.81×10^6
1:1.93:3.23	7 28	3.53×10^6 4.11×10^6
1:2.6:4.05	7 28	4.32×10^6 3.96×10^6
1:3.76:3.0	7 28	3.10×10^6 3.67×10^6
1:4.65:6.18	7 28	2.95×10^6 3.31×10^6

under investigation longitudinal or transverse vibrations of varying frequency. In this way the frequency of natural vibrations of the sample is determined, from which the modulus of elasticity can be easily established. In the course of recent years, this method has been widely used for the determination of elastic constants of very different materials.

Table VI-2 presents results of one such determination of the moduli of elasticity of various types of concrete. It is seen from the table that the modulus of elasticity does not change much with changes in the composition of concrete. At the same time, the test results show that the modulus of elasticity increases with an increase in the age of concrete. Absolute values of the modulus of elasticity at an age of 28 days were in no case smaller than 3.0×10^6 tons/m²; the average value established from six determinations was 3.78×10^6 tons/m².

Thus the experimental data show that the actual modulus of elasticity of concrete is much larger than 2×10^6 tons/m^2; i.e., it is larger than the value established by the official *Technical Rules and Construction Code.*

The foregoing discussion makes it possible to consider that the actual value of Young's modulus for concrete (at an age not less than 28 to 30 days) is not less than 3×10^6 tons/m^2. This value of the modulus should be taken for design computations.

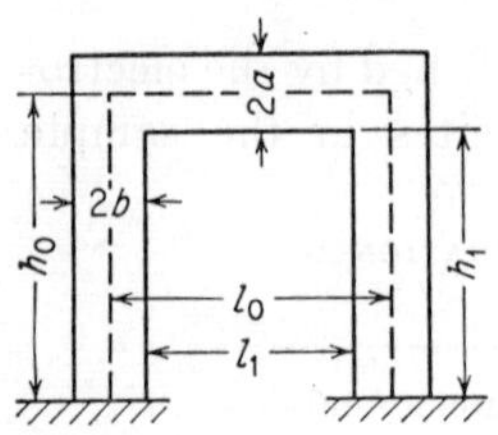

FIG. VI-4. Design values for computations of frame vibration frequencies.

c. Design Dimensions of the Upper Part of the Foundation. The cross-sectional dimensions of foundation units are usually much larger than the spans between them; hence the influence of the rigidity of corner sections of the frame should be taken into account. If one is to consider the corner sections as being absolutely flexible, then in the determination of deflections and bending moments of separate frame elements, the span l_0 (Fig. VI-4) and the height h_0 should be introduced into the computations.

However, if one considers the frame corner sections as being absolutely rigid, it becomes necessary to use in computations the inside free span l_1 and the inside free height h_1 of the frame. For usual foundation sizes, the value of l_0 often exceeds the value of l_1 by 25 per cent. Formulas for deflection computations contain the value of the span in the third or fourth power. Therefore design values of the span and height considerably affect the results of computations, in particular the value of the natural vibrations of the frame. For example, if one is to calculate frequencies of natural vibrations of a frame having $l_0 = 5.50$ m and $l_1 = 4.00$ m, then for different design values of the span (from $l = l_0$ to $l = l_1$), the frequency of vibrations computed for $l_0 = 5.50$ m will be (for the case in which the frame is loaded only by its own weight) approximately two times smaller than the frequency for the case where $l = 4.00$ m.

As a matter of fact, the frame corner sections are neither absolutely flexible nor absolutely rigid; therefore design values of the span and height of the frame should be smaller than l_0 and h_0, and larger than l_1 and h_1. They should be determined from the following expressions:

$$\begin{aligned} l &= l_0 - 2\alpha b \\ h &= h_0 - 2\alpha a \end{aligned} \qquad \text{(VI-2-4}a\text{)}$$

where $2\,a$ = height of frame beam (Fig. VI-4)

$2\,b$ = width of column

The value of coefficient α is taken from the graph (Fig. VI-5). When this graph is used, intermediate values of h_0 and l_0 should be determined by interpolation.

If haunches are provided in frame or beam corner sections, the values of a and b are taken as shown in Fig. VI-6.

d. Rigidity of the Upper Platform of the Foundation. The upper foundation platform on which the machine is placed is formed by longitudinal beams which tie the transverse frames together and a reinforced-

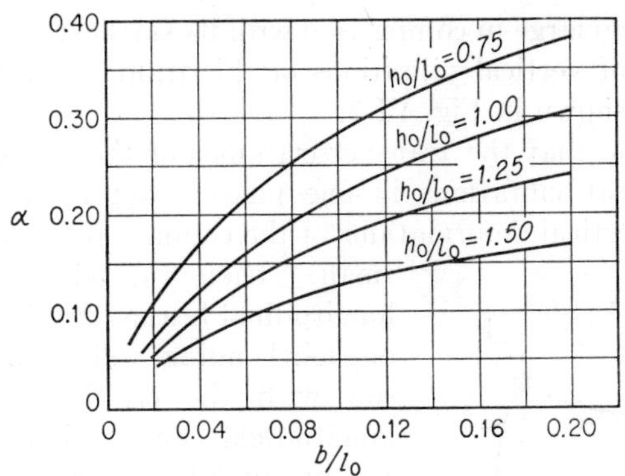

FIG. VI-5. Graph for determination of coefficient α in Eqs. (VI-2-4*a*).

concrete slab provided with openings required for pipes, machine parts, condensers, and so on. The rigidity of this platform depends essentially on the rigidity and relative position of the bedframe of the machine.

The upper foundation platform together with the bedframe of the machine represents a structure which is extremely rigid in the horizontal direction; therefore, for an approximate determination of the amplitude of horizontal vibrations of the foundation, the upper platform may be considered to be absolutely rigid. This assumption simplifies computations of horizontal vibrations of the frame foundation but does not involve large errors in the results of computations.

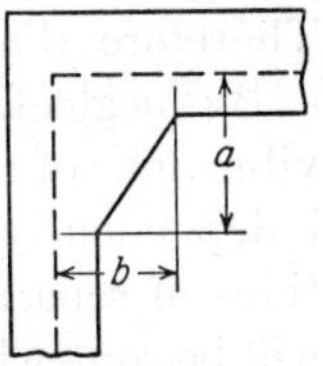

FIG. VI-6. Values of a and b to be used in Eqs. (VI-2-4*a*) if a frame has haunches.

e. Computation of Forced Vertical Vibrations. The foregoing assumptions reduce a dynamic computation of the frame foundation to computations of the amplitudes of forced vibrations of a three-dimensional frame system consisting of thin beams and columns embedded in an absolutely rigid slab. The latter rests on an elastic base, i.e., on soil.

Although it is possible to obtain rigorous solutions to the problem of forced vibrations of such a system,[30] the solutions obtained are so cumbersome and lead to such complicated calculations that they are of little use for practical purposes. Therefore several assumptions are necessary which will simplify the solution of the problem without influencing its accuracy and will make the solution practical.

When vertical vibrations occur, the most intensive deformations are observed in the beams of the transverse frames. The columns of the frames are deformed less, and the corner haunch sections of the plane transverse frames do not deform at all. If a separate frame vibrates at a frequency not too large in comparison with its smallest natural frequency, then the form of vertical vibrations of this frame approximately corresponds to that shown in Fig. VI-7.

Let us assume that the transverse frames of the foundation are subjected to vertical vibrations in one phase. Let us also assume that differences in vertical deformations of the columns of separate frames are small. Then the elastic resistance of longitudinal beams, developing as a result of their bending, will be small in comparison with the elastic resistance against longitudinal compression of the columns. Longitudinal beams are usually fixed only at the corner sections of the transverse frames. Therefore the vertical vibrations of a transverse frame are also affected by the resistance to torsion of the longitudinal beams. This resistance is also small, as is the resistance resulting from the bending of the longitudinal beams. Therefore the influence of longitudinal beams may be disregarded.

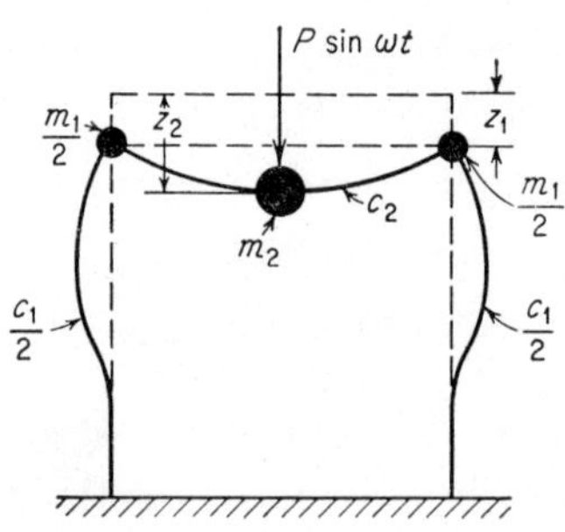

FIG. VI-7. Vibrations of a separate frame.

By neglecting the influence of longitudinal beams on the vertical vibrations of the transverse frames, it is possible to consider the vibrations independently of each other. The natural frequencies of vertical vibrations of separate frames, calculated on the basis of such an assumption, will be somewhat smaller than the actual values. In order to compensate for the influence of the longitudinal beams, we shall disregard the actions of other factors which affect natural frequencies in an opposite manner. These factors include the shearing forces and the inertia of rotation of cross sections of the units of the frame.

A. I. Lur'ye[29] showed that the frequencies of natural vibrations of the frames may be strongly influenced by the elasticity of the base under the foundation. Computations show that the frequencies of natural vertical vibrations, computed with consideration of the elasticity of the base under the foundation, may differ by 10 to 20 per cent from the values computed without taking this factor into account.

The frequency of natural vertical vibrations f_{nz} of the frame considered to be a rigid solid resting on an elastic base is usually much smaller than the operational frequency of the turbodynamo. We denote by f_{n1} the smallest frequency of natural vertical vibrations of the frame; assuming that the base is absolutely rigid, this frequency depends only on the elastic and inertial properties of the frame. For turbodynamos there usually exists the following interrelationship:

$$f_{nz} < \omega < f_{n1} \tag{VI-2-5}$$

where ω is the frequency of rotation of the machine.

For low-speed electrical machines (for example, motor generators) the frequency of rotation may also be smaller than f_{nz}; therefore the following interrelationship may exist:

$$\omega < f_{nz} < f_{n1} \tag{VI-2-6}$$

If one is to consider the elasticity of the base under the foundation, then the two smallest frequencies of vertical vibrations of the system "frame–rigid-slab–elastic-base" will have the following interrelationship with the frequencies f_{nz}, ω, and f_{n1}:

$$f^*_{nz} < f_{nz} < \omega < f_{n1} < f^*_{n1}$$

The latter inequalities show that consideration of the elasticity of soil leads to an increase in the fundamental frequency of natural vibrations based only on the elastic and inertial properties of the frame.

If requirement (VI-2-5) or (VI-2-6) is satisfied in the design of a foundation, then neglecting the elastic properties of soil in the computation of vertical vibrations of the foundation contributes to an increase in the safety factor of the dynamic stability of the foundation. Consequently, the computation of forced vibrations of a frame foundation may be reduced to the computation of vibrations of plane frames resting on an absolutely rigid base. Then the columns of each frame may be considered as being rigidly embedded in an immovable foundation slab.

Let us consider the loads acting on a transverse foundation frame. Each frame usually supports one of the machine bearings. The width of bearings supporting the rotating machine parts usually is small in comparison with the length of the beams. Therefore the load transmitted by the bearings may be considered concentrated and located in the middle of the frame beam. The static load is the part of the rotor weight resting on this bearing. In addition, the same bearing transmits to the frame an exciting vertical force $P \sin \omega t$. The frame beam is also subjected to the action of a uniform load imposed by its own weight. Let us replace it by an equivalent concentrated load located at the center of the frame beam span. In order that this change should not influence the results of dynamic computations, the magnitude of this equivalent load should be selected so that the kinetic energy of the system will not change. Without furnishing proof here, let us note that the magnitude of an equivalent mass, selected on the basis of the above condition, corresponds to 45 per cent of the dead weight of the frame beam.

The frame columns are subjected to the following loads:

1. The loads imposed by adjacent longitudinal frame beams, including their own weight. These loads, falling on each frame, are computed according to the laws of statics. Loads imposed by longitudinal beams may be considered concentrated at the tops of the columns, since the longitudinal beams are supported by the corner sections of the transverse frames.

2. The loads imposed by the weight of the transverse beam, also concentrated at the tops of the columns. On the basis of the same considerations which governed the selection of the equivalent concentrated mass at the center of the beam span, the dead weight loads imposed by the adjacent transverse beam on each column are taken to equal 25.5 per cent of the weight of this beam.

3. The weight of the column, replaced by an equivalent weight load concentrated at the top of the column. It follows from the theory of longitudinal vibrations of prismatic bars that the value of this load should be equal to 33 per cent of the column weight.

As a result of reducing the dead weight loads, we come to the consideration of vibrations of a plane frame whose elements are weightless and whose masses are concentrated in two places (Fig. VI-7): one mass m_2 at the center of the frame beam span, and two other masses, each equaling $m_1/2$, at the tops of the columns. Vibrations of all frame units are determined by the vertical displacements z_1 and z_2 of these masses. In this manner, the problem of vertical forced vibrations of transverse frames of the foundation is reduced to the problem of vibrations of a system with two degrees of freedom. It is assumed that the exciting vertical force acts on mass m_2.

Let us denote by c_2 the coefficient of rigidity of the frame beam; this coefficient represents a vertical force which should be applied to the center of the frame beam span in order to cause a deflection of unit length, i.e.,

$$c_2 = \frac{1}{l_2} \tag{VI-2-7}$$

The value of l_2 is determined by the formula

$$l_2 = \frac{l^3(1 + 2k)}{96EI_l(2 + k)} + \frac{3l}{8GA_l} \tag{VI-2-8}$$

$$k = \frac{hI_l}{lI_h} \tag{VI-2-9}$$

where E, G = Young's modulus, modulus of elasticity in shear of material of frame beam

A_l, I_l = cross-sectional area, moment of inertia of frame beam

A_h, I_h = cross-sectional area, moment of inertia of column

Let us denote by $c_1/2$ the coefficient of rigidity of a column; this coefficient represents the vertical force which must be applied to the column in order to cause a unit change in its length; it is evident that

$$\frac{c_1}{2} = \frac{EA_h}{h} \qquad \text{(VI-2-10)}$$

The differential equations of forced vibrations of the system shown in Fig. VI-7 will be exactly the same as Eqs. (IV-6-1), and the solutions for the amplitudes are determined by Eqs. (IV-6-4) and (IV-6-5).

The differential equations (IV-6-1) do not take into account the influence of damping reactions; so the computations of amplitudes of forced vibrations, presented in Chap. IV, will produce adequate results only in cases in which the fundamental frequencies of the system (Fig. VI-7) differ by at least ±30 per cent from the frequency of machine rotation. If this condition is not satisfied, and forced vibrations of the foundation occur in the resonance zone, then the use of the foregoing equations leads to large errors in the determination of the amplitudes of forced vibrations.

The natural frequencies of vertical vibrations of the foundation will be determined as roots of Eq. (IV-6-8). The solution of this equation will provide the two natural frequencies f_{n1} and f_{n2} of the frame under consideration. Let us assume that the frequency of excitation is close to one of these frequencies.

The following relationship usually exists between the limiting frequencies f_l and f_{nz}:

$$f_l < f_{nz}$$

Therefore the frequencies of natural vertical vibrations lie in the range

$$f_{n2} < f_l < f_{nz} < f_{n1}$$

If the frequency of excitation lies close to the lower natural frequency, then the form of the frame vibrations does not differ much from the form of vibrations of frequency f_l, and it can be considered with sufficient accuracy that the system under consideration has one degree of freedom; therefore the amplitude of forced vibrations of the system is determined by the expression

$$A_2^* = \frac{P}{m_2\sqrt{(f_{n2}{}^2 - \omega^2 + 4c^2\omega^2}} \qquad \text{(VI-2-11)}$$

where c is the damping constant. If the difference $f_{n2}{}^2 - \omega^2$ is small, then, approximately,

$$A_2^* = \frac{P}{2c\omega m_2} \tag{VI-2-12}$$

An approximate value of the damping constant is 5 to 10 per cent of f_{n2}.

If the frequency of excitation lies close to the higher natural frequency f_{n1}, and the form of frame vibrations does not differ much from the form of vibrations of frequency f_{nz}, then A_2 is determined from formula (VI-2-11) or (VI-2-12), in which f_{n2} and m_2 are replaced by f_{n1} and $(m_1 + m_2)$.

f. Computation of Horizontal Transverse Vibrations. In the computation of forced horizontal vibrations of the frame foundation we neglect the elasticity of the upper slab and the soil; i.e., we assume that the slab is absolutely rigid and the frame columns are embedded in an unyielding foundation.

Let us consider, for example, a foundation having six columns and three transverse frames. We replace all vibrating masses of the foundation and machine by three equivalent masses m_1, m_2, and m_3, each located at the center of one transverse frame beam span. Each of these masses is computed by adding the following:

1. The mass of the concentrated and distributed deadweight load on the frame beam, including its own weight
2. The mass formed by 30 per cent of the weight of the two columns of the transverse frame
3. The mass of the deadweight transferred by longitudinal frame beams adjacent to the transverse frame under consideration, their own weight included

The amplitudes of forced transverse vibrations depend on the sizes of these masses and on values characterizing the rigidity of transverse frames.

Let us replace each transverse frame by an equivalent spring (Fig. VI-8), and the upper slab by a prismatic bar which is assumed to be absolutely rigid. The motion of the system (Fig. VI-8) is determined by x, the lateral displacement of the center of mass of the prismatic bar, and by φ, the angle of its rotation with respect to the center of mass. Consequently the system has two degrees of freedom; i.e., it has two natural frequencies of vibration.

The differential equations describing the transverse horizontal vibra-

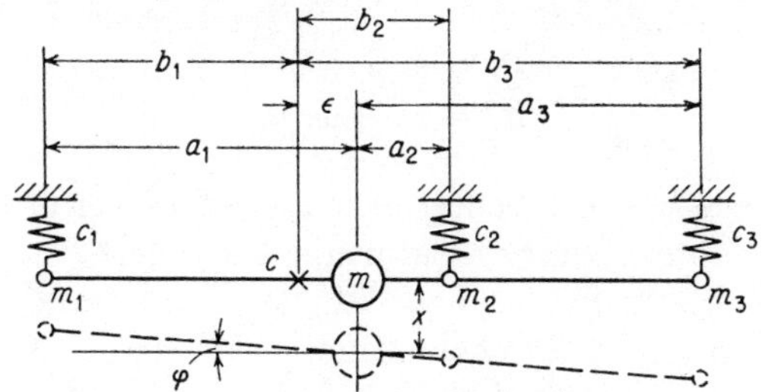

FIG. VI-8. Derivation of Eq. (VI-2-13).

tions of the foundation in the case under consideration read as follows:

$$m\ddot{x} + \sum_{i=1}^{n} R_i = P \sin \omega t$$
$$M_m\ddot{\varphi} + \sum_{i=1}^{n} M_i = M \sin \omega t \qquad \text{(VI-2-13)}$$

where $m = \sum_{i=1}^{n} m_i$ = sum of all equivalent masses

$M_m = \sum_{i=1}^{n} m_i a_i^2$ = total moment of inertia of all equivalent masses with respect to common center of mass

a_i = distance between common center of mass and mass m; we consider these distances to be positive in one direction and negative in opposite direction

R_i = elastic force acting on mass m_i during its forward displacement up to value x_i

M_i = moment of inertia of force R_i with respect to an axis passing through center of mass

P, M = exciting force, moment

The summation should be performed for all transverse frames, so that n denotes the number of transverse foundation frames (usually n equals 3 or 4).

When the mass m_i is displaced by the value x_i, then the elastic force of the equivalent springs acting thereon equals

$$R_i = c_i x_i$$

where c_i is the rigidity of the ith spring.

The moment of the elastic force is

$$M_i = c_i a_i x_i$$

Since

$$x_i = x + a_i \varphi$$

we may write:

$$R_i = c_i(x + a_i\varphi)$$

and

$$R_i = (x + \epsilon\varphi)C \qquad \text{(VI-2-14)}$$

where ϵ = distance between center of mass and center of rigidity

$C = \Sigma c_i$ = total rigidity of equivalent springs

To determine the rigidity c_i of the ith transverse frame, let us apply a horizontal force to the center of the frame beam span. This force equals unity and is directed along the axis of the frame beam. It is known that the lateral displacement δ caused by this force equals

$$\delta = \frac{h^3(2 + 3k)}{12EI_h(1 + 6k)} \qquad \text{(VI-2-15)}$$

Here, as before,

$$k = \frac{hI_l}{lI_h}$$

where h = height of column

l = length of frame beam

I_h, I_l = moments of inertia of cross sections of column, frame beam

Having determined the deflection δ for the ith transverse frame, we find the rigidity c_i characterizing the equivalent spring corresponding to this frame:

$$c_i = \frac{1}{\delta} \qquad \text{(VI-2-16)}$$

In the same way we find

$$\Sigma M_i = \Sigma c_i a_i x_i = C(x + \epsilon\varphi)\epsilon + \gamma\varphi \qquad \text{(VI-2-17)}$$

where

$$\gamma = \Sigma c_i b_i^2$$

is the moment of a couple causing the rotation of the prismatic bar through a unit angle.

Substituting the computed values of ΣR_i and ΣM_i into Eqs. (VI-2-13), we obtain two differential equations of forced vibration of the foundation:

$$\begin{aligned} m\ddot{x} + Cx + c\epsilon\varphi &= P \sin \omega t \\ M_m\ddot{\varphi} + C\epsilon x + (C\epsilon^2 + \gamma)\varphi &= M \sin \omega t \end{aligned} \qquad \text{(VI-2-18)}$$

or

$$\begin{aligned} \ddot{x} + f_{nx}^2 x + f_{nx}^2\epsilon\varphi &= p \sin \omega t \\ \ddot{\varphi} + \frac{\epsilon}{r^2} f_{nx}^2 x + \left(\frac{\epsilon^2}{r^2} f_{nx}^2 + f_{n\varphi}^2\right)\varphi &= R \sin \omega t \end{aligned} \qquad \text{(VI-2-19)}$$

The following designations were used in the foregoing formulas:

$$f_{nx}^2 = \frac{C}{m} \qquad \text{(VI-2-20)}$$

where f_{nx} is the limiting frequency of natural lateral vibrations of the foundation when the center of rigidity of the foundation coincides with the center of mass, i.e., when $\epsilon = 0$;

$$f_{n\varphi}{}^2 = \frac{\gamma}{M_m} \tag{VI-2-21}$$

where $f_{n\varphi}$ is the limiting frequency of natural rocking vibrations of the foundation when $\epsilon = 0$;

$$r^2 = \frac{M_m}{m}$$

where r is the radius of gyration; finally

$$p = \frac{P}{m} \qquad R = \frac{M}{M_m}$$

We seek solutions of the system of Eqs. (VI-2-19) corresponding only to the forced vibrations of foundations in the form

$$x = A_x \sin \omega t \qquad \varphi = A_\varphi \sin \omega t$$

Inserting these values of x and φ into Eq. (VI-2-19), we obtain the following two equations containing the amplitudes of forced vibrations A_x and A_φ of the foundation as unknown values:

$$(f_{nx}{}^2 - \omega^2)A_x + f_{nx}{}^2 \epsilon A_\varphi = p$$

$$\frac{\epsilon^2}{r^2} f_{nx}{}^2 A_x + \left(\frac{\epsilon^2}{r^2} f_{nx}{}^2 + f_{n\varphi}{}^2 - \omega^2\right) A_\varphi = R$$

Solving these equations for A_x and A_φ, we obtain

$$A_x = \frac{[(\epsilon^2/r^2)f_{nx}{}^2 + f_{n\varphi}{}^2 - \omega^2]p - f_{nx}{}^2 R}{\Delta(\omega^2)} \tag{VI-2-22}$$

$$A_\varphi = \frac{(\epsilon^2/r^2)f_{nx}{}^2 p - (f_{nx}{}^2 - \omega^2)R}{\Delta(\omega^2)} \tag{VI-2-23}$$

where $$\Delta(\omega^2) = \omega^4 - (\alpha f_{nx}{}^2 + f_{n\varphi}{}^2)\omega^2 + f_{nx}{}^2 f_{n\varphi}{}^2 \tag{VI-2-24}$$

$$\alpha = 1 + \frac{\epsilon^2}{r^2}$$

The natural frequencies f_{n1} and f_{n2} of foundation vibrations are determined as roots of the equation

$$\Delta(\omega^2) = 0 \tag{VI-2-25}$$

It is clear that the above formulas for the determination of amplitudes of forced vibration of the foundation may be applied only when the frequency of excitement ω differs from the fundamental frequencies of the foundation. Otherwise, the amplitudes of vibration should be computed analogously to the computation of the amplitudes of vertical vibrations when one of the natural frequencies is close to the engine frequency.

For example, if ω lies close to the lower frequency f_{n2} determined as the root of Eq. (VI-2-25), and if f_{n2} does not differ much from f_{nx}, then the foundation will be subjected predominantly to vibrations accompanied by lateral displacements as a rigid body. The amplitude of vibrations may be approximately established from Eq. (VI-2-11) where m_2 is assumed to equal m. If resonance occurs because the frequency of machine rotation lies close to the second frequency f_{n1} and the latter does not differ much from f_φ, then the foundation will undergo chiefly rocking vibrations.

In this case, an approximate value of the amplitude of vibrations may be found from the same Eq. (VI-2-11), in which P, m_2, and f_{n2} should be replaced respectively by M, M_m, and f_{n1}.

g. Design Values of the Permissible Amplitude of Vibrations. Design values of the permissible amplitude of vibrations of foundations for turbodynamos and electrical machines should be established on the basis of data derived from the study of operating machines. It is hardly possible to establish these limits on the basis of any theoretical premises.

As a matter of fact, if a permissible amplitude of vibrations is established on the basis of permissible stresses for the foundation materials, it is found that the computed amplitudes are tens of times larger than those permissible for normal machine operation. Therefore a selection of the design value of vibration amplitude should be based on the amplitudes accepted by machine operators as permissible for a given machine type. Table VI-3 presents data on the permissible values of amplitudes of vibrations of turbogenerator bearings.

The absolute values of permissible amplitudes of vibrations may be much larger for machines running at 1,500 rpm than for 3,000-rpm machines. If one admits that the permissible amplitude of vibrations of bearings may be taken as the arithmetic mean of the values given in Table VI-3 for machines with certain speeds, then the following values may be accepted:

For 3,000-rpm machines:	
Vertical vibrations:	0.02 to 0.03 mm
Horizontal vibrations:	0.04 to 0.05 mm
For 1,500-rpm machines:	
Vertical vibrations:	0.04 to 0.06 mm
Horizontal vibrations:	0.07 to 0.09 mm

Vibration investigations of 36 foundations under turbodynamos established that actual vibration amplitudes do not exceed the above permissible values. Only in 1 foundation was an amplitude found as high as 0.016 mm. In 4 foundations, the amplitude of vibrations of the upper part of the foundation lay within the range 0.010 to 0.016 mm. In all other foundations the amplitudes of vibrations were within the range

TABLE VI-3. PERMISSIBLE VIBRATION AMPLITUDE VALUES OF TURBOGENERATOR BEARINGS

Type of vibrations	Location of measurements	Evaluation of the engine	Amplitudes† of vibrations, mm, corresponding to speeds, rpm, of:				
			3,000	2,500	2,000	1,500	1,000 and less
Vertical	Extreme bearings	Is fit for operation	0.02	0.03	0.04	0.06	0.08
		No adjustment is needed	0.03	0.05	0.06	0.09	0.11
		An adjustment is desirable	0.04	0.08	0.09	0.13	0.15
	Central bearings	Is fit for operation	0.01	0.02	0.03	0.04	0.05
		No adjustment is needed	0.02	0.03	0.05	0.06	0.08
		An adjustment is desirable	0.03	0.04	0.08	0.09	0.13
Horizontal and transverse	Extreme bearings	Is fit for operation	0.05	0.07	0.08	0.09	0.12
		No adjustment is needed	0.08	0.10	0.11	0.12	0.15
		An adjustment is desirable	0.13	0.14	0.15	0.17	0.20
	Central bearings	Is fit for operation	0.03	0.04	0.05	0.07	0.09
		No adjustment is needed	0.05	0.06	0.08	0.10	0.12
		An adjustment is desirable	0.08	0.09	0.13	0.14	0.17

† The largest permissible values of amplitudes are presented.

0.004 to 0.010 mm; i.e., they were considerably smaller than permissible values.

Vibrations with an amplitude of 0.016 mm, had no influence on the normal operation of the turbogenerator.

Permissible values of foundation vibration amplitudes under electric machines having speeds close to 1,500 or 3,000 rpm may be the same as those for turbogenerators. It is very difficult to select even tentative design values of permissible amplitudes of vibrations for electric machines with speeds lower than 1,500 rpm, since scarcely any data are available on the results of vibration investigations of these machines. For low-frequency electrical machines (less than 500 rpm) such as motor generators and Leonard generators, a design value of permissible vibration amplitudes may be selected on the basis of values established for reciprocating engines (around 0.20 mm).

VI-3. Examples of Dynamic Computations of Foundations of the Frame Type

Example 1. *Illustrative design of a frame foundation under a* 1,500-*kw turbogenerator*

1. DATA. The speed of the turbogenerator is $N = 3{,}000$ rpm; i.e., the frequency of excitation is $\omega \approx 300\ \text{sec}^{-1}$; $\omega^2 \approx 9 \times 10^4\ \text{sec}^{-2}$.

The foundation to be designed will have six columns and three transverse frames. Table VI-4 gives the initial data required for the dynamic computations of the foundation which are taken from the Design Assigment. Figure VI-9 shows the geometry of the foundation with indication of the loads imposed by the stationary and rotating parts of the machine.

TABLE VI-4. DESIGN DATA FOR COMPUTATIONS OF EXAMPLE VI-3-1 AND FIG. VI-9

Dimensions and design parameters	Frame I	Frame II	Frame III
Height of transverse frames h_0, m	4.30	4.30	4.30
Span of transverse frames l_0, m	3.20	3.20	3.20
Height of cross section of transverse frame $2a$, m	0.95	1.00	1.00
Height of cross section of column $2b$, m	0.80	0.83	0.80
Area of cross section of column A_h, m²	0.76	0.78	0.76
Moment of inertia of cross-sectional area of column I_h, m⁴	0.039	0.041	0.039
Cross-sectional area of frame beam A_l, m²	0.58	0.83	0.64
Moment of inertia of cross-sectional area of beam I_l, m⁴	0.0425	0.069	0.053
Weight of frame beam W_l, tons	4.26	6.10	4.70
Weight of column W_h, tons	6.53	6.67	6.53

The loads imposed by the rotating parts of the turbogenerator act only on the beams of the transverse frames. The loads imposed by the stationary parts (the stator of the generator and the cover of the turbine) are transmitted to the longitudinal beams.

The design value of the modulus of elasticity for the material of the upper part of the foundation (concrete type 110†) we assume to equal $E = 3 \times 10^6$ tons/m².

2. COMPUTATIONS. The computation of forced vertical vibrations of the foundation is begun by some preliminary computations. Table VI-5 gives their results. Table VI-6 presents results of computations of equivalent masses m_1 and m_2 for each of the transverse frames.

† See footnote, Art. IV-1-c, p. 132.

Results of computations of equivalent rigidities are given in Table VI-7, and those of limiting frequencies of each frame in Table VI-8.

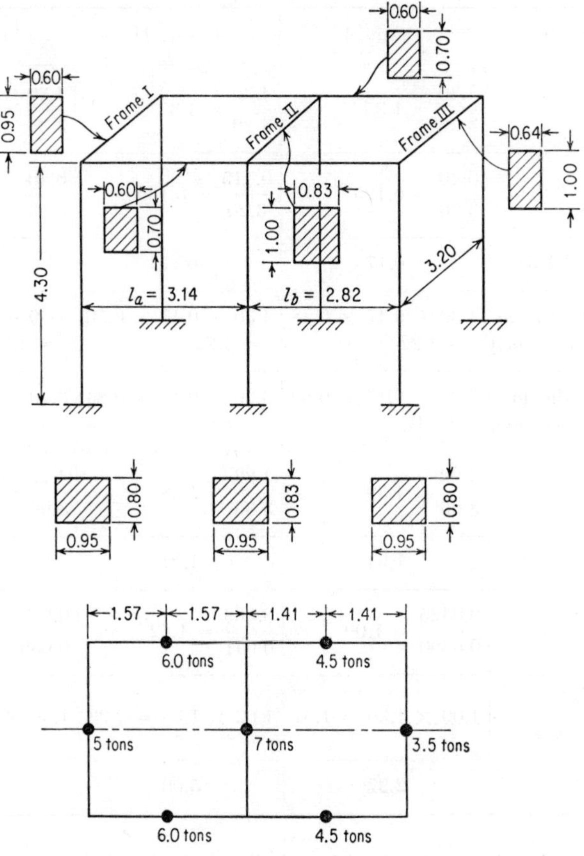

FIG. VI-9. Foundation of Example VI-3-1.

Let us now compute the amplitudes of forced vertical vibrations of the foundation. According to the foregoing data, the design value of the eccentricity of rotating masses of the turbogenerator is taken to equal

$$r_0 = 0.05 \times 10^{-3} \text{ m}$$

The weights of rotating parts falling on each frame equal

$$W_1 = 1.5 \text{ tons} \qquad W_2 = 2.0 \text{ tons} \qquad W_3 = 1.0 \text{ ton}$$

The magnitude of exciting vertical load acting on each transverse frame of the foundation is:

Frame I: $P_1 = 0.05 \times 10^{-3} \times 1.5/9.81 \times 9 \times 10^4 = 0.69$ ton
Frame II: $P_2 = 0.05 \times 10^{-3} \times 2.0/9.81 \times 9 \times 10^4 = 0.92$ ton
Frame III: $P_3 = 0.05 \times 10^{-3} \times 1.0/9.81 \times 9 \times 10^4 = 0.46$ ton

TABLE VI-5. RESULTS OF PRELIMINARY COMPUTATIONS FOR EXAMPLE VI-3-1, FIG. VI-9

Design parameters	Frame I	Frame II	Frame III
$\frac{h_0}{l_0}$	$\frac{4.30}{3.20} = 1.35$	$\frac{4.30}{3.20} = 1.35$	$\frac{4.30}{3.20} = 1.35$
$\frac{b}{l_0}$	$\frac{0.40}{3.20} = 0.125$	$\frac{0.415}{3.20} = 0.130$	$\frac{0.40}{3.20} = 0.125$
α (from Fig. VI-5)	0.17	0.17	0.17
Reduced height h, m [from Eq. (VI-2-4*a*)]	$4.30 - 0.17 \times 0.48 = 4.22$	$4.30 - 0.17 \times 0.50 = 4.22$	$4.30 - 0.17 \times 0.50 = 4.22$
Reduced length , m [from Eq. (VI-2-4*a*)]	$3.20 - 0.17 \times 0.80 = 3.06$	$3.20 - 0.17 \times 0.83 = 3.06$	$3.20 - 0.17 \times 0.80 = 3.06$
$l_s = \frac{h}{l}$	$\frac{4.22}{3.06} = 1.39$	$\frac{4.22}{3.06} = 1.38$	$\frac{4.22}{3.06} = 1.39$
l_s^2	1.94	1.91	1.94
$\frac{I_l}{I_h}$	$\frac{0.0425}{0.0390} = 1.09$	$\frac{0.069}{0.041} = 1.62$	$\frac{0.053}{0.039} = 1.36$
$k = \frac{hI_l}{lI_h}$	$1.09 \times 1.39 = 1.51$	$1.62 \times 1.38 = 2.23$	$1.36 \times 1.39 = 1.89$
k^2	2.27	5.00	3.57

We then determine the amplitude of forced vibrations of each frame.

FRAME I. We find the value of coefficient $\Delta(\omega^2)$ from Eq. (IV-6-8):

$$\Delta(\omega^2) = 81 \times 10^8 - (1 + 0.55)(30.2 + 75.0)9 \times 10^8 + (1 + 0.55)30.2 \times 75.0 \times 10^8 = 21.2 \times 10^{10}$$

The amplitude of longitudinal vibrations of the column we find from Eq. (IV-6-4):

$$A_1 = \frac{30.2 \times 10^4 \times 0.69}{1.28 \times 21.2 \times 10^{10}} = 0.71 \times 10^{-6} \text{ m}$$

The amplitude of vertical vibrations of the center of the frame beam span we find from Eq. (IV-6-5):

$$A_2 = \frac{(1 + 0.55) \times 75.0 \times 10^4 + 0.55 \times 30.2 \times 10^4 - 9 \times 10^4}{0.71 \times 21.2 \times 10^{10}} 0.69 = 4.7 \times 10^{-6} \text{ m}$$

TABLE VI-6. COMPUTATIONS OF EQUIVALENT MASSES FOR EXAMPLE VI-3-1, FIG. VI-9

Loads and equivalent masses	Frame I	Frame II	Frame III
Concentrated load imposed on the frame beam, tons..	5.0	7.0	3.5
Equivalent load from frame beam weight, $0.45 \times W_l$, tons...	$0.45 \times 4.26 = 1.91$	$0.45 \times 6.10 = 2.74$	$0.45 \times 4.70 = 2.12$
Equivalent mass m_2 reduced to center of frame beam span, tons $\times$ sec²/m	$\frac{5.0 + 1.91}{9.81} = 0.71$	$\frac{7.0 + 2.74}{9.81} = 1.00$	$\frac{3.5 + 2.12}{9.81} = 0.58$
Equivalent load of columns' weight, $0.33 \times W_h$, tons...	$0.33 \times 6.53 = 2.20$	$0.33 \times 6.67 = 2.25$	$0.33 \times 6.53 = 2.20$
Load imposed by longitudinal beams, tons	$6.0 + 2.4 \times 0.42 \times 3.14 = 9.16$	$6.0 + 4.5 + 2.4 \times 0.42 \times (3.14 + 2.82) = 16.50$	$4.5 + 2.4 \times 0.42 \times 2.82 = 7.36$
Equivalent load imposed on columns by transverse beams of frames, tons.......	1.09	1.55	1.20
Equivalent mass m_1 reduced to top of column, tons $\times$ sec²/m	$\frac{2.20 + 9.16 + 1.09}{9.81} = 1.28$	$\frac{2.25 + 16.5 + 1.55}{9.81} = 2.08$	$\frac{2.20 + 7.36 + 1.20}{9.81} = 1.11$

Thus the amplitude of total vertical vibrations of the frame under consideration equals

$$A_t = A_1 + A_2 = (0.71 + 4.7)10^{-6} \text{ m} = 0.005 \text{ mm}$$

FRAME II

$$\Delta(\omega^2) = 81 \times 10^8 - (1 + 0.48)(30.8 + 49.8)9 \times 10^8 + (1 + 0.48)30.8 \times 49.8 \times 10^8 = 12.8 \times 10^{10}$$

$$A_1 = \frac{30.8 \times 10^4 \times 0.92}{2.08 \times 12.8 \times 10^{10}} = 1.0 \times 10^{-6} \text{ m}$$

$$A_2 = \frac{(1 + 0.48)49.8 \times 10^4 + 0.48 \times 30.8 \times 10^4 - 9 \times 10^4}{1.0 = 12.8 \times 10^{10}} 0.92$$

$$= 5.7 \times 10^{-6} \text{ m}$$

$$A_t = A_1 + A_2 = (1.0 + 5.7)10^{-6} \text{ m} = 0.007 \text{ mm}$$

TABLE VI-7. RESULTS OF COMPUTATIONS OF EQUIVALENT RIGIDITIES FOR EXAMPLE VI-3-1, FIG. VI-9

Parameters and equivalent rigidities	Frame I	Frame II	Frame III
$\dfrac{l^3(1 + 2k)}{96EI_l(2 + k)}$	$\dfrac{3.06^3(1 + 2 \times 1.51)}{96 \times 3 \times 10^6 \times 0.0425(2 + 1.51)}$ $= 2.70 \times 10^{-6}$	$\dfrac{3.06^3(1 + 2 \times 2.23)}{96 \times 3 \times 10^6 \times 0.069(2 + 2.23)}$ $= 1.88 \times 10^{-6}$	$\dfrac{3.06^3(1 + 2 \times 1.89)}{96 \times 3 \times 10^6 \times 0.053(2 + 1.89)}$ $= 2.35 \times 10^{-6}$
$\dfrac{3l}{8GA_l}$	$\dfrac{3 \times 3.06}{8 \times 10^6 \times 0.58} = 1.98 \times 10^{-6}$	$\dfrac{3 \times 3.06}{8 \times 10^6 \times 0.83} = 1.38 \times 10^{-6}$	$\dfrac{3 \times 3.06}{8 \times 10^6 \times 0.64} = 1.80 \times 10^{-6}$
Deflection of frame beam at center of span under action of unit force, m........	$(2.70 + 1.98)10^{-6} = 4.68 \times 10^{-6}$	$(1.88 + 1.38)10^{-6} = 3.26 \times 10^{-6}$	$(2.35 + 1.80)10^{-6} = 4.15 \times 10^{-6}$
Rigidity c_2 of frame beam, tons/m.......	21.4×10^4	30.8×10^4	24.0×10^4
Rigidity c_1 of column, tons/m $\left(\dfrac{2EA_h}{h}\right)$	$\dfrac{2 \times 3 \times 10^6 \times 0.76}{3.06} = 149 \times 10^4$	$\dfrac{2 \times 3 \times 10^6 \times 0.78}{3.06} = 153 \times 10^4$	$\dfrac{2 \times 3 \times 10^6 \times 0.76}{3.06} = 149 \times 10^4$

TABLE VI-8. RESULTS OF COMPUTATIONS OF LIMITING FREQUENCIES FOR EXAMPLE VI-3-1 AND FIG. VI-9

Ratios between masses and the limiting frequencies	Frame I	Frame II	Frame III
$\alpha = \dfrac{m_2}{m_1}$	$\dfrac{0.71}{1.28} = 0.55$	$\dfrac{1.00}{2.08} = 0.48$	$\dfrac{0.58}{1.11} = 0.52$
Square of frequency of natural vibrations of frame beam, considering columns to be absolutely rigid f_l^2, sec^{-2}	$\dfrac{21.4 \times 10^4}{0.71} = 30.2 \times 10^4$	$\dfrac{30.8 \times 10^4}{1.00} = 30.8 \times 10^4$	$\dfrac{24.0 \times 10^4}{0.58} = 41.5 \times 10^4$
Square of frequency of natural vibrations of columns, considering frame beams to be absolutely rigid f_{nz}^2, sec^{-2}	$\dfrac{149 \times 10^4}{1.28 + 0.71} = 75.0 \times 10^4$	$\dfrac{153 \times 10^4}{2.08 + 1.00} = 49.8 \times 10^4$	$\dfrac{149 \times 10^4}{1.11 + 0.58} = 88.3 \times 10^4$

FRAME III

$$\Delta(\omega^2) = 81 \times 10^8 - (1 + 0.48)(41.5 + 88.3)9 \times 10^{10} + (1 + 0.48)41.5 \times 88.3 \times 10^8 = 37.8 \times 10^{10}$$

$$A_1 = \frac{41.5 \times 10^4 \times 0.46}{1.1 \times 37.8 \times 10^{10}} = 0.46 \times 10^{-6} \text{ m}$$

$$A_2 = \frac{(1 + 0.48)88.3 \times 10^4 \times 0.48 \times 41.5 \times 10^4 - 9 \times 10^4}{0.58 \times 37.8 \times 10^{10}} 0.46$$

$$= 1.6 \times 10^{-6} \text{ m}$$

$$A_t = A_1 + A_2 = (0.46 + 1.6)10^{-6} \text{ m} \cong 0.002 \text{ mm}$$

It is clear that the computed amplitudes of vertical vibrations of the foundation are much smaller than the permissible values.

We compute the forced horizontal (transverse) vibrations of the foundation. First we determine the equivalent mass of each frame. As has been indicated, the magnitude of each of these masses is determined by the following:

1. Concentrated and distributed loading imposed on the frame beam, including its own weight
2. 30 per cent of the column weights
3. Loads imposed by longitudinal frame beams adjacent to the transverse frame under consideration, the deadweight of longitudinal beams also included

Thus we have, for frame I:

$$m_1 = \frac{5.0 + 4.26 + 0.30 \times 6.53 \times 2 + 9.16}{9.81} = 2.13 \text{ tons} \times \text{sec}^2/\text{m}$$

for frame II:

$$m_2 = \frac{7.00 + 6.10 + 0.30 \times 6.67 \times 2 + 16.50}{9.81} = 3.25 \text{ tons} \times \text{sec}^2/\text{m}$$

and for frame III:

$$m_3 = \frac{3.5 + 4.70 + 0.30 \times 6.53 \times 2 + 3.5}{9.81} = 1.40 \text{ tons} \times \text{sec}^2/\text{m}$$

The total equivalent mass equals

$$m + m_1 + m_2 + m_3 = 2.13 + 3.25 + 1.40 = 6.78 \text{ tons} \times \text{sec}^2/\text{m}$$

We determine the distance from the axis of frame I, along the foundation, to the total equivalent mass:

$$a_1 = \frac{3.14 \times 3.25 + (3.14 + 2.82)1.40}{6.78} = 2.95 \text{ m}$$

The distances from the common center of mass to the axes of frames II and III are:

$$a_2 = 2.95 - 3.14 = -0.19 \text{ m}$$
$$a_3 = 2.95 - 5.96 = -3.01 \text{ m}$$

The moment of inertia of all the equivalent masses with respect to the common center of mass equals

$$M_m = m_1 a_1^2 + m_2 a_2^2 + m_3 a_3^2$$
$$= 2.13 \times 2.95^2 + 3.25 \times 0.19^2 + 1.40 \times 3.01^2$$
$$= 31.2 \text{ tons} \times \text{m} \times \text{sec}^2$$

From Eqs. (VI-2-15) and (VI-2-16) we find the equivalent rigidities: for frame I, the displacement δ_1 caused by a unit horizontal force directed along the axis of the frame beam is

$$\delta_1 = \frac{4.25^3(2 + 3 \times 1.51)}{12 \times 3 \times 10^6 \times 0.039(1 + 6 \times 1.51)} = 3.44 \times 10^{-3}\ \text{m}$$

The equivalent rigidity equals

$$c_1 = \frac{1}{\delta_1} = \frac{1}{3.44 \times 10^{-3}} = 0.29 \times 10^3\ \text{tons/m}$$

For frame II:

$$\delta_2 = \frac{4.22^3(2 + 3 \times 2.23)}{12 \times 3 \times 10^6 \times 0.041(1 + 6 \times 2.23)} = 3.12 \times 10^{-3}\ \text{m}$$

$$c_2 = \frac{1}{3.12 \times 10^{-3}} = 0.32 \times 10^3\ \text{tons/m}$$

and for frame III:

$$\delta_3 = \frac{4.22^3(2 + 3 \times 1.89)}{12 \times 3 \times 10^6 \times 0.039(1 + 6 \times 1.89)} = 3.85 \times 10^{-3}\ \text{m}$$

$$c_3 = \frac{1}{3.85 \times 10^{-3}} = 0.26 \times 10^3\ \text{tons/m}$$

The total rigidity of all three frames equals

$$C = (0.29 + 0.32 + 0.26)10^3 = 0.87 \times 10^3\ \text{tons/m}$$

We determine the distance to the center of rigidity, along the foundation, from the axis of frame I:

$$b_1 = \frac{3.14 \times 0.32 \times 10^3 + (3.14 + 2.82)0.26 \times 10^3}{0.87 \times 10^3} = 2.95\ \text{m}$$

The distance between the center of mass and the center of rigidity of the frame is

$$\epsilon = 2.95 - 2.95 = 0$$

Consequently, the center of inertia and the center of rigidity coincide. For this particular case, Eqs. (VI-2-12) and (VI-2-13) for determining the amplitudes of vibration are simplified.

The amplitude of lateral horizontal (transverse) vibrations of the foundation equals

$$A_x = \frac{P}{m(f_{nx}^2 - \omega^2)} \tag{VI-2-26}$$

The amplitude of rocking vibrations around the common center of gravity in the plane of the upper platform is

$$A_\varphi = \frac{M}{M_m(f_{n\varphi} - \omega^2)} \tag{VI-2-27}$$

We determine the limiting natural frequencies of the foundation. From Eq. (VI-2-20) we have:

$$f_{nx}^2 = \frac{0.87 \times 10^3}{6.78} = 0.128 \times 10^3\ \text{sec}^{-2}$$

The total exciting force equals

$$P = 0.69 + 0.92 + 0.46 = 2.07\ \text{tons}$$

The amplitude of transverse vibrations of the foundation is

$$A_x = \frac{2.07}{6.78(0.128 \times 10^3 - 9 \times 10^3)} = 0.034 \text{ mm}$$

To determine the rigidity of the foundation against torsion, we have

$$b_1 = 2.95 \text{ m} \qquad b_2 = a_2 = -0.19 \text{ m} \qquad b_3 = a_3 = -3.01 \text{ m}$$

It follows that

$$\begin{aligned} \gamma &= c_1b_1^2 + c_2b_2^2 + c_3b_3^2 \\ &= (0.29 \times 2.95^2 + 0.32 \times 19^2 + 0.26 \times 3.01^2)10^3 \\ &= 4.89 \times 10^3 \text{ tons} \times \text{m} \end{aligned}$$

From Eq. (VI-2-21) we compute the frequency of natural rocking vibrations of the foundation:

$$f_{n\varphi}^2 = \frac{\gamma}{M_m} = \frac{4.89 \times 10^3}{31.2} = 0.157 \times 10^3 \text{ sec}^{-2}$$

Let us assume that exciting forces in the generator and the turbine act in the same direction at each instant of time. Then the magnitude of the exciting moment will equal

$$\begin{aligned} M &= P_1a_1 + P_2a_2 + P_3a_3 \\ &= 0.69 \times 2.95 - 0.92 \times 0.19 - 0.46 \times 3.01 \\ &= 0.48 \text{ tons} \times \text{m} \end{aligned}$$

The amplitude of rocking vibrations determined from Eq. (VI-2-27) is

$$A_\varphi = \frac{0.48}{31.2(0.157 - 9.0)10^3} = 0.175 \times 10^{-5} \text{ radians}$$

The largest horizontal displacement as a result of rocking vibrations is

$$a_3A_\varphi = 3.01 \times 0.175 \times 10^{-5} = 0.005 \times 10^{-3} \text{ m} = 0.005 \text{ mm}$$

Thus the total maximum amplitude of horizontal displacement of the foundation in the direction perpendicular to the axis of the main shaft of the turbine equals

$$a_x = A_x + a_3A_\varphi = 0.034 + 0.005 = 0.039 \text{ mm}$$

The latter value lies within the range of permissible design values.

Example 2. Illustrative design of a foundation for a 500-kw generator

1. DATA. The motor generator runs at 750 rpm; consequently the frequency of forced vibrations will be

$$\omega \approx 75 \text{ sec}^{-1} \qquad \omega^2 \approx 5.6 \times 10^3 \text{ sec}^{-2}$$

The foundation is designed to consist of six columns and three transverse frames. Figure VI-10 gives a diagram of the foundation and the loads acting thereon. The weight of the rotating part of the motor generator is 5.9 tons; its total weight is 13.2 tons.

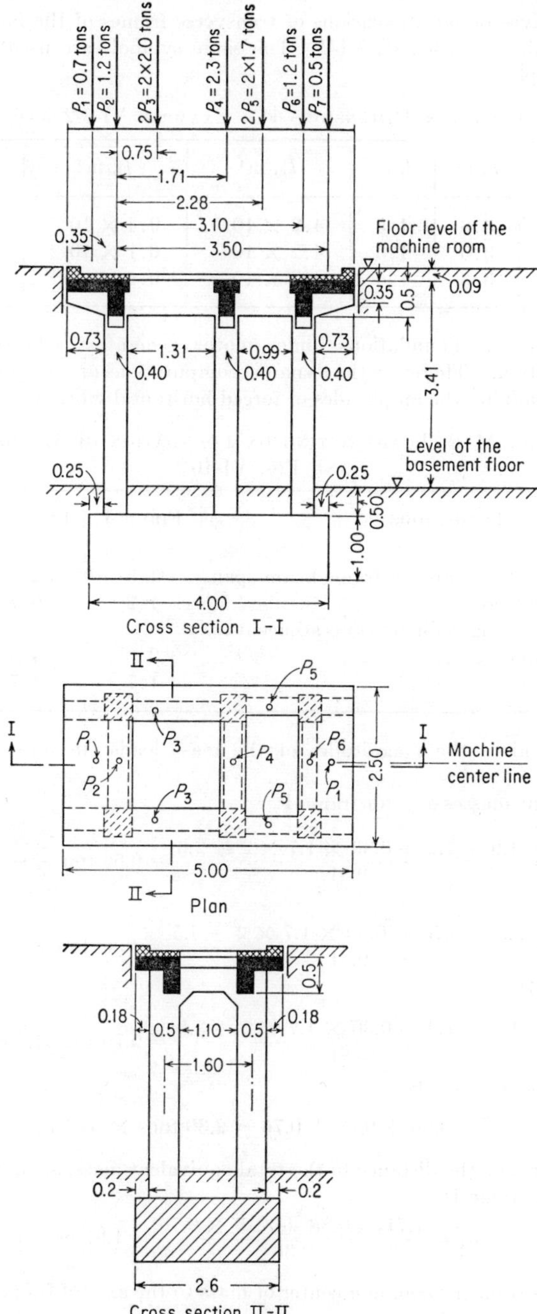

FIG. VI-10. Foundation of Example VI-3-2.

Table VI-9 gives design dimensions of transverse frames of the foundation determined in accordance with Art. VI-2. The same symbols are used here as in the preceding example.

TABLE VI-9. DESIGN DIMENSIONS FOR EXAMPLE VI-3-2 AND FIG. VI-10

Frame	h, m	l, m	I_h, m^4	I_l, m^4	k
I	3.6	1.5	4.2×10^{-3}	9.4×10^{-3}	5.4
II	3.6	1.5	4.2×10^{-3}	6.1×10^{-3}	3.5
III	3.6	1.5	4.2×10^{-3}	9.4×10^{-3}	5.4

2. COMPUTATIONS. Foundations under motor generators vibrate mostly in a transverse direction. Therefore the dynamic computations of the foundation may be limited to determining the amplitudes of forced horizontal vibrations.

TABLE VI-10. MAIN LOADS ACTING ON FOUNDATION OF EXAMPLE VI-3-2 AND FIG. VI-10

Loads, tons	Frame I	Frame II	Frame III
Concentrated load in center of frame beam span	1.9	2.3	1.7
Weight of frame beam	1.2	0.7	1.2
Loads imposed by longitudinal beams adjacent to transverse frame	2.6	4.5	3.6
Weight of columns	1.7	1.7	1.7

Table VI-10 gives the magnitudes of the mass loads (in tons) acting on the foundation.

The equivalent masses are: for frame I:

$$m_1 = \frac{1.9 + 1.2 + 0.30 \times 1.7 \times 2 + 2.6}{9.81} = 0.66 \text{ ton} \times \text{sec}^2/\text{m}$$

for frame II:

$$m_2 = \frac{2.3 + 0.70 + 0.30 \times 1.7 \times 2 + 4.5}{9.81} = 0.88 \text{ ton} \times \text{sec}^2/\text{m}$$

and for frame III:

$$m_3 = \frac{1.7 + 1.2 + 0.30 \times 1.7 \times 2 + 3.6}{9.81} = 0.76 \text{ ton} \times \text{sec}^2/\text{m}$$

The total equivalent mass is

$$m = 0.66 + 0.88 + 0.76 = 2.30 \text{ tons} \times \text{sec}^2/\text{m}$$

We now determine the distance to the total equivalent mass, along the foundation, from the axis of frame I:

$$a_1 = \frac{1.71 \times 0.88 + 3.10 \times 0.76}{2.30} = 1.67 \text{ m}$$

The distances from the common center of mass to the axes of frames II and III are

$$a_2 = 1.67 - 1.71 = -0.04 \text{ m}$$
$$a_3 = 1.67 - 3.50 = -1.83 \text{ m}$$

The moment of inertia of all equivalent masses with respect to the common center of mass equals

$$\begin{aligned} M_m &= m_1a_1^2 + m_2a_2^2 + m_3a_3^2 \\ &= 0.66 \times 1.67^2 + 0.88 \times 0.04^2 + 0.76 \times 1.83^2 \\ &= 4.3 \text{ tons} \times \text{m} \times \text{sec}^2 \end{aligned}$$

We compute the equivalent rigidities of each frame: for frame I:

$$\delta_1 = \frac{3.6^3(2 + 3 \times 5.4)}{12 \times 3 \times 10^6 \times 4.2 \times 10^{-3}(1 + 6 \times 5.4)} = 1.68 \times 10^{-4} \text{ m}$$

$$c_1 = \frac{1}{1.68 \times 10^{-4}} = 0.59 \times 104 \text{ tons/m}$$

for frame II:

$$\delta_1 = \frac{3.6^3(2 + 3 \times 3.5)}{12 \times 3 \times 10^6 \times 4.2 \times 10^{-3}(1 + 6 \times 3.5)} = 1.74 \times 10^{-4} \text{ m}$$

$$c_2 = \frac{1}{1.74 \times 10^{-4}} = 0.57 \times 10^{-4} \text{ tons/m}$$

and for frame III:

$$\delta_1 = \frac{3.6^3(2 + 3 \times 5.4)}{12 \times 3 \times 10^6 \times 4.2 + 10^{-3}(1 + 6 \times 5.4)} = 1.68 \times 10^{-4} \text{ m}$$

$$c_3 = \frac{1}{1.68 \times 10^{-4}} = 0.59 \times 10^4 \text{ tons/m}$$

The total rigidity of all frames is

$$C = (0.59 + 0.57 + 0.59)10^4 = 1.75 \times 10^4 \text{ tons/m}$$

We determine the distance to the center of rigidity, along the foundation, from the axis of frame I:

$$b_1 = \frac{(1.71 \times 0.57 + 3.10 \times 0.59)10^4}{1.75 \times 10^4} = 1.61 \text{ m}$$

The distance between the centers of mass and rigidity of the foundation is

$$\epsilon = 1.67 - 1.61 = 0.05 \text{ m}$$

To determine the rigidity of the foundation against torsion, we have:

$$\begin{aligned} b_1 &= 1.61 \text{ m} \\ b_2 &= a_2 + \epsilon = -0.04 + 0.06 = 0.02 \text{ m} \\ b_3 &= a_3 + \epsilon = -1.83 + 0.06 = -1.77 \text{ m} \end{aligned}$$

The rigidity of the foundation against torsion is then

$$\begin{aligned} \gamma &= (0.59 \times 1.61^2 + 0.57 \times 0.02^2 + 0.59 \times 1.77^2)10^4 \\ &= 3.65 \times 10^4 \text{ tons} \times \text{m} \end{aligned}$$

From Eqs. (VI-2-20) and (VI-2-21) we compute the limiting frequencies of natural vibrations of the foundation:

$$f_{nx}^2 = \frac{1.75 \times 10^4}{2.30} = 7.6 \times 10^3 \text{ sec}^{-2}$$

$$f_{n\varphi}^2 = \frac{3.65 \times 10^4}{4.30} = 8.5 \times 10^3 \text{ sec}^{-2}$$

The square of the radius of gyration equals

$$r^2 = \frac{M_m}{m} = 4.30/2.30 = 1.86$$

We find the value

$$\alpha = 1 + \frac{2}{r^2} = 1 + 0.06^2/1.86^2 \approx 1$$

In the case under consideration, α does not differ much from unity; therefore the natural frequencies f_{n1} and f_{n2} of the foundation, determined as roots of Eq. (VI-2-35), differ very little from the limiting frequencies f_{nx} and $f_{n\varphi}$.

According to Eq. (VI-2-4) we determine the design value of the radius of unbalance (eccentricity) of the rotating masses of the motor generator:

$$r_0 = 500/750^2 = 0.9 \times 10^{-3} \text{ m}$$

The exciting force imposed by all rotating masses of the motor generator equals

$$P = 0.9 \times 10^{-3} \times 5.9/9.81 \times 9 \times 10^3 = 4.8 \text{ tons}$$

Assuming, as in the preceding example, that the unbalanced state refers only to the static loads, we obtain for the exciting moment

$$M = 1.9 \times 1.67 - 2.3 \times 0.04 - 1.7 \times 1.83 = 0.06 \text{ ton} \times \text{m}$$

The influence of this moment on the amplitudes of forced vibrations of the foundation may be neglected.

The amplitude of horizontal displacement is obtained from Eq. (VI-2-26):

$$A_x = \frac{4.8}{2.3(7.6 - 6.1)10^3} = 1.4 \times 10^{-3} \text{ m} = 1.4 \text{ mm}$$

The foregoing computations show that the value of one of the natural vibration frequencies is close to the operational frequency of the engine; it follows that the amplitudes of vibrations considerably exceed permissible values. The foundation should be brought out of the zone of resonance; this may be done by increasing the cross sections of the columns to 0.6 by 0.7 m instead of the dimensions 0.4 by 0.5 m accepted in the design.

VII

MASSIVE FOUNDATIONS

VII-1. Massive Foundations for Motor Generators

General Remarks in Regard to Design and Design Computations. Motor generators usually operate at much lower speeds than turbogenerators. Therefore according to Eq. (VI-2-4) the eccentricity of rotating masses in motor generators and analogous electric machines is considerably larger than in turbogenerators. This is indirectly confirmed by the results of measurements of foundation vibrations.

If the eccentricities in motor generators were indeed of the same order as in turbogenerators, the exciting forces imposed by motor generators would be so small that they could not induce appreciable vibrations even under conditions of resonance, when the frequency of vibrations approaches one of the natural frequencies of the foundation. Then no dynamic computations of foundation vibrations under motor generators would be needed. Observations show, however, that the foundations under low-speed motor generators (up to 300 to 400 rpm) often undergo vibrations with amplitudes of the order of 0.1 to 0.3 mm. Foundations weighing several hundred tons may undergo forced vibrations with such amplitudes only when exciting loads are large. If one is to take the eccentricity of rotating masses in the motor generator as having the same value as in turbogenerators (0.2 mm), then for a 75-ton weight of rotating masses of a motor generator with a speed of 300 rpm, the value of exciting forces generated will equal only 1.5 tons. Such an exciting load, even under conditions of resonance, cannot induce vibrations with an amplitude of the order of 0.1 to 0.3 mm in a foundation weighing several hundred tons.

Thus the results of instrumental measurements of vibrations of low-frequency motor generators provide a basis for the assumption that in these machines the eccentricity of rotating masses is much larger than in turbogenerators.

Stress analysis of a massive foundation is not required because of the small magnitude of stresses imposed by static and dynamic external loads. In addition to the computation of amplitudes of transverse vibrations, it is necessary to center the foundation and check the magnitude of the pressure imposed on the soil by its weight. The permissible pressure on the soil may be taken to equal the permissible pressure under conditions of static loading only.

The foundation is built of concrete; the upper part is made of concrete type 100,† and the lower slab may be made of concrete type 60† or cyclopean concrete. In portions weakened by openings and grooves, the foundation is reinforced to fit the field conditions; approximately 20 to 30 kg of steel are used for 1 m^3.

Example. Dynamic computation

1. DATA. A dynamic computation of the massive foundation under a 3,000-kw motor generator running at 300 rpm is to be performed. Figure VII-1 shows the main dimensions of the foundation selected on the basis of construction assignments from the engine manufacturer and the client who ordered the design. The static loads and points of application are also shown in the figure.

Geologic conditions are as follows: loessial clay with some sand extends to a depth of 28 m; its moisture content is about 9 to 10 per cent; it is underlaid by dense brown clays. The ground-water level is at a depth of 14 m. The following coefficients of elasticity of the soil have been established from investigations of its elastic properties: Coefficient of elastic uniform compression:

$$c_u = 5 \times 10^3 \text{ tons/m}^3$$

Coefficient of elastic nonuniform compression:

$$c_\varphi = 10 \times 10^3 \text{ tons/m}^3$$

Coefficient of elastic uniform shear:

$$c_\tau = 3.5 \times 10^3 \text{ tons/m}^3$$

The foundation is to be erected in a machine room with several operating motor generators. The width of the building does not permit increasing the width of the foundation beyond that shown in Fig. VII-1. The distance between footings under columns and motor generators is 25 cm.

The following values necessary for dynamic computations were established from calculations:

Weight of the foundation (taking into account all cavities) and engine:

$$W = 1{,}136 \text{ tons}$$

Mass of the foundation and engine:

$$m = 115.7 \text{ tons} \times \text{sec}^2/\text{m}$$

† See footnote, Art. IV-1-c, p. 132.

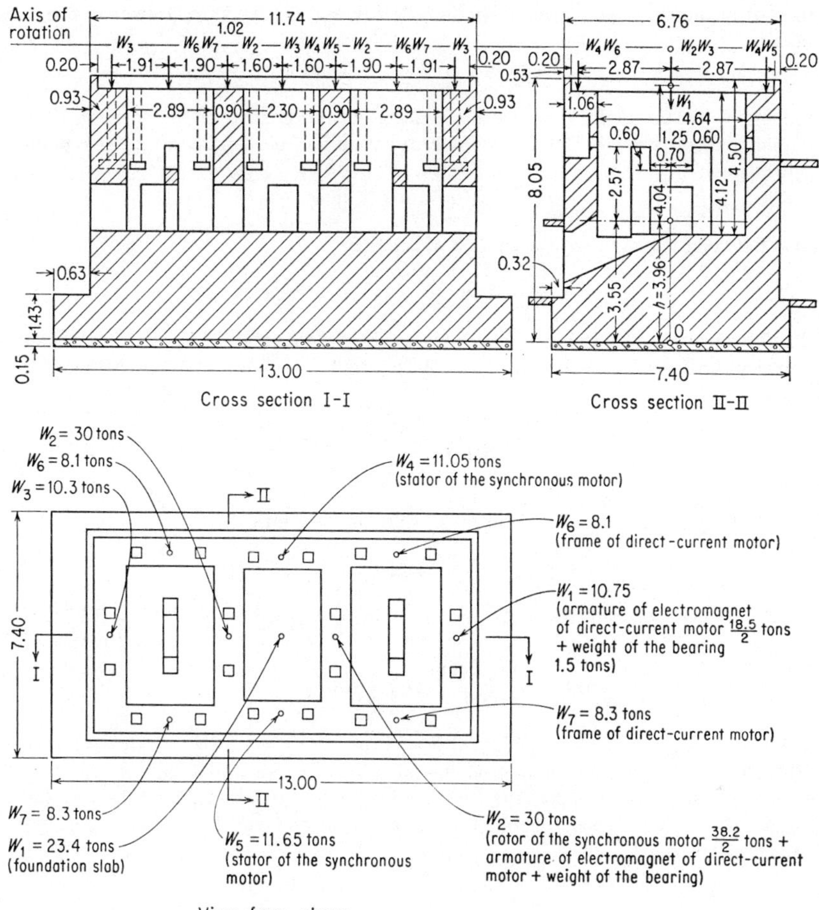

FIG. VII-1. Massive foundation for 3,000-kw 300-rpm motor generator of example of Art. VII-1.

Foundation area in contact with soil:

$$A = 96.0 \text{ m}^2$$

Distance between the level of the foundation contact area with soil and the common center of mass:

$$h = 4.6 \text{ m}$$

Moment of inertia of the foundation contact area with respect to the longitudinal axis passing through the centroid of the contact area:

$$I = 440 \text{ m}^4$$

Moment of inertia of the mass of the foundation and engine with respect to the same axis:

$$I_m = 3{,}974.6 \text{ tons} \times \text{m} \times \text{sec}^2$$

Moment of inertia of the mass of the foundation and engine with respect to the axis which passes through the center of gravity and is perpendicular to the plane of vibrations:

$$I_0 = 1{,}510.6 \text{ tons} \times \text{m} \times \text{sec}^2$$

Ratio between the moments of inertia of masses:

$$\gamma = 0.38$$

2. Computations. Using these data, we begin by establishing the frequency of natural vibrations of the foundation. The frequency of natural vertical vibrations (from Eq. III-1-5) is

$$f_{nz}^2 = \frac{5 \times 10^3 \times 96.0}{115.7} = 4.4 \times 10^3 \text{ sec}^{-2}$$

$$f_{nz} = 64.3 \text{ sec}^{-1}$$

The number of natural vertical vibrations of the foundation is

$$N_z = 9.55 f_{nz} = 9.55 \times 64.3 = 614 \text{ min}^{-1}$$

The difference between the numbers of natural vibrations and forced vibrations equals per cent

$$\eta_z = \frac{614 - 300}{300} 100 = 105 \text{ per cent}$$

Hence, the design of the foundation is satisfactory in regard to vertical vibrations.

In order to determine the frequencies of natural vibrations of the foundation f_{n1} and f_{n2} in a transverse plane, the limiting frequencies $f_{n\varphi}$ and f_{nx} of the foundation should first be established. From Eqs. (III-2-6) and (III-3-2), we have

$$f_{n\varphi}^2 = \frac{10 \times 10^3 \times 440}{3974.6} = 1.11 \times 10^3 \text{ sec}^{-2}$$

$$f_{nx}^2 = \frac{3.5 \times 10^3 \times 96}{115.7} = 2.91 \times 10^3 \text{ sec}^{-2}$$

Then we obtain

$$\frac{f_{n\varphi}^2 + f_{nx}^2}{2\gamma} = \frac{1.11 \times 10^3 + 2.91 \times 10^3}{2 \times 0.38} = 5.3 \times 10^3$$

$$\frac{f_{n\varphi}^2 f_{nx}^2}{\gamma} = \frac{1.11 \times 10^3 \times 2.91 \times 10^3}{0.38} = 8.5 \times 10^6$$

According to Eq. (III-4-8),

$$f_{n1,2}^2 = 5.3 \times 10^3 \pm \sqrt{28.0 \times 10^6 - 8.5 \times 10^6} = (5.3 \pm 4.43)10^3 \text{ sec}^{-2}$$

Hence

$$f_{n2}^2 = 0.87 \times 10^3 \text{ sec}^{-2}$$

$$f_{n2} = 29.5 \text{ sec}^{-1}$$

Thus the minimum number of natural vibrations of the foundation is

$$N_2 = 9.55 \times 29.5 = 282 \text{ min}^{-1}$$

This differs from the operational speed of the engine by only

$$\eta_2 = \frac{282 - 300}{300} 100 = -3 \text{ per cent}$$

Therefore it can be assumed that if the rotating masses of the motor generator are only slightly out of balance, considerable transverse vibrations of the foundation may develop. It follows that the design of the foundation is not satisfactory in regard to these vibrations.

3. Modification. The dimensions of the machine and the building do not permit any considerable changes in foundation width. Changes in the depth of the foundation or an increase in the foundation length has very little influence on the frequencies of natural vibrations of the foundation; therefore the only way to increase their value is to increase artificially the rigidity of the base under the foundation. In the case under consideration, the best way to achieve an increase in the rigidity of the base would be the provision of short conical precast reinforced-concrete piles. However, the use of these piles would require a comparatively long time for their casting and curing. In addition, the driving of piles inside a building with machines in operation would cause considerable inconvenience.

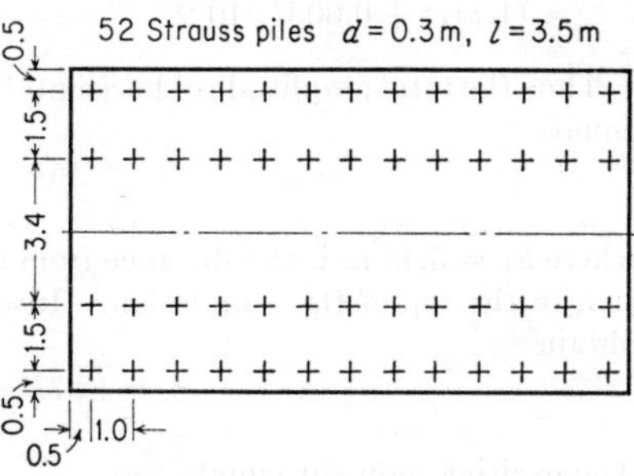

Fig. VII-2. Distribution of situ-cast concrete bore piles under the foundation of Fig. VII-1.

Therefore the decision was taken to increase the rigidity of the base by the installation of 52 situ-cast bore piles system Strauss, each 3.5 m long. Figure VII-2 shows the distribution of these piles in plan.

The coefficients of elasticity of such a pile base are about three times larger than those of the natural base under the foundation. One can take for the pile base:

$$c_\varphi = 30 \times 10^3 \text{ tons/m}^3$$
$$c_\tau = 10.5 \times 10^3 \text{ tons/m}^3$$

Let us compute the forced vibrations of the foundation in a transverse plane when piles are used:

$$f_\varphi^2 = 3 \times 1.11 \times 10^3 = 3.3 \times 10^3 \text{ sec}^{-2}$$
$$f_x^2 = 3 \times 2.91 \times 10^3 = 8.72 \times 10^3 \text{ sec}^{-2}$$

From Eq. (III-4-8),

$$f_{1,2}^2 = \left(\frac{3.3 + 8.72}{2 \times 0.38} \pm \sqrt{\frac{3.3 + 8.72^2}{2 \times 0.38} - \frac{3.3 \times 8.72}{0.38}}\right) 10^3$$
$$= (15.7 \pm 12.3)10^3$$

Hence we have

$$f_1^2 = 29.0 \times 10^3 \text{ sec}^{-2}$$
$$f_2^2 = 3.4 \times 10^3 \text{ sec}^{-2}$$
$$f_2 = 58 \text{ sec}^{-1}$$

We find the multiplier,

$$\Delta(\omega^2) = mI_0(f_1^2 - \omega^2)(f_2^2 - \omega^2)$$
$$= 115.7 \times 1510.6(29 - 0.9)(3.4 - 0.9)10^6 = 12.3 \times 10^{12}$$

According to Eqs. (III-4-1) and (III-4-12), the amplitudes of forced vibrations of the foundation will equal

$$A_x = \frac{30 \times 10^3 \times 440 - 1136 \times 4.6 + 10.5 \times 10^3 \times 96 \times (4.6)^2 - 1510.6 \times 0.9 \times 10^3}{12.3 \times 10^{12}} P + \frac{10.5 \times 10^3 \times 96 \times 4.6}{12.3 \times 10^{12}} M_i = (2.7P + 0.37M_i)10^{-6} \text{ m}$$

$$A_\varphi = \frac{10.5 \times 10^3 \times 96 \times 4.6}{12.3 \times 10^{12}} P + \frac{10.5 \times 10^3 \times 96 - 115.7 \times 0.9 \times 10^3}{12.3 \times 10^{12}} M_i = (1.71P + 0.00M_i)10^{-6}$$

Thus the total amplitude of horizontal vibrations of the upper part of the foundation equals

$$A = A_x + h_1 A_\varphi$$

where $h_1 = 3.45$ m is the distance from the common center of the engine and foundation to the top of the foundation. Inserting the numerical values of A_x and A_φ, we obtain

$$A = (2.7P + 6.27M_i)10^{-6} \text{ m}$$

The exciting moment equals

$$M_i = PH$$

where $H = 5.2$ m is the distance from the axis of the machine shaft to the common center of gravity; thus

$$A = 27.7 \times 10^6 P$$

According to the data from the plant, the design value of the rotor weight is around 60 tons; an approximate value of the eccentricity r_0 of rotating masses may be taken as ten times that for turbogenerators with speeds of 1,500 rpm; i.e., $r_0 = 2$ mm. Then the design value of the exciting force equals

$$P = 2 \times 10^{-3} \times 60/9.81 \times 0.9 \times 10^3 = 10.8 \text{ tons}$$

Inserting this value of P into the expression for the amplitude, we obtain

$$A = 27.7 \times 10^{-6} \times 10^8 = 0.3 \times 10^{-3} \text{ m} = 0.3 \text{ mm}$$

For low-frequency motor generators, the permissible design value of amplitude of vibrations may be taken to equal 0.30 mm. It follows that the foundation under consideration satisfies the conditions of dynamic stability.

VII-2. Massive Foundations under Turbodynamos

Basically, massive foundations under turbodynamos are blocks with cavities and grooves for individual parts of the machine or for mounting auxiliary equipment. Such a foundation consists of an upper part designed as a very rigid box or as two walls with grooves and openings, and a lower foundation slab transmitting the load to the soil. Special design features of massive foundations for turbodynamos are seen in

Fig. VII-3*a* and *b*, which shows a general view and a longitudinal section of a foundation for a 1,200-kw turbodynamo.

The upper part of this foundation consists of a complicated combination of individual structural units: girders, columns, walls, slabs, and others. Dynamic and static computations of this part of a foundation therefore involve a high degree of approximation. The dimensions of the upper part of the foundation and its individual units are usually determined by the construction assignment prepared by the machine manufacturer. Thus a designer's task is limited to determining the dimensions of the lower foundation slab and designing the reinforcement

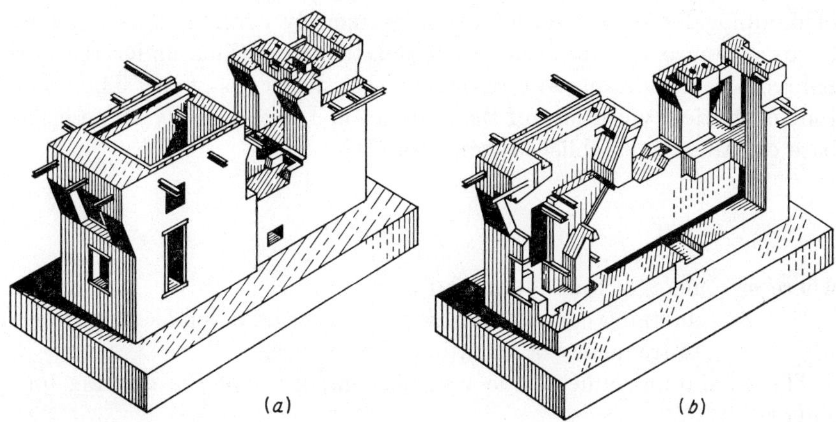

Fig. VII-3. Foundation for a 1,200-kw turbodynamo: (*a*) general view; (*b*) longitudinal section.

for the foundation. Essential points of instruction for the construction of frame foundations given in Art. VI-1 should be followed in the design of massive foundations for turbodynamos. Concrete type 150† is used for the upper part of a massive foundation, and concrete type 100† for the lower foundation slab.

All structural units of the upper part of the foundation are designed so that their numbers of natural vibrations should not be smaller than 3,000 min-1.

Unbalanced inertial forces of turbodynamos may induce vibrations of the foundation as a rigid body on an elastic base, as well as vibrations of the separate structural elements constituting the foundation.

Experience in operating high-frequency turbodynamos has not revealed any cases of significant vibrations of massive foundations acting as rigid bodies on elastic bases for the following reasons:

† See footnote, Art. IV-1-c, p. 132.

The natural frequencies of foundation vibrations are usually smaller than the operational frequencies of high-frequency turbodynamos, and it is hardly possible that these two frequencies will coincide. High-frequency turbodynamos are well balanced in regard to both static and dynamic loads. Their actual eccentricities (Art. VI-2) do not exceed 2 mm. Therefore the exciting loads inducing foundation vibrations are relatively small.

Even under the most unfavorable conditions, the exciting loads cannot induce vibrations with impermissible amplitudes, because the foundation mass is large in comparison with the mass of rotating machine parts. In the case of high-frequency vibrations, there is considerable influence of damping forces. In order to approximately evaluate this influence, let us compute the amplitude of foundation vibrations under the most unfavorable conditions—at resonance (i.e., when $\omega = f_{nz}$). The amplitude of vertical vibrations of the foundation as a rigid body on an elastic base can then be established from Eq. (III-1-21):

$$A_z = \frac{P_z}{2mc\omega}$$

where m = mass of foundation and machine

c = damping constant whose value may be taken as proportional to frequency of vibrations, i.e., $c = \eta\omega$

The maximum value of the vertical component of the exciting force equals

$$P_z = r_0 m_0 \omega^2$$

where r_0 = eccentricity

m_0 = mass of rotating parts of machine

Inserting expressions for P_z and c in the formulas for A_z, we obtain the following expression for the amplitude of vertical vibrations of the foundation at resonance:

$$A_z = \frac{1}{2} \frac{r_0 \mu}{\eta}$$

where $\mu = m_0/m$ is the ratio between the rotating machine masses and the total mass of the foundation and machine.

For turbodynamos μ equals approximately 0.05; the value of the coefficient of proportionality η may be taken as 0.5. For these values of μ and η, we obtain

$$A_z = 5 \times 10^{-2} r_0$$

Even for $r_0 = 0.2$ mm, the maximum amplitude of vertical vibrations of the foundation to be expected under conditions of resonance will be on the order of only 0.010 mm. Vibrations of such an amplitude are not dangerous. Actual amplitudes of foundation vibrations are much smaller, because it is hardly likely that the frequencies of natural vertical vibrations and the frequencies of the machine will coincide. Therefore in design computations of massive foundations under turbodynamos with speeds greater than 1,000 rpm dynamic computation of the foundation as a rigid body on an elastic base is not required.

In order to prevent the vibration of individual units constituting the foundation, it is advisable to check them as to danger of resonance and then design them so that their natural frequencies will be larger than the operational frequencies of the machine.

Investigations of resonance should be performed on transverse girders of the foundation which support the machine bearings, because it is these girders which carry the dynamic loads imposed by the machine. In the computation of frequencies of natural vibrations of these elements, formulas for single-degree-of-freedom systems may be used in most cases with a sufficient degree of approximation. When computing deflections one should consider only the dead loads carried directly by the element studied.

High frequencies are characterized by large damping forces developing as a result of vibrations of individual units. Therefore, for foundations under turbodynamos with speeds greater than 3,000 rpm, there is no necessity to check individual units as to danger of resonance. It suffices to perform a static computation of the foundation elements directly supporting the loads. The same computations should be made also for machines having speeds below 3,000 rpm.

Dynamic stresses in bases under foundations are very small because the amplitudes of vibrations of massive foundations under turbodynamos are very small. Therefore the permissible bearing value of soils under foundations for high-frequency engines may be taken to equal about 80 to 100 per cent of the permissible bearing value for static load only.

VII-3. Foundations for Rolling Mills

In the process of hot-rolling operations, in addition to constant (with respect to time) loads acting on the foundation, there appear also variable loads. These loads may induce foundation oscillations and dynamic stresses in both the soil and the foundation.

The larger the rolling mill, the larger the alternate loads imposed on the foundation and soil. Of the heavy rolling mills, reversible double-level mills are the ones most commonly used in engineering metallurgy. Therefore the computations outlined below refer to this type of mill. However, the data presented here may be easily applied to other types of rolling mills, such as three-level types and nonreversible types.

Modern rolling mills consist of the following main units:

1. Driving roll motor, whose foundation in some cases is rigidly tied to the foundation of the stands; occasionally no tie exists between these foundations
2. Ilgner power system, which is always mounted on a separate foundation
3. Operating and drive-gear stands, usually having a common foundation

a. Dynamic Loads Imposed on the Foundation by the Driving Roll Motor. Reversible direct-current motors are commonly used for the operation of rolling mills. The operational speed of these motors is rather low—around 58 rpm. The maximum (switching-off) moment at the shaft of the motor may reach several hundred tons × meters. The power is supplied from the Ilgner power system; the motor is mounted on a massive foundation. It will be assumed in further discussions that the motor is rigidly tied to the foundation, which is considered to be an absolutely rigid body resting on a fully elastic soil whose essential constants are known.

If a torsional moment M is applied to the rotor shaft, then the stator, and consequently the foundation, will be under the action of a moment whose magnitude equals $|M|$, but whose direction is opposite to that of M. This moment is the only alternating load acting on the foundation.

Changes in the torsional moment M applied to the rotor shaft, and consequently changes in the alternating moment acting on the foundation and soil, are a complicated function of many independent variables whose influence is difficult to evaluate. Therefore calculations are usually based on several assumptions. First of all some assumptions should be made concerning the distribution of the so-called reduced pressure of metal on the rolls. The magnitude of this pressure essentially affects the magnitude of M.

In computations of power consumed by the rolling mill, it is customarily assumed that the reduced pressure along the arc of contact between the ingot and the rolls remains constant. Usually it is assumed that the angular speed of the rotor is constant. Under these conditions, the magnitude of the moment of rotation may be expressed at any instant as a linear function of time.

Figure VII-4 shows a graph of changes in M for one of the first passes of the ingot through the rolls, plotted on the basis of the preceding assumptions concerning the reduced pressure of metal on the rolls and the angular speed of the mill. The horizontal axis of the graph shows periods of time t corresponding to successive stages in the passage of the ingot.

These stages are as follows: (1) no-load speeding up of the rolls; (2) the ingot is gripped and forced through; (3) rolling with acceleration; (4) rolling at constant speed; (5) slowing down of the rolls; (6) exit of the ingot; (7) stoppage of the mill. Figure VII-5 gives a graph of changes in M during the whole process of rolling of an ingot.

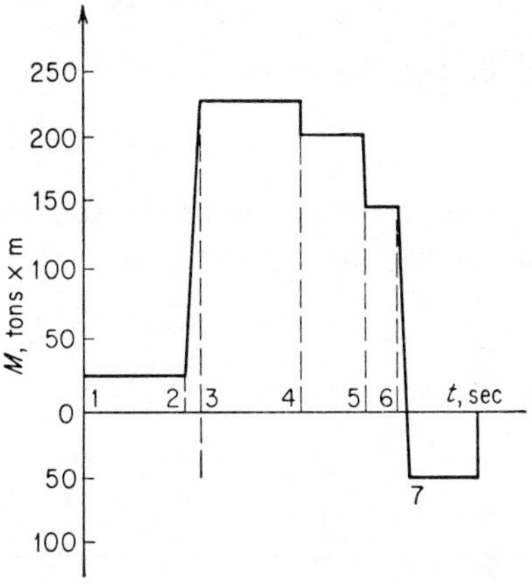

FIG. VII-4. Graph of changes in the torsional moment of the shaft during one passage of an ingot on a rolling mill.

It is seen from graphs VII-4 and VII-5 that the external torsional moment, and consequently the exciting moment acting on the foundation of the motor, do not change much in the course of a pass of the ingot, except for the periods of its entry and exit. Therefore, instead of the diagram of changes in M shown in Fig. VII-4, one can assume that changes will occur according to the diagram in Fig. VII-6a.

When the ingot emerges from the rolls, the absolute value of change occurring in M in practice may be considered to equal the change occur-

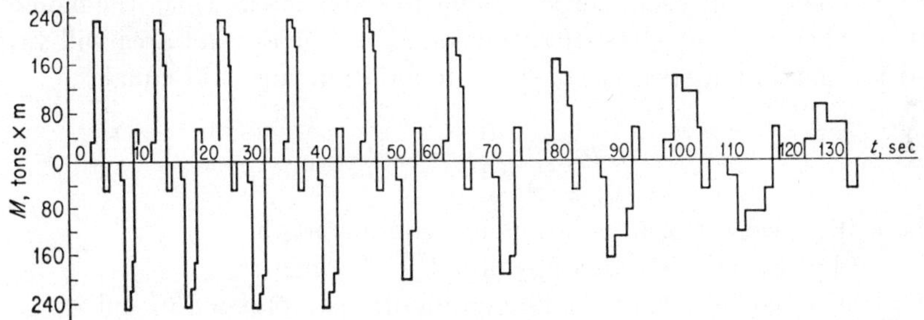

FIG. VII-5. Graph of changes in the torsional moment of the shaft during entire rolling process of an ingot.

ring when the ingot is gripped by the rolls. The exit of an ingot from the rolls is accompanied by foundation vibration. Due to a decrease in loading, the magnitudes of stresses induced by the vibration will not exceed those observed during the steady process of rolling. This makes it possible to base calculations not on Fig. VII-6a, but on Fig. VII-6b.

In conformity with this diagram, let us set up the following conditions for the exciting moment M:

1. When $t = 0$, $M = 0$.

2. At the time the ingot is gripped by the rolls ($t \leq \tau$),

$$\frac{d}{dt} M = \text{constant} > 0$$

3. For the steady process of rolling ($t > \tau$),

$$M = M_{\max} = \text{constant}$$

Under the action of the torsional moment, the foundation will rotate around an axis passing through the center of gravity of the foundation

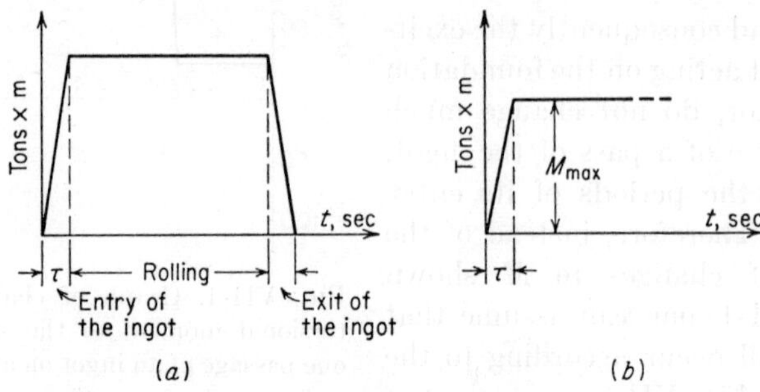

FIG. VII-6. Simplified design diagram of changes in the torsional moment of a rolling mill shaft.

area in contact with soil, perpendicular to the plane in which the moment acts. Therefore the stresses in the soil along the contact area will vary, and the maximum stress $p_{\max}$ at the foundation edge will equal

$$p_{\max} = \frac{W}{A} + c_\varphi a \varphi_{\max} \tag{VII-3-1}$$

where W = weight of foundation and equipment thereon
A = foundation area in contact with soil
c_φ = coefficient of elastic nonuniform compression of soil
$2a$ = foundation width in plane of action of moment

Let us compute $\varphi_{\max}$ for the interval of time corresponding to the gripping of the ingot by the rolls. The equations of motion of the foundation are as follows:

$$\begin{aligned} \ddot{x} + \alpha_{11}x + \alpha_{12}\varphi &= 0 \\ \ddot{\varphi} + \alpha_{21}x + \alpha_{22}\varphi &= M_i t \end{aligned} \tag{VII-3-2}$$

where $$M_i = \frac{M_{\max}}{M_\tau}$$

and M_τ is the moment of inertia of the installation mass (machine and foundation).

The coefficients of this system of equations depend on the elastic properties of soil and the dimensions and mass of the foundation and motor [Eqs. (III-4-5)].

Assuming that at time zero the displacement x_0 and the angle of rotation φ of the center of gravity of the foundation equal zero, we obtain the general solution for φ:

$$\varphi = \frac{\alpha_{11} - f_{n1}^2}{f_{n1}^2(f_{n2}^2 - f_{n1}^2)} M \sin f_{n1}t - \frac{\alpha_{11} - f_{n2}^2}{f_{n2}^2(f_{n2}^2 - f_{n1}^2)} M \sin f_{n2}t + \varphi_s \qquad \text{(VII-3-3)}$$

where f_{n1} and f_{n2} are the natural frequencies of the foundation established from Eq. (III-4-8). The expression

$$\varphi_s = \frac{\alpha_{11}}{f_{n1}^2 f_{n2}^2} M_i t \qquad \text{(VII-3-4)}$$

gives the value of φ for the condition that the torsional moment has only a static effect. The other terms in Eq. (VII-3-3) evaluate the dynamic action of the external torsional moment applied to the foundation. Assuming

$$\varphi = \mu\varphi_s$$

we obtain for the dynamic coefficient

$$\mu = 1 + \frac{(\alpha_{11} - f_{n1}^2)f_{n2}^2}{\alpha_{11}(f_{n2}^2 - f_{n2}^2)f_{n1}t} \sin f_{n1}t - \frac{(\alpha_{11} - f_{n1}^2)f_{n1}^2}{\alpha_{11}(f_{n2}^2 - f_{n1}^2)f_{n2}t} \sin f_{n2}t \qquad \text{(VII-3-5)}$$

If the gripping period is small in comparison with the periods T_1 and T_2 of natural vibrations of the foundation, then, assuming that

$$\sin f_{n1}t = f_{n1}t$$
$$\sin f_{n2}t = f_{n2}t$$

we obtain

$$\mu = 2$$

In this case the maximum rotation of the foundation under the action of the alternating torsional moment will not exceed the twofold value of displacement caused by the static action of the same moment.

As the ingot gripping time increases, the value $\mu - 1$ decreases, approaching zero for high values of t. Consequently, if the gripping time is large in comparison with the periods T_1 and T_2, then the action of the alternating torsional moment upon the foundation does not differ much from the static pressures.

The value of μ may be computed with a comparatively high degree of accuracy as soon as one knows the periods of natural vibrations of the foundation in the plane of action of the alternating torsional moment;

then one will also know the time required to grip the ingots. In practice this time varies within a comparatively wide range. Only approximate values of the periods T_1 and T_2 may be established by computations. The calculation of T_1 and T_2 involves laborious arithmetic operations; therefore practical design computations of the foundation under the driving roll motor should be based on the most unfavorable conditions by setting $\mu = 2$.

Evidently, for $t > \tau$, i.e., for the steady process of rolling, φ will not exceed the maximum value characterizing the times of gripping of the ingot by the rolls and its emergence from the rolls. Therefore the stresses in the soil along the foundation contact area during the process of rolling will not exceed the stresses developing at the time of entry of the ingot. The values of these stresses should be used in calculations of the base under the foundation for the rolling mill.

b. Dynamic Action on the Foundation by the Ilgner Power System. The purpose of the Ilgner power system is to feed power to the motor driving the rolls. The power system consists of one or several direct-current generators and a flywheel, mounted on the same shaft. The generators are set in motion by an electric motor.

The power W_1 of an asynchronous motor, taken from the line, remains almost constant during power-system operation. It equals the average quadratic power required for rolling during one cycle (15 to 19 passes). The power supplied by the direct-current generators is almost constant. The power W_2 taken from these two generators by a motor driving the rolls undergoes extremely sharp changes. The range of these changes is from zero, which corresponds to a pause in the rolling, to the maximum power required by the motor. The maximum power may be considerably larger than the power which at a given instant is supplied to the generators by the asynchronous motor. The flywheel and other rotating masses increase the amount of energy which is yielded by the generators, since their kinetic energy changes as a result of changes in the speed of the shaft. Thus a deficiency in energy required for rolling is made up. In addition, during the operation interval in which the generators do not supply energy to the drive motor, the flywheel and other rotating masses accumulate the energy which is taken by the asynchronous motor from the line.

Let us investigate the dynamic loads acting on the foundation during the operation of the power system. If

W_1 = power taken by motor from line
ω = angular speed of aggregate shaft (a varying value)
M_1 = torsional moment of motor shaft

then

$$W_1 = M_1\omega$$

In electrical motors, the stators (and consequently the foundation) are under the action of a reactive moment whose absolute value equals that of the torsional moment M_1, but whose direction is opposite.

In addition to this moment M_1, a moment M_2 is also acting on the foundation, induced by the generators. This moment has the same sign as the moment of the generator shaft. If W_2 is the power yielded by the generators, then

$$W_2 = M_2\omega$$

The resulting external moment M_i acting on the foundation evidently will equal the difference between the moments; i.e.,

$$M_i = M_2 - M_1$$

Neglecting power losses in the engine, we obtain

$$W_2 = W_1 - \frac{d\omega}{dt}\,\omega \sum I_i$$

where $\Sigma I_i = I$ is the sum of the moments of inertia of all the rotating masses of the power system, i.e., of the flywheel, motor generators, and armatures of electromagnets. Since

$$W_2 - W_1 = \omega(M_2 - M_1) = M_i\omega$$

it follows that

$$M_i = \frac{d\omega}{dt}\,I \qquad \text{(VII-3-6)}$$

when the energy yielded by the power system equals the energy taken from the line, i.e., when

$$\frac{d\omega}{dt} = 0$$

the external moment acting on the foundation also equals zero. At the same time, the foundation will be subjected to internal moments tending to produce torsion in it.

Let us assume that the Ilgner power system consists of two alternating-current generators, an asynchronous motor, and a flywheel. The total flywheel moment GD^2 of all the rotating masses of the aggregate is about 870 tons $\times$ m^2; hence

$$I = \frac{GD^2}{4g} = \frac{870}{4 \times 9.81} = 22.3 \text{ tons} \times \text{m} \times \text{sec}^2$$

If N is the number of rpm of the power system, then

$$\omega = \frac{2\pi}{60} N \qquad \frac{d\omega}{dt} = \frac{2\pi}{60} \frac{dN}{dt}$$

Therefore,
$$M_i = \frac{2\pi}{60} I \frac{dN}{dt} = 2.24 \frac{dN}{dt}$$

The rate of change of N, i.e., dN/dt, varies within the range 2.8 to 10.4. The magnitude of the external moment acting on the power-system foundation during the whole cycle of rolling of one ingot changes from 2.6 tons × m (periods of running idle) up to 24 tons × m (periods of rolling).

The design value of the exciting moment should be taken to equal $2M_i$ for the most unfavorable case. The angle of foundation rotation, induced by this moment, is determined from Eq. (VII-3-3).

In addition to the exciting moments caused by changes in the kinetic energy of the power system, the foundation may be subjected to periodic exciting loads caused by the unbalanced state of the engine with respect to magnetic forces and static equilibrium. The computation of forced vibrations of the foundation caused by these loads is performed in the same way as for foundations under motor generators.

c. Dynamic Loads on the Common Foundation of Working and Gear Stands. In the process of the rolling-mill operation, the frame of the driving-gear stand, and consequently its foundation, are subjected to the action of a varying exciting moment equal in magnitude and sign to the moment of the shaft of the driving roll motor.

The forces appearing as a result of the acceleration of the ingot may be neglected because of their minute magnitudes; hence it may be considered that stresses occur only in the working stand during the rolling operations. These stresses have a tendency to rupture the stand. The sum of all the external alternating loads equals zero.

The drive-gear and working stands may be mounted on a separate foundation, not tied to that under the driving roll motor. In this case, the dynamic influences of external loads on the foundations are evaluated separately but similarly.

If the drive-gear stand, working stand, and driving roll motor are mounted on a common foundation, then the drive-gear stand is subjected to the action of a torsional moment whose sign is opposite that of the moment acting on the stator of the driving roll motor. Therefore the sum of all the external dynamic loads transmitted to the foundation and soil equals zero. The foundation will be under the action of internal torsional moments whose magnitude equals the moment of the shaft of the motor, as well as under the action of the equipment weight. These loads should be considered in the stress analysis of the foundation and its separate elements. The dynamic nature of the internal moments is taken into account by introducing in the calculations the twofold magnitude of the maximum torsional moment of the shaft of the driving roll motor.

In the case under consideration, stresses in the soil are determined for a design load consisting of the combined weight of the foundation and the equipment mounted thereon.

d. Remarks concerning Design. The foundations for the principal rolling equipment (stands, reducer, gear) are always built as massive

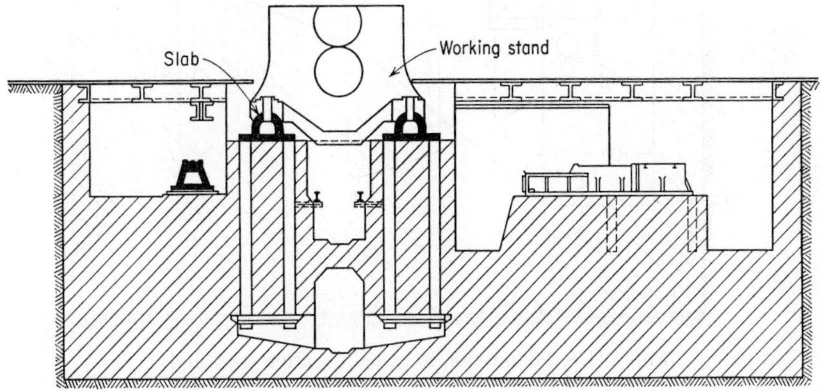

Fig. VII-7. Foundation for a stand of a sheet-rolling mill.

units which either are monolithic or are provided with deformation joints. As illustration, diagrams of massive foundations are shown as follows: Fig. VII-7: a foundation for the stand of a sheet-rolling mill; Fig. VII-8: a foundation for a light-section steel mill; Fig. VII-9: a foundation for a drive-gear stand.

The main part of the foundation under the drive-gear and working stands is always designed as one block. This part of the foundation usually has two tunnels, located along the axis of the stand at different heights. The upper tunnel serves for the removal of mill scale and for the runoff of cooling water under the working stand, as well as for the inspection of equipment and the runoff of lubricant under the drive-gear stand. The lower tunnel serves for the inspection of anchor bolts; it is provided with several recesses to facilitate access to anchor plates. In the central part of the foundation are located spindle benches which are provided with wells for counterweights and an appliance for changing the first roller of the stand.

The foundation under the driving roll motor (Fig. VII-10) is built as a separate massive block or as a block forming one monolith with the

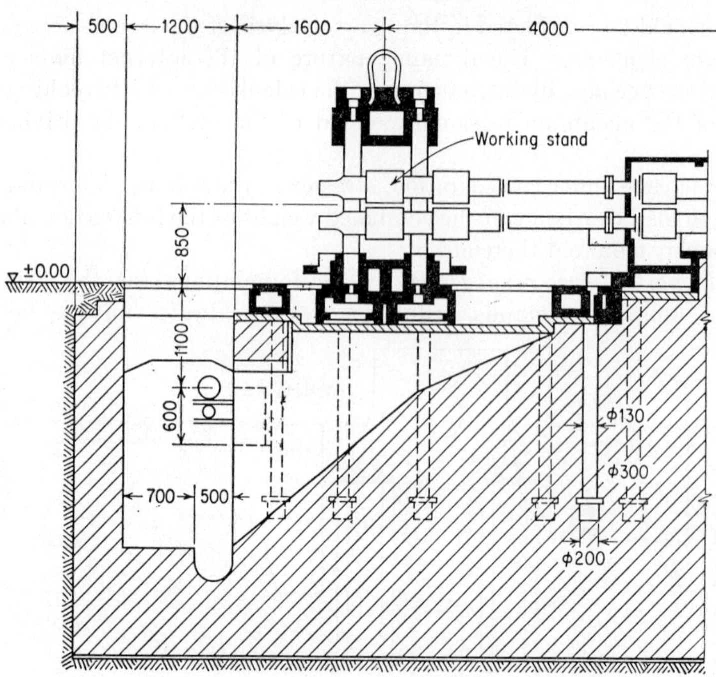

FIG. VII-8. Foundation for a light-section steel mill.

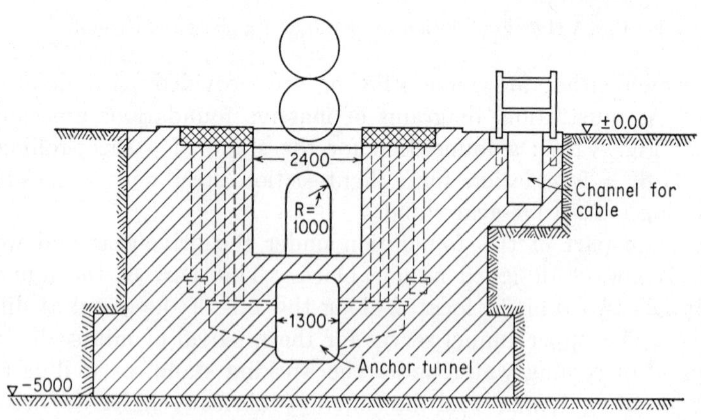

FIG. VII-9. Foundation for a drive-gear stand.

foundation under the drive-gear stand. The foundation has a deep groove for the inspection and mounting of the equipment and a channel for the air-cooling of the motor.

On both sides of the working stand, along the rolling-mill axis, are located the foundations for manipulators and roller conveyors. Usually

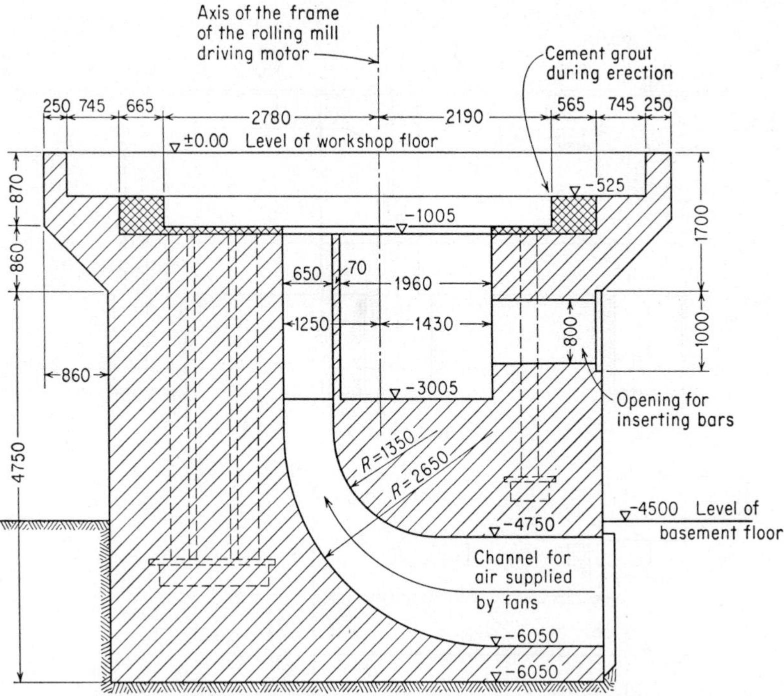

FIG. VII-10. Foundation for the motor driving the rolls.

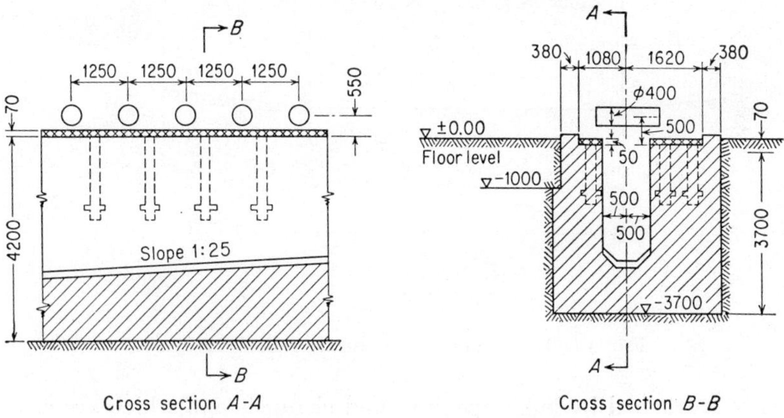

FIG. VII-11. Block foundation for roller conveyors.

these foundations are also built as massive blocks with required channels and grooves (Fig. VII-11). Sometimes they are designed as frame foundations (Fig. VII-12).

Foundations under rolling-mills equipment are made of concrete and reinforced concrete. Concrete is employed for massive foundations which

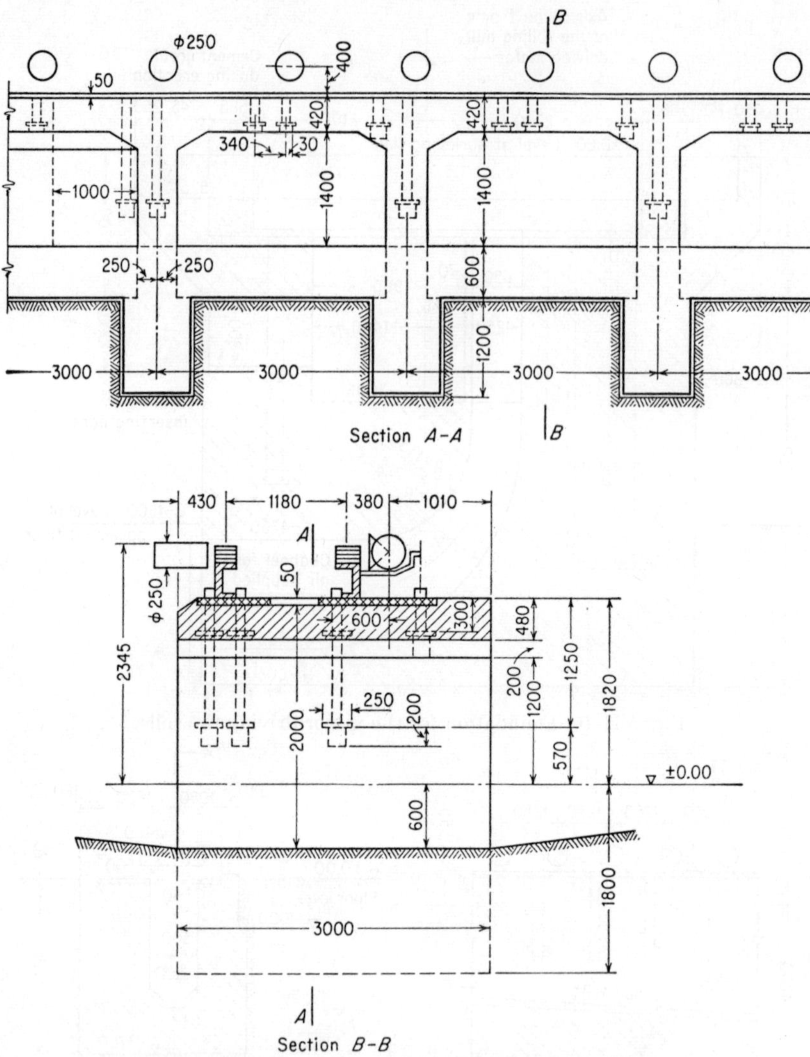

FIG. VII-12. Frame foundation for roller conveyors.

are not weakened by large openings and channels and which are erected on sufficiently rigid and homogeneous bases. Otherwise, reinforced concrete is used. As a rule, concrete type 100† is employed.

The foundation area in contact with soil should be, as far as possible, all on the same level. Large differences in depths of separate sections of

† See footnote, Art. IV-1-c, p. 132.

the foundation should not be permitted. If locating all foundation contact areas on the same level leads to a considerable overexpenditure of material, then deformation joints may be provided between sections lying at different depths.

The location of expansion, shrinkage, and settlement joints in foundations under rolling-mill equipment is determined by the distribution of the equipment, the depth of separate foundation sections, the soil bearing value, and the temperature regime of rolling. Distances between deformation joints are selected according to the official *Technical Rules and Construction Code.* Joints should be located so that they divide the foundation into separate sections which support units of equipment not connected with each other. For example, in order to avoid uneven settlement, the foundations under working and drive-gear stands should not be separated.

Continuous footings longer than 20 to 30 m and foundation sections under stands larger than 15 by 15 m or 20 by 20 m should be provided with deformation joints. If a large section of the foundation cannot be divided by deformation joints, then, in order to prevent the appearance of shrinkage cracks, such a foundation may be divided by temporary joints with reinforcement extending beyond the joints. Later these joints are filled with concrete of the same type. The projecting reinforcement is overlapped and welded.

e. Design Loads. For the analysis of stresses within the foundation and for the determination of pressure on the base, the following loads should be considered:

Weight of the rolling-mill equipment
Weight of the driving roll motor
Maximum disconnection moment at the motor shaft
Horizontal force transmitted to the footings under manipulators and tilting devices
Erection loads
Foundation weight

Static computations of the foundation may be limited to:

1. Stress analysis of separate units of the foundation, such as units weakened by openings, cantilevers, and others
2. Computation of local stresses under supporting slabs
3. Analysis of stresses within the foundation
4. Computation of pressures transmitted to the soil

The foundation is considered to be a girder of varying rigidity resting on an elastic base.

For calculations listed in points 1 and 2, a value of the dynamic coefficient equaling 2 is introduced in the calculations of the weight of the rolling mill and of the driving roll motor. For calculations listed in points 3 and 4 the actual weight of the same machines is taken, without introducing a dynamic coefficient.

If a foundation is treated in design computations as a beam resting on an elastic base, then, in order to simplify operations, it is permissible to consider separate comparatively rigid units of the foundation as being absolutely rigid. An uneven settlement at the contact of the foundation under the roller conveyers with the foundation under the rolling mill leads to the appearance of stresses along this contact. To determine these stresses, it is permitted to consider the foundation under the rolling mill to be an absolutely rigid unit.

The permissible pressure on the soil under the foundations of rolling mills and driving roll motors for dynamic loads may be taken to equal the corresponding permissible pressure for static loads only.

In concrete or lightly reinforced foundations, the soil pressure imposed by separate machinery units and established for conditionally separated foundation sections without considering the influence of other foundation units should not exceed the permissible bearing value of soil.

Foundations subjected to horizontal impacts, such as those under manipulators and tilting devices, should be designed for the double value of the maximum horizontal force.

f. Data on Performance of Existing Foundations under Rolling Mills. The author and B. M. Terenin investigated several foundations under rolling mills at one of the Soviet plants. These foundations were built of concrete, and each consisted of a single massive block supporting the driving roll motor as well as the drive-gear and working stands.

The foundations investigated were not reinforced at places weakened by recesses, openings, and channels. Results of laboratory tests showed that concrete had been used which, at the age of 28 days, had a temporary compressive strength of 90 kg/cm^2, with slight deviations in some parts of the foundation. Concrete type 60† was used for the foundations under lifting platforms of rolling mill "750," and concrete type 130† for the foundation under the first working stand of the same mill.

The foundations were placed on loessial clays with some sand. Owing to the wetting of the soil, for different reasons the foundations underwent uneven settlements resulting in the appearance of cracks. In the block of the central part of the foundation under rolling mill "750" several

† See footnote, Art. IV-1-c, p. 132.

cracks were observed in the tunnel under the driving gear and operation stands, in the tunnels under the lifting platforms, in wells at the contact between the foundations of roller conveyors and the foundations of rolling mills, and in the foundation unit under the driving roll motor. The appearance of these cracks was due to two causes:

1. A horizontal foliation of the foundation under the drive-gear and operating stands developed at the level of the anchor plates. The most distinct crack was observed in the tunnel under the drive-gear stand. Under the operating stands were found slightly developed small horizontal cracks coinciding with working joints. These cracks indicate that a long interruption had occurred in the concreting of the foundation and that no measures were taken to secure the monolithic character of the foundation.

2. There was a differential settlement of the foundation under the rolling mill and the foundation under adjacent auxiliary equipment. This settlement was caused by the wetting of soil and resulted in cracks in the tunnels of the rear and front lifting platforms, in the wall of the middle platform of the staircase, in the arch near the lifting platform of the second operating stand, and under the decelerator of the driving roll motor.

A vertical crack was observed approximately in the middle of the tunnel of rolling mill "450." This crack ran along the walls in places where they were weakened by niches, and along the arch.

A vertical crack was found in the tunnel of rolling mill "360" near the inlet opening; two vertical cracks were found in niches, one of them running along the arch.

In the tunnel of rolling mill "280" a vertical crack was found under the operating stand through which water was flowing abundantly. Channels of rolling mills "360" and "280," especially in their lower sections, were filled with water.

An instrumental investigation of vibrations of foundations under the rolling mills was performed at several points along the foundation axis and along its height: on the slabs of the operating and drive-gear stands, at the level of niches where anchor slabs of the foundation were located, and at points on the floor of the tunnels.

Results of the measurements are shown in Table VII-1. It is seen from this table that the largest amplitudes of vibrations were found directly on the slab under the drive-gear stand of rolling mill "750." The measurements performed here showed that the foundation underwent extremely irregular high-frequency vibrations with amplitudes of the order of 0.006 to 0.010 mm, caused by impacts of the gear.

These measured values of vibration amplitudes under rolling mills show that the additional pressure on the soil and the stresses within the foundation caused by dynamic loads are small in comparison with stresses imposed by the weight of equipment and foundation. Therefore a value of 3 for the dynamic coefficient, often taken in design computations of such foundations, is exaggerated.

TABLE VII-1. RESULTS OF VIBRATION MEASUREMENTS ON ROLLING-MILL FOUNDATIONS

Rolling mill	Vibrations measured at:	Nature of vibrations and amplitude
"750"	Station (1 at slab of gear stand)	Extremely irregular high-frequency vibrations with amplitudes 0.005–0.010 mm. At time of entry and exit of ingot, vibrometer records impacts inducing vertical and horizontal vibrations with amplitude 0.030–0.050 mm.
"750"	Station 2 (at edge of foundation near slab of gear stand)	High-frequency vibrations with amplitude less than 0.003 mm. At time of entry and exit of ingot, vertical impacts are recorded with amplitudes of some 0.006–0.010 mm.
"750"	Station 3 (housing under gear stand)	The same as for Station 2
"750"	Station 4 (on floor of tunnel under gear stand)	The same as for Station 2
"750"	Station 5 (at surface of foundation near rolling-mill driving motor)	High-frequency vibrations with amplitude 0.003 mm. At time of entry and exit of ingot, impacts are recorded inducing vibrations with amplitude of 0.010 mm.
"450"	Station 6 (at surface of foundation near rolling-mill driving motor)	Quickly damped vibrations were recorded, with amplitudes on the order of 0.0015 mm. Impacts at time of entry and exit of ingot are only slightly noticeable.
"360"	Station 7 (at surface of foundation near rolling-mill driving motor)	Vibrations of same nature as those at Station 6
"280"	Station 8 (at surface of foundation near rolling-mill driving motor)	The same as for Station 6
Slabbing	Near working stand	The same as for Station 2

Measurements of vibrations of the foundation under rolling mill "750," performed on the upper and lower parts of the foundation divided by a horizontal crack, established that these two sections underwent vibrations of the same character with the same amplitude. This indicates that the complete foundation vibrated as one block. It followed that foliation of the foundation is not dangerous for rolling-mill operations.

VII-4. Foundations for Crushing Equipment

a. Design Computations of Foundations under Jaw Crushers. There are many different arrangements of jaw-crusher operating mechanisms. However, one common feature of these crushers is that, analogously to reciprocating engines, they create unbalanced inertial forces varying with time. These inertial forces form exciting loads which induce forced vibrations of the foundation.

The most common arrangement of the operating units of the jaw crusher is one in which the motion of the mechanism is due to the action of so-called lower couples of rotation. Some typical arrangements of jaw crushers of this group are shown in Fig. VII-13. Approximate formulas for the determination of unbalanced inertia forces are also given. Accurate methods of computation of the exciting loads imposed by jaw crushers may be found in specialized publications.[2]

It follows from the equations in Fig. VII-13 that exciting loads imposed by jaw crushers are of the same nature as exciting loads imposed by reciprocating engines. Therefore all directives outlined in Chap. IV concerning the design of foundations for reciprocating engines may be applied to the dynamic computation and design of foundations for jaw crushers.

b. Computations of a Foundation under a Gyratory Crusher. In gyratory crushers the ore is pulverized between the crushing head of the main shaft, undergoing a rocking motion along a circle, and the armored jacket of the upper stationary part.

Under the action of frictional forces, the crushing cone moves around the axis of the crusher and develops an angular velocity whose value is close to that of the movement but has opposite sign. As a result of this, the frame of the machine, and consequently the foundation, is subjected to the action of gyroscopic and inertial loads which may be approximately expressed by one resultant exciting force:

$$R = (m_1 r_1 - m_2 r_2)\omega^2 \qquad \text{(VII-4-1)}$$

where m_1 = total mass of main shaft and crushing cone attached to it
m_2 = mass of camshaft and units rigidly connected with it (gears, counterweights, and others)
r_1 = distance between crusher axis and center of gravity of main shaft
r_2 = distance between another axis and center of gravity of eccentric shaft
ω = frequency of rotation of crusher

This force, rotating at a constant angular speed, acts in a horizontal plane

No.	Diagram of the crusher	Approximate values of inertia forces	Designations
1	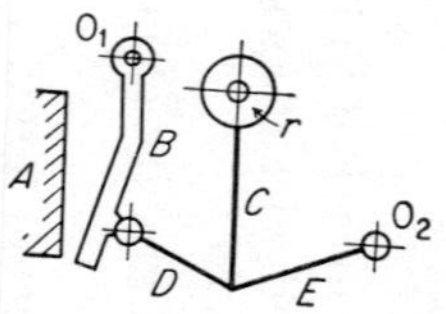	$P_z = (M_0 + M_c)r\omega^2 \sin \omega t$ $P_x = (M_0 + 0.8M_c)r\omega^2 \cos \omega t$ } I $P_z = [(M_0 + M_c)r - M_d r_1]\omega^2 \sin \omega t$ $P_x = 0.25M_B r\omega^2 \sin \omega t$ } II	M_B = mass of moving (crushing) jaw M_c = mass of connecting rod
2	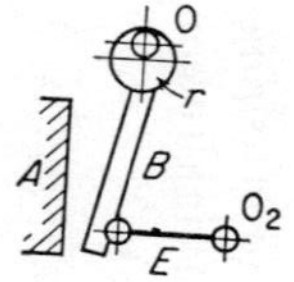	$P_z = (M_0 + M_B)r\omega^2 \sin \omega t$ $P_x = (M_0 + 0.5M_B)r\omega^2 \sin \omega t$ } I $P_z = [(M_0 + M_B)r - M_d r_1]\omega^2 \sin \omega t$ $P_x = [(M_0 + 0.5M_B)r - M_d r_1]\omega^2 \cos \omega t$ } II	M_0 = mass of eccentric (or 50% of crank-shaft mass) M_d = total mass of counterweights
3	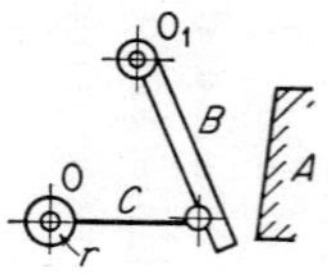	$P_z = (M_0 + 0.7M_c)r\omega^2 \sin \omega t$ $P_x = (M_0 + M_c + 0.5M_B)r\omega^2 \cos \omega t$ } I $P_z = 0$ $P_x = [(M_0 + M_c + 0.5M_B)r - M_d r_1]\omega^2 \cos \omega t$ } II	r = eccentricity r_1 = distance from axis of rotation to center of gravity of counterweights center
4	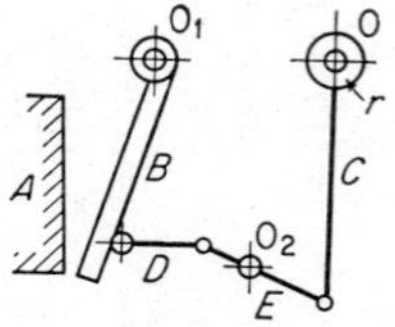	$P_z = (M_0 + M_c)r\omega^2 \sin \omega t$ $P_x = (M_0 + 0.8M_c)r\omega^2 \cos \omega t$ } $P_z = [(M_0 + M_c)r - M_d r_1]\omega^2 \sin \omega t$ $P_x = 0.25M_B r\omega^2 \sin \omega t$ } II	ω = angular speed

FIG. VII-13. Data on jaw crushers.

No.	Diagram of the crusher	Approximate values of inertia forces	Designations
5	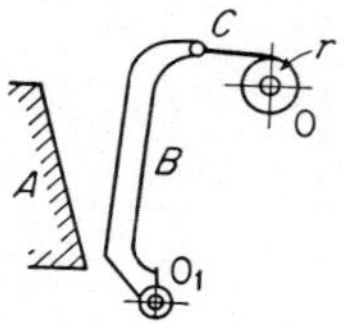	$P_z = (M_0 + 0.7M_c)r\omega^2 \sin \omega t$ $P_x = (0.5M_B + M_c + M_0)r\omega^2 \cos \omega t$ } I $P_z = 0$ $P_x = [(0.5M_B + M_c + M_0)r - M_d r_1]\omega^2 \cos \omega t$ } II	P_z = vertical component of resultant inertia force P_x = horizontal component of resultant inertia force
6	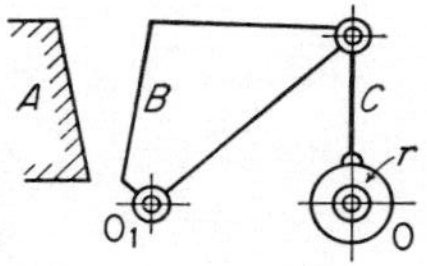	$P_z = (M_0 + M_c + 0.5M_B)r\omega^2 \sin \omega t$ $P_x = (M_0 + 0.7M_c + 0.5M_B)r\omega^2 \cos \omega t$ } I $P_z = [(M_0 + M_c + 0.5M_B)r - M_d r_1)]\omega^2 \sin \omega t$ $P_x = [(M_0 + 0.7M_c + 0.5M_B)r - M_d r_1]\omega^2 \cos \omega t$ } II	*Notes:* 1. Forces P_z and P_x are applied to **axis** of main shaft. 2. Equations I refer to crushers without counterweights.
7	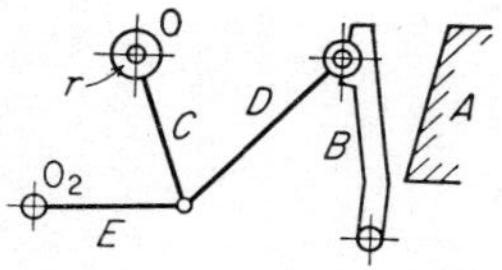	$P_z = (M_0 + M_c)r\omega^2 \sin \omega t$ $P_x = (M_0 + 0.8M_c)r\omega^2 \cos \omega t$ } I $P_z = [(M_0 + M_c)r - M_d r_1]\omega^2 \sin \omega t$ $P_x = 0.25M_B r\omega^2 \sin \omega t$ } II	Equations II refer to crushers with counterweights.

FIG. VII-13 (*Continued*)

passing through the center of the main shaft (in crushers with a sharp cone) or through the point of rest (in crushers with a flat cone).

Resolving R into components along the horizontal axes x and y, the principal inertial axes of the installation, we obtain

$$P_x = R \sin \omega t$$
$$P_y = R \cos \omega t$$

The dynamic computation of a foundation under a gyratory crusher is reduced to the determination of amplitudes of forced vibrations imposed on principal vertical planes of the foundation by exciting forces P_x and P_y and moments $P_x h_1$ and $P_y h_1$.

Thus, dynamic computation of a foundation under a gyratory crusher in principle does not differ at all from the dynamic computation of a foundation under a jaw crusher.

VIII

PROPAGATION OF ELASTIC WAVES IN SOIL

INTRODUCTION

As stated in Chap. I, there are several reasons why the application of Hooke's law to soils is limited. For example, it has been indicated that the elastic constants of soil depend on normal stresses and that elastic deformations may affect the initial internal stresses which always exist in soil. It should also be noted that the solution of problems related to the propagation of waves may be greatly influenced by dissipative properties of soil which govern the absorption of wave energy.

When solving problems related to the propagation of waves in soils, one has to start with models of the phenomenon, which are very far from reality. For example, the investigation of waves emanating from machine foundations leads to a composite dynamic theory-of-elasticity problem which starts with displacements in a certain section of the soil surface—while the rest of the soil is free of stresses. In the simplest case the soil is considered to be a semi-infinite elastic solid. The solution of such a composite problem involves considerable mathematical difficulties. Therefore, a source of waves is represented as an alternating force, either concentrated or distributed over the given soil surface area. This model of the source of waves is far from reality, and the results of such a solution may differ (sometimes considerably) from the results of experimental investigations of wave propagation from an actual source of waves such as a vibrating foundation.

However, in spite of the indicated limitations, the development of the theory of propagation of waves in soils on the basis of the theory of elasticity, even for highly abstract conditions, gives us a chance to investigate several very important specific features of wave propagation in soils. For example, the application of methods of the theory of elasticity made it possible to establish the influence of a free surface or a layer on wave propagation in soil. These investigations led to the discovery of new waves, whose theoretically established properties were repeatedly confirmed by experimental data. Experiments also verified the principal conclusions concerning the dispersion of these waves, i.e., the relationship between the velocity of their propagation and their length. As a result, it even became possible to apply the phenomena of dispersion of surface waves to the study of the geology of construction sites. Similarly, amplitudes of vertical vibrations of soil, obtained for vertical sources of waves on the basis of theoretical considerations, are in good agreement with the results of experiments.

By considering the influence of wave-energy absorption by soil, it becomes possible to correlate experimental data with the theory of amplitude changes in the vertical component of soil vibrations as a function of distance. However, some theoretical conclusions concerning the propagation of elastic waves through soils do not agree with experimental data, and even contradict them. For example, according to the theory of surface waves, the orbits of motion of particles on the soil surface have the form of ellipses with one axis perpendicular to the surface. Experiments show that the orbits of motion of soil particles often deviate from the elliptic shape and that the inclination of the axes of the orbit with respect to the soil surface depends on distance from the source. According to theory, for a vertical source of waves the amplitudes of horizontal longitudinal components of vibrations should change monotonously according to the same law which governs changes in the vertical components of vibrations. Actually this is not the case. Most of the experiments show that in cases of vertical sources of waves, the amplitudes of horizontal (longitudinal) components of vibrations of the soil surface change with distance according to a much more complicated law. Similarly, theoretical data related to the distribution of amplitudes with depth (for a surface source) do not agree with the results of some experiments.

The above discussion shows that the development of the theory of wave propagation, based on the theory of elasticity, should be carried out simultaneously with numerous experiments which would provide data for the adjustment of theoretical conclusions.

VIII-1. Elastic Waves in an Infinite Solid Body

a. Longitudinal and Transverse Waves. Assuming that soil satisfies all the conditions of an absolutely isotropic homogeneous elastic body, it is possible to use for the study of wave propagation in soil the general differential equations of motion of an absolutely elastic body; these equations are as follows:[28]

$$(\lambda + \mu)\frac{\partial \Delta}{\partial x} + \mu\nabla^2 u + X = \rho\frac{\partial^2 u}{\partial t^2}$$
$$(\lambda + \mu)\frac{\partial \Delta}{\partial y} + \mu\nabla^2 v + Y = \rho\frac{\partial^2 v}{\partial t^2} \qquad \text{(VIII-1-1)}$$
$$(\lambda + \mu)\frac{\partial \Delta}{\partial z} + \mu\nabla^2 w + Z = \rho\frac{\partial^2 w}{\partial t^2}$$

where u, v, w = components of elastic displacement along x, y, z axes
Δ = relative change in volume

$$\Delta = \frac{\partial u}{\partial x} + \frac{\partial v}{\partial y} + \frac{\partial w}{\partial z}$$

∇^2 = Laplace operator

$$\nabla^2 = \frac{\partial^2}{\partial x^2} + \frac{\partial^2}{\partial y^2} + \frac{\partial^2}{\partial z^2}$$

ρ = density of soil
X, Y, Z = components of body forces
λ, μ = Lamé coefficients, interrelated with Young's modulus E and Poisson's ratio ν

$$\lambda = \frac{\nu}{(1 + \nu)(1 - 2\nu)} E$$
$$\mu = \frac{1}{2(1 + \nu)} E = G \qquad \text{(VIII-1-2)}$$

The influence of the body forces, as in static problems, is small and therefore may be neglected.

If one assumes that soil is a solid elastic medium extending to infinity in any direction, then it follows that two types of waves independent of each other are propagated from the source of vibrations; therefore the displacement at any point in the soil is the sum of the displacements caused by each wave; i.e.,

$$u = u_1 + u_2 \qquad v = v_1 + v_2 \qquad w = w_1 + w_2$$

The components of displacements u, v, and w also satisfy the conditions

$$\frac{\partial w_1}{\partial y} - \frac{\partial v_1}{\partial z} = 0 \qquad \frac{\partial u_1}{\partial z} - \frac{\partial w_1}{\partial x} = 0 \qquad \frac{\partial v_1}{\partial x} - \frac{\partial u_1}{\partial y} = 0 \qquad \text{(VIII-1-3)}$$

which will be valid if u_1, v_1, and w_1 have potentials; hence

$$u_1 = \frac{\partial \varphi}{\partial x} \qquad v_1 = \frac{\partial \varphi}{\partial y} \qquad w_1 = \frac{\partial \varphi}{\partial z} \qquad \text{(VIII-1-4)}$$

The components of displacements u_2, v_2, and w_2 satisfy the equation

$$\Delta^2 = \frac{\partial u_2}{\partial x} + \frac{\partial v_2}{\partial y} + \frac{\partial w_2}{\partial z} = 0 \qquad \text{(VIII-1-5)}$$

Thus, the components of the total displacement of soil equal

$$u = \frac{\partial \varphi}{\partial x} + u_2 \qquad v = \frac{\partial \varphi}{\partial y} + v_2 \qquad w = \frac{\partial \varphi}{\partial z} + w_2 \qquad \text{(VIII-1-5}a\text{)}$$

Substituting these values of u, v, and w into Eqs. (VIII-1-1), we obtain the following equations which should be satisfied by φ, u_2, v_2, and w_2:

$$\frac{\partial^2 \varphi}{\partial t^2} = a^2 \nabla^2 \varphi \qquad \text{(VIII-1-6)}$$

$$\frac{\partial^2 u_2}{\partial t^2} = b^2 \nabla^2 u_2$$
$$\frac{\partial^2 v_2}{\partial t^2} = b^2 \nabla^2 v_2 \qquad \text{(VIII-1-7)}$$
$$\frac{\partial^2 w_2}{\partial t^2} = b^2 \nabla^2 w_2$$

where
$$a^2 = \frac{\lambda + 2\mu}{\rho} \qquad \text{(VIII-1-8)}$$

$$b^2 = \frac{\mu}{\rho} \qquad \text{(VIII-1-9)}$$

Equations (VIII-1-6) and (VIII-1-7) differ only in the coefficients of ∇^2.

Expressions (VIII-1-3) represent the components of the vector of rotation; therefore it is evident that a wave propagating in soil, as described by Eq. (VIII-1-6), will not induce shear deformations. Deformations corresponding to this wave are of such order that soil undergoes only a relative change in volume which equals

$$\Delta_1 = \nabla^2 \varphi$$

Hence, the wave corresponding to Eq. (VIII-1-6) is a wave of compression and expansion. However, condition (VIII-1-5) shows that no changes in volume occur for waves whose components satisfy Eqs. (VIII-1-7); when these waves are propagated, the soil elements undergo only relative displacements whose components equal

$$2w_x = \frac{\partial w_2}{\partial y} - \frac{\partial v_2}{\partial z} \qquad 2w_y = \frac{\partial u_2}{\partial z} - \frac{\partial w_2}{\partial x} \qquad 2w_z = \frac{\partial v_2}{\partial x} - \frac{\partial u}{\partial y}$$

Hence, the waves satisfying differential Eqs. (VIII-1-7) are waves of shear.

When waves of compression and expansion are propagated, soil displacements are parallel to the direction of transmission; therefore waves of compression and expansion are called longitudinal waves.

Shear waves may be called transverse waves, because their propagation induces soil displacements only in a direction perpendicular to the direction of transmission.

Equations (VIII-1-6) and (VIII-1-7) are not interrelated. Hence, in an infinite solid body longitudinal and transverse waves propagate independently of each other. If the conditions of appearance of the waves are such that the components of soil displacement at the initial moment correspond only to a change in volume, then transverse waves will not appear, and vice versa.

b. Propagation Velocities of Longitudinal and Transverse Waves. The integral of a differential equation of the type of Eqs. (VIII-1-6) or (VIII-1-7) describing waves may be taken in the form

$$\frac{1}{r} F\left(t - \frac{r}{c}\right)$$

where r = radius vector of point under consideration
F = arbitrary function satisfying initial and boundary conditions
c = velocity of wave propagation; for waves of compression and expansion $c = a$; for waves of shear $c = b$

Using Eqs. (VIII-1-2), it is possible to bring Eqs. (VIII-1-8) and (VIII-1-9), describing velocities of waves of compression and expansion, to the form

$$a = \sqrt{\frac{1 - \nu}{(1 + \nu)(1 - 2\nu)} \frac{E}{\rho}} \qquad \text{(VIII-1-10)}$$

$$b = \sqrt{\frac{1}{2(1 + \nu)} \frac{E}{\rho}} \qquad \text{(VIII-1-11)}$$

Hence it follows that the velocities of propagation of longitudinal and transverse waves depend only on the elastic properties and density of the soil.

The ratio between the velocities of the compression and shear waves equals

$$\frac{a}{b} = \sqrt{\frac{2(1 - \nu)}{1 - 2\nu}} \qquad \text{(VIII-1-12)}$$

and always exceeds unity. Consequently, longitudinal waves have a higher velocity of propagation than transverse waves. The difference between the two velocities is directly proportional to the value of the Poisson ratio ν. The relative value of the velocity of compression waves is smaller in sandy soils than in clayey soils, because the Poisson ratio is smaller for the former. Table VIII-1 gives numerical values of velocities of waves of compression and shear.

TABLE VIII-1. VELOCITIES OF COMPRESSION WAVES a AND SHEAR WAVES b

Soil	ρ, kg $\times$ sec²/cm⁴	a, m/sec	b, m/sec
Moist clay	1.8×10^{-6}	1,500	150
Loess at natural moisture	1.67×10^{-6}	800	260
Dense sand and gravel	1.70×10^{-6}	480	250
Fine-grained sand	1.65×10^{-6}	300	110
Medium-grained sand	1.65×10^{-6}	550	160
Medium-sized gravel	1.8×10^{-6}	750	180

c. Propagation of Waves Induced by Underground Explosion: Camouflet.† If an explosion occurs at a considerable depth (for example, in a mine), then to establish soil displacements at points located some distance away from the surface the waves emanating from the explosion center may be considered to be either longitudinal or transverse waves propagating in an infinite solid body.

When a camouflet takes place, the initial zone of excitement of the soil approaches the shape of a sphere. Then the radial components of displacements, i.e., the components in the direction of wave propagation, are large in comparison with the tangential components. Therefore, for a camouflet explosion the boundary conditions of wave propagation when $r = r_0$ may be taken to be as follows:

$$u = f(t)\frac{x}{r} \qquad v = f(t)\frac{y}{r} \qquad w = f(t)\frac{z}{r} \tag{VIII-1-13}$$

where $f(t)$ is the assigned function of time

Conditions (VIII-1-13) show that there occur only radial displacements of the soil on the camouflet surface. Therefore only longitudinal waves will be propagated from the explosion center. The following relation may be accepted for these waves:

$$\varphi = \frac{1}{r}F\left(t - \frac{r}{a}\right) \tag{VIII-1-14}$$

Since the radial component of displacements equals

$$u_r = \sqrt{u^2 + v^2 + w^2}$$

† TRANSLATION EDITOR'S NOTE: "Camouflet" is a term of French and British military origin and refers to an underground blast, the effects of which do not produce visible displacements of the soil surface.

we obtain, according to Eq. (VIII-1-14),

$$u_r = -\frac{1}{r^2}\left[\frac{r}{a}\frac{\partial F\left(t-\frac{r}{a}\right)}{\partial t} + F\left(t-\frac{r}{a}\right)\right] \qquad \text{(VIII-1-15)}$$

In order to find the function F, let us use boundary conditions (VIII-1-13), which may be rewritten as follows:

$$u_r = f(t)$$

Setting $r = r_0$ in Eq. (VIII-1-15), we have

$$\frac{r_0}{a}\frac{\partial F(t - r_0/a)}{\partial t} + F\left(t-\frac{r_0}{a}\right) = -r_0^2 f(t) \qquad \text{(VIII-1-16)}$$

or

$$\frac{r_0}{a}\frac{\partial F(t_1)}{\partial t_1} + F(t_1) = -r_0^2 f\left(t_1+\frac{2}{a}\right) \qquad \text{(VIII-1-17)}$$

where

$$t_1 = t - \frac{r_0}{a}$$

The solution of Eq. (VIII-1-17) has the form

$$F = e^{-(a/r_0)t_1}\left[C - r_0 a\int_0^{t_1} f\left(t_1+\frac{r_0}{a}\right)e^{(a/r_0)_1}\,dt_1\right] \qquad \text{(VIII-1-18)}$$

The displacement of the camouflet surface created by the explosion may be expressed by the following function:

$$f(t) = \alpha t e^{-\beta t} \qquad \text{(VIII-1-19)}$$

Values of α and β depend on the properties of the explosive charge. Since at the initial moment there are no displacements on the camouflet surface, C should be zero.

Substituting the expression for $f(t)$ from Eq. (VIII-1-19) under the integral of Eq. (VIII-1-18) and integrating, we obtain

$$F(t) = \frac{r_0^2 a\alpha}{\beta r_0 - a}\left(t + \frac{r_0}{\beta r_0 - a}\right)e^{-\beta t} \qquad \text{(VIII-1-20)}$$

According to Eq. (VIII-1-15), we find the following expression for the magnitude of displacement at a distance $r > at$ from the center of the explosion:

$$u_r = -\frac{r_0^2 a\alpha}{r^2(\beta r_0 - a)}\left\{\frac{r}{a}\left[1-\left(t-\frac{r}{a}-\frac{r_0}{r_0-a}\right)\beta\right] + t - \frac{z}{a}\frac{r_0}{\beta r_0 - a}\right\}\exp\left[-\beta\left(t-\frac{r}{a}\right)\right] \qquad \text{(VIII-1-21)}$$

This expression makes it possible to evaluate the effect of an explosion at any distance from the charge. It is easy to see that at small distances from the explosion center the displacements of the soil decrease rapidly; the decrease is approximately inversely proportional to the square of the distance. At large distances the decrease of the amplitudes is inversely proportional to the distance.

The results of computations depend on the assigned dimensions of the camouflet, as well as on the constants α and β which characterize changes with time of the soil displacement at the camouflet boundary. These constants can be determined experimentally.

VIII-2. Changes with Depth in Amplitudes of Soil Vibrations Produced by Surface Waves

a. General Expressions for Components of Soil Displacements Induced by Waves Caused by a Vertical Exciting Force Acting along a Surface Line. With the exception of sources located in deep mines, practically all the industrial sources of wave excitement in soil lie close to the surface, as do all receivers of waves, such as foundations under buildings. Therefore the study of waves propagating in a zone close to the surface is of practical interest in the study of the dynamics of bases and foundations. In this connection, we must establish how the propagation of waves in soil is influenced by the presence of the free surface of soil.

Let us confine ourselves to the case in which $w = 0$ and

$$\frac{\partial u}{\partial z} = \frac{\partial v}{\partial z} = 0$$

Then if the action of body forces is neglected, the differential equations of wave propagation are as follows:

$$\begin{aligned}(\lambda + \mu)\frac{\partial \Delta}{\partial x} + \mu\nabla^2 u &= \rho\frac{\partial^2 u}{\partial t^2} \\ (\lambda + \mu)\frac{\partial \Delta}{\partial y} + \mu\nabla^2 v &= \rho\frac{\partial^2 v}{\partial t^2}\end{aligned} \qquad \text{(VIII-2-1)}$$

Let us take a general solution of these equations in the form

$$\begin{aligned}u &= \frac{\partial \Phi^*}{\partial x} + \frac{\partial \Psi^*}{\partial y} \\ v &= \frac{\partial \Phi^*}{\partial y} - \frac{\partial \Psi^*}{\partial x}\end{aligned} \qquad \text{(VIII-2-2)}$$

where Φ^* and Ψ^* are analytic functions.

We direct the y axis perpendicular to the soil surface; then for all points in soil, $y > 0$. Considering steady propagation of waves with frequency

ω, we take

$$\begin{aligned} \Phi^* &= e^{i\omega t}\Phi(x,y) \\ \Psi^* &= e^{i\omega t}\Psi(x,y) \end{aligned} \qquad \text{(VIII-2-3)}$$

Then the equations which should be satisfied by the functions Φ and Ψ are as follows:

$$\begin{aligned} (\nabla^2 + h^2)\Phi &= 0 \\ (\nabla^2 + k^2)\Psi &= 0 \end{aligned} \qquad \text{(VIII-2-4)}$$

where $$h = \frac{\omega}{a} \qquad k = \frac{\omega}{b} \qquad \text{(VIII-2-5)}$$

Since $$\omega = \frac{2\pi}{T}$$

where T is the period of the propagated waves, it is clear that

$$h = \frac{2\pi}{aT} \qquad \text{and} \qquad k = \frac{2\pi}{bT}$$

However, aT and bT represent the lengths of longitudinal and transverse waves. Hence, h and k are reciprocal values of the wavelengths, and k is always larger than h.

Particular solutions of Eqs. (VIII-2-4) are taken as follows:

$$\begin{aligned} \Phi &= Ae^{-\alpha y}e^{i\xi x} \\ \Psi &= Be^{-\beta y}e^{i\xi x} \end{aligned} \qquad \text{(VIII-2-6)}$$

where $$\begin{aligned} \alpha^2 &= \xi^2 - h^2 \\ \beta^2 &= \xi^2 - k^2 \end{aligned} \qquad \text{(VIII-2-7)}$$

A, B, and ξ are arbitrary constants determined by boundary conditions.

Let us assume that the border plane $y = 0$ is subjected to the action of external normal forces distributed continuously over the entire plane; these forces induce normal stresses equaling

$$\sigma_y = \sigma_0 e^{i\xi x} \qquad \text{(VIII-2-8)}$$

We assume that the tangential stresses on the border plane equal zero; i.e.,

$$\tau_{yx} = 0 \qquad \text{(VIII-2-9)}$$

Stresses σ_y and τ_{yx} are expressed through functions Φ and Ψ as follows:

$$\begin{aligned} \frac{\tau_{yx}}{\mu} &= 2\frac{\partial^2\Phi}{\partial x \partial y} - k^2\Psi - 2\frac{\partial^2\Psi}{\partial x^2} \\ \frac{\sigma_y}{\mu} &= -k^2\Phi - 2\frac{\partial^2\Phi}{\partial x^2} - 2\frac{\partial^2\Psi}{\partial x \partial y} \end{aligned} \qquad \text{(VIII-2-10)}$$

Substituting, into the right-hand parts of these equations, expressions for Φ and Ψ taken from Eqs. (VIII-2-6) when $y = 0$, we obtain the equations for calculating the constants A and B:

$$\begin{aligned} -2i\xi\alpha A + (2\xi^2 - k^2)B &= 0 \\ (2\xi^2 - k^2)A + 2i\xi\beta B &= \frac{\sigma_0}{\mu} \end{aligned} \tag{VIII-2-11}$$

Hence

$$\begin{aligned} A &= \frac{2\xi^2 - k^2}{F(\xi)} \frac{\sigma_0}{\mu} \\ B &= \frac{2i\xi\alpha}{F(\xi)} \frac{\sigma_0}{\mu} \end{aligned} \tag{VIII-2-12}$$

where

$$F(\xi) = (2\xi^2 - k)^2 - 4\xi^2\alpha\beta \tag{VIII-2-13}$$

Using Eqs. (VIII-2-2), (VIII-2-6), and (VIII-2-12), we obtain

$$\begin{aligned} u &= i\xi \frac{(2\xi^2 - k^2)e^{-\alpha y} - 2\alpha\beta e^{-\beta y}}{F(\xi)} e^{i\xi x} \frac{\sigma_0}{\mu} \\ v &= \alpha \frac{-(2\xi^2 - k^2)e^{-\alpha y} + 2\xi^2 e^{-\beta y}}{F(\xi)} e^{i\xi x} \frac{\sigma_0}{\mu} \end{aligned} \tag{VIII-2-14}$$

In order to transform the exciting force into one acting along the line

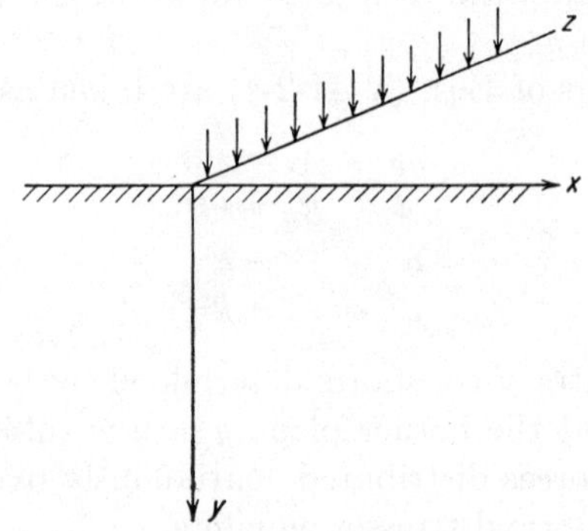

FIG. VIII-1. Exciting force [Eqs. (VIII-2-15)].

$x = 0$, $y = 0$ (Fig. VIII-1), we assume

$$\sigma_0 = -P \frac{d\xi}{2\pi}$$

Substituting this expression into the right-hand parts of Eqs. (VIII-2-14) and integrating with respect to ξ from $-\infty$ to $+\infty$, we obtain the following expressions for the displacements u and v:

$$\begin{aligned} u &= -\frac{iP}{2\pi\mu} \int_{-\infty}^{+\infty} \frac{\xi[(2\xi^2 - k^2)e^{-\alpha y} - 2\alpha\beta e^{-\beta y}]e^{i\xi x}}{F(\xi)} d\xi \\ v &= -\frac{P}{2\pi\mu} \int_{-\infty}^{+\infty} \frac{\alpha[-(2\xi^2 - k^2)e^{-\alpha y} + 2\xi^2 e^{-\beta y}]e^{i\xi x}}{F(\xi)} d\xi \end{aligned} \tag{VIII-2-15}$$

Equations (VIII-2-15) correspond to the forced waves induced by an exciting force acting along the line $x = 0$, $y = 0$.

b. Propagation Velocity of Surface Waves. Free surface waves occur where the border surface is not subjected to the action of forces and the waves are induced by some initial excitement. Assuming for this case that $\sigma_0 = 0$, we obtain equations for determining the constants A and B:

$$\begin{aligned} -2i\xi\alpha A + (2\xi^2 - k^2)B &= 0 \\ (2\xi^2 - k^2)A + 2i\xi\beta B &= 0 \end{aligned} \qquad \text{(VIII-2-16)}$$

Equations (VIII-2-16) will give solutions other than zero for A and B only when the determinant of this system equals zero. Hence we obtain equations for determining ξ:

$$F(\xi) = 0 \qquad \text{(VIII-2-17}a\text{)}$$

Instead of this equation, which contains irrational expressions, let us consider the following equation, which does not contain radical signs:

$$\begin{aligned} F(\xi)f(\xi) &= (2\xi^2 - k^2)^4 - 16(\xi^2 - h^2)(\xi^2 - k^2)\xi^4 \\ &= k^8\left[1 - 8\frac{\xi^2}{k^2} + \left(24 - 16\frac{h^2}{k^2}\right)\frac{\xi^4}{k^4} - 16\left(1 - \frac{h^2}{k^2}\right)\frac{\xi^6}{k^6}\right] = 0 \end{aligned} \qquad \text{(VIII-2-17)}$$

where

$$f(\xi) = (2\xi^2 - k^2)^2 + 4\xi^2\alpha\beta$$

Since $k > h$, one of the roots of Eq. (VIII-2-17) lies between 1 and $+\infty$. It is easy to show that the other two roots, if they are real, lie between 0 and h^2/k^2.

The first root corresponds to positive values of α and β; therefore this root does not satisfy the condition $f(\xi) = 0$. The last two roots make α and β positive and imaginary; therefore they do not satisfy the equation $F(\xi) = 0$. The latter equation has only one root ($\xi^2 = \varkappa^2$) which is larger than 1. Therefore $\varkappa^2 > k^2$. For a Poisson ratio of 0.5, the real root of Eq. (VIII-2-17) is

$$\frac{\varkappa}{k} = 1.04678$$

For $\nu = 0.25$, all roots of Eq. (VIII-2-17) are real; they are equal to

$$\frac{\xi^2}{k^2} = \frac{1}{4}; \frac{1}{4}(3 - \sqrt{3}); \frac{1}{4}(3 + \sqrt{3})$$

Of these roots, only the last satisfies the conditions of the problem; its value is

$$\frac{\varkappa}{k} = \frac{1}{2}\sqrt{3 + \sqrt{3}} = 1.087664\ldots$$

Analogously to Eqs. (VIII-2-5), let us designate

$$\varkappa = \frac{\omega}{c}$$

where c is the velocity of propagation of the surface waves under consideration; it is clear that

$$c = \frac{k}{\varkappa} b$$

For $\nu = 0.5$, $\quad c = 0.9553b$
for $\nu = 0.25$, $\quad c = 0.9194b$

Thus it is seen that surface waves propagate with somewhat smaller velocity than transverse waves.

c. Approximate Expressions for Soil Surface Displacements Induced by Waves Excited by a Concentrated Vertical Force. Let us return to the consideration of the general Eqs. (VIII-2-15) for displacements induced by waves excited by a vertical concentrated force acting on the surface. Integrals on the right-hand sides of Eqs. (VIII-2-15), as they are written, are indeterminate, since the function $F(\xi)$ for some values of ξ (e.g., $\xi = \pm\varkappa$) equals zero. In the general case, at $y = 0$, it is very difficult to disclose this indeterminateness. Therefore let us confine ourselves to the investigation of soil displacements at points on the surface ($y = 0$); then we obtain

$$u_0 = \frac{iP}{2\pi\mu} \int_{-\infty}^{+\infty} \frac{\xi(2\xi^2 - k^2 - 2\alpha\beta)e^{i\xi x}}{F(\xi)} d\xi$$
$$v_0 = -\frac{P}{2\pi\mu} \int_{-\infty}^{+\infty} \frac{k^2\alpha e^{i\xi x}}{F(\xi)} d\xi \qquad \text{(VIII-2-18)}$$

These integrals remain indeterminate; however, it is possible to single out their principal value. Skipping details of this operation, which includes integrating in the area of the complex variables along the contour containing special points, we present here the following final equations:

$$u_0 = -\frac{P}{\mu} H_0 e^{i\omega t - \varkappa x} + S_u$$
$$v_0 = -i\frac{P}{\mu} K_0 e^{i\omega t - \varkappa x} + S_v \qquad \text{(VIII-2-19)}$$

where $\quad H_0 = \dfrac{-\varkappa(2\varkappa^2 - k^2 - 2\alpha_1\beta_1)}{F'(\varkappa)} \qquad K_0 = -\dfrac{\alpha_1 k^2}{F'(\varkappa)}$

$$\alpha_1{}^2 = \varkappa^2 - h^2 \qquad \beta_1{}^2 = \varkappa^2 - k^2 \qquad F'(\varkappa) = \left.\frac{F(\xi)}{\partial\xi}\right|_{\xi=\varkappa} \qquad \text{(VIII-2-20)}$$

For large $k\varkappa$ and $h\varkappa$, asymptotic expansions of S_u and S_v are as follows:

$$S_u = \frac{P}{\mu}\sqrt{\frac{2}{\pi}}\sqrt{1-\frac{h^2}{k^2}}\,\frac{e^{i(\omega t-kx-\pi/4)}}{(kx)^{3/2}} - \frac{P}{\mu}\sqrt{\frac{2}{\pi}\frac{h^3k^2\sqrt{k^2-h^2}}{(k^2-2h^2)^3}}\,\frac{ie^{i(\omega t-hx-\pi/4)}}{(hx)^{3/2}} + \cdots$$

$$S_v = \frac{2P}{\mu}\sqrt{\frac{2}{\pi}}\left(1-\frac{h^2}{k^2}\right)\frac{ie^{i(\omega t-kx-\pi/4)}}{(kx)^{3/2}} + \frac{P}{2\mu}\sqrt{\frac{2}{\pi}\frac{h^2k^2}{(k^2-h^2)^2}} - \frac{ie^{i(\omega t-hx-\pi/4)}}{(hx)^{3/2}} + \cdots \qquad \text{(VIII-2-21)}$$

Equations (VIII-2-21) show that not one wave, but a series of waves will be propagated from the site of the excitement of vibrations; the velocity of propagation of these waves equals the velocity of propagation of the longitudinal, transverse, and surface waves. The amplitudes of surface waves [described by the first terms in Eqs. (VIII-2-19)] do not change with increase in distance from the source of waves. However, it follows from Eqs. (VIII-2-21) that the amplitudes of waves, propagating with the velocities a and b, decrease very rapidly (at least inversely proportionally to $x^{3/2}$) with an increase in the distance from the source. Therefore at a sufficient distance from the source of the waves, the amplitudes of the additional spectrum of waves are small in comparison with the amplitudes of the surface waves.

Owing to interference in the area close to the source of the separate waves, the wave amplitudes do not change monotonously in proportion to distance from the source, but, as this distance grows, maximums and minimums of amplitudes are observed; the values of these maximums and minimums become smaller as the distance grows.

d. The Variation with Depth of Surface-wave Amplitudes. If one assumes that x is not very small, then the additional spectrum of waves may be neglected and it may be considered that only surface waves are propagated which are described by the first terms in Eqs. (VIII-2-19).

Let us investigate the changes of amplitudes of these waves with depth; according to Eqs. (VIII-2-2) and (VIII-2-6) we have, substituting x for $-x$:

$$u = (-i\varkappa Ae^{-\alpha_1 y} - \beta_1 Be^{-\beta_1 y})e^{i\varkappa x}$$
$$v = (-\alpha_1 Ae^{-\alpha_1 y} + i\varkappa B^{-\beta_1 y})e^{i\varkappa x} \qquad \text{(VIII-2-22)}$$

where $\quad \alpha_1{}^2 = \varkappa^2 - h^2 \qquad \beta_1{}^2 = \varkappa^2 - k^2$

Assuming $y = 0$, we obtain

$$u_0 = (-i\varkappa A - \beta_1 B)e^{i\varkappa x}$$
$$v_0 = (-\alpha_1 A + i\varkappa B)e^{i\varkappa x} \qquad \text{(VIII-2-23)}$$

Equating the right-hand parts of Eqs. (VIII-2-19) and (VIII-2-23), and disregarding a temporary multiple and S_u and S_v, we obtain two equations for determining constants A and B corresponding only to the surface waves:

$$i\varkappa A + \beta_1 B = \frac{P}{\mu} H_0$$

$$A - i\varkappa B = i\frac{P}{\mu} K_0$$

Solving these equations for A and B, we obtain

$$A = i\frac{\varkappa H_0 + \beta_1 K_0}{\alpha_1\beta_1 + \varkappa^2}\frac{P}{\mu}$$

$$B = \frac{\alpha_1 H_0 + \varkappa K_0}{\alpha_1\beta_1 - \varkappa^2}\frac{P}{\mu}$$

Substituting these values of A and B into Eqs. (VIII-2-22), introducing a temporary multiple $e^{i\omega t}$, and disregarding imaginary terms, we finally obtain

$$u = \left(\varkappa\frac{\varkappa H_0 + \beta_1 K_0}{\alpha_1\beta_1 - \varkappa^2}e^{-\alpha_1 y} - \beta_1\frac{\alpha_1 H_0 + \varkappa K_0}{\alpha_1\beta_1 - \varkappa^2}e^{-\beta_1 y}\right)\frac{P}{\mu}\cos(\omega t - \varkappa x)$$

$$v = \left(-\alpha_1\frac{\varkappa H_0 + \beta_1 K_0}{\alpha_1\beta_1 - \varkappa^2}e^{-\alpha_1 y} + \varkappa\frac{\alpha_1 H_0 + \varkappa K_0}{\alpha_1\beta_1 - \varkappa^2}e^{-\beta_1 y}\right)\frac{P}{\mu}\sin(\omega t - \varkappa x) \qquad \text{(VIII-2-24)}$$

These expressions for components of soil displacement show that for surface waves the orbit of motion of soil particles is an ellipse whose axes coincide with the coordinate axes. The interrelationship between the amplitudes of the vertical and horizontal components depends on the depth.

Equations (VIII-2-24) make it possible to evaluate changes with depth in the amplitudes of waves. We present below such an evaluation for Poisson ratios of 0.5 and 0.25. Table VIII-2 presents values of several coefficients contained in Eqs. (VIII-2-24) for these two cases.

Substituting data of Table VIII-2 into Eqs. (VIII-2-24), we obtain the following amplitudes of waves: for Poisson's ratio $\nu = 0.5$:

$$u = (-0.1298e^{-2\pi y/L_c} + 0.0706e^{-(0.2958)2\pi y/L_c})\frac{P}{\mu}$$

$$v = (0.1298e^{-2\pi y/L_c} - 0.2387e^{-(0.2958)2\pi y/L_c})\frac{P}{\mu} \qquad \text{(VIII-2-25)}$$

for Poisson's ratio $\nu = 0.25$:

$$u = (-0.2958e^{-(0.8474)2\pi y/L_c} + 0.1707e^{-(0.3933)2\pi y/L_c}) \frac{P}{\mu}$$
$$v = (0.2507e^{-(0.8474)2\pi y/L_c} - 0.4341e^{-(0.3933)2\pi y/L_c}) \frac{P}{\mu} \qquad \text{(VIII-2-26)}$$

where L_c is the wavelength.

TABLE VIII-2. VALUES OF COEFFICIENTS IN EQS. (VIII-2-24)

Poisson's ratio ν	E	λ	$\frac{a}{b}$	$\frac{h^2}{k^2}$	H_0	K_0	α_1	β_1	$\frac{\varkappa H_0 - \beta K_0}{\alpha_1\beta_1 - \varkappa^2}$	$\frac{\alpha_1 H_0 + \varkappa K_0}{\alpha_1\beta_1 - \varkappa^2}$
0.5	3	∞	∞	0	0.05921	0.10890	$\varkappa$	$0.2958\varkappa$	$\frac{-0.1298}{\varkappa}$	$\frac{-0.2387}{\varkappa}$
0.25	2.5	μ	$\sqrt{3}$	⅓	0.12500	0.18349	$0.8474\varkappa$	$0.3933\varkappa$	$\frac{-0.2958}{\varkappa}$	$\frac{-0.4341}{\varkappa}$

Figure VIII-2 gives graphs of u and v for $\nu = 0.5$ and $P/\mu = 1$. It is seen from these graphs that the amplitudes of the vertical components of waves increase with an increase in the depth to $y = 0.2L_c$; further on they decrease, attaining, at a depth of the order of $0.35L_c$, values corresponding to those at the surface. With further increase in depth, the amplitudes decrease. Thus it can be considered as an approximation that the amplitudes of vertical components of surface waves change relatively little to a depth of the order of $0.4L_c$.

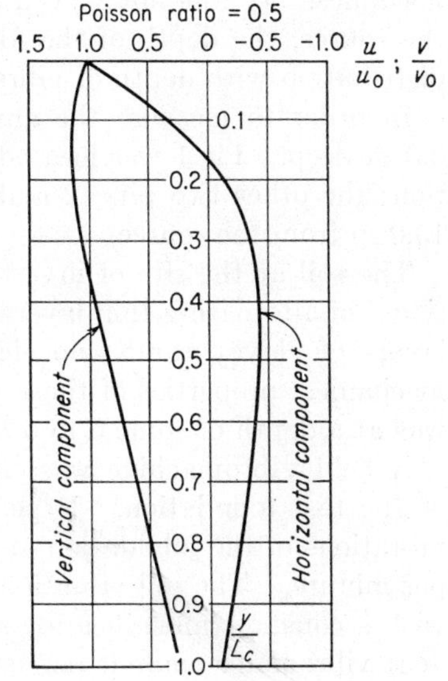

FIG. VIII-2. Variations of wave amplitudes with depth.

Amplitudes of the horizontal longitudinal component of a wave change with depth in an opposite manner: to a depth of the order of $0.15L_c$ the amplitudes rapidly decrease, approaching zero. Then, to a depth of the order of $0.45L_c$ they increase; with further increase in depth, the amplitudes gradually decrease.

e. Experimental Investigations of Variation with Depth of Surface-wave Amplitudes. The foregoing theoretical data concerning the distribution of amplitudes of soil vibration near the surface refer to surface waves whose influence prevails at distances from the wave source which are considerably larger than the wavelength. At relatively small distances from the source, the distribution of vibration amplitudes with depth may be influenced by soil vibrations produced by waves of the supplementary spectrum; therefore the actual distribution of soil vibrations with depth may greatly differ from that established by Eqs. (VIII-2-24). The distribution may also be influenced by the fact that sources of waves actually greatly differ from the concentrated vertical force assumed in the above discussion. Thus, the assumption that waves produced by foundations undergoing vertical vibrations are excited by vertical exciting forces may lead to considerable error in the determination of soil vibration amplitudes at small distances from the source.

Great difficulties are involved in theoretical investigations of the influence of these factors on the distribution of soil vibration amplitudes with depth. Therefore, experimental investigations are of special importance. The author performed such investigations under diverse soil conditions in which foundations for machinery acted as sources of waves.

In one of the experiments, a foundation with a 1.0- by 1.0-m area in contact with soil acted as a source of waves. The foundation was placed 2.5 m below the soil surface, the usual depth of foundations under machines. It was not placed on the surface since, in an area close to the source, the depth of the foundation may considerably influence the distribution with depth of vibration amplitudes.

In order to measure the amplitudes, three test pits were dug, each 2.5 m deep. Pit 1 was located at a distance of 3.0 m from the foundation; the other two pits (2 and 3) were placed at distances of 7.30 and 15.0 m from the source.

The soil at the site of investigation was fairly homogeneous and consisted of alternating thin layers of clay with some sand and silt and thin layers of clayey sands; no significant difference was observed in the mechanical properties of these varieties of soil. The ground-water level was at a depth of some 5 to 6 m.

A field vibromachine was employed for the excitement of vibrations of the test foundation. Experiments were performed only for vertical vibrations of the foundation at rates of 800, 1,000, and 1,200 oscillations per minute. The soil vibrations were recorded by an optical vibrometer with a constant magnification of about 500 times. This double-component vibrometer made it possible not only to measure the amplitudes of the horizontal and vertical components of waves, but also to fix the Lissajous figures.

Measurements were performed only in a longitudinal plane at eight points of different depths: the first point was approximately 0.10 m below the soil surface; the others were located each 0.30 m lower than the preceding one; the last point was at a depth of 2.2 m from the surface. The

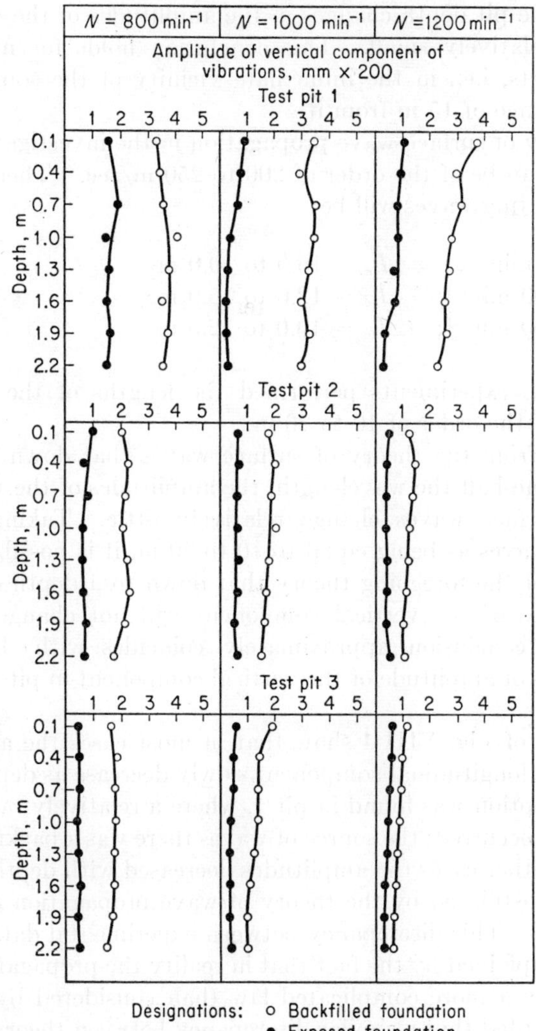

FIG. VIII-3. Experimentally determined variation with depth of the vertical component of oscillations induced by a vibratory machine.

measurements in each pit were performed for two cases: (1) side faces of the foundation were completely relieved from soil pressure; (2) the foundation was backfilled. In the first case, only the foundation area in contact with the soil served as a source of waves; in the second case, both the contact area and the side faces acted as a source of waves.

Figures VIII-3 and VIII-4 give the results of the measurements as graphs of changes in amplitudes with depth. It is seen from Fig. VIII-3 that within the pit depth changes in the amplitudes of the vertical component are relatively small. This conclusion holds for measurements in all three pits, i.e., in the immediate vicinity of the source of waves and at a distance of 15 m from it.

The velocity of surface-wave propagation in the investigated soils may be considered to be of the order of 200 to 250 m/sec. Then, the length of the propagating waves will be:

For $N = 800\ \text{min}^{-1}$: $L_c = 15.0$ to 19.0 m
For $N = 1{,}000\ \text{min}^{-1}$: $L_c = 12.0$ to 15.0 m
For $N = 1{,}200\ \text{min}^{-1}$: $L_c = 10.0$ to 12.5 m

Thus in the experiments performed the lengths of the propagating waves were of the order of 10 to 20 m.

It appears from the theory of surface waves that down to depths of the order of one-half the wavelength, the amplitudes of the vertical component of surface waves change relatively little. Taking lengths of propagating waves as being equal to 10 to 20 m, it is possible to assume on the basis of the foregoing theory that down to a depth of 5 to 10 m the amplitudes of the vertical component will not change much with depth. This conclusion approximately coincides with the results of measurements of amplitude of the vertical component in pits with depths to 2.5 m.

The graphs of Fig. VIII-4 show that in most cases the amplitudes of the horizontal longitudinal component slowly decrease as depth increases. The only exception was found in pit 1, where a relatively rapid damping of amplitudes occurred; the source of waves there was a backfilled foundation. In all other cases the amplitudes decreased with depth much more slowly than postulated by the theory of wave propagation near the free surface of soil. This discrepancy between experimental data and theory is evidently explained by the fact that in reality the propagation of waves is governed by a more complicated law than considered by the theory. It should be noted that a similar discrepancy between theory and experimental data was disclosed by studies of changes with the distance from the source of the same horizontal longitudinal component of a wave in the case of vertical vibrations of the foundation acting as a source of waves (see Art. VIII-3).

A comparison of soil vibration amplitudes in the two types of foundations shows that the amplitudes are much larger in backfilled foundations than in exposed foundations. This is explained as follows: in the first case, the energy of the vibrating foundation is transferred to the soil through the side surfaces and through the foundation area in contact with soil; and in the second case, only through the base contact area.

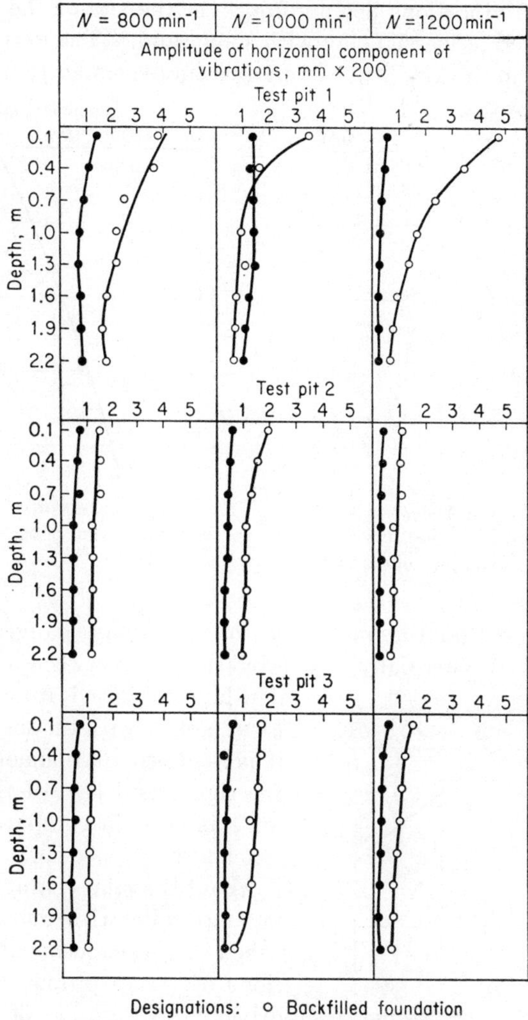

FIG. VIII-4. Experimentally determined variation with depth of the horizontal component of oscillations induced by a vibratory machine.

Graphs obtained as a result of experiments carried out under other soil conditions show that the changes of amplitudes with depth differed somewhat from the changes outlined in the preceding discussion. Figure VIII-5 gives a graph showing the variation of vertical amplitude with depth along the side face of a 4-m-deep ditch dug at a distance of 2.5 m from a test foundation whose area in contact with soil equaled 0.8 m^2. This foundation was placed on the surface and was excited vertically at the rate of 670 vibrations per minute. On the site of the investigation the soil consisted of loessial clay with some sand. The graph shows that at a depth of 4 m the amplitudes were two times smaller than those on the

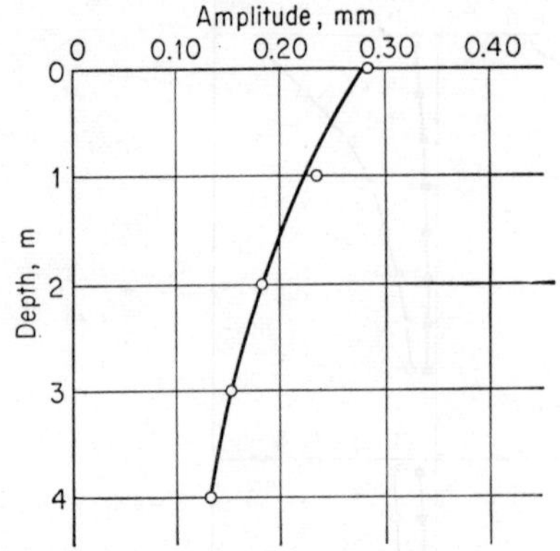

Fig. VIII-5. Variation with depth of the vertical component of oscillations induced by a vibratory machine operating at 670 rpm.

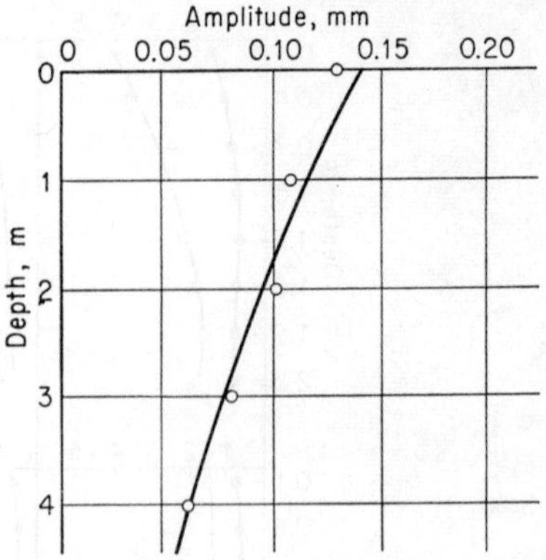

Fig. VIII-6. Variation with depth of the vertical component of oscillations induced by a pile driver.

soil surface, and that the amplitudes of vibrations changed with depth approximately exponentially. The relationship between amplitude and depth, established from this experiment, agrees with the theory, provided it is assumed that one of the terms in the right-hand part of the expression for v is large in comparison with the other term.

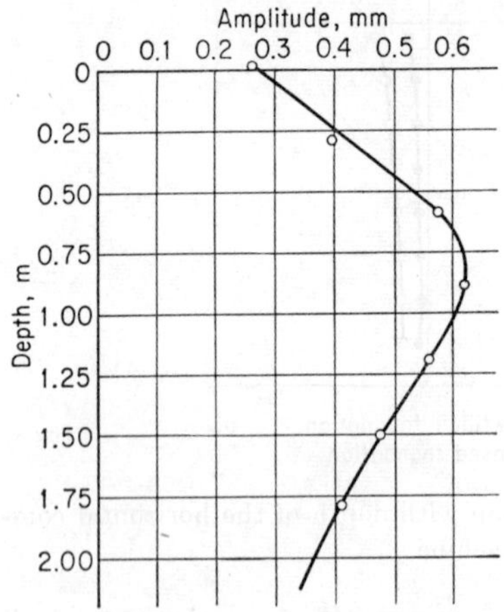

Fig. VIII-7. Variation with depth of the vertical component of oscillations induced by a forge hammer.

About the same character of changes in amplitude with depth was established from experimental investigations of waves propagated by a pile driver.† A graph of the results is given in Fig. VIII-6. Loessial soil was found on the site of the investigation.

Finally, Fig. VIII-7 gives graphs of the variation with depth of the vertical component of soil vibration amplitudes, plotted from the results of the author's investigations of distribution of waves produced by the foundation under a forge hammer whose dropping parts weighed 7.5 tons.

† The investigations were performed by Ya. N. Smolikov.

Measurements were performed at a distance of 7 m from the source of waves. At this site were found layers of clays with some sand and silt alternating with clayey sands.

In spite of some contradictions between the experimental data discussed above and the theoretical data, the conclusion is possible that at small depths, approximately in the range 0.2 to 0.5 wavelength, changes in vibration amplitudes are relatively small. This finding is of a certain practical importance, because an increase in the depth of wave receivers, such as the footings under columns of a building, may have significant influence on the vibration amplitudes of the receivers only at considerable depths. A small deepening of the wave receiver in practice will not influence the amplitudes of its vibrations. Similarly, the vibration amplitudes of a receiver will not be significantly affected by a small change in its depth in relation to the depth of the source of waves. Therefore it is not necessary to place foundations under machines deeper than foundations under adjacent structures (walls, columns, etc.).

The depth of the foundation under an engine may be selected without consideration of the effect of waves produced by this foundation on footings under walls and other structures. This condition in many cases (especially where there is a high level of ground water and in the presence of previously constructed foundations) eliminates complications which may develop in the process of construction of a machinery foundation located near footings under walls or columns.

VIII-3. Dependence of the Amplitude of Soil Vibrations on Distance from the Source of Waves

a. General Expressions for Soil Displacements Produced by Concentric Waves. Of practical interest are waves emanating from an area of excitement concentrated on a small section or distributed over a limited area of the soil surface. Of special importance is the problem of wave propagation with reference to displacements of a section of the soil surface not subjected to external loads. As indicated above, the solution of a problem with such boundary conditions is very difficult and has not yet been found. Therefore in the following discussion we will confine ourselves to known exciting forces distributed over a section of the free surface of the soil, including the case in which the section is infinitely small.

Assuming that the source of waves is a harmonic function of time, we shall seek solutions of Eqs. (VIII-1-6) and (VIII-1-7) in the form

$$\varphi = e^{i\omega t}\Phi(x,y,z) \tag{VIII-3-1}$$

$$u_2 = e^{i\omega t}U \qquad v_2 = e^{i\omega t}V \qquad w_2 = e^{i\omega t}W \tag{VIII-3-2}$$

where ω is the frequency of excitement.

Substituting Eqs. (VIII-3-1) and (VIII-3-2) into Eqs. (VIII-1-6) and (VIII-1-7), we find that Φ, U, V, and W should satisfy the following equations:

$$(\nabla^2 + h^2)\Phi = 0 \tag{VIII-3-3}$$

$$(\nabla^2 + k^2)U = 0 \qquad (\nabla^2 + k^2)V = 0 \qquad (\nabla^2 + k^2)W = 0 \tag{VIII-3-4}$$

In addition, the following relationship should be satisfied:

$$\frac{\partial U}{\partial x} + \frac{\partial V}{\partial y} + \frac{\partial W}{\partial z} = 0 \tag{VIII-3-5}$$

The solution of Eqs. (VIII-3-4) is sought in the form

$$U = \frac{\partial^2 \Psi}{\partial x \partial y} \qquad V = \frac{\partial^2 \Psi}{\partial y \partial z} \qquad W = \frac{\partial^2 \Psi}{\partial z^2} + k^2 \Psi \tag{VIII-3-6}$$

Substituting Eqs. (VIII-3-6) into Eqs. (VIII-3-4), we find that the function Ψ should satisfy the equation

$$(\nabla^2 + k^2)\Psi = 0 \tag{VIII-3-7}$$

Let us introduce a cylindrical system of coordinates, placing the origin of coordinates on the soil surface; then we have:

$$x = r \cos \theta \qquad y = r \sin \theta \qquad z = z$$

In this system of coordinates,

$$\nabla^2 = \frac{\partial^2}{\partial r^2} + \frac{1}{r}\frac{\partial}{\partial r} + \frac{\partial^2}{\partial z^2}$$

Designating by q a displacement in the direction of the radius vector r, we obtain

$$u = \frac{x}{r} q \qquad v = \frac{y}{r} q$$

According to Eqs. (VIII-1-5*a*) and (VIII-3-6), we have

$$q = \frac{\partial \Phi}{\partial r} + \frac{\partial^2 \Psi}{\partial r\, \partial z} \qquad w = \frac{\partial \Phi}{\partial z} + \frac{\partial^2 \Psi}{\partial z^2} + k^2 \Psi \tag{VIII-3-8}$$

Let us assume that an exciting load of the type

$$P_z = -PJ_0(\xi r) \tag{VIII-3-9}$$

is acting on the soil surface, i.e., at $z = 0$; in the above expression $J_o(\xi r)$ is the Bessel function of the first type, of zero order. No tangential stresses are found on the surface;

therefore,

$$\tau_r = 0 \tag{VIII-3-10}$$

We take the solution of Eqs. (VIII-3-3) and (VIII-3-7) in the form

$$\Phi = Ae^{-\alpha z}J_0(\xi r) \qquad \Psi = Be^{-\beta z}J_0(\xi r) \tag{VIII-3-11}$$

where A, B, and ξ are arbitrary constants.

Using the boundary conditions of Eqs. (VIII-3-9) and (VIII-3-10), we obtain

$$A = \frac{2\xi^2 - k^2}{F(\xi)}\frac{P}{\mu} \qquad B = \frac{2\alpha}{F(\xi)}\frac{P}{\mu} \tag{VIII-3-12}$$

where the function $F(\xi)$ is determined by Eq. (VIII-2-13).

The displacement components on the soil surface (when $y = 0$) corresponding to Eqs. (VIII-3-11) and (VIII-3-12) are as follows:

$$\begin{aligned} q_0 &= \frac{-\xi(2\xi^2 - k^2 - 2\alpha\beta)}{F(\xi)} J_1(\xi r)\frac{P}{\mu} \\ w_0 &= \frac{k^2\alpha}{F(\xi)} J_0(\xi r)\frac{P}{\mu} \end{aligned} \tag{VIII-3-13}$$

where J_1 is the Bessel function of the first type and the first order.

In order to consider the excitation of waves by a vertical exciting force $P_v e^{i\omega t}$, acting at the origin of coordinates, let us assume that

$$P = -\frac{P_v}{2\pi}\xi\, d\xi$$

Substituting this expression for P into the right-hand parts of Eqs. (VIII-3-13) and integrating in the range from 0 to $+\infty$, we obtain for the displacements on the soil surface:

$$q_0 = \frac{P_v}{2\pi\mu}\int_0^\infty \frac{\xi^2(2\xi^2 - k^2 - 2\alpha\beta)}{F(\xi)} J_1(\xi r)\, d\xi \tag{VIII-3-14}$$

$$w_0 = -\frac{P_v}{2\pi\mu}\int_0^\infty \frac{k^2\xi\alpha}{F(\xi)} J_0(\xi r)\, d\xi \tag{VIII-3-15}$$

b. Displacement of Soil at Small Distances from the Source of Waves. For a vertical exciting source, the vertical component of soil vibrations is of practical interest. Therefore let us confine ourselves to the investigation of the value of w_0 only, assuming that the value $\rho = kr$ is small. We assume that

$$\bar{J} = \frac{1}{k}\int_0^\infty \frac{k^2\xi\alpha J_0(\xi r)}{F(\xi)}\, d\xi$$

Let us introduce a new variable of integration, assuming that

$$\xi = k\theta$$

where k is determined in accordance with Eqs. (VIII-2-5).

According to Eqs. (VIII-2-7) and (VIII-2-13), we have:

$$F(\xi) = (2\xi^2 - k^2)^2 - 4\xi^2\alpha\beta = k^4[(2\theta^2 - 1)^2 - 4\theta^2 \sqrt{\theta^2 - \vartheta^2} \sqrt{\theta^2 - 1}]$$

Hence
$$\bar{J} = \int_0^\infty \frac{\theta \sqrt{\theta^2 - \vartheta^2}\, J_0(\rho\theta)\, d\theta}{(2\theta^2 - 1)^2 - 4\theta^2 \sqrt{\theta^2 - \vartheta^2} \sqrt{\theta^2 - 1}} \qquad \text{(VIII-3-16)}$$

where
$$\vartheta = \frac{h}{k}$$

and h is determined by Eqs. (VIII-2-5).

The velocity of longitudinal wave propagation is always higher than the velocity of transverse wave propagation; therefore $h < k < 1$, and we have

$$\bar{J} = \int_0^\vartheta + \int_\vartheta^1 + \int_1^\infty = \bar{J}_\vartheta + \bar{J}_1 + \bar{J}_\infty$$

Let us consider

$$\bar{J} = \int_0^\vartheta \frac{\theta \sqrt{\theta^2 - \vartheta^2}\, J_0(\rho\theta)\, d\theta}{2(\theta^2 - 1)^2 - 4\theta^2 \sqrt{\theta^2 - \vartheta^2} \sqrt{\theta^2 - 1}}$$

According to the meaning of the problem, the coefficients α and β should have positive values for all values of θ. During changes of θ within the range 0 to ϑ, $\theta < \vartheta$; hence, in order to fulfill the condition $\alpha > 0, \beta > 0$ in the integral expression for $\bar{J}_\vartheta$, it is necessary to substitute $\sqrt{\theta^2 - \vartheta^2}$ for $+i\sqrt{\vartheta^2 - \theta^2}$ and $\sqrt{\theta^2 - 1}$ for $+i\sqrt{1 - \theta^2}$. Then we have

$$\bar{J} = i\int_0^\infty \frac{\theta \sqrt{\vartheta^2 - \theta^2}\, J_0(\rho\theta)\, d\theta}{(2\theta^2 - 1)^2 + 4\theta^2 \sqrt{\vartheta^2 - \theta^2} \sqrt{1 - \theta^2}} = i\bar{J}_2 \qquad \text{(VIII-3-17)}$$

For $\bar{J}_1$, $1 > \theta > \vartheta$; hence

$$\bar{J}_1 = \int_\vartheta^1 \frac{\theta \sqrt{\theta^2 - \vartheta^2}\, J_0(\rho\theta)\, d\theta}{(2\theta^2 - 1)^2 - 4i\theta^2 \sqrt{\theta^2 - \vartheta^2} \sqrt{1 - \theta^2}} \qquad \text{(VIII-3-18)}$$

Transforming the integral function, we obtain

$$\bar{J}_1 = \bar{J}_1{}^{(a)} + i\bar{J}_1{}^{(b)}$$

where
$$\bar{J}_1{}^{(a)} = \int_\vartheta^1 \frac{(2\theta^2 - 1)^2\theta \sqrt{\theta^2 - \vartheta^2}\, J_0(\rho\theta)\, d\theta}{(2\theta^2 - 1)^4 + 16\theta^4(\theta^2 - \vartheta^2)(1 - \theta^2)} \qquad \text{(VIII-3-19)}$$

$$\bar{J}_1{}^{(b)} = \int_\vartheta^1 \frac{4\theta^3(\theta^2 - \vartheta^2) \sqrt{1 - \theta^2}\, J_0(\rho\theta)\, d\theta}{(2\theta^2 - 1)^4 + 16\theta^4(\theta^2 - \vartheta^2)(1 - \theta^2)} \qquad \text{(VIII-3-20)}$$

As θ changes from 1 to ∞, it takes the value $\varkappa k$, where $\varkappa$ is the root of the equation $F(k\theta) = 0$, corresponding to a surface wave. Therefore the integral function J_∞, for some values of θ, increases without limit, and the integral J_∞ becomes indeterminate. As indicated in Art. VIII-2, in order to disclose this indeterminateness, only the principal part of the Cauchy integral should be considered. Let us denote this part of the integral by CJ_∞; then

$$J_\infty = C \int_1^\infty \frac{\theta \sqrt{\theta^2 - \vartheta^2}\, J_0(\rho\theta)\, d\theta}{(2\theta^2 - 1)^2 - 4\theta^2 \sqrt{\theta^2 - \vartheta^2}\sqrt{\theta^2 - 1}} \qquad \text{(VIII-3-21)}$$

It is clear that J_∞ is a real value. Using the results obtained with reference to $\bar{J}$, we have

$$w_0 = \frac{Pk}{\mu} e^{i\omega t}(f_1 + if_2) \qquad \text{(VIII-3-22)}$$

where

$$f_1 = \frac{1}{2\pi} (\bar{J}_1{}^{(a)} + C\bar{J}_\infty)$$

$$f_2 = \frac{1}{2\pi} (\bar{J}_2 + \bar{J}_1{}^{(b)}) + \frac{k_0}{2} J_0\left(\rho \frac{\varkappa}{k}\right) \varkappa$$

Thus the determination of displacements on the soil surface at a small distance from the source of the waves, e.g., for a small value of ρ, is reduced to the computation of the foregoing integrals.

The easiest way to compute these integrals is by expanding them into a series. Omitting the rather involved intermediate operations for their computation, let us use the equation obtained by O. Ya. Shekhter[41] for f_1 and f_2. For the Poisson ratio $\nu = 0.5$, we have

$$f_1 = -0.0796 \frac{1}{\rho} + 0.0598\rho - 0.00607\rho^3 + 0.000243\rho^5 - 0.00000517\rho^7 + \cdots$$

$$f_2 = 0.0571J_0(1.047) + 0.0474 - 0.00647\rho^2 + 0.000264\rho^4 - 0.00000517\rho^6 + \cdots$$

For the Poisson ratio $\nu = 0.25$, we have

$$f_1 = -0.119 \frac{1}{\rho} + 0.0895\rho - 0.0104\rho^3 + 0.000466\rho^5 - 0.0000109\rho^7 + \cdots$$

$$f_2 = 0.0998J_0(1.08777\rho) + 0.0484 - 0.00595\rho^2 + 0.000240\rho^4 - 0.00000484\rho^6 + \cdots$$

For the Poisson ratio $\nu = 0$, we have

$$f_1 = -0.159 \frac{1}{\rho} + 0.1392 - 0.0185\rho^3 + 0.000937\rho^5 - 0.0000246\rho^7 + \cdots$$

$$f_2 = 0.163J_0(1.1441\rho) + 0.0512 - 0.00585\rho^2 + 0.000228\rho^4 - 0.00000445\rho^6 + \cdots$$

Returning to Eq. (VIII-3-22), let us separate the real and imaginary parts; neglecting the latter, we obtain the following simple equation for the vertical component of soil displacement at small distances from the source of waves:

$$w = -A_0\Psi(\rho)\sin(\omega t - \gamma) \qquad \text{(VIII-3-23)}$$

where

$$A_0 = \frac{P\omega}{b\mu} \qquad \text{(VIII-3-24)}$$

$$\Psi(\rho) = \sqrt{f_1^2 + f_2^2} \qquad \text{(VIII-3-25)}$$

$$\tan\gamma = \frac{f_1}{f_2} \qquad \text{(VIII-3-26)}$$

Figure VIII-8 gives graphs of the function $\Psi(\rho)$ plotted for different values of the Poisson ratio. The use of these graphs in design computa-

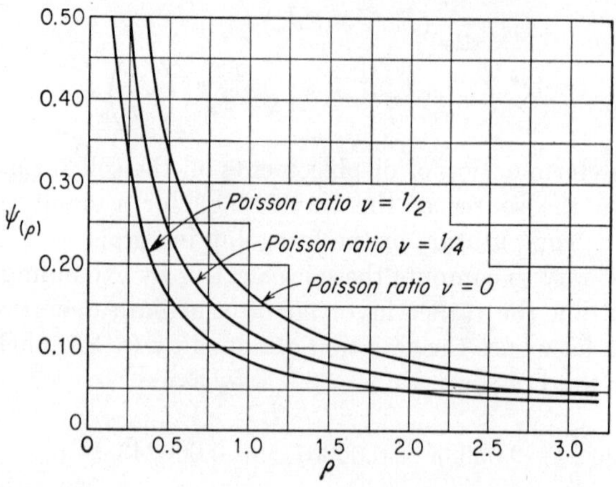

FIG. VIII-8. Graph facilitating the use of Eq. (VIII-3-25).

tions makes it possible to avoid calculations of f_1 and f_2 from the foregoing formulas.

c. Displacement of Soil at Large Distances from the Source of Waves. In order to find approximate expressions for soil displacements at large distances from the source of waves, the principal part according to Cauchy should be singled out of integrals (VIII-3-14) and (VIII-3-15). Integrating along the contour of the area of the complex variable gives us the following principal values of q_0 and w_0:

$$Cq_0 = -\frac{\varkappa P}{2\mu} H_0 Y_1(\rho) + \frac{ik^2P}{\pi\mu}\int_h^k \frac{\xi^2(2\xi^2 - k^2)\alpha\beta D_1(\xi r)}{F(\xi)f(\xi)}\,d\xi \qquad \text{(VIII-3-27)}$$

$$Cw_0 = \frac{i\varkappa P}{2\mu} K_0 J_0(\rho) - \frac{ik^2P}{2\pi\mu}\int_h^k \frac{\xi(2\xi^2 - k^2)^2\alpha}{F(\xi)f(\xi)} D_0(\xi r)\,d\xi - \frac{ik^2P}{2\pi\mu} C\int_k^\infty \frac{\xi\alpha}{F(\xi)} D_0(\xi r)\,d\xi \qquad \text{(VIII-3-28)}$$

where

$$D_0(\varkappa r) = -Y_0(\varkappa r) - J_0(\varkappa r) \qquad \text{(VIII-3-29)}$$

$$D_1(\varkappa r) = -Y_1(\varkappa r) - iJ_1(\varkappa r) \qquad \text{(VIII-3-30)}$$

Y_0 and Y_1 are Bessel functions of the second type, of the zero and first orders; the coefficients H_0 and K_0 are determined by Eqs. (VIII-2-20); all other symbols have already been defined.

Equations (VIII-3-27) and (VIII-3-28) include standing waves. In order to obtain a system of transient waves, it is necessary to impose on Eqs. (VIII-3-27) and (VIII-3-28) the solutions with reference to free surface waves; these may be taken as follows:

$$q_0^* = \frac{i\varkappa P}{2\mu} H_0 J_1(\varkappa r) \qquad w_0^* = -\frac{i\varkappa P}{2\mu} K_0 J_0(\varkappa r) \quad \text{(VIII-3-31)}$$

Adding these solutions to the preceding ones, and taking into account a temporary factor $e^{i\omega t}$, we obtain

$$q_0 = -\frac{\varkappa P}{2\mu} H_0 D_1(\varkappa r) e^{i\omega t} + \frac{Pik^2}{\pi\mu} \int_h^k \frac{\xi^2(2\xi^2 - k^2)\alpha\beta D_1(\xi r) e^{i\omega t}}{F(\xi) f(\xi)} d\xi \quad \text{(VIII-3-32)}$$

$$w_0 = -\frac{i\varkappa^2 P}{2\pi\mu} C \int_k^\infty \frac{\xi\alpha}{F(\xi)} D_0(\xi r) e^{i\omega t}\, d\xi - \frac{ik^2 P}{2\pi\mu} \int_h^k \frac{\xi(2\xi^2 - k^2)^2 \alpha D_0(\xi r) e^{i\omega t}}{F(\xi) f(\xi)} d\xi \quad \text{(VIII-3-33)}$$

Since the above expressions consist only of transient waves, they give a solution of the problem for the case of the application of a normal periodic force $Pe^{i\omega t}$ at the origin of coordinates.

If the distance from the source of waves is such that the value $\varkappa r$ is large, expressions (VIII-3-32) and (VIII-3-33) may be reduced to the form

$$q_0 = -\frac{\varkappa P}{2\mu} H_0 D_1(\varkappa r) e^{i\omega t} + S_q \quad \text{(VIII-3-34)}$$

$$w_0 = \frac{\varkappa P}{2\mu} K_0 D_0(\varkappa r) e^{i\omega t} + S_w \quad \text{(VIII-3-35)}$$

Here S_q and S_w contain terms which are inversely proportional at least to $(kr)^2$ and $(hr)^2$. Since $D_1(\varkappa r)$ and $D_0(\varkappa r)$ are inversely proportional to $\varkappa r^{1/2}$, it is clear that, at sufficiently large distances from the source of waves, the values of S_q and S_w will be small and may be neglected. Replacing $D_1(\varkappa r)$ and $D_0(\varkappa r)$ by their asymptotic values, we obtain the following final equations:

$$q_0 = -\frac{i\varkappa P}{2\mu} H_0 \sqrt{\frac{2}{\pi\varkappa r}}\, e^{i(\omega t - \varkappa r - \pi/4)} \quad \text{(VIII-3-36)}$$

$$w_0 = \frac{\varkappa P}{2\mu} K_0 \sqrt{\frac{2}{\pi\varkappa r}}\, e^{i(\omega t - \varkappa r - \pi/4)} \quad \text{(VIII-3-37)}$$

It is seen from Eqs. (VIII-3-36) and (VIII-3-37) that, at sufficiently large distances, the vibration amplitudes decrease at a rate inversely proportional to the square root of distance from the source.

Neglecting the imaginary parts of Eqs. (VIII-3-36) and (VIII-3-37), we have

$$q_0 = -\frac{\varkappa P}{2\mu} H_0 \sqrt{\frac{2}{\pi \varkappa r}} \sin\left(\omega t - \varkappa r - \frac{\pi}{4}\right) \qquad \text{(VIII-3-38)}$$

$$w_0 = \frac{\varkappa P}{2\mu} K_0 \sqrt{\frac{2}{\pi \varkappa r}} \cos\left(\omega t - \varkappa r - \frac{\pi}{4}\right) \qquad \text{(VIII-3-39)}$$

These expressions show that the orbit of motion of soil particles is an ellipse in which the ratio between the axes, as in the case of plane waves, equals K_0/H_0.

By equating the vibration amplitudes computed by Eqs. (VIII-3-23) and (VIII-3-39), we obtain the distance to points at which the amplitudes of vertical displacements of soil particles as computed by each of these equations have the same value; as a result we obtain an equation for $k \approx \varkappa$:

$$\Psi(\rho) \sqrt{\rho} = K_0 \sqrt{\frac{2}{\pi}} \qquad \text{(VIII-3-40)}$$

For the Poisson ratio $\nu = 0.5$, $K_0 = 0.109$; hence

$$\Psi(\rho) \sqrt{\rho} = 8.7 \times 10^{-2}$$

Using the graph of the function $\Psi(\rho)$ for $\nu = 0.5$ (Fig. VIII-8), we obtain

$$\rho = \frac{2\pi}{L_c} r = 1$$

It therefore follows that for $r = r_0 = L_c/2\pi$ (for the Poisson ratio $\nu = 0.5$), Eqs. (VIII-3-23) and (VIII-3-39) will give the same value for the amplitude of the vertical component of soil vibrations. For distances $r > r_0$, amplitudes of vibrations should be calculated from Eq. (VIII-3-39); for smaller distances, from Eq. (VIII-3-23).

By the use of Eq. (VIII-3-40) it is also easy to determine limiting distances for other values of the Poisson ratio.

d. Experimental Investigations. The author performed experimental investigations of propagation of waves under various soil conditions. Water-saturated sands were a subject of the most detailed investigations. The results of these investigations permit some conclusions concerning the validity limits of theoretically established relationships between soil displacements and distances from the source of waves.

A test foundation was used as a source. The area of this foundation in contact with soil was 1 m². Vibrations were excited by a special vibrator with a controllable speed. Investigations were mostly performed in cases in which the foundation (the source of waves) underwent vertical vibrations. Since the area of the foundation in contact with soil is small, it may be considered as representing a concentrated source.

The foregoing theoretical investigations took as the source of waves a vertical exciting force whose magnitude was determined by a certain assigned value P. If the foundation area in contact with soil serves as a source of waves, then the value of P may be taken as follows:

$$P = c_u A A_z \tag{VIII-3-41}$$

where c_u = coefficient of elastic uniform compression of base under foundation

A = foundation area in contact with soil

A_z = amplitude of vertical vibrations of foundation

During the experiments under discussion, the value of c_u was established from the resonance curve of the vertical vibrations of the foundation; it equaled 5 kg/cm³. Soil vibrations were measured by means of an oscillograph which made it possible to record the vertical and horizontal components of vibrations and the orbits of motion of soil particles.

According to the theory of surface waves, soil particles located on the soil surface move along vertical ellipses in which the ratio between the vertical and horizontal axes is constant and equals K_0/H_0. For soils satisfying the Poisson hypothesis ($\nu = 0.25$), this ratio equals 1.47; for incompressible soils ($\nu = 0.5$), $K_0/H_0 = 1.85$.

Figure VIII-9 shows the orbits of motion of soil particles recorded at different distances from the wave source and at vibromachine speeds equaling 800 (*a*) and 1,200 rpm (*b*). The vibration amplitude of the foundation (source of waves) remained constant and equaled 0.30 to 0.32 mm. Experimental orbits of motion of the soil particles differ considerably from theoretical orbits, especially for relatively small distances from the source. At small distances the orbits of motion of soil particles may have a very complicated shape (a figure eight or a deformed ellipse). With an increase in distance, the orbit approaches an ellipse; however, the angle of its inclination with respect to the vertical axis does not equal $\pi/2$, as should be the case according to the theory of surface waves.

It appears that the considerable distortions of the orbit observed at small distances from the source of waves are explained by the influence of the supplementary spectrum of waves propagated through the soil simultaneously with the surface waves. It will be shown later that, at large distances from the source of waves, the displacement of soil particles is affected by the absorption of the wave energy by soil. Consequently, the phase angle between the horizontal and vertical components does not remain constant and equal to $\pi/2$, but changes from point to point; therefore the inclination of the axes of the ellipse depends on the distance of the point under consideration from the source.

In order to verify Eq. (VIII-3-23), amplitudes of vertical vibrations of the soil surface were measured at small distances from the source of

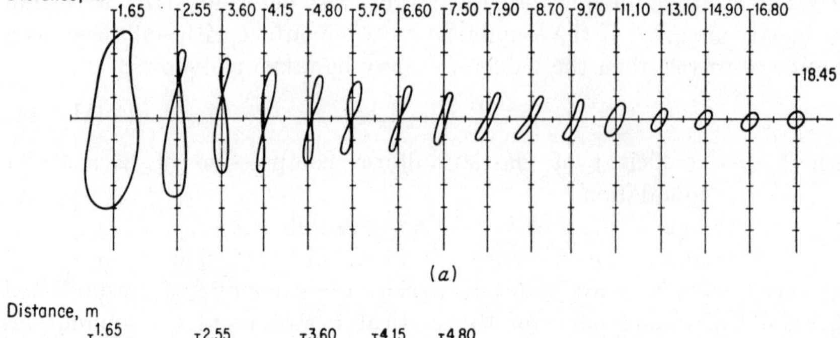

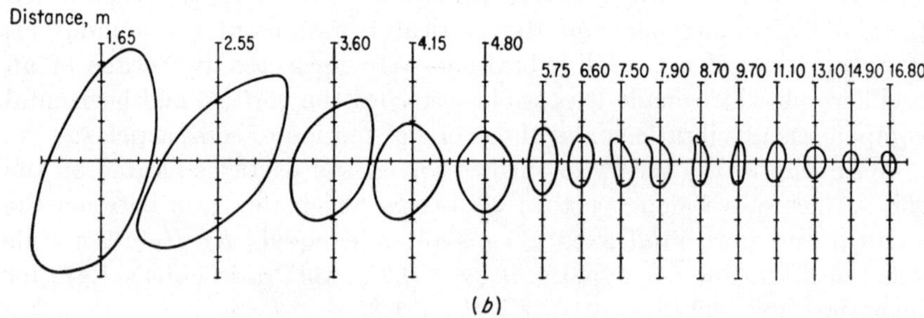

FIG. VIII-9. Measured orbits of motion of surface particles at varying distances from vibromachines operating at: (a) 800 rpm; (b) 1,200 rpm.

waves. Table VIII-3 gives the results of these measurements performed at different vibromachine speeds. The amplitude of vibrations of the foundation (source) was equal to 0.30 mm.

TABLE VIII-3. COMPARISON OF THEORETICAL AND EXPERIMENTAL VIBRATION-AMPLITUDE VALUES OF SOIL SURFACE

N, min^{-1}	ω, sec^{-1}	r, m	A_0, cm	ρ	$\psi(\rho)$	Amplitudes, mm		Design values	
						Computed	Measured	b, m/sec	μ, kg/cm^2
600	62.5	1.6	2.24×10^{-2}	0.90	0.115	2.5×10^{-2}	2.5×10^{-2}	150	374
810	84.0	1.6	4.3×10^{-2}	1.12	0.090	3.9×10^{-2}	3.9×10^{-2}	120	245
1,000	104.0	2.4	5.3×10^{-2}	2.10	0.057	3.10×10^{-2}	3.25×10^{-2}	120	245
1,500	156.0	1.6	6.25×10^{-2}	1.92	0.060	3.75×10^{-2}	3.7×10^{-2}	130	287
1,800	187.0	1.6	6.70×10^{-2}	2.20	0.053	3.50×10^{-2}	3.6×10^{-2}	135	312
2,100	218.0	1.6	8.80×10^{-2}	2.70	0.050	4.4×10^{-2}	4.2×10^{-2}	130	287

When amplitudes of soil vibrations were computed from Eq. (VIII-3-23), the velocity of transverse waves was taken to equal 120 to 150 m/sec². This value is fairly accurate, since it was confirmed by other experiments on the site of the investigation. The values of the modulus in shear corresponding to these velocities of transverse waves were found to equal 245 to 374 kg/cm². The value of $\Psi(\rho)$ was found from the graphs of Fig. VIII-8 for the arithmetical mean between $\nu = 0.5$ and $\nu = 0.25$.

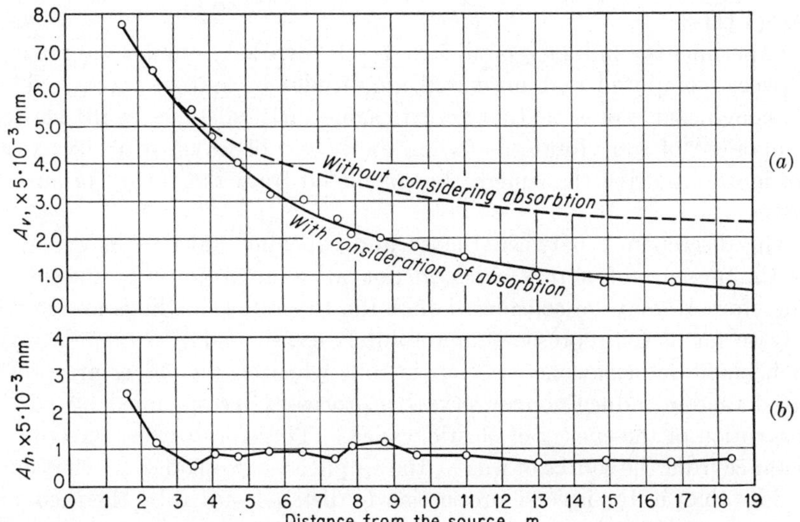

FIG. VIII-10. (a) Experimental check of Eqs. (VIII-3-42) and (VIII-3-43) with reference to the vertical component A_v of vibration amplitude on soil surface; (b) measured variation of horizontal component A_h of vibration amplitude. In both cases vibrations at the source are vertical.

Amplitudes of vertical soil vibrations computed from Eq. (VIII-3-23) agreed with measured amplitudes. It follows that Eq. (VIII-3-23) may be used with confidence for small distances from the source.

In accordance with Eqs. (VIII-3-38) and (VIII-3-39), the amplitudes of surface waves at relatively large distances from the source of waves decrease at a rate which is inversely proportional to the square root of the distance. Hence, if A_r is the amplitude at a distance r from the source, and A_0 is the amplitude at a distance r_0, then

$$A_r = A_0 \sqrt{\frac{r_0}{r}} \qquad \text{(VIII-3-42)}$$

Figure VIII-10 presents a graph of experimentally established values of vertical amplitudes of vibrations at different distances from a foundation subjected to 810 vibrations per minute. The experimental data are shown by little circles. The same figure gives a graph (dashed line) of changes in amplitude in accordance with Eq. (VIII-3-42); the initial values of r and A_r are taken from Table VIII-2. Comparison of the theoretical curve with the experimental data shows that only within the range of small values of r (as compared with r_0) do the amplitudes computed from Eq. (VIII-3-42) coincide with the experimental data. With an increase in the ratio r/r_0, the discrepancy between the two increases. The measured amplitudes are always smaller than those computed from Eq. (VIII-3-42).

Experiments conducted on different soils revealed a similar discrepancy between computed and measured amplitudes of surface waves. Thus the conclusion is possible that actual changes with distance of the vertical component of soil vibrations (when the source of waves is also vertical) are more intensive than might be concluded from the theory of surface waves.

The discrepancy between theory and experimental data is explained by the presence of factors which are not taken into account by the theory. Therefore it may be considered that the theoretical findings are correct only so far as soil represents an absolutely elastic body. In reality, even such small deformations as occur in soils when elastic waves are propagated therein induce nonconservative processes accompanied by partial absorption of the energy of elastic waves. Therefore with an increase in distance from the source of waves, the amplitudes drop off somewhat more rapidly than in the inverse proportion to distance stated by the theory of wave propagation in an absolutely elastic body.

Some assumptions should be made in order to take into account the absorption of wave energy by soil and to introduce a correction into the relationship between amplitude and distance. These assumptions concern the dependence of wave absorption on the properties of the propagating waves and the distance from the source of waves. Hereafter we shall assume that, when a spherical wave is propagated over the soil surface from a concentrated source, the absorption of the energy of this wave is proportional to the amount of energy entering a given soil layer as well as to its thickness; however, the absorption will not depend on the radius of curvature of the wave. Then an approximate formula for amplitudes of spherical waves propagating over the soil surface may be written as follows:

$$A_r = A_0 \sqrt{\frac{r_0}{r}}\, e^{-\alpha(r-r_0)} \qquad \text{(VIII-3-43)}$$

where A_r, A_0 = amplitudes of soil vibrations at distances r, r_0

α = coefficient of wave energy absorption, having dimensions meters^{-1} or centimeters^{-1}

If the soil were an absolutely elastic medium, no energy would be absorbed; in this case α would equal zero. Hence the coefficient α is a soil constant determining the deviation of its physical properties from the properties of an absolutely elastic body.

The continuous line in Fig. VIII-10 shows a relationship between the vertical component of soil vibrations and distance as computed from Eq. (VIII-3-43) for $\alpha = 0.100$ m^{-1}. It is seen from this graph that calculated and experimental amplitude values fully agree for all distances from the source of waves. Several analogous graphs obtained as results of experiments with other frequencies and other sources of waves are given in Art. VIII-8. These graphs also show good agreement between the results of computations from Eq. (VIII-3-43) and the experimental data, as do investigations carried out under different soil conditions.

Hence the conclusion is possible that the equations for the computation of the vertical component of vibration amplitudes as a function of distance (in the case of a vertical source) are in good agreement with experiments if the absorption of wave energy by the soil is taken into account.

Energy absorption by the soil is an irreversible process, similar to the formation of loops during elastic hysteresis. Therefore it may be assumed that these two processes are caused by similar factors inducing deviations in the behavior of soil (in cases of small deformations) from the behavior of an absolutely elastic body. Very little is known about the nature of these factors, not only for such complex material as soil, but for such materials as concrete, wood, and steel.

There are two main points of view in regard to these phenomena. According to the first, observed deviations in the behavior of materials are explained by the existence of forces of internal friction or resistance, which are dissipative forces; i.e., they depend not on the magnitude of deformation, but on the rate thereof. The influence of these forces on free and forced vibrations is fairly well established in systems with one degree of freedom, such as the vertical vibrations of massive foundations under machines. It is usually assumed that dissipative forces are directly proportional to the velocities of deformation. For continuous elastic systems it may also be considered that in addition to stresses depending on deformation, there appear stresses depending on the rate of deformation at a given moment. In the simplest case (for low velocities) it may be considered that there is a linear relationship between dissipative stresses and the rates of deformation. Elastic systems which permit the existence of stresses depending on the magnitude and rate of deformation are called viscoelastic systems.

We shall not deal here in detail with the theory of wave propagation in a viscoelastic body, but shall only indicate the principal conclusions of this theory concerning the influence of the viscosity of the medium on the velocity of propagation and the damping of three-dimensional and surface waves.

Let us use the following symbols:

α = damping constant of wave which depends on distance
v = velocity of propagation of wave in viscous medium
a, b = velocities of propagation of longitudinal, transverse waves in perfectly elastic medium
$\eta' = \eta/\rho$ = kinematic viscosity of medium
η = coefficient of viscosity
ρ = density of medium
ω = frequency of wave

Assuming that the medium is incompressible and is characterized by a low viscosity, i.e., assuming that η/a^2T and η/b^2T are small, we obtain the following expressions which establish the dependence of v and α on the viscosity of the medium:

For longitudinal waves:

$$v = a \qquad \alpha_a = \frac{4\eta'\omega^2}{3a^3}$$

For transverse waves:

$$v = b \qquad \alpha_b = \frac{\eta'\omega^2}{b^3}$$

For surface waves:

$$v = c \qquad \alpha_c = \alpha_b$$

Hence, in soil of low viscosity, surface waves are damped in the same way as transverse three-dimensional waves; the ratio between the constant α_a of a longitudinal wave and the constant α_b of a transverse wave equals

$$\frac{\alpha_a}{\alpha_b} = \frac{4b^3}{3a^3}$$

In soils, usually $3a^3/4 > b^3$; therefore the damping of longitudinal waves is smaller than the damping of transverse or surface waves.

It follows from the expressions for damping constants that in all three types of waves they increase proportionally to the square of the frequency of propagating waves. Hence, according to the theory of wave propagation in soils characterized by viscosity, waves produced by high-frequency machines are damped in the soil much more rapidly than waves created by low-frequency machines.

According to the second point of view, the damping of waves in soil is due to the influence of elastic recovery. The application of the theory of elastic recovery to the question of damping of waves in soil was considered by B. Deryagin,[12,13] V. G. Gogoladze,[19] and others. The principal conclusions arrived at by B. Deryagin in regard to the damping of seismic waves in soils follow.

For three-dimensional waves of not too small periods,

$$\alpha_a = \frac{2}{3}\pi^2 \frac{b^2}{a^3 T}\frac{B}{G}$$

$$\alpha_b = \frac{\pi^2}{2bT}\frac{B}{G}$$

where B = constant of elastic recovery
G = modulus of rigidity

Hence it follows that

$$\frac{\alpha_a}{\alpha_b} = \frac{4}{3}\left(\frac{b}{a}\right)^3$$

If $a = 1.8b$, then $\alpha_a = 0.23\alpha_b$.

Thus it follows also from the theory of elastic recovery that the damping of longitudinal three-dimensional waves occurs much more slowly than the damping of transverse waves.

The theory of wave propagation in a viscoelastic body leads to the conclusion that damping constants of waves are proportional to the squares of their frequencies. However, the theory of wave propagation in a medium characterized by elastic recovery leads to the conclusion that the damping of waves is proportional to the first power of the frequency. This means that the rate of damping for one wavelength is constant.

For surface waves, the damping constant determined by B. Deryagin on the basis of the theory of elastic recovery does not differ much from α_b. In this regard, conclusions of the elastic-recovery and viscoelastic theories agree. It is seen that in the question of damping of waves in soil the theories differ only in regard to the relationship between the damping constant and the period of propagating waves. How far either theory corresponds to reality may be shown only by experimental investigations of the propagation in soil of waves with different periods.

The author experimentally investigated under winter conditions the propagation of waves with periods of vibrations of 0.111 sec (N = 540 min^{-1}), 0.083 sec (N = 720 min^{-1}), and 0.067 sec (N = 900 min^{-1}) in water-saturated brown clays with some silt and sand. He found[4] that α decreases at a rate which is inversely proportional to the square of the period; this finding agrees with the conclusions of the theory for a viscoelastic medium. However, it should be noted that theoretical curves computed from Eq. (VIII-3-43) differ considerably from experimental values of soil vibration amplitudes. Therefore the values of α obtained for waves with the periods indicated above are approximate, and the dependence of α on frequency is confirmed only with a high degree of approximation, especially as the investigations were conducted for three wave periods only.

For the determination of the relationship between damping constant and frequency, one can use data obtained from investigations of vibration-amplitude changes on the soil surface as a function of distance. These data were obtained from experimental studies of screening of waves by line and closed sheet-pile rows (see Art. VIII-8). Data on amplitudes of the vertical component of soil vibrations were obtained during vertical vibrations of the foundation (source of waves) before the sheetpiling was installed.

From measured amplitudes of soil vibrations, values of α were computed for different numbers of oscillations per minute, and then curves of vertical soil vibration amplitudes were plotted. Comparison of the theoretical and experimental curves of soil vibration amplitudes shows satisfactory agreement.

TABLE VIII-4. SOIL DAMPING CONSTANTS COMPUTED FROM TWO TESTS

Foundation	Oscillations, min^{-1}	Damping constant, m^{-1}
7	600	0.100
	800	0.100
	1,200	0.100
	1,500	0.100
	1,800	0.090
8	600	0.040
	800	0.050
	1,000	0.050
	1,200	0.040
	1,620	0.040

Table VIII-4 gives approximate computed values of a damping constant for wave propagation from two test foundations. This table shows that in both cases, i.e., for waves emanating from foundation 7 (with the foundation area in contact with soil equaling 1 m^2) and for waves propagating from foundation 8 (contact area equaling 4 m^2), the damping constant does not depend on the period of the propagated waves. Hence, at least for the case of wave propagation in water-saturated fine-grained sands, the conclusions of the theory of wave propagation in a medium with elastic recovery agree more closely with the experimental data than the conclusions of the theory of wave propagation in a viscoelastic medium.

The hypothesis that dissipative forces are linearly dependent on the velocity of deformation does not seem to be supported by experimental data. It is possible that the assumption concerning the dependence of the forces of internal resistance on deformation,† when extended to continuous media (which include soil), gives more satisfactory results in regard to damping of vibrations than the theory of a viscoelastic body or a body characterized by so-called recovery.

Values of the constant α in waves propagated from foundation 7 were about two times larger than in waves propagated from foundation 8. Apparently this can be explained by the fact that for foundation 7 the measurements of wave propagation were made along a ridge of yellow sand, and for foundation 8 along gray medium sands containing organic silt.

These experimental investigations of wave propagation from foundation 7 were made partly after the bulk of the soil had already thawed; others were made at a time when the soil was still frozen. From these experiments an average value of α equaling 0.058 m^{-1} was obtained for frozen soils; this is about one-half the value for a thawed soil. Hence, other conditions being equal, waves from machine foundations propagate to larger distances in winter than in summer.

Table VIII-5 gives approximate values of the coefficient of absorption of wave energy (damping constant) obtained by the author as a result of investigations of wave propagation in different soils.

TABLE VIII-5

No.	Soil	Coefficient of absorption, m^{-1}
1	Yellow water-saturated fine-grained sand	0.100
2	Yellow water-saturated fine-grained sand in a frozen state	0.060
3	Gray water-saturated sand with laminae of peat and organic silt	0.040
4	Clayey sands with laminae of more clayey sands and of clays with some sand and silt, above ground-water level	0.040
5	Heavy water-saturated brown clays with some sand and silt	0.040–0.120
6	Marly chalk	0.100
7	Loess and loessial soil	0.100

Figure VIII-10*b* shows a graph—experimentally obtained—of amplitude changes in the horizontal longitudinal components of vibrations. It is seen from this graph that, while the amplitude of the vertical com-

† This assumption was advanced by I. L. Korchinskiy.[25]

ponent (Fig. VIII-10a) changes with distance more or less monotonously, no regularity is observed in changes of the amplitude of the horizontal longitudinal component. Graphs of relationship between the amplitude of the horizontal longitudinal component and distance, given in Art. VIII-8, show this same lack of regularity.

Thus experiments demonstrate that the application of the theory of surface waves to the computation of amplitudes of the horizontal longitudinal component of vibrations is not possible in the case of a vertical source of waves. The factors indicated above (i.e., the supplementary spectrum and the absorption of wave energy by soil) greatly affect the distribution of amplitudes of the horizontal component of vibrations;

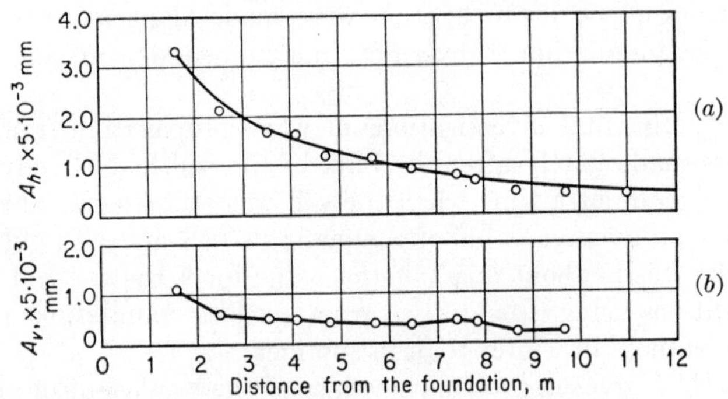

FIG. VIII-11. Vibrations at the source are induced by a rocking motion: (a) variation with distance of horizontal component A_h of vibrations at soil surface; (b) the same for the vertical component A_v.

therefore the results of computations conducted without taking into account these factors cannot be applied to the investigation of wave propagation in soil.

From the principle of reciprocity, the assumption is possible that an analogous conclusion will be justified for the vertical component of soil vibrations if the source of waves consists of a foundation undergoing horizontal or close to horizontal vibrations. This is confirmed by experiments. Figure VIII-11a shows an experimental graph of changes in the amplitudes of the horizontal longitudinal component of vibrations A_h; Fig. VIII-11b shows an experimental graph of changes of amplitudes of the vertical vibration component when the source of waves was the same foundation undergoing rocking vibrations with a frequency of 30 sec^{-1}. These graphs show that changes in the horizontal longitudinal component follow Eq. (VIII-3-43); the amplitudes of the vertical component change with distance in the same complicated way as the horizontal longitudinal component when the foundation (source of waves) undergoes vertical vibrations.

VIII-4. Influence of the Area of the Source of Waves on the Amplitude of Vibrations of the Soil Surface

Let us assume that the distance r of the point under consideration from the source is large by comparison with the lengths of propagating waves. In this case the influence of the supplementary spectrum may be neglected, and it may be considered that displacements on the soil surface are created by surface waves only. Let us also assume that the exciting force producing waves in soil is uniformly distributed along a limited surface having the shape of a circle with radius R. Assuming that the exciting force imposes pressure p on the soil, we may, in Eqs. (VIII-3-34) and (VIII-3-35), replace the value P of the concentrated force by a force acting on an element of the area of the circle and equaling

$$ps\,ds\,d\varphi$$

According to Eqs. (VIII-3-34) and (VIII-3-35), displacements on the soil surface at a distance r from the point of application of the concentrated force P equal

$$\begin{aligned} w_0 &= AD_0(\varkappa r)ps\,ds\,d\varphi \\ q_0 &= BD_1(\varkappa r)ps\,ds\,d\varphi \end{aligned} \qquad \text{(VIII-4-1)}$$

where

$$A = \frac{\varkappa}{2\mu} K_0 e^{i\omega t}$$

$$B = -\frac{\varkappa}{2\mu} H_0 e^{i\omega t}$$

All other symbols have been previously defined.

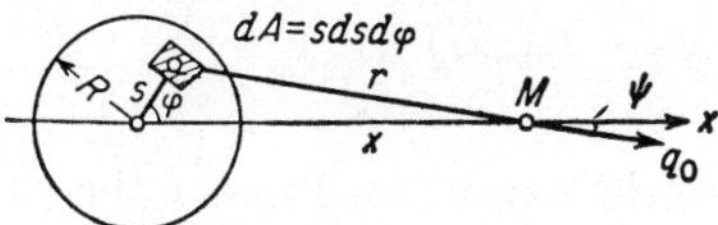

FIG. VIII-12. Diagram illustrating Eqs. (VIII-4-1) and (VIII-4-2).

It follows directly from Fig. VIII-12 that

$$\begin{aligned} r &= \sqrt{x^2 + s^2 - 2xs\cos\varphi} \\ x - s\cos\varphi &= r\cos\psi \end{aligned} \qquad \text{(VIII-4-2)}$$

The horizontal component of soil displacement along the x axis equals

$$u_0 = q_0 \cos\psi$$

Displacements produced at the point M located at a distance x from the center of a circle whose whole area is subjected to the action of an evenly distributed exciting force of intensity p are as follows:

$$\begin{aligned} w_0 &= \int_0^{2\pi}\int_0^R ApD_0(\varkappa r)s\,ds\,d\varphi \\ u_0 &= \int_0^{2\pi}\int_0^R BpD_1(\varkappa r)\cos\psi s\,ds\,d\varphi \end{aligned} \qquad \text{(VIII-4-3)}$$

From the theory of Bessel's functions, under the condition that $x > s > 0$, we have

$$
\begin{aligned}
J_0(\varkappa r) &= J_0(\varkappa s)J_0(\varkappa x) + 2\sum_{n=1}^{\infty} J_n(\varkappa s)J_n(\varkappa x)\cos n\varphi \\
Y_0(\varkappa r) &= J_0(\varkappa s)Y_0(\varkappa x) + 2\sum_{n=1}^{\infty} J_n(\varkappa s)Y_n(\varkappa x)\cos n\varphi
\end{aligned}
\tag{VIII-4-4}
$$

In addition,

$$J_0'(\xi) = -J_1(\xi) \qquad -Y_0'(\xi) = -Y_1(\xi)$$

Differentiating Eqs. (VIII-4-4) with respect to x and taking into account the fact that

$$\frac{\partial r}{\partial x} = \frac{x - s\cos\varphi}{r}$$

we obtain

$$-J_1(\varkappa r)\varkappa\frac{x - s\cos\varphi}{r} = -\varkappa J_0(\varkappa s)J_1(\varkappa s) + 2\sum_{n=1}^{\infty}\frac{dJ_n(\varkappa x)}{dx}J_n(\varkappa s)\cos n\varphi$$

$$-Y_1(\varkappa r)\varkappa\frac{x - s\cos\varphi}{r} = -\varkappa J_0(\varkappa s)Y_1(\varkappa x) + 2\sum_{n=1}^{\infty}\frac{dY_n(\varkappa x)}{dx}J_n(\varkappa s)\cos n\varphi$$

Or, using Eqs. (VIII-4-2), we have

$$J_1(\varkappa r)\cos\psi = J_0(\varkappa s)J_1(\varkappa x) + \sum_{n=1}^{\infty} C_n\cos n\varphi$$

$$Y_1(\varkappa r)\cos\psi = J_0(\varkappa s)Y_1(\varkappa x) + \sum_{n=1}^{\infty} C_n^{(1)}\cos n\psi$$

where C_n and $C_n^{(1)}$ are the expressions under the summation signs in the two penultimate equations.

Substituting these expressions into Eqs. (VIII-3-29) and (VIII-3-30), which give expressions for $D_0(\varkappa r)$ and $D_1(\varkappa r)$, and integrating with respect to φ, we have

$$w_0 = -2\pi Ap\int_0^R J_0(\varkappa s)[Y_0(\varkappa x) + iJ_0(\varkappa x)]s\,ds$$
$$u_0 = -2\pi Bp\int_0^R J_0(\varkappa s)[Y_1(\varkappa x) + iJ_1(\varkappa x)]s\,ds$$

After integrating with respect to s, we obtain

$$
\begin{aligned}
w_0 &= 2\pi ApD_0(\varkappa x)\frac{\varkappa RJ_1(\varkappa R)}{\varkappa^2} \\
u_0 &= 2\pi BpD_1(\varkappa x)\frac{\varkappa RJ_1(\varkappa R)}{\varkappa^2}
\end{aligned}
\tag{VIII-4-5}
$$

It follows from a comparison of Eq. (VIII-4-5) with Eqs. (VIII-3-38) and (VIII-3-39) that, if a source of waves is distributed over a circular area with radius R, changes in amplitude of surface waves are related to changes in distance from the source with the same regularity as was observed in the case of a concentrated exciting force. In addition, however, for the same value of the intensity p of the exciting force, the amplitude will be greatly influenced by the dimensions of the area of the source. The size of this area influences the amplitudes of both vertical and horizontal vibration components.

Let us consider two cases:

1. The source of waves is the foundation under a low-frequency engine; consequently X is small, and the wavelength L_c is large. If one assumes that the dimensions of the foundation area in contact with soil are small in comparison with the wavelength, then $\varkappa R$ will also be small. By expanding the function J_1 into a series with respect to the argument for this boundary condition, we obtain approximately

$$J_1(\varkappa R) = \frac{\varkappa R}{2}$$

Then

$$\begin{aligned} w_0 &= ApD_0(\varkappa X)A' \\ u_0 &= BpD_1(\varkappa X)A' \end{aligned} \qquad \text{(VIII-4-6)}$$

where A' is the area of the source of waves in contact with soil.

Thus for foundations under low-frequency engines (where the value of $\varkappa R$ is small) the amplitudes of soil vibrations grow in proportion to the area of the source of waves in contact with soil.

2. The source of waves is a foundation under a high-frequency engine (turbogenerator, forge hammer, and others) for which the product $\varkappa R$ is relatively large. For these values of the independent variable, the function J_1 may be replaced by its asymptotic value; i.e.,

$$J_1(\varkappa R) = \frac{1}{\sqrt{2\pi\varkappa R}} \sin\left(\varkappa R - \frac{\pi}{4}\right)$$

Therefore, for this case,

$$\begin{aligned} w_0 &= \frac{ApD_0(\varkappa x)}{2} \sqrt{2\pi\varkappa R} \sin\left(\varkappa R - \frac{\pi}{4}\right) \\ u_0 &= \frac{BpD_1(\varkappa x)}{2} \sqrt{2\pi\varkappa R} \sin\left(\varkappa R - \frac{\pi}{4}\right) \end{aligned} \qquad \text{(VIII-4-7)}$$

Hence the amplitudes of surface waves radiated from high-frequency sources will undergo periodic changes when the radius R continuously varies. Maximum magnitudes of amplitudes will be attained when

$$\sin\left(\varkappa R - \frac{\pi}{4}\right) = \pm 1$$

i.e., when $$\varkappa R - \frac{\pi}{4} = \frac{2n+1}{2}\pi \qquad n = 1, 2, 3, \ldots$$

or when $$\frac{R}{L_c} = \frac{2n+3}{8}$$

Amplitudes at all points on the soil surface equal zero if $\sin\left(\varkappa R - \frac{\pi}{4}\right) = 0$, i.e., if

$$\frac{R}{L_c} = \frac{2n+1}{8}$$

If a source of waves, for example, a vibrator placed on the soil, vibrates as a solid body with one degree of freedom, then theoretically the resonance curve of forced vertical vibrations has one maximum. However, if

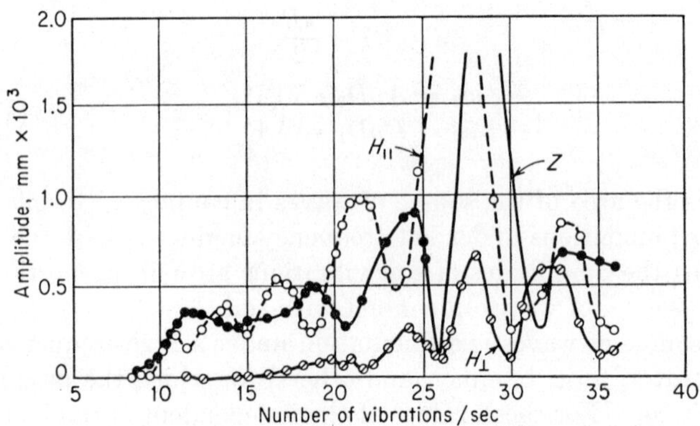

FIG. VIII-13. Resonance peaks established by Koehler.[24]

one simultaneously measures the resonance curves of the source of waves and the soil surface, the curve obtained will have several maximums in addition to the principal maximum corresponding to the maximum on the resonance curve of forced vibrations of the source of waves. Some investigators explain these additional maximums by coincidences of the frequency of natural vibrations of the soil layer and the frequency of propagating waves.

Figure VIII-13 shows a resonance curve plotted by R. Koehler,[24] who suggests that the presence of several maximums and minimums can be explained by the phenomena of interference and resonance (with the exception of a large maximum lying between 21 and 31 hertz and produced by natural vibrations of the engine). It is seen from curve Z that the first maximum for the vertical component of vibrations corresponds to 11.2 to 11.5 hertz. Resonance curves recorded at distances of 60 to 90 m from the source also demonstrated the presence of this maximum. On the basis of this observation Koehler came to the conclusion that the recorded frequency of vibrations corresponding to the first maximum of the resonance curve represents the frequency of natural vibrations of a layer of carbonaceous gray clay having a thickness of 5.8 m and located at a depth of 4.8 to 10.6 m. Koehler does not explain the natures of the other maximums on the resonance curve.

Let us assume that a resonance curve is recorded in the soil by a device installed at a constant distance r from the source of waves, and that the radius of the source remains constant. When the frequency ω of the source changes, $\varkappa$ will also change, since the latter is proportional to ω:

$$\varkappa = \frac{\omega}{c}$$

where c is the velocity of the propagating waves (independent of the frequency).

We assume that the amplitude of vibrations of the source is constant; consequently p is constant. Thus the only variable will be ω or $\varkappa$.

Let us consider the influence of changes in the frequency of the source on the amplitudes of vibrations (for example, on the amplitude of the vertical component of the soil vibrations). Using Eq. (VIII-4-5), we have

$$w_0 = -\frac{\pi}{\mu} K_0 p R J_1(\varkappa R) \sqrt{J_0{}^2(\varkappa x) + Y_0{}^2(\varkappa x)}$$

If the distance from the source of waves is large, then, approximately,

$$Y_0{}^2(\varkappa x) + J_0{}^2(\varkappa x) = \frac{2}{\pi \varkappa x}$$

and consequently,

$$w_0 = a \frac{J_1(\varkappa R)}{\sqrt{\varkappa}} \tag{VIII-4-8}$$

where

$$a = -\frac{K_0 p}{\mu} R \sqrt{\frac{2\pi}{x}}$$

It follows from Eq. (VIII-4-8) that when soil vibrations are excited by a source with a constant intensity p and varying frequency, the amplitudes of soil vibrations change with the frequency of vibrations. It is noteworthy that the amplitude of soil vibrations does not change in a monotonous way, but has several maximums and minimums analogous to those occurring in the experimental curves presented in Fig. VIII-13.

Of the greatest practical importance is the computation of the vertical component of soil vibrations produced by a vertical source. Assuming that in Eq. (VIII-4-5) the value of $\varkappa x$ is large enough so that we may replace $D_0(\varkappa x)$ by its asymptotic value, we obtain the following expression for the amplitude of vertical soil vibrations:

$$A_w = \frac{\pi p R K_0}{\mu} J_1(\varkappa R) \sqrt{\frac{2}{\pi \varkappa x}} \qquad \text{(VIII-4-9)}$$

Values of the function $J_1(\varkappa R)$ are taken from the tables.

Example. Computation of amplitudes of the vertical component of soil oscillations produced by shocks to a foundation under a forge hammer

1. DATA. The foundation under the hammer is a block; the following specifications apply to it and the soil:
Foundation contact area:

$$A = 65.6 \text{ m}^2$$

Coefficient of elastic uniform soil compression:

$$c_u = 6.0 \times 10^3 \text{ tons/m}^3$$

Amplitude of vertical vibrations of the foundation:

$$A_z = 0.85 \times 10^{-3} \text{ m}$$

Frequency of natural vertical vibrations of the foundation:

$$\omega = 70 \text{ sec}^{-1}$$

Velocity of propagation of transverse waves:

$$b = 100 \text{ m/sec}$$

Modulus of rigidity of soil:

$$G = 17 \times 10^3 \text{ tons/m}^2$$

2. COMPUTATIONS. Let us determine the equivalent radius of the circle:

$$R = \sqrt{\frac{A}{\pi}} = \sqrt{\frac{65.6}{3.14}} = 4.46 \text{ m}$$

We find coefficient $\varkappa$:

$$\varkappa = \frac{\omega}{c} \cong \frac{\omega}{b} = \frac{70}{100} = 0.7$$

Hence,

$$\varkappa R = 0.7 \times 4.46 = 3.13$$

We find from the tables,

$$J_1(\varkappa R) = 0.30$$

The dynamic pressure on the soil will be

$$p = c_u A_z = 6 \times 10^3 \times 0.85 \times 10^{-3} = 5.1 \text{ tons/m}^2$$

We take the value of the coefficient K_0 as the arithmetic mean of its values for Poisson ratios of 0.5 and 0.25:

$$K_0 = \frac{0.108 + 0.183}{2} = 0.145$$

Substituting all the above values into Eq. (VIII-4-9), we obtain

$$A = \frac{3.14 \times 5.10 \times 4.46 \times 0.145}{1.7 \times 10^3} \times 0.30 \sqrt{\frac{2}{3.14 \times 0.7x}}$$
$$= 1.79 \times 10^{-3} \sqrt{\frac{1}{x}}$$

As indicated in Art. VIII-3, the influence of energy absorption by soil is not great at relatively small distances from the source of waves. Therefore for small values of $x > R$, the influence of energy absorption may be neglected. For example, assuming $x = 6$ m, we obtain an amplitude of the vertical component of soil vibrations of

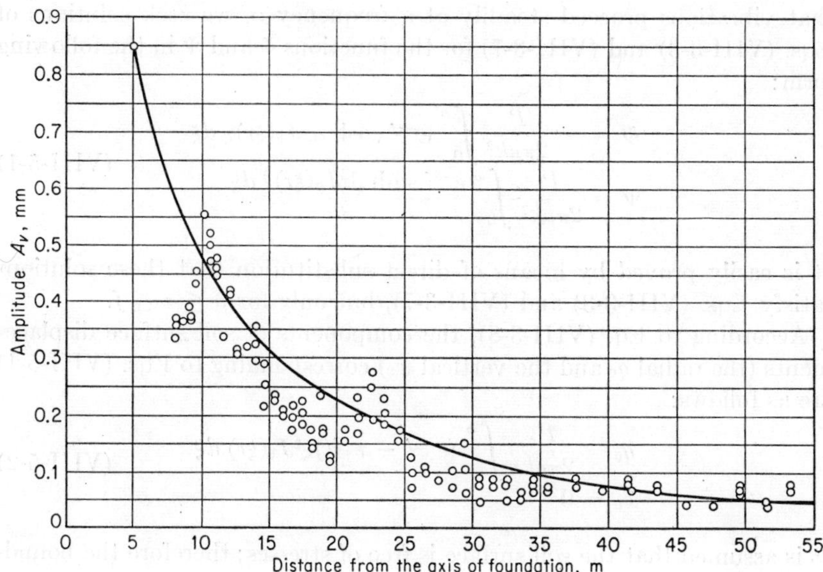

FIG. VIII-14. Comparison of measured and computed vertical components of vibration amplitudes.

0.73 mm. Taking this as the initial amplitude A_0, we can determine from Eq. (VIII-3-42) the amplitudes at distances $r > r_0$. For the given soil conditions, the value of the coefficient of energy absorption by soil may be taken as 0.04 m^{-1}. Then for the amplitudes at distances $r > 6$ m from the foundation axis, we have

$$A_r = \frac{1.79}{r} e^{-0.04(r-6)}$$

Figure VIII-14 presents a graph of amplitude changes with distance, computed from the above equation. The same figure shows measured amplitudes at different distances from the hammer foundation. If one does not consider the interference maximums and minimums in the experimentally obtained distribution of amplitudes, then a comparison of the graph of computed values with the measured values of amplitudes shows good agreement.

VIII-5. Dependence of Soil Vibration Amplitudes on the Depth of the Source of Waves

Foundations under engines, as well as other industrial sources of seismic waves, are usually located below the soil surface. Therefore it is interesting to investigate the influence of the depth of the source on the amplitudes of soil vibrations.

Let us consider as a source of waves a vertical exciting force acting at a depth f below the soil surface. We shall place the origin of coordinates on the surface, directing the z axis in the downward direction. Assuming that vibrations proceed steadily at a frequency ω, we seek solutions of Eqs. (VIII-3-3) and (VIII-3-7) for the functions Φ and Ψ in the following form:

$$\begin{aligned} \Phi &= -\frac{P}{2\pi\mu k^2}\int_0^\infty e^{-\alpha f}\cosh \alpha z J_0(\xi r)\xi\, d\xi \\ \Psi &= \frac{P}{2\pi\mu k^2}\int_0^\infty \frac{e^{-\beta f}\sinh \beta z J_0(\xi r)\xi\, d\xi}{\beta} \end{aligned} \qquad \text{(VIII-5-1)}$$

It is easily proved by means of direct substitution that these solutions satisfy Eqs. (VIII-3-3) and (VIII-3-7), but only for $0 \leq z \leq f$.

According to Eq. (VIII-3-8), the components of soil surface displacements (the radial q_0 and the vertical w_0) corresponding to Eqs. (VIII-5-1) are as follows:

$$\begin{aligned} \bar{q}_0 &= \frac{P_v}{2\pi\mu k^2}\int_0^\infty (e^{-\alpha f} - e^{-\beta f})\xi^2 J_1(\xi r)\, d\xi \\ \bar{w}_0 &= 0 \end{aligned} \qquad \text{(VIII-5-2)}$$

It is assumed that the soil surface is free of stresses; therefore the boundary conditions are:

$$z = 0 \qquad \tau_r = 0 \qquad \sigma_z = 0 \qquad \text{(VIII-5-2}a\text{)}$$

τ_r, the shear stress, and σ_z, the normal stress, are expressed through Φ and Ψ by the following equations:

$$\tau_r = \mu\left(\frac{\partial q}{\partial z} + \frac{\partial w}{\partial r}\right) = \mu\left(2\frac{\partial^2\Phi}{\partial r\partial z} + 2\frac{\partial^3\Psi}{\partial r\partial z^2} + k^2\frac{\partial\Psi}{\partial r}\right) \qquad \text{(VIII-5-3)}$$

$$\begin{aligned} \sigma_z = \lambda\theta + 2\mu\frac{\partial w}{\partial z} &= \lambda\left(\frac{\partial^2\Psi}{\partial r^2} + \frac{1}{r}\frac{\partial\Phi}{\partial r} + \frac{\partial^2\Phi}{\partial z^2}\right) \\ &\quad + 2\mu\left(\frac{\partial^2\Phi}{\partial z^2} + \frac{\partial^3\Psi}{\partial z^3} + k^2\frac{\partial\Psi}{\partial z}\right) \end{aligned} \qquad \text{(VIII-5-4)}$$

Substituting the values of Φ and Ψ which correspond to $z = 0$ into the right-hand parts of Eqs. (VIII-5-3) and (VIII-5-4), we find that, when

Eq. (VIII-5-3) satisfies the conditions of Eqs. (VIII-5-2*a*), the normal stresses do not equal zero, but are equal to

$$P\Big|_{z=0} = \frac{P_v}{2\pi k^2}\int_0^\infty [(k^2 - 2\xi^2)e^{-\alpha f} + 2\xi^2 e^{-\beta f}]J_0(\xi r)\xi\, d\xi \quad \text{(VIII-5-5)}$$

In order to obtain a solution for a surface free of normal stresses, it is necessary to superimpose on solutions (VIII-5-2) a solution obtained for a surface under the action of stresses of magnitudes equal to those of Eq. (VIII-5-5), but of opposite sign. For this purpose, we take, in Eq. (VIII-3-13),

$$P = \frac{P_v}{2\pi k}[(k^2 - 2\xi^2)e^{-\alpha f} + 2\xi^2 e^{-\beta f}]\xi\, d\xi \quad \text{(VIII-5-6)}$$

Integrating from 0 to ∞, we obtain the following expressions for the components of displacements of the soil surface produced by this force:

$$q_0^* = \frac{P_v}{2\pi\mu k^2} \int_0^\infty \frac{\xi^2(2\xi^2 - k^2 - 2\alpha\beta)[(k^2 - 2\xi^2)e^{-\alpha f} + 2\xi^2 e^{-\beta f}]J_1(\xi r)\, d\xi}{F(\xi)} \quad \text{(VIII-5-7)}$$

$$w_0^* = -\frac{P_v}{2\pi\mu}\int_0^\infty \frac{\alpha\xi[(k^2 - 2\xi^2)e^{-\alpha f} + 2\xi^2 e^{-\beta f}]J_0(\xi r)\, d\xi}{F(\xi)}$$

Thus the total displacements of soil at the soil surface equal

$$q_0 = \bar{q}_0 + q_0^* = \frac{P_v}{2\pi\mu} \int_0^\infty \frac{\xi^2[(2\xi^2 - k^2)e^{-\beta f} - 2\alpha\beta e^{-\alpha f}]J_1(\xi r)\, d\xi}{F(\xi)} \quad \text{(VIII-5-8)}$$

$$w_0 = \bar{w}_0 + w_0^* = -\frac{P_v}{2\pi\mu} \int_0^\infty \frac{\alpha\xi[2\xi^2 e^{-\beta f} - (2\xi^2 - k^2)e^{-\alpha f}]J_0(\xi r)\, d\xi}{F(\xi)} \quad \text{(VIII-5-9)}$$

For $f = 0$, Eqs. (VIII-5-8) and (VIII-5-9) coincide with Eqs. (VIII-3-14) and (VIII-3-15).

The integrals in the right-hand parts of Eqs. (VIII-5-8) and (VIII-5-9) and the integrals of Eqs. (VIII-3-14) and (VIII-3-15) are to a certain degree indeterminate; in order to single out their principal values, it is necessary to integrate in the area of the complex variable along the determinate contour. If, after determining the principal values of the integrals in Eqs. (VIII-5-8) and (VIII-5-9), we superimpose on this result free surface waves, then taking into account a temporary factor, we will obtain, for sufficiently large values of r, the following formulas for the displacement components q_0 and w_0 on the soil surface:

$$q_0 = -\frac{\varkappa P}{2\mu} H_f D_1(\varkappa r) e^{i\omega t}$$
$$w_0 = \frac{P}{2\mu} K_f D_0(\varkappa r) e^{i\omega t} \tag{VIII-5-10}$$

where the coefficients H_f and K_f are as follows:

$$H_f = -\frac{\varkappa}{F'(\varkappa)} (2\varkappa^2 - k^2) e^{-\beta_1 f} - 2\alpha_1 \beta_1 e^{-\alpha_1 f} \tag{VIII-5-11}$$

$$K_f = \frac{\alpha_1}{F'(\varkappa)} (2\varkappa^2 - k^2) e^{-\alpha_1 f} - 2\varkappa^2 e^{-\beta_1 f} \tag{VIII-5-12}$$

If $f = 0$, then $H_f = H_0$, $K_f = K_0$ [see Eq. (VIII-2-20)], and Eqs. (VIII-5-10) identically coincide with Eqs. (VIII-3-34) and (VIII-3-35) for the displacement components when the exciting force is located on the soil surface.

The author conducted special investigations to verify Eqs. (VIII-5-10). The soils on the site of the investigation consisted of water-saturated gray sands. Three foundations were built for these experiments; all had the same weight (W = 6.8 tons) and the same area in contact with soil (1.0 m^2), but they were placed at different depths. The base area of foundation 5 was placed 2.0 m below the soil surface, that of foundation 6 was 1.0 m below the soil surface, and that of foundation 7 was on the soil surface. The soil vibrations were measured on the surface of the sand layer; the overlying peat layer was removed.

During all these investigations, the foundations which served as sources of waves were subjected to vertical vibrations only, with the vibrator running at about 950 rpm. The average amplitude of forced vertical vibrations of the foundations was about 0.42 mm; soil vibration amplitudes recorded for other amplitudes of vibrations were reduced to this magnitude. Soil vibrations were measured by a vibrometer in a vertical longitudinal plane; in almost all cases only the vertical component of vibrations was recorded. The investigations were conducted on frozen soil.

Small circles plotted in Fig. VIII-15 represent experimental values of reduced double amplitudes of soil vibrations at varying distances from foundation 7; the dashed line is a computed curve plotted on the basis of the assumption that the wave is cylindrical and is characterized by damping. For a damping constant of 0.045 m^{-1}, the computed curve is in fairly good agreement with experimental values of amplitudes. Measurements were taken twice at each point: while moving the device away from and toward the foundation.

A section of the profile of some 7.5 m length was cleared of the peat layer in autumn. In the remainder the peat was removed immediately before measurement; therefore the sand was much less frozen here. The boundary between the two sections of the profile lies between points 11 and 12. In the first section (points 4 to 11) there is some regularity in the distribution of amplitudes; this is disturbed at point 12. There is a certain irregularity in the distribution of amplitudes of the second section.

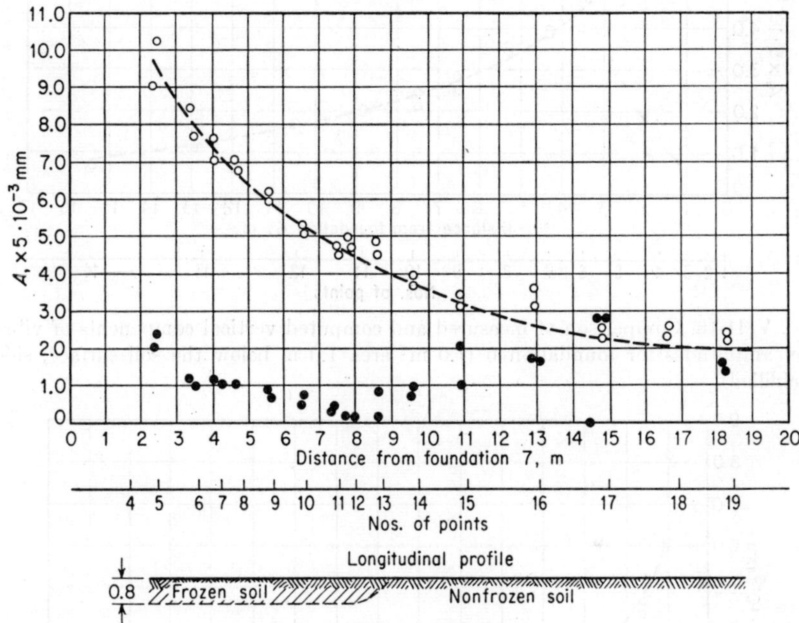

FIG. VIII-15. Comparison of measured and computed vertical components of vibration amplitudes for foundation 7 (1.0 m² area on soil surface).

Values of the horizontal component of soil vibrations are shown in Fig. VIII-15 by black circles. It is noteworthy that between points 4 and 11, where the soil was uncovered in the autumn, the amplitudes of the horizontal component of vibrations decrease fairly smoothly, but in the unfrozen section the changes in amplitude have a different character: with a decrease in distance from the source of waves, the amplitudes increase up to point 17 (at a distance of 14.9 m from the foundation); then, at points 18 and 19, the amplitudes decrease. It follows that during vertical vibrations of the foundation on a soil which was frozen to a certain depth, one wave is propagated through the soil. Its vertical component is much larger than its horizontal component. When this wave reaches the section of unfrozen soil, there appear supplementary waves which interfere with the principal wave and distort the distributions of amplitudes of both the vertical and (especially) the horizontal components of soil vibrations.

Figure VIII-16 presents a distribution curve of vertical vibration

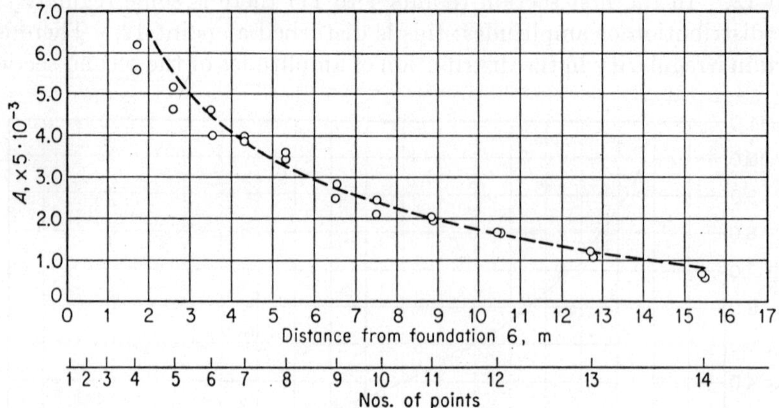

FIG. VIII-16. Comparison of measured and computed vertical components of vibration amplitudes for foundation 6 (1.0 m² area 1.0 m below the soil surface, sides backfilled).

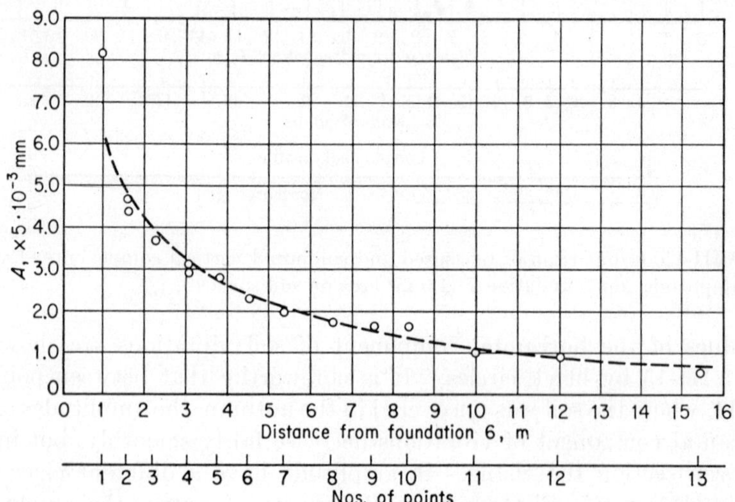

FIG. VIII-17. Comparison of measured and computed vertical components of vibration amplitudes for foundation 6 (1.0 m² area 1.0 m below the soil surface, *not* backfilled).

amplitudes of soil during the vertical vibration of foundation 6, whose base area in contact with soil lies at a depth of 1.0 m and whose sides are backfilled with soil. Figure VIII-17 presents an analogous curve for the same foundation, but for the case in which its sides were exposed along their whole length; the ground water was not pumped out of the excavation. The dashed curves indicate amplitudes obtained by computations taking into account a damping constant of 0.050 to 0.060 m^{-1}. These curves are in good agreement with the experimental data; in both cases the interference of waves is slightly noticeable. The peat was removed from the soil surface in the fall; therefore the soil froze to a depth of about 1 m.

Figure VIII-18 presents analogous experimental data on the distribution of waves from foundation 5, placed at a depth of 2 m and backfilled. The damping constant of soil was found to equal 0.064 m^{-1}. Figure VIII-19 gives experimental data on the distribution of vertical vibration

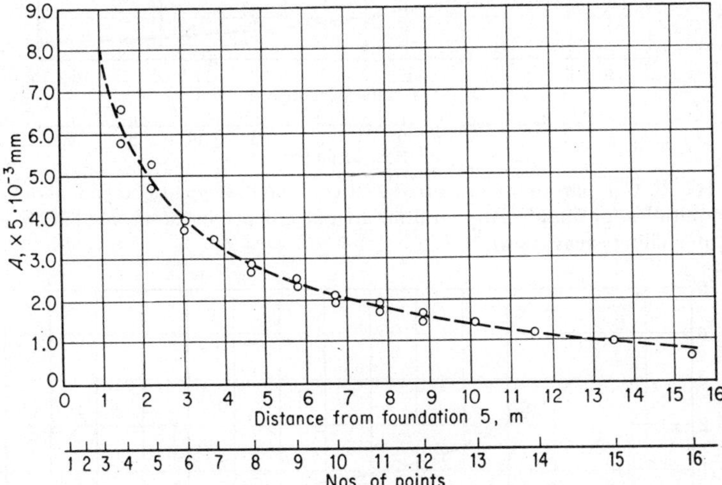

FIG. VIII-18. Comparison of measured and computed vertical components of vibration amplitudes for foundation 5 (1.0 m^2 area at a depth of 2.0 m and backfilled).

amplitudes of soil for the same foundation, but not backfilled and with water filling the excavation; in this case, the damping constant of soil was found to equal 0.060 m^{-1}. Figure VII-20 gives the results of measurements when the water was pumped out of the excavation and its level was kept constantly some 20 cm higher than the foundation area in contact with soil. In this case the damping constant was found to equal 0.121 m^{-1}; i.e., it was approximately two times larger than in all the foregoing cases.

The profile along which the propagation of waves from foundation 5 was measured is identical to that of foundation 6. It was cleared of the peat layer early in the fall; therefore at the time when measurements were started along this profile the soil was frozen to a depth of some 1 m.

If one is to exclude the influence of damping of vibrations on the propagation of waves from nonbackfilled foundation 5 when the water was pumped out of the excavation, then the average value of the damping constant for the other five measurements equals 0.058 m^{-1}.

In order to find out how the amplitudes of soil vibrations are affected by the depth of the foundation area in contact with soil, a comparison should be made of amplitude values obtained at points located at the

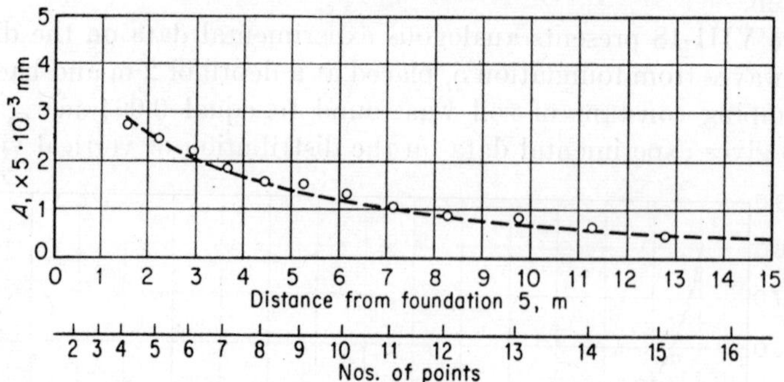

FIG. VIII-19. Comparison of measured and computed vertical components of vibration amplitudes for foundation 5 (1.0 m² area at a depth of 2.0 m, *not* backfilled and with water filling excavation).

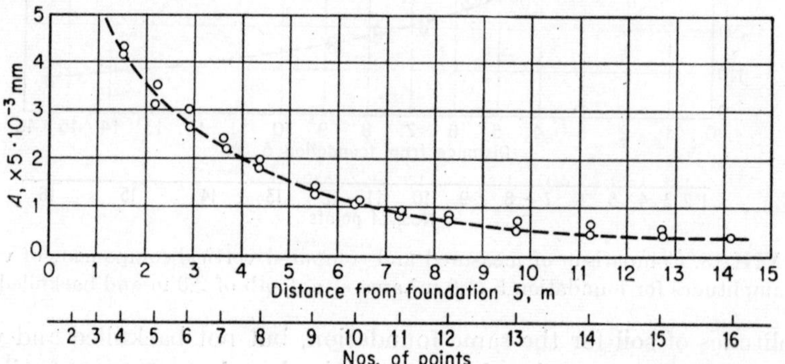

FIG. VIII-20. Comparison of measured and computed vertical components of vibration amplitudes for foundation 5 (1.0 m² area at a depth of 2.0 m, *not* backfilled and with water pumped out of excavation).

same distances from the foundations but at different depths. For this purpose, the graphs of the distribution of amplitudes at varying distances from the source of waves may be used. On the basis of these graphs, one can plot curves of the relationship between vibration amplitudes and depths of the foundation base contact area. These curves are given in Fig. VIII-21*a* for backfilled foundations and in Fig. VIII-21*b* for nonbackfilled foundations.

Let us denote by A_0 and A_f the amplitudes of soil vibrations when the source is placed on the soil surface and the foundation is placed at a depth f. According to Eqs. (VIII-2-20) and (VIII-5-12), we have

$$\frac{A_f}{A_0} = \left(2\frac{\varkappa^2}{k^2} - 1\right)e^{-\alpha_1 f} - 2\frac{\varkappa^2}{k^2}e^{-\beta_1 f} \qquad \text{(VIII-5-13)}$$

We shall compute the ratio A_f/A_0 for the conditions of the experiment.

According to the graphs of measured changes in amplitudes with distance, the velocity of transverse wave propagation for the given soil conditions is about $b = 100$ m/sec. Taking for sand the Poisson ratio $\nu = 0.35$, we obtain from Eq. (VIII-1-12) the velocity of longitudinal

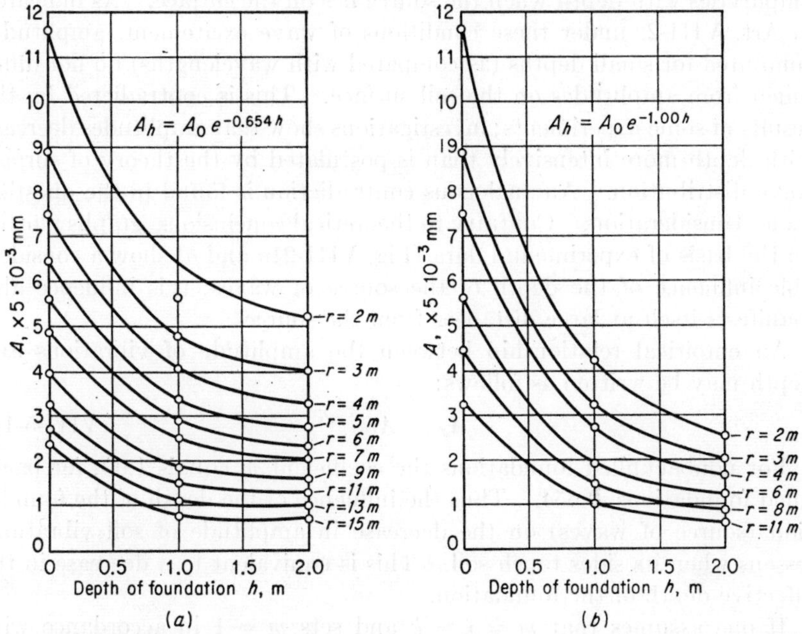

FIG. VIII-21. Effect of depth of foundation on the vibration amplitudes at varying distances r from the foundation: (*a*) backfilled foundation; (*b*) nonbackfilled foundation.

wave propagation $a = 210$ m/sec. The velocity of surface wave propagation may be taken as $c = 0.93b = 93$ m/sec. The foundation (source of waves) vibrates at 950 oscillations per minute; therefore the frequency of propagating waves $\omega = 100$ sec^{-1}. Thus we have

$$\varkappa^2 = \frac{\omega^2}{(0.93)^2 b^2} = 1.15 \text{ m}^{-2}$$

$$k^2 = \frac{\omega^2}{b^2} = 1.00 \text{ m}^{-2}$$

$$h^2 = \frac{\omega^2}{a^2} = 0.25 \text{ m}^{-2}$$

$$\alpha_1 = \sqrt{\varkappa^2 - h^2} = 0.95 \text{ m}^{-1}$$

$$\beta_1 = \sqrt{\varkappa^2 - k^2} = 0.39 \text{ m}^{-1}$$

Consequently for the soil conditions under consideration, we obtain, according to Eq. (VIII-5-13),

$$\frac{A_f}{A_0} = 1.33e^{-0.95f}$$

Calculations from this formula show that lowering the foundation 2 m (a small change compared to the wavelength) will not essentially influence the amplitudes of vibrations of the soil surface. This conclusion has much in common with theoretical data concerning the distribution of amplitudes with depth when the source lies on the surface. As indicated in Art. VIII-2, under these conditions of wave excitement, amplitudes computed for small depths (as compared with wavelengths) do not differ much from amplitudes on the soil surface. This is contradicted by the results of some experiments; investigations show that amplitudes decrease with depth more intensively than is postulated by the theory of surface wave distribution. An analogous contradiction is found in the question under consideration. Contrary to theoretical conclusions, graphs plotted on the basis of experimental data (Fig. VIII-21*a* and *b*) show a considerable influence of the depth of the source of waves; this influence also manifests itself at large distances from the source.

An empirical relationship between the amplitude of vibrations and depth may be written as follows:

$$A_f = A_0e^{-mf} \tag{VIII-5-14}$$

For nonbackfilled foundations the coefficient m equals 1.00; for backfilled foundations, 0.654. Thus the influence of the depth of the foundation (source of waves) on the decrease in amplitude of soil vibrations lessens when its sides touch soil. This is equivalent to a decrease in the effective depth of the foundation.

If one assumes that $m \approx \varkappa \approx k$ and sets $m = 1$ in accordance with experimental data for nonbackfilled foundations, then the velocity of transverse waves will equal 100 m/sec; i.e., it will be close to the magnitude obtained from the analysis of the graphs of relationship between vibration amplitudes and distances. Then Eq. (VIII-5-14) may be written as follows:

$$A_f = A_0e^{-(2\pi/L_b)f} \tag{VIII-5-15}$$

This expression coincides with the theoretical formula if the vibrating foundation is considered not as an exciting force acting at a depth f below the surface, but as a local exciter of soil vibrations. These vibrations are determined by the following values of Φ and Ψ:[27]

$$\Phi = 2\int_0^\infty \frac{\cosh \alpha z}{\alpha} e^{-\alpha f} J_0(\xi)\xi\, d\xi$$

$$\Psi = 0$$

Depths of machine foundations usually are such that the ratio f/L_b is considerably smaller than unity. For example, for low-frequency machines (reciprocating compressors, diesels, etc.) the depth of the foundation usually does not exceed 4.0 to 5.0 m, while in soft soils the lengths of waves propagated from these foundations are usually of the order of 40 to 100 m; consequently, for these engines f/L_b will be of the order of 0.10 to 0.05, and

$$e^{-2\pi f/L_b} = 0.50 \text{ to } 0.75$$

For high-frequency sources (for example, hammer foundations) the depths of foundations are approximately of the same order, but the waves propagated are considerably shorter. For soft soils, the lengths of waves propagated by hammers are 6 to 12 m; for these values of L_b, the ratio f/L_b will be of the order 0.3 to 0.8, and

$$e^{-2\pi f/L_b} = 0.15 \text{ to } 0.007$$

Thus it is clear that even a small increase in the depth of the source has considerable effect on the decrease of amplitudes of soil surface vibrations, especially for high-frequency sources.

VIII-6. The Spreading of Surface Waves in Layered Soils

a. The Influence of a Layered Soil Structure on the Distribution of Surface Waves. Many soils are formed of layers characterized by different mechanical properties. The layered structure of soil is responsible for many peculiarities in wave propagation which cannot be explained if the soil is considered to be a homogeneous body.

For example, the layered structure of soil may essentially affect the propagation of surface waves. Theoretical analyses can handle mainly cases in which a soil layer overlies a mass of homogeneous soil. The influence of a layered structure on the propagation of surface waves has been studied by many authors (see, for example, Ref. 15). As a result of these investigations it was established that in the process of surface-wave propagation in a layered medium, there is a dispersion of waves; thus the velocity of wave propagation is not constant and is determined not only by the elastic and inertial properties of the medium, but also by the lengths of propagating waves and their frequencies.

In addition, it was established that the presence of a layer overlying a homogeneous soil mass and having properties different from this layer does not produce the monotonous change of amplitude with depth which takes place when surface waves are propagated in a homogeneous semi-infinite mass. If such a layer is present, graphs of the distribution of amplitudes with depth may have several maximums and minimums whose values depend on the ratio between the wavelength and the layer thickness. For a certain ratio between these values, the possibility is not excluded that the amplitude may increase in the layer with an increase in depth. All of the above may partly explain the discrepancies between experimental and theoretical graphs of the distribution of amplitudes, as indicated in Art. VIII-2.

To simplify all calculations, let us confine ourselves to establishing the influence of the layer on the distribution of free plane surface waves.

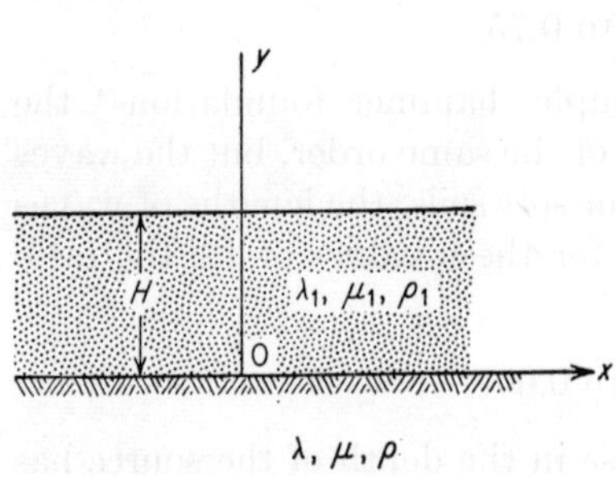

FIG. VIII-22. Coordinate system for Eqs. (VIII-6-1) and Fig. VIII-23.

We assume that the elastic properties and densities of the layer and the underlying soil mass are determined respectively by the constants λ_1, μ_1, ρ_1 and λ, μ, ρ. Let us place the origin of coordinates on the bottom surface of the layer (Fig. VIII-22) and direct the y axis vertically upward. We take the solutions of Eqs. (VIII-2-4) for the underlying soil mass as follows:

$$\begin{aligned} \Phi &= Ae^{\alpha y}e^{i\xi x} \\ \Psi &= Be^{\beta y}e^{i\xi x} \end{aligned} \qquad \text{(VIII-6-1)}$$

According to Eq. (VIII-2-2) the components of displacements corresponding to these values of Φ and Ψ are as follows:

$$\begin{aligned} u &= (i\xi Ae^{\alpha y} + \beta Be^{\beta y})e^{i\xi x} \\ v &= (\alpha Ae^{\alpha y} - i\xi Be^{\beta y})e^{i\xi x} \end{aligned} \qquad \text{(VIII-6-2)}$$

where α and β are determined by Eqs. (VIII-2-7); A and B are arbitrary constants.

For waves propagating in the layer, the solutions of Eqs. (VIII-2-4) will be as follows:

$$\begin{aligned} \Phi_1 &= (C \cosh \alpha_1 y + D \sinh \alpha_1 y)e^{i\xi x} \\ \Psi_1 &= (E \cosh \beta_1 y + F \sinh \beta_1 y)e^{i\xi x} \end{aligned} \qquad \text{(VIII-6-3)}$$

where

$$\alpha_1{}^2 = \xi^2 - h_1{}^2 \qquad \beta_1{}^2 = \xi^2 - k_1{}^2$$
$$h_1 = \frac{\omega}{a_1} \qquad k_1 = \frac{\omega}{b_1} \qquad \text{(VIII-6-4)}$$

and a_1 and b_1 are propagation velocities of longitudinal and transverse waves in the layer; it is clear that h_1 and k_1 are the reciprocals of the wavelengths.

For the displacement components in the layer we obtain

$$\begin{aligned} u_1 &= i\xi(C \cosh \alpha_1 y + D \sinh \alpha_1 y)e^{i\xi x} \\ &\qquad + \beta_1(E \sinh \beta_1 y + F \cosh \beta_1 y)e^{i\xi x} \\ v_1 &= \alpha_1(C \sinh \alpha_1 y + D \cosh \alpha_1 y)e^{i\xi x} \\ &\qquad - i\xi(E \cosh \beta_1 y + F \sinh \beta_1 y)e^{i\xi x} \end{aligned} \qquad \text{(VIII-6-5)}$$

The arbitrary constants $A, B, \ldots, F$ should be selected in such a way as to satisfy the boundary conditions of the problem.

It follows from the conditions of continuity of stress and deformation that deformations and stresses in the layer equal deformations and stresses in the underlying soil at their boundary. When $y = 0$, these contact conditions may be written as follows:

$$\begin{aligned} u &= u_1 \qquad & \sigma_y &= \sigma_{y1} \\ v &= v_1 \qquad & \tau_{yx} &= \tau_{yx1} \end{aligned} \tag{VIII-6-6}$$

It is assumed that the upper surface of the layer is free of stresses; therefore boundary conditions for $y = H$ will be

$$\sigma_y = 0 \qquad \tau_{yx} = 0 \tag{VIII-6-7}$$

The components of stresses are expressed through Φ and Ψ by Eqs. (VIII-2-10).

Using the solutions of Eqs. (VIII-6-2) and (VIII-6-5) and the boundary conditions of Eqs. (VIII-6-6) and (VIII-6-7), we obtain six equations relating the constants ξ, A, B, C, D, E, and F:

$$\begin{aligned} i\xi A + \beta B &= i\xi C + \beta_1 F \\ \alpha A - i\xi B &= \alpha_1 D - i\xi E \\ (2\xi^2 - k^2)A - 2i\xi\beta B &= \frac{\mu_1}{\mu}[(2\xi^2 - k_1^2)C - 2i\xi\beta_1 F] \\ 2i\xi\alpha A + (2\xi^2 - k^2)B &= \frac{\mu_1}{\mu}[(2\xi^2 - k_1^2)E + 2i\xi\alpha_1 D] \end{aligned} \tag{VIII-6-8}$$

$$(2\xi^2 - k_1^2)(C \cosh \alpha_1 H + D \sinh \alpha_1 H) - 2i\xi\beta_1(E \sinh \beta_1 H + F \cosh \beta_1 H) = 0$$

$$(2\xi^2 - k_1^2)(E \cosh \beta_1 H + F \sinh \beta_1 H) + 2i\xi\alpha_1(C \sinh \alpha_1 H + D \cosh \alpha_1 H) = 0$$

The solution of the system of Eqs. (VIII-6-8) for the constants $A, \ldots, F$ will be other than zero only if the determinant equals zero. Hence we obtain an equation for the determination of the velocity of propagation of surface waves in a layer overlying a soil mass:

$$\Omega(\xi) = 0 \tag{VIII-6-9}$$

In the general case this equation is very complicated; it contains irrational expressions and is very difficult to solve. It is, however, solved[40] for the case in which the densities of the soil mass and the overlying layer do not differ much ($\rho = \rho_1$); in addition, both the layer and the underlying soil mass must satisfy the Poisson hypothesis ($\nu = 0.25$). Figure VIII-23 presents graphs of the relationship between c_a/c and the ratio between layer thickness and wavelength for several values of μ/μ_1 (c_a is the velocity of surface waves with overlying layer, and c is the velocity of the same waves without overlying layer). The wavelength depends on the frequency of propagating waves; therefore the graphs show the dependence of the propagation velocity on the frequency in a case in which waves propagate in a layer of constant thickness.

It is seen from Fig. VIII-23 that with a decrease in the wave frequency (i.e., with a decrease in H/L) the velocity of surface waves increases. A particularly intensive change in velocity depending on changes in H/L takes place when $H/L < 0.5$. For these values of H/L, the velocity of waves increases approximately in proportion to the wavelength, i.e.,

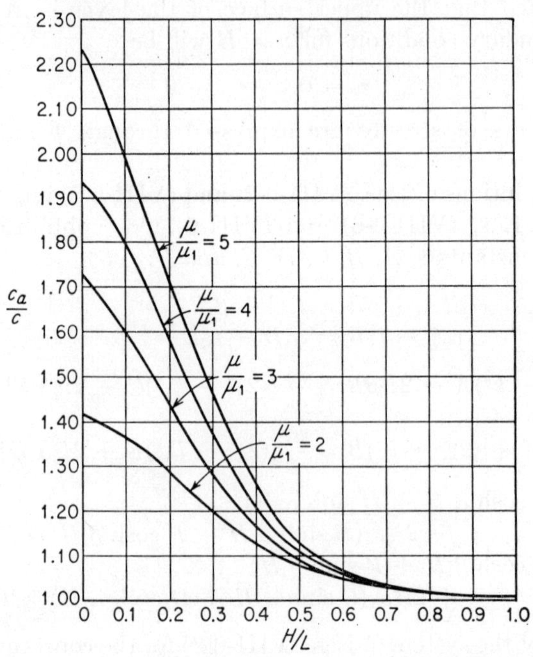

Fig. VIII-23. Effect of the ratio of the surface layer thickness H to the wavelength L on the velocity ratio c_a/c for different ratios of the Lamé coefficients μ (of the underlying soil mass) to μ_1 (of the upper layer).

inversely proportionally to the frequency. The coefficient of proportionality depends on the ratio μ/μ_1. The larger the difference between the mechanical properties of the overlying layer and the underlying soil mass, the larger the value of the coefficient of proportionality and, consequently, the larger the velocity of wave propagation. The waves are propagating mostly in the overlying layer, and the influence of the underlying soil on the velocity of wave propagation is insignificant when the frequency of vibrations is so high that the lengths of propagating waves are equal to or smaller than the layer thickness.

As indicated in Art. VIII-2, when waves are propagated in a homogeneous, elastic, semi-infinite mass, the ratio between the amplitudes of the horizontal and vertical components of displacement of particles on the soil surface remains constant and equals 0.6811. If one sets $y = H$ in Eqs. (VIII-6-5) and determines the ratio between the horizontal and vertical components of vibrations, an expression is obtained which shows that the ratio depends on H/L.

Figure VIII-24 presents graphs of the relationship between u_a/v_a on the soil surface and L/H; it is seen from these graphs that for short waves ($L/H \geq 1$) the ratio between the horizontal and vertical components of vibration does not differ much from a value typical for waves propagating in a homogeneous semi-infinite mass. Within a range of frequencies

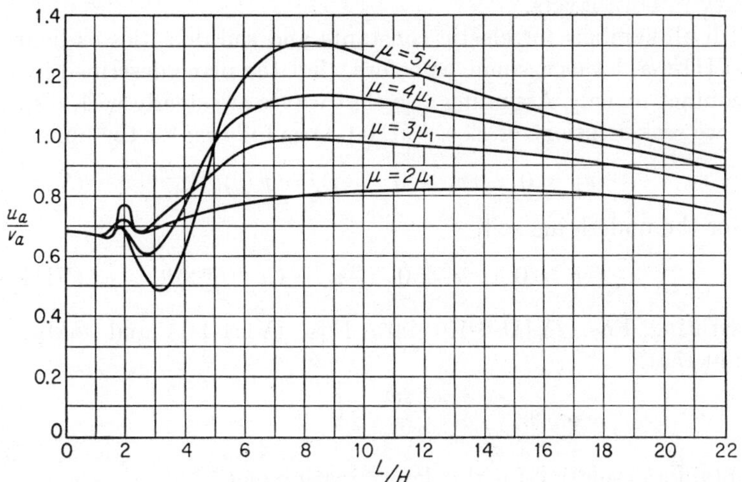

FIG. VIII-24. Effect of the ratio of the wavelength L to the upper layer thickness H on the ratio of horizontal u to vertical v displacements on the surface of the upper layer.

corresponding to wavelengths approximately two to four times larger than the layer thickness, the ratio u_a/v_a becomes smaller than that for waves propagating in a soil that is not layered; i.e., the ellipsis of vibrations becomes elongated in the vertical direction. When the frequency decreases so that the wavelength is four to five times larger than H, the relative value of the horizontal component of vibrations increases and the ellipsis of vibrations expands in the horizontal direction; for $\mu_1 > 3\mu$, at a certain value of the frequency, it becomes a circle.

The foregoing data concern the influence of one layer, overlying a soil mass with properties different from its own, on surface waves of the Raleigh type. An investigation of wave propagation in a many-layered soil is much more difficult, and not much has been done in this area. However, there is no doubt that the presence of several soil layers showing a diversity in mechanical properties may influence considerably the properties of propagating waves. A theoretical treatment of problems of wave propagation in many-layered soils is too difficult; therefore experimental investigations should be conducted for the solution of problems of this type.

b. Transverse Surface Waves in Two-layered Soil Systems. A layered soil structure may change the laws governing wave propagation in an infinite or finite solid body; in addition, the layered system may cause the formation of new waves characterized by specific properties. It was found, for example, that where a soil layer is underlaid by a more rigid base, specific transverse waves may occur in the soil, with the same velocity in both layers.

With all symbols for elastic constants and soil densities as defined in Art. VIII-6-a, let us assume, as before, that the plane described by $t = 0$ is a contact plane. Assuming also that waves are steady with respect to time, we seek solutions of Eqs. (VIII-1-1) as follows: for the upper layer:

$$u_1 = 0 \qquad v_1 = 0 \qquad w_1 = F(y)e^{i(\xi x - \omega t)} \tag{VIII-6-10}$$

and for the underlying soil:

$$u = 0 \qquad v = 0 \qquad w = Ce^{\beta y + i(\xi x - \omega t)} \tag{VIII-6-11}$$

Substituting Eqs. (VIII-6-10) into Eqs. (VIII-1-1) and taking into account that

$$\frac{\partial w}{\partial z} = 0$$

we obtain an equation for the determination of $F(y)$:

$$\frac{d^2F}{dy^2} = (k_1^2 - \xi^2)F = 0 \tag{VIII-6-12}$$

As before, $\qquad k_1^2 = \dfrac{\omega^2}{b_1^2}$

(b_1 is the velocity of transverse waves in the upper layer). The solution of Eq. (VIII-6-12) is as follows:

$$F(y) = A \sin \alpha_1 y + B \cos \alpha_1 y \tag{VIII-6-13}$$

where $\qquad \alpha_1^2 = k_1^2 - \xi^2$

Substituting Eqs. (VIII-6-11) into (VIII-1-1) for the underlying soil, we find that this solution will satisfy the equations of motion if

$$\beta^2 = \xi^2 - k^2 \tag{VIII-6-14}$$

where $\qquad k^2 = \dfrac{\omega^2}{b^2}$

and b is the velocity of transverse waves in the underlying soil. The requirements of continuity of deformations and stresses provide us with the boundary and contact conditions; therefore for $y = 0$, it is necessary that

$$\begin{aligned} w &= w_1 \\ \tau_{yz} &= \tau_{yz1} \end{aligned} \tag{VIII-6-15}$$

In addition, since the upper surface of the overlying soil layer is not subjected to the action of stresses, the following condition should be also observed:† If $v = H$,

$$\tau_{yz1} = 0 \tag{VIII-6-16}$$

Since

$$\tau_{yz1} = \mu \frac{\partial w_1}{\partial y} = \mu_1 \alpha_1 (A \cos \alpha_1 y - B \sin \alpha_1 y) e^{i(\xi x - \omega t)}$$

we obtain, according to Eq. (VIII-6-16),

$$A \cos \alpha H - B \sin \alpha H = 0 \tag{VIII-6-17}$$

According to Eqs. (VIII-6-15) we have

$$B = C \qquad \mu_1 \alpha A = \mu \beta C \tag{VIII-6-18}$$

Excluding from Eqs. (VIII-6-17) and (VIII-6-18) the constants A, B, and C, we obtain an equation for determining the constant ξ:

$$\tan \alpha H = \frac{\mu \beta}{\mu_1 \alpha} \tag{VIII-6-19}$$

In order that the wave amplitudes in the underlying soil may decrease with depth, the real part of coefficient β should be positive. To satisfy this condition, the following interrelationship should exist:

$$\xi > k \tag{VIII-6-20}$$

But

$$\xi = \frac{k_b}{c_b} \tag{VIII-6-21}$$

where c_b is the velocity of transverse wave propagation. Substituting this expression for ξ into the left-hand part of Eq. (VIII-6-20), we obtain

$$c_b < b \tag{VIII-6-22}$$

i.e., the velocity of propagation of transverse surface waves is lower than the velocity of propagation of transverse waves in the underlying soil. It is easy to prove that $c_b > b_1$. As a matter of fact, if we assume that

† The equality of other stress component magnitudes, i.e., of τ_{yx} and σ_y, is satisfied automatically, because for both the upper layer and the underlying soil, the displacement components in the direction of the x and y axes equal zero.

$c_b < b_1$, then according to (VIII-6-21) we would obtain

$$\frac{\omega}{\xi} < b_1 \qquad \text{(VIII-6-23)}$$

or

$$k_1 < \xi \qquad \text{(VIII-6-24)}$$

Then $\alpha^2 < 0$, which would be possible if α were an imaginary value. For example, let us assume that $\alpha = i\gamma$. Substituting this expression into Eq. (VIII-6-19), we obtain:

$$\tan h(\gamma H) = -\frac{\mu\beta}{\mu_1\alpha} < 0$$

which is impossible; hence $c_b > b_1$.

Thus it is clear that the transverse waves under discussion may appear and propagate only when $b > b_1$. Since the densities ρ and ρ_1 usually do not differ much from one another, the condition governing the propagation of surface transverse waves will be as follows:

$$\mu > \mu_1 \qquad \text{(VIII-6-25)}$$

i.e., the underlying soil should have a higher elastic rigidity in shear than the overlying layer. The limiting values of propagation velocity of the waves under consideration will be

$$b_1 < c_b < b$$

We now transform Eq. (VIII-6-19); taking into account that $\mu = b^2\rho$ and $\mu_1 = b_1{}^2\rho$, we have

$$\tan H = \frac{b^2}{b_1{}^2}\frac{\rho}{\rho_1}\frac{\beta}{\alpha} \qquad \text{(VIII-6-26)}$$

Substituting into Eq. (VIII-6-26) the expressions for β and α,

$$\beta = \sqrt{\xi^2 - k_1{}^2} = \omega\sqrt{\frac{1}{c_b{}^2} - \frac{1}{b^2}}$$

$$\alpha = \sqrt{k_1{}^2 - \xi^2} = \omega\sqrt{\frac{1}{b_1{}^2} - \frac{1}{c_b{}^2}}$$

and assuming that $\rho = \rho_1$, after various transformations we obtain Eq. (VIII-6-26) in the following form:

$$\tan 2\pi\frac{H}{L}\sqrt{\frac{c_b{}^2}{b_1{}^2} - 1} = \frac{b^2}{b_1{}^2}\sqrt{\frac{1 - c_b{}^2/b^2}{c_b{}^2/b_1{}^2 - 1}} \qquad \text{(VIII-6-27)}$$

This equation establishes the dependence of the velocity c_b of the surface transverse waves not only on b and b_1, but also on the ratio between the layer thickness and the length of propagating waves. It follows that the roots of Eq. (VIII-6-27) depend on the frequency of the waves.

The expression governed by tan in Eq. (VIII-6-27) is always positive, but, depending on the magnitude of H, it may attain any value. For example, it may equal $\pi/2$ or may even be larger, in consequence of which the tangent may acquire negative values; but the right-hand part of Eq. (VIII-6-27) will always remain positive. Therefore all the values of αH lying within the range

$$\frac{2n-1}{2}\pi < \alpha H < n\pi \qquad n = 1, 2, 3, \ldots$$

should be eliminated. As a consequence of the periodicity of the tangent, Eq. (VIII-6-27) may be written as follows:

$$2\pi \frac{H}{L}\sqrt{\frac{c_b^2}{b_1^2} - 1} = k\pi + \arctan \frac{b^2}{b_1^2}\sqrt{\frac{1 - c_b^2/b^2}{c_b^2/b_1^2 - 1}} \qquad \text{(VIII-6-28)}$$

where $k = 0, 1, 2, 3, \ldots$.

Hence it follows that the dispersion of the waves under consideration is determined by a multiterm expression; consequently, theoretically there are several curves of dispersion which correspond to each set of values of b, b_1, and H. Figure VIII-25 presents a graph of wave dispersion. This graph gives the relationship between c_b/b_1 and H/L for $0 < H < \pi/2$. It is seen that surface transverse waves, as in the case of waves of the Raleigh type,† are characterized by a normal dispersion; i.e., their velocity increases with an increase in wavelength.

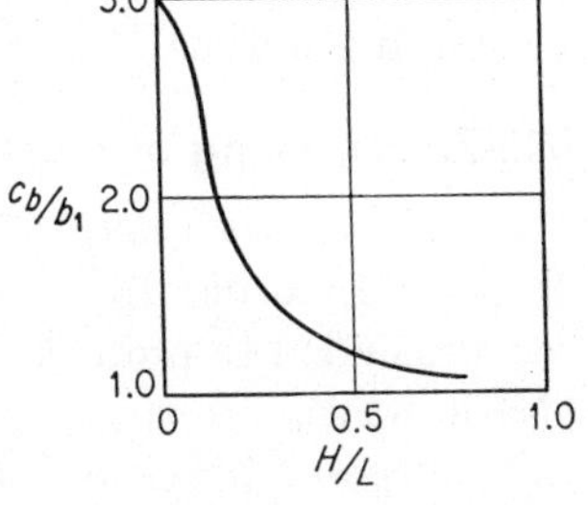

Fig. VIII-25. Effect of the ratio of the upper layer thickness H to the wavelength L on the ratio of transverse wave velocities in the upper layer (c_b at its surface and b_1 within the layer).

Let us consider the nature of changes in wave amplitudes with depth, confining ourselves to cases in which $y < H$, i.e., considering points located only in the overlying layer.

The distribution of amplitudes along the depth of the overlying layer depends essentially on the ratio between layer thickness and wavelength. In the general case the amplitudes are not distributed in a monotonous manner, but form nodes and protuberances. In the nodes the amplitudes of vibrations equal zero; therefore by equating the right-hand part of Eq. (VIII-6-13) to zero, we obtain an equation for the determination of the location of the node line along the depth of the overlying layer:

$$A \sin \alpha y_0 + B \cos \alpha y_0 = 0$$

or

$$\tan \alpha y_0 = -\frac{B}{A}$$

† In the presence of a soil layer overlying a soil mass characterized by different properties.

According to Eq. (VIII-6-17) we have

$$\frac{A}{B} = \cot \alpha H$$

hence
$$\tan \alpha H = -\cot \alpha y_0 = \tan\left(\frac{\pi}{2} + \alpha y_0\right)$$

From the above expression we obtain

$$\alpha y_0 = \alpha H - \frac{\pi}{2} \qquad \text{(VIII-6-29)}$$

Thus it is seen that nodes in the distribution of amplitudes may appear along the depth of the overlying layer only when $\alpha H > \pi/2$. The depth of the node, computed from the contact surface, equals

$$y_0 = H - \frac{\pi}{2\alpha}$$

Specifically, when $\alpha H = \pi/2$, $y_0 = 0$ and the surface transverse waves are propagated only in the overlying layer and do not penetrate into the underlying soil mass.

VIII-7. Screening of Elastic Waves Propagating through Soil

a. Elementary Concepts of Physics concerning the Diffraction of Elastic Waves. In order to decrease vibrations of structures it is sometimes recommended to protect the structure from the influence of the energy of elastic waves propagating in soils. The following methods are used for this purpose: provision of backfilled or open trenches, employment of sheetpiling or barriers of special design, and provision of compacted zones. There is a widely shared opinion that the best effect is obtained by placing a barrier around the source of vibrations. For example, if it is necessary to protect the walls of a forge shop from the harmful influence of waves propagating from a hammer foundation, it is recommended to place barriers around the foundation. If a vibrating diesel foundation induces considerable vibrations of a structure located at a distance of even several tens of meters from it, then it is also recommended to place a barrier around the source of the waves. Only when the source of vibrations is moving (transport) is it recommended to provide barriers near the receiver.

Experience with various types of barriers has shown that they are often of no use at all, or their effect is very small. However, cases are on record in which barriers were very effective. It appears that the use of sheetpiling, trenches, etc., is not always effective; these installations are of use only under certain conditions rarely encountered in practice.

This is explained by the nature of the propagation of elastic waves encountering an obstacle in the form of a screen. In such cases the concept of "wave beams" cannot be used, since the propagation is no longer linear. In the presence of an obstacle such as a screen (or a slit in a screen) the propagation of elastic waves is accompanied by a phenomenon known as the diffraction of waves. It appears that in cases in which barriers were of no use, designers did not take into account elementary features of the theory of wave propagation in the presence of a screen.

If plane steady waves propagating over the earth's surface meet a screen placed parallel to the wave front, then (in the case of linear wave propagation) a screened zone ("shadow") is formed behind the screen, in which no vibrations are observed. There is a sharp boundary between this screened zone and the area of wave propagation behind the screen. It looks as if vibrations are interrupted at this boundary. At the same time, the character of wave propagation behind the screen remains the same and does not depend on the wavelength or the screen width.

As a matter of fact, the character of propagation changes after the waves have passed the screen. For example, no sharp decrease is observed in the amplitude of vibrations at the boundaries of the screened zone. The amplitudes decrease as the waves approach the screened zone; inside the zone they decrease in such a way that the sharp contour of the zone vanishes. The depth of the screened zone is limited; as the distance from the screen grows, the waves penetrate more and more into it. At some distance from the screen, the waves which were propagated separately in areas to the right and left of the zone merge. After that the process of wave propagation basically does not differ from propagation where no screen is present. The depth of the screened zone is finite; it increases with an increase in the ratio of screen length to wavelength. If screen length is large in comparison with wavelength, then the nature of the propagation approaches that of linear propagation.

As the screen dimensions decrease in relation to the wavelength, the depth of the screened zone decreases. If the screen dimensions are small in comparison with the length of the propagating waves, no screened zone at all is formed, or else this zone is very small. Then the screen just dissipates the waves falling upon it and partly reflects them.

The phenomenon of wave propagation in the screened zone represents one of the special features of wave diffraction. However, the phenomenon of diffraction is not limited to the distortion of the boundaries of the screened zone; it also changes the nature of the distribution of amplitudes. At the zone boundary the amplitudes do not change in jumps as they should in the case of linear propagation of waves, but, as indicated above, the amplitudes change continuously, and the character of the changes in amplitudes behind the screen may be very complicated. As the distance from the screen grows, the average amplitude of the wave within the screened zone increases and, at some distance from the screen, attains the amplitude of a wave that has not been diffracted.

The screen creates a screened zone of a certain finite depth; therefore the screen affects the distribution of amplitudes over the surface only within the screened zone. The largest effect of the screen is observed directly behind it. As the distance from the screen grows, the propagation of waves in the area behind it increases, and the effect of the screen on the amplitudes of vibrations gradually decreases.

Thus, if one evaluates the effect of the screen on the basis of relative changes in amplitudes,† then the following relationship may be established for the propagation of a plane undamped wave: as the distance from the screen grows, the relative change in amplitude continuously increases from a minimum in the section directly behind the screen to unity at points located at distances larger than the depth of the screened zone. It appears that the amplitude of soil vibrations directly behind the screen depends mainly on the relationship between the screen dimensions and the length of the propagating waves. As the screen dimensions increase in comparison with the wavelength, the initial value of the ratio between the amplitudes decreases; if the screen dimensions are large in comparison with the wavelength, then the waves directly behind the screen will not get into the central portion of the screened zone, and consequently the amplitudes of soil vibrations there will equal zero.

If the waves were propagating only along the surface and not penetrating into the soil, then, under conditions of wave diffraction, it would be possible to isolate the source of waves. For this purpose, it would suffice to place an enclosing screen around the source. However, even surface waves penetrate to depths on the order of one wavelength; therefore the diffraction phenomenon is observed not only on the surface, but in depth as well. The placing of an enclosing barrier around a source would be accompanied by the creation of a continuous screened zone behind it only on the condition that the barrier cut through the entire soil zone in which waves are propagating.

It is practically impossible to satisfy this condition. Therefore in the design of a screen (barrier), it should be remembered that the dimensions of the screened zone formed by the barrier depend not only on the dimensions of the barrier in plan, but also on its depth. Thus, if one provides a barrier with a large length along the surface but with a small depth in comparison with the wavelength, its effect will be negligible.

The foregoing considerations concerning the diffraction of elastic waves by a screen are based exclusively on some elementary concepts borrowed from experimental investigations of diffraction of sound waves in air and of waves propagating in water.

† The relative change in amplitudes is defined as the ratio of the amplitudes at a point when the screen is present to the amplitudes at the point when no screen is present.

There is no doubt that the diffraction of elastic waves in soil is somewhat different from the diffraction of sound waves or waves traveling through water. The differences could be established, especially quantitatively, if the theory of diffraction of elastic waves by means of a screen were worked out. However, no theory is available now which would give a basis for quantitative computations. Several mathematical investigations have appeared in this field, but solutions obtained on the basis of these investigations are of a form which does not permit the practical evaluation of the influence of the screen on the decrease in soil vibrations.[26,42,43]

Obstacles in the direction of wave propagation produce not only the diffraction of waves, but also their dissipation and reflection. The dissipation may occur even if the dimensions of the obstacle are very small in comparison with the wavelength. This phenomenon is used in special branches of physics, particularly in the study of minute particles within the field of light or sound waves.

However, in comparison with other processes taking place in the soil when elastic waves are propagating (for example, by comparison with energy absorption by the soil), it appears that dissipation is of secondary importance for the protection of structures from waves in soil. Nevertheless, this phenomenon, combined with the reflection of waves, may contribute somewhat to the efficiency of a barrier, although the amount of energy dissipated and reflected by the obstacle apparently makes up only a small part of the total energy which produces vibrations of structures.

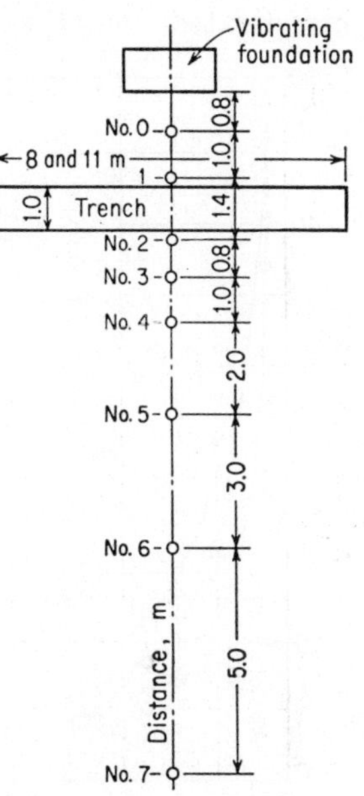

Fig. VIII-26. Plan of experiment on wave-screening action of trench.

b. Experimental Screening of Elastic Waves in Loess. The author, in cooperation with P. A. Saichev and Ya. N. Smolikov, investigated the screening of waves by means of a trench. The source of waves was a reinforced-concrete foundation with an area of 0.81 m^2 in contact with the soil. A vibrator installed on this foundation produced forced vertical vibrations at the desired frequency and amplitude. In all the experiments, although the frequencies changed, the amplitude of vibrations remained more or less constant, equaling some 0.19 mm.

Figure VIII-26 shows the location of the source and the points at which amplitudes were measured. Two profiles were selected for the investigation of the distribution of amplitudes: (1) a longitudinal profile, perpendicular to the trench; and (2) a transverse profile, parallel to the trench. The distance from the trench to the source of vibrations was 2 m.

In the first series of experiments the length of the trench was 8 m. At the surface the trench was some 1 m wide; the width somewhat decreased with depth and at a depth of 4.0 m equaled 0.7 m; the walls of the trench were not braced.

The distribution of amplitudes over the surface was studied in the following order:

Before the trench was excavated, the distribution of amplitudes was investigated along the longitudinal and transverse profiles for frequencies of 730 and 930 oscillations per minute.

Then the trench was excavated to a depth of 0.8 m, and for the same frequencies the vibration amplitudes were measured along the longitudinal profile. These measurements were successively repeated when the trench reached depths of 1.8, 2.5, and 4 m. Further excavation without bracing was dangerous, but the use of bracing could introduce new factors into the phenomenon under investigation, and it would be difficult to take these factors into account. After the trench reached a depth of 4.0 m, the distribution of amplitudes was studied not only for the above-mentioned frequencies, but also for 810 and 1,030 oscillations per minute.

Then the length of the trench was increased by 3 m, and in addition to the investigation of amplitude distribution along the longitudinal profile, measurements along the transverse profile were conducted for 670, 730, 930, and 1,030 oscillations per minute.

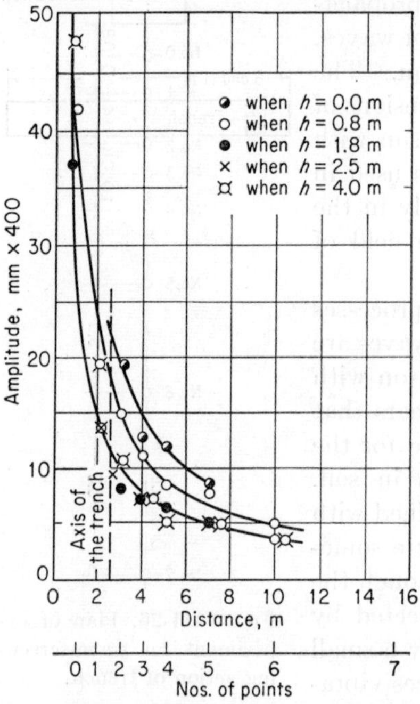

FIG. VIII-27. Vibration amplitudes measured during experiment illustrated by Fig. VIII-26, for different depths h of the 8.0-m-long trench and a 730-rpm vibrator speed.

Figure VIII-27 presents graphs of amplitude distribution along the longitudinal profile for 730 rpm at different depths of the 8.0-m-long trench. The real amplitude of vibrations was one four-hundredth of the values given. It is seen from Fig. VIII-27 that there are no sharp changes in the amplitude distribution for 730 oscillations per minute. The amplitude distribution graphs are smooth curves, generally typical for the amplitude distribution along the soil surface when no screening is present. Therefore the conclusion is possible that for this frequency, a trench of any of the depths investigated has no significant influence on the distribution of amplitudes over the soil surface. Analogous results were obtained for other frequencies and for trenches shallower than 4 m.

Figure VIII-28 shows graphs of the amplitude distribution when the trench was 4.0 m deep and 11.0 m long. It is seen from these graphs that at the trench the amplitudes underwent a sharp change at 930 oscillations per minute. Changes in amplitudes in the area beyond the trench are so insignificant that in the first approximation the amplitudes may be considered as being invariable over the entire area investigated.

It is noteworthy that even at a fairly large distance from the trench the soil vibration amplitudes are considerably smaller than the vibration amplitudes at the same points when the trench was 4 m deep and the frequency was less than 930 oscillations per minute. This leads to the assumption that in this case the influence of the trench on the decrease of the amplitudes of soil vibrations is observed at distances much larger than the depth of the trench.

An investigation of the screening properties of a trench of 4 m depth and 8 m length was also carried out for 810 and 1,030 vibrations per minute. Figure VIII-29 shows graphs of amplitude distribution for these frequencies and, for comparison, the graphs for 730 and 930 oscillations per minute.

A comparison of the amplitude distributions beyond the trench during 930 and 1,030 oscillations per minute leads to the conclusion that an increase in the frequency by 100 oscillations per minute has almost no effect on the amplitude distribution. However, an increase from 730 to 810 oscillations per minute resulted in considerable decrease in amplitudes beyond the trench. It appears that when the trench had a depth of 4 m and the waves propagated at 810 oscillations per minute, the trench screened the vibrations, but not to such a degree as to cause a rupture in the amplitude distribution as in the case of 930 and more oscillations per minute.

To illustrate the relationship between vibration amplitudes at the soil surface on opposite sides of the trench, Fig. VIII-30 shows the seismograms recorded at points 1 and 2. Let us note that while vibrations were propagated at a frequency lower than 930 vibrations per minute, they were felt without any instrument, by "sensory perception," on both sides of the trench; when the frequency exceeded 930, the vibrations were felt only on the side of the trench nearest the source.

An increase in the trench length by 3 m did not have a significant effect on the amplitude distribution beyond the trench for frequencies of 930 and 1,030 oscillations per minute. However, an increase in the trench length had considerable influence on the decrease of amplitudes of vibra-

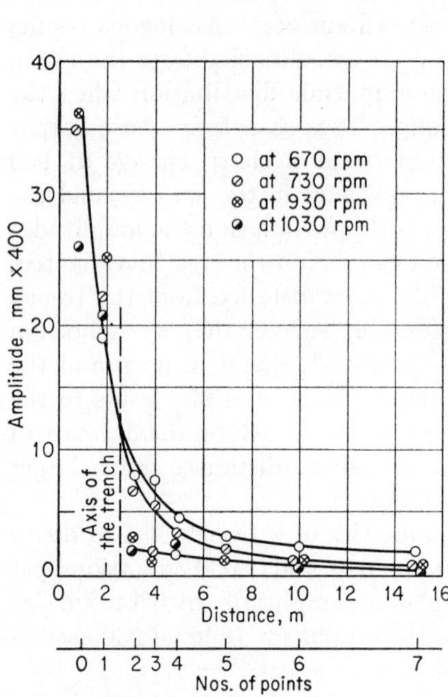

FIG. VIII-28. Vibration amplitudes measured during experiment illustrated by Fig. VIII-26 for a 4.0-m depth h of the 11.0-m-long trench and varying vibrator speeds.

FIG. VIII-29. Vibration amplitudes measured during experiment illustrated by Fig. VIII-26 for a 4.0-m depth h of the 8.0-m-long trench and varying vibrator speeds.

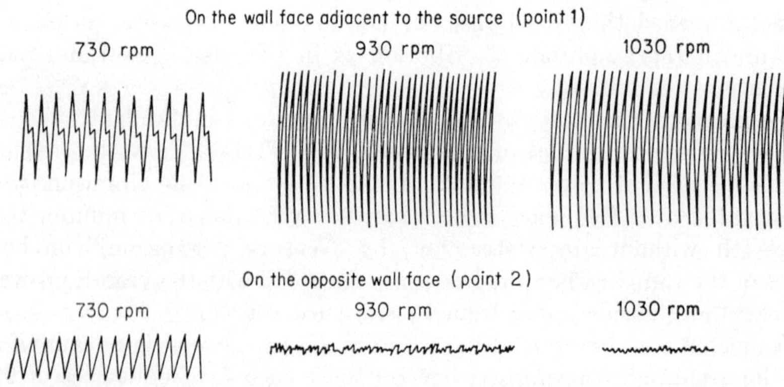

FIG. VIII-30. Seismograms recorded at the soil surface on both sides of the trench of Fig. VIII-26.

tions beyond the trench for waves of 730 oscillations per minute (see Figs. VIII-28 and VIII-29).

In addition to the investigation of the distribution of amplitudes over the soil surface, an investigation of the distribution of amplitudes along the depth of the trench was conducted. For this purpose, cavities were dug in both walls of the trench at distances of 1 m from each another. Measurements of the vertical component of vibrations were conducted by a seismograph and recorder. These measurements were performed at 1,050 and 670 vibrations per minute. A distinct screening effect of the trench was observed at 1,050 vibrations per minute, but not at 670

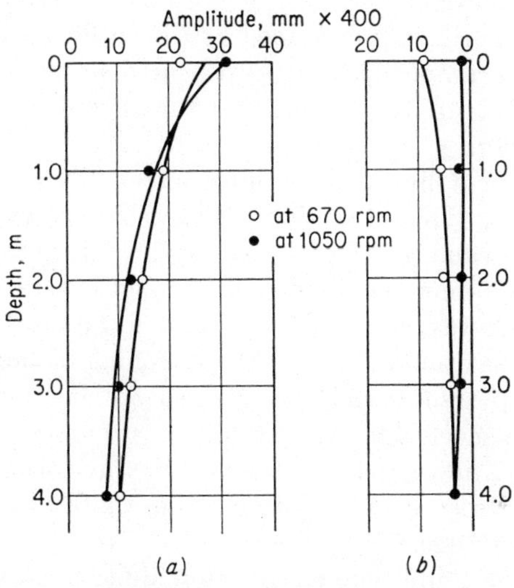

FIG. VIII-31. Variation of vibration amplitudes along the depth of the trench of Fig. VIII-26: (*a*) trench wall adjoining vibrating foundation; (*b*) opposite trench wall.

vibrations per minute. Figure VIII-31 shows the graphs of the distribution of amplitudes. Figure VIII-31*a* refers to the wall of the trench nearest the source of vibrations; Fig. VIII-31*b* refers to the opposite wall.

At the trench wall adjacent to the source of waves, vibrations at the two frequencies do not differ much one from the other; as was to be expected, the vibration amplitudes at this wall decrease with an increase in depth in approximately the same manner as vibrations on the soil surface. On the wall opposite the source, the distribution of amplitudes at 670 vibrations per minute basically does not differ from the distribution of amplitudes on the wall adjacent to the source. On the other hand, the distribution of amplitudes for the screened frequency (1,050 vibrations per minute) is distorted by diffraction. This is explained by the fact that on the side of the trench opposite the source, two reciprocally opposed processes influence the amplitude distribution: (1) the screening effect of the trench, causing a decrease in amplitudes; this effect increases with depth; (2) a decrease in amplitudes due to the increase in depth of the point under consideration. Depending on the relative significance of the screening effect in comparison with the depth effect, the distribution of amplitudes in depth may be of a different nature. For example, it may happen that the amplitudes will not decrease, but will even somewhat increase with depth.

If no screening effect is observed, then the amplitudes should decrease with increasing depth; this was so when waves were propagated at 670 oscillations per minute.

After all the experiments described had been finished, the trench was backfilled and the amplitudes were measured at points 0 to 4 of the longitudinal profile. It turned out that the backfilled trench lost all its screening properties.

Investigations of a linear sheetpiling screen were conducted on the same site, but on a different section. The decision was made first to dig a trench 0.8 m deep and then to drive sheetpiling along the trench axis to a depth of 2.8 m; 25-mm planks of 3.0 m length were used. Then the depth of the trench was increased by 1.0 m and the sheetpiling was driven 1 m. Thus the maximum depth of the screening installation, consisting in the 1.8-m-deep trench and the 2.8-m-deep sheetpiling, was 4.6 m. The length of the sheetpiling was 6.0 m.

The same foundation which was used for the investigation of the trench was taken as the source of vibrations; the amplitude of excitement of vibrations was the same.

Measurements of amplitude distribution along the longitudinal profile were made prior to driving the sheet piles (when the trench was 0.8 m deep), at the time the sheetpiling was driven to a depth of 3.6 m, and finally when the trench was 1.8 m deep and the sheetpiling was at a depth of 4.6 m from the soil surface. Measurements were made at 730 and 1,030 oscillations per minute.

Figures VIII-32 and VIII-33 show graphs of the amplitude distribution along the longitudinal profile for the experimental frequencies. In spite of some scattering of points, it may be seen from Fig. VIII-32 that prior to sheet-pile driving and when the sheet piles were at a depth of 3.6 m, no rupture occurred in the distribution of amplitudes at the barrier (this refers to the waves characterized by 730 and 1,030 vibrations per minute). An analogous conclusion may be drawn in regard to the distribution of amplitudes when the sheet pile was 4.6 m deep and the waves were characterized by 730 vibrations per minute (Fig. VIII-33). However, at 1,030 vibrations per minute, at a sheet-pile depth of 4.6 m, a rupture in the distribution of amplitudes occurred near the barrier (points 1 and 2).

The amplitude distribution at this depth of the barrier and at 1,030 vibrations per minute clearly shows the presence of the screening effect of the barrier, analogous to the case in which the wave frequency was the same and the trench was 4.0 m deep (without sheetpiling).

c. Experimental Screening of Waves in Water-saturated Soils. It is clear that the employment of a screen in the form of a trench is not possible in some soils, for example, in water-saturated sands. Therefore sheetpiling should be used in such soils. The author experimentally

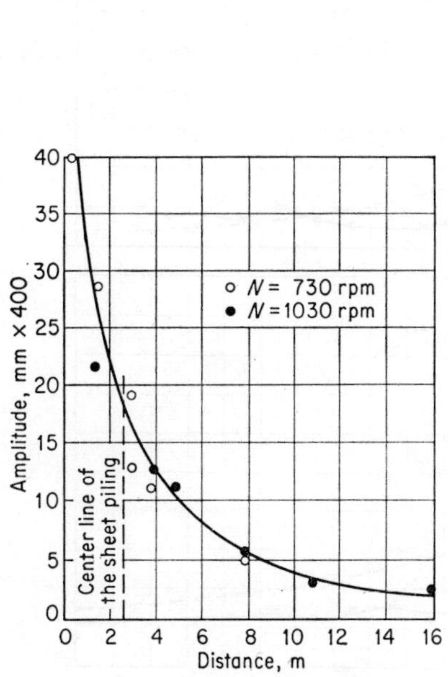

Fig. VIII-32. Vibration amplitudes measured at the soil surface on both sides of a trench and 3.6-m-deep sheet-pile wall.

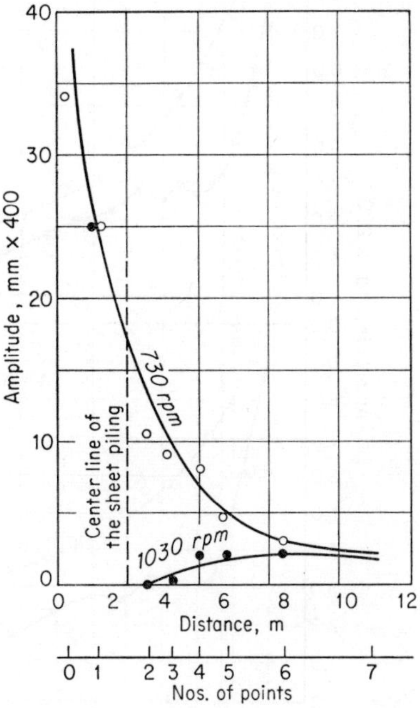

Fig. VIII-33. Vibration amplitudes measured at the soil surface on both sides of a trench and 4.6-m-deep sheet-pile wall.

investigated such screens[10] in the field, where a water-saturated dense sand was found down to a depth of some 8 m. The sand was underlaid by a thick layer of organic silt, below which sands were again found.

Foundation 7 was used as a source of waves in the investigations of the screening effect of linear sheetpiling. The foundation area in contact with soil was 1.0 m². Prior to the sheet-pile driving, amplitudes of vertical soil vibrations were measured at several different frequencies of forced vertical vibrations of the foundation. The vibrometer recorded soil vibrations in a vertical plane which passed through the foundation and the point where vibrations were measured. Lissajous curves

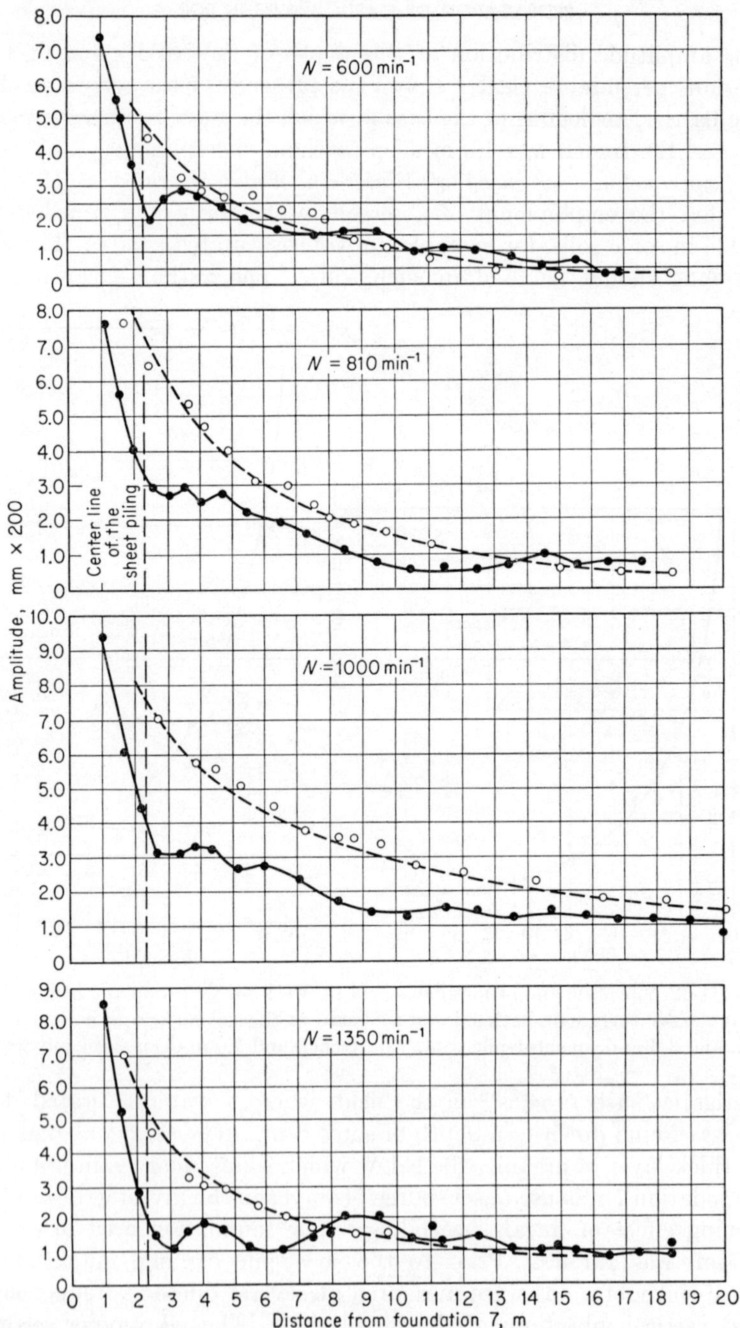

FIG. VIII-34*a*. Variation along a longitudinal profile of amplitudes of the vertical component of vertical surface soil vibrations before (open circles) and after (black circles) sheet-pile driving (see Fig. VIII-37, foundation 7).

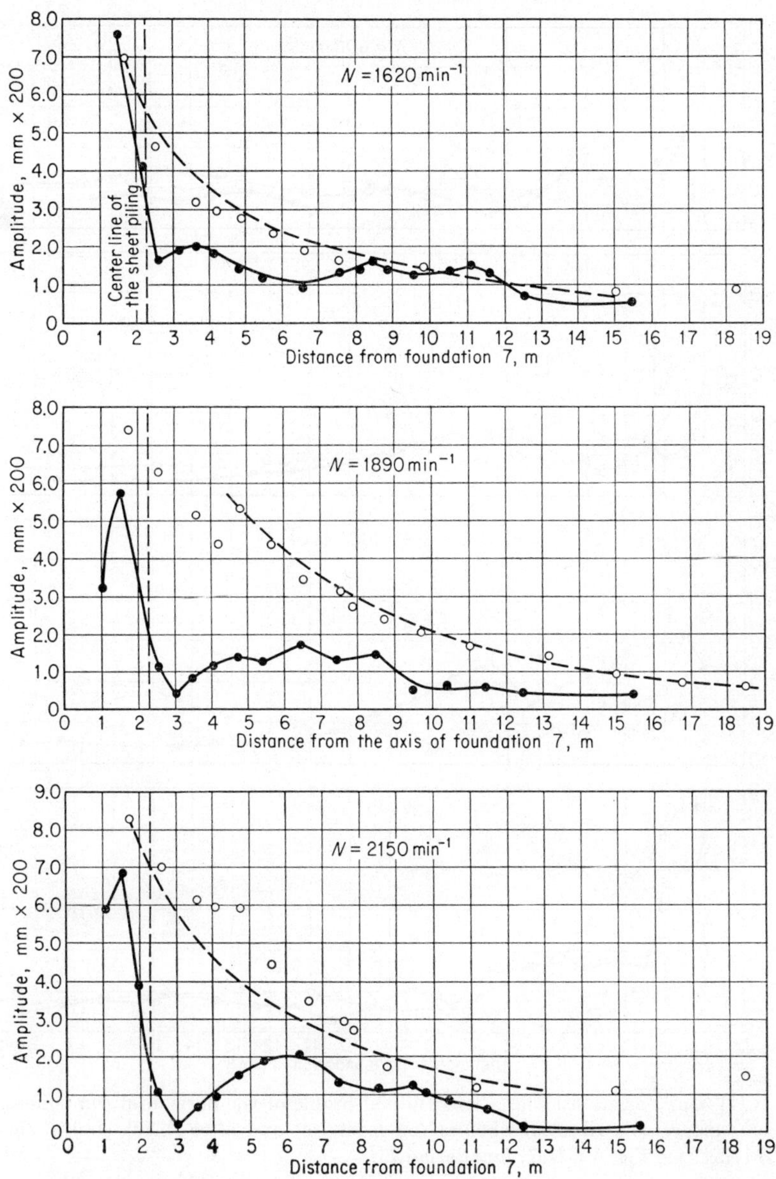

FIG. VIII-34*b*. Same as Fig. VIII-34*a*, but for other vibrator speeds.

obtained by means of the vibrometer were recorded at each point. Thus data were collected not only on the values of the vibration amplitudes, but also on the shape of the orbit of particle motion at a given point on the soil surface. The results of the measurements are shown in Figs. VIII-34 and VIII-35, where small open circles show the amplitudes of the

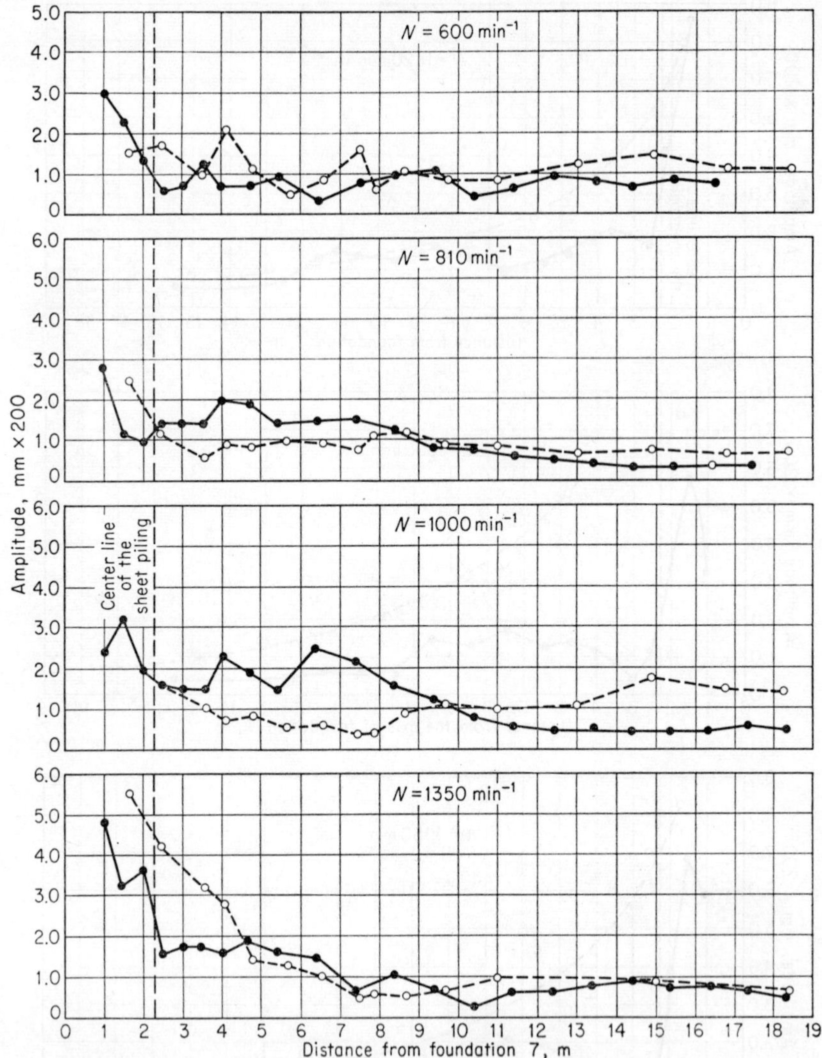

FIG. VIII-35*a*. Variation along a longitudinal profile of the horizontal component of vertical surface soil vibrations before (open circles) and after (black circles) sheet-pile driving (see Fig. VIII-37, foundation 7).

vertical and horizontal longitudinal vibration components obtained prior to sheet-pile driving.

The same figures indicate by dashed lines the theoretical curves computed from Eq. (VIII-3-41) for a given value of the damping constant α of the waves.

It is seen from these graphs that the theoretical curves coincide fairly

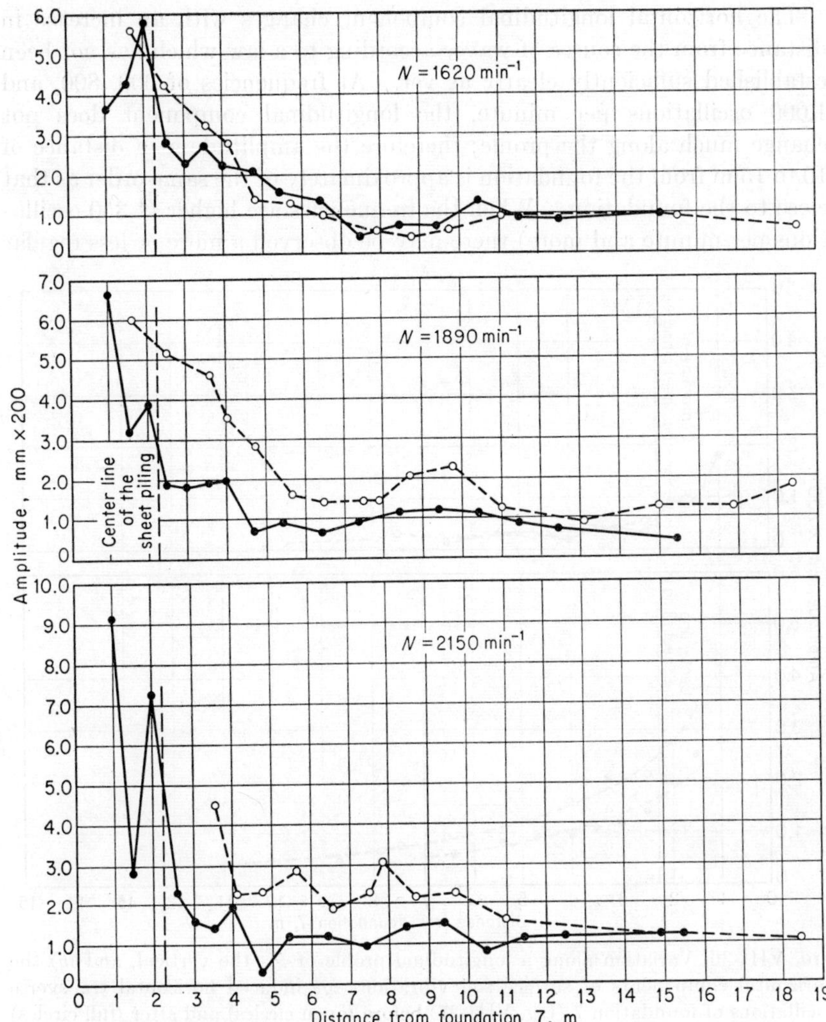

Fig. VIII-35*b*. Same as Fig. VIII-35*a*, but for other vibrator speeds.

well with experimentally established points, with the exception of measurements at a frequency of 2,150 oscillations per minute, when the computed amplitudes for average distances from the source of waves differ considerably from the experimentally established values. The decrease in amplitudes of the vertical component of vibrations with an increase in the distance from the source of waves has a more or less monotonous character which in some places is distorted, apparently by the interference of waves.

The horizontal longitudinal component changes with an increase in distance from the source of waves according to a law which has not been established sufficiently clearly as yet. At frequencies of 600, 800, and 1,000 oscillations per minute, the longitudinal component does not change much along the profile; therefore the amplitude at a distance of 10 to 15 m from the foundation is approximately of the same order as that close to the foundation. When the frequencies are higher (1,350 oscillations per minute and more) there may be observed a more or less regular

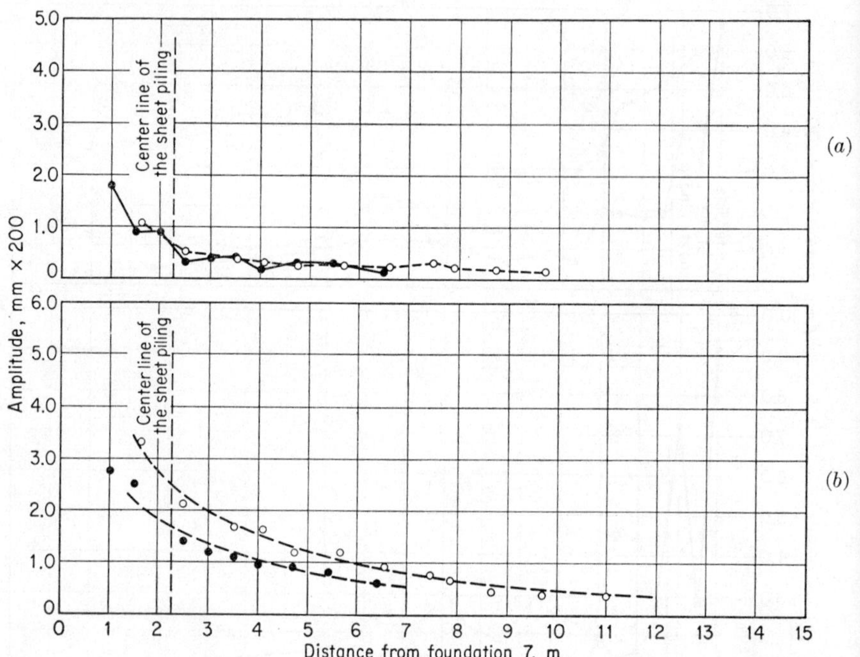

FIG. VIII-36. Variation along a longitudinal profile of (*a*) the vertical, and (*b*) the horizontal components of surface soil vibrations for induced horizontal transverse oscillations of foundation 7 (Fig. VIII-37) before (open circles) and after (full circles) sheet-pile driving.

decrease in the horizontal vibration amplitudes, especially at relatively small distances from the source of waves. In some cases the amplitudes do not decrease with an increase in distance, but on the contrary they grow (for example, see data for $N = 1{,}000 \text{ min}^{-1}$ and $N = 1{,}890 \text{ min}^{-1}$).

In addition, an investigation was made of the distribution of amplitudes along the same profile when the foundation underwent transverse forced vibrations in a plane parallel to the profile. The frequency of these oscillations was taken to equal the frequency of natural transverse oscillations of the foundation (some 300 min^{-1}). Under these conditions the form of the foundation vibrations did not differ much from rocking vibrations.

Graphs of Fig. VIII-36 present results of these vibration-amplitude measurements over the soil surface. Small open circles indicate the experimentally obtained amplitudes of the vertical (Fig. VIII-36a) and horizontal (Fig. VIII-36b) components of soil vibrations. Unlike the case of vertical vibrations of the foundation (source of waves), a monotonous change with distance of the horizontal component of soil vibrations

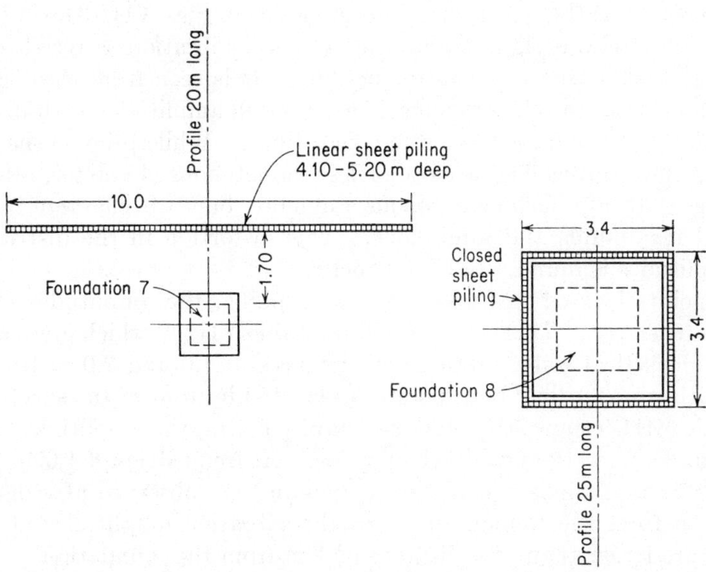

FIG. VIII-37. Plans of foundation 7 (used in the experiments described by Figs. VIII-34 to VIII-36) and foundation 8 (used in the experiments described by Figs. VIII-39 and VIII-40).

was observed here; the vertical component remained constant at almost all points.

After the distribution of amplitudes had been measured over the soil surface for vertical and transverse vibrations of the foundation, the sheetpiling was driven, for which 6- to 7-m-long timber beams having a cross section of 20 by 20 cm^2 and more were used. The sheetpiling was driven at a distance of 1.7 m from the foundation (source of waves). In plan it was located perpendicular to the profile along which the amplitude distribution was measured. Figure VIII-37 shows the arrangement of the sheetpiling; the total length of the linear sheetpiling in plan was 10 m (5 m in each direction from the source of waves). The depth to which individual piles were driven varied within the range of 4.10 to 5.2 m.

Jetting was used during the sheet-pile driving. The space between individual piles at the soil surface reached 5.0 cm. Of 44 piles, 1 was driven to a depth of 6.10 m, and 2 to a depth of 2.7 m.

Measurements of the distribution of amplitudes for vertical vibrations of the foundation were performed along the same profile before and after sheet-pile driving. The frequencies in both cases were approximately the same. The amplitudes of soil vibrations were also measured when the foundation underwent transverse vibrations under conditions of the first resonance—close to 300 min^{-1}.

The results of these measurements are given in Figs. VIII-34 to VIII-36. Small full circles designate amplitudes of soil vibrations after sheet-pile driving; their curve is a continuous line. It is seen from these graphs that the curves describing changes in vibration amplitudes with distance from the foundation undergo some distortions. While prior to sheet-pile driving these curves had a more or less monotonous character, after the driving they are not monotonous anymore but are characterized by several maximums and minimums. This distortion in the distribution of amplitudes is found at all frequencies.

At points located in front of the sheetpiling, the amplitudes change very intensively. This is illustrated by Table VIII-6, which gives amplitudes of vertical soil vibrations at distances of 1.0 and 2.0 m from the foundation center line; both points are located in front of the sheetpiling.

Table VIII-6 shows that over a distance of 1 m, the amplitudes on the average decrease to one-half their value. At frequencies of 1,620, 1,890, and 2,150 oscillations per minute, maximums are observed at a distance of 1.5 m from the foundation; thus the vibration amplitudes at these points are larger than at a distance of 1 m from the foundation.

Table VIII-6. Decrease of Vibration Amplitudes over a Distance of 1 m

N, min^{-1}	Amplitudes, mm × 200, when		Ratio between amplitudes
	$r = 1.0$ m	$r = 2.0$ m	
600	7.4	3.6	2.1
810	7.6	4.0	1.9
1,000	8.6	4.1	2.1
1,350	8.5	2.8	3.0
1,620	7.4	4.1	1.8
1,890	3.1	3.1	1.0
2,150	5.0	3.9	1.5

The smallest vibration amplitudes beyond the sheetpiling were recorded at a point located at a distance of 2.5 m from the source, i.e., at a distance of 0.5 m from the sheetpiling, and a distance of 3 m from the foundation center line. This indicates that during wave propagation sheetpiling undergoes vertical vibrations at a smaller amplitude than the surrounding soil. To show how the vibration amplitudes decrease immediately beyond the sheetpiling, Table VIII-7 gives amplitude values before and after sheet-pile driving.

It is seen from Table VIII-7 that with an increase in the number of oscillations, i.e., with a decrease in the wavelength, the sheetpiling became more effective for points located directly behind it. A decrease in the amplitude of vertical vibrations, caused by the installation of the sheetpiling, was observed for all frequencies investigated in the course of the experiments. Therefore it may be stated that linear sheetpiling of the dimensions studied decreases the amplitudes of vertical soil vibrations during vertical vibrations of the foundation, at least for $N = 600$ min^{-1} and greater.

TABLE VIII-7. EFFECTIVENESS OF SHEET-PILE SCREEN (FIG. VIII-37) AT HIGHER FREQUENCIES OF VIBRATION (FOUNDATION 7)

N, min^{-1}	Amplitudes, mm × 20		Ratio between amplitudes
	Prior to sheet-pile driving	After sheet-pile driving	
600	4.5	2.0	2.2
810	6.5	2.9	2.2
1,000	6.5	2.9	2.2
1,350	4.6	1.5	3.1
1,620	4.6	1.6	2.9
1,890	6.3	1.1	5.7
2,150	7.0	1.1	6.4

After the amplitudes of vertical soil vibrations attain a minimum, they begin to increase with an increase in the distance from the sheetpiling, and hence from the source of waves. For $N = 600$ to 1,620 min^{-1} they attain a maximum at a distance of 1.0 to 1.2 m; for $N = 1{,}890$ min^{-1} and $N = 2{,}150$ min^{-1}, at 4 m.

It is seen from the graphs of amplitude distribution that at a distance of 1.0 to 1.2 m behind the sheetpiling (corresponding to the maximum), the smallest difference between the vibration amplitudes measured before and after sheet-pile driving was observed at the smallest number of oscillations, i.e., when $N = 600$ min^{-1}. At wavelengths corresponding to this number of oscillations, a decrease in the amplitudes of vertical soil vibrations after (as compared to before) sheet-pile driving was observed only immediately behind the piling. At a distance of 1.0 m from the sheetpiling and further, the difference in amplitudes before and after driving is so small that it may be considered that at these distances the sheetpiling does not have a noticeable effect on the decrease in amplitudes. Thus the zone of decreased amplitudes, or "screened zone," formed by the sheet piling when $N = 600$ min^{-1} extends behind the screen only for some 0.2 m.

When higher frequencies are studied, the depth of the screened zone is much larger. Table VIII-8 gives values of the depth of the screened zone for different values of N. The depth of the screened zone is taken as the distance from the sheetpiling at which the amplitudes after sheet-pile driving are no larger than one-half the amplitudes at the same points measured prior to sheet-pile driving.

TABLE VIII-8. INCREASED DEPTH OF SCREENED ZONE BEHIND SHEETPILING AT HIGHER FREQUENCIES FOR FOUNDATION 7 (FIG. VIII-37)

N, min^{-1}	*Depth of screened zone, m*
600	0.2–0.3
810	0.5–0.7
1,000	0.5–0.7
1,350	2.5–2.7
1,620	2.7–3.0
1,890	10.0–12.0

When $N = 2{,}150$ min^{-1}, the amplitudes of soil vibrations decrease in the area directly behind the sheetpiling; for example, at a distance of 0.7 m, the amplitude of soil vibrations after sheet-pile driving was one-fourteenth the amplitude before driving. As the distance from the sheetpiling grows, amplitudes increase, and within the range of distances of 4.2 to 9.5 m, the ratio between the amplitudes before and after sheet-pile driving changes within the range of 1.0 to 1.75. Thus, at these distances, when the number of oscillations equals 2,150 min^{-1}, the sheetpiling does not much affect the decrease in the amplitudes of soil vibrations. However, as the distance from the sheetpiling increases further, a new sharp decrease in amplitudes occurs after the driving, so that at a distance of 10 to 13 m from the piling the ratio of amplitudes before and after sheet-pile driving is of the order of 10 to 12.

Thus, due to a varying distribution of soil vibration amplitudes after sheet-pile driving, the following zones are observed in the area adjacent to the sheetpiling: first, directly behind the sheetpiling, a zone is found where a noticeable decrease in amplitudes is observed; then follows a zone of relatively small influence, and further again a zone of considerable influence. The presence of similar zones of decreased and increased screening effect is noticeable not only when $N = 2{,}150$ min^{-1}, but also at lower frequencies. However, the differences in amplitudes then are smaller.

Table VIII-8 shows that with an increase in the number of oscillations, i.e., with a decrease in the length of propagating waves, the depth of the screened zone increases, and consequently the sheetpiling is effective at larger distances.

Summarizing the foregoing discussion with regard to the influence of line sheetpiling of the dimensions studied for vertical vibrations of the source of waves, the following may be stated concerning the distribution of amplitudes of the vertical component of soil vibrations at different numbers of oscillations, i.e., at different wavelengths: the sheetpiling causes a decrease in amplitudes at points located directly behind it for all frequencies studied, i.e., $N = 600$ min^{-1} and greater. However, when $N = 600$ min^{-1}, the depth of the screened zone is very small, being of the order of 20 to 30 cm. As the number of oscillations increases, the depth of the screened zone also increases (at the same time, the amplitude directly behind the sheetpiling decreases), and when $N = 1{,}000$ to $1{,}620$ m^{-1}, the depth of the screened zone is some 2.5 to 3.0 m; when $N = 810$ to $1{,}000$ min^{-1}, the depth is some 0.5 to 0.7 m; and when $N = 1{,}890$ min^{-1}, the depth of the screened zone extends some 10 to 12 m behind the sheetpiling. Hence, if a foundation which should be protected from propagating waves has a width exceeding 0.5 to 0.7 m, its total area in contact with soil will lie within the screened zone if N equals or exceeds 800 to 1,000 min^{-1}.

If the oscillation numbers of the propagating waves are smaller than those indicated, then the sheetpiling will screen only part of the foundation base contact area which is to be protected from waves. If oscillation numbers are less than 1,620 min^{-1}, then the contact dimensions of the protected foundation should not be larger than 2.5 to 3.0 m.

All that has been said above concerning the influence of linear sheetpiling on the distribution of amplitudes over the soil surface refers to the vertical component of soil vibrations under vertical vibrations of the foundation; the influence of sheetpiling on the horizontal longitudinal component is much more complicated. This is why no regularity has been established in the distribution of amplitudes of the horizontal component after sheet-pile driving. When N was equal to 600 min^{-1}, the amplitudes directly behind the sheetpiling were reduced to approximately one-half the previously observed values; then, to a distance of some 8 m from the sheetpiling, vibration amplitudes were about the same before and after sheet-pile driving. At distances larger than 8 m, there followed again a zone of relative decrease, where amplitudes of vibrations were 1.5 to 2 times larger before than after sheet-pile driving.

When N was 810 min^{-1}, in the 8-m zone, the sheetpiling induced, not a decrease in amplitudes, but even an increase in the amplitudes of horizontal soil vibrations; at distances larger than 8 m from the sheetpiling, there followed a relative decrease in the vibration amplitudes: prior to driving the amplitudes were approximately 1.5 to 2 times larger than the amplitudes at the same points after driving.

When N was 1,000 min^{-1}, the sheetpiling influence on horizontal vibrations was about the same as when N was 810 min^{-1}. When oscillation numbers were larger than 1,000 min^{-1}, in all cases, the sheetpiling caused a decrease in the horizontal component in a zone extending directly behind it for 1.5 to 2.5 m; then followed a zone of increase in amplitudes, where absolute values exceeded the values of amplitudes prior to sheetpile driving; finally came a zone where the sheetpiling had only an insignificant influence on the amplitudes of horizontal longitudinal soil vibrations.

In comparison with the influence of sheetpiling on the amplitude of the vertical component, its influence on the horizontal longitudinal component is small; therefore the use of sheetpiling as a screening installation may be recommended mainly for the purpose of decreasing the vertical component of soil vibrations when the exciting source of waves also undergoes vertical vibrations.

Such a decrease of vertical vibration amplitudes is necessary in some cases in which the source of waves causes vertical vibrations of foundations of adjacent structures† at impermissible amplitudes.

In addition to the investigation of the distribution of amplitudes of the horizontal and vertical vibration components along a profile perpendicular to the line of sheetpiling, an investigation was also performed of the distribution of amplitudes induced by vertical vibrations of the source of waves along profiles *parallel* to the sheetpiling. Measurements were made for oscillation numbers equaling 1,620, 1,890, and 2,150 min^{-1} at points directly adjacent to the sheetpiling, both in front of and behind it. The results of these measurements are presented in the form of graphs (Fig. VIII-38) which characterize changes in soil vibration amplitudes in front of and behind the sheetpiling. These graphs reveal the complexity of the law governing amplitude changes around the piling. It can be seen that the screening effect of the sheetpiling changes along its length.

In all cases, the ratio between the soil vibration amplitudes in front of and behind the sheetpiling has the largest value along the two central quarters of its length. Toward the edges of the piling, this ratio decreases rapidly and its value differs relatively little from unity, which indicates a decrease in the screening effect close to the edges. On the basis of these data, the conclusion is possible that the design of sheetpiling to protect a structure from waves propagating through the soil should provide that the length of the piling is greater than the length of the protected structure.

† See Art. IX-3.

The influence of line sheetpiling on the distribution of soil amplitudes was also studied under conditions in which the foundation underwent transverse vibrations at the same frequency before and after sheet-pile driving. Amplitudes of soil vibrations, obtained as a result of these

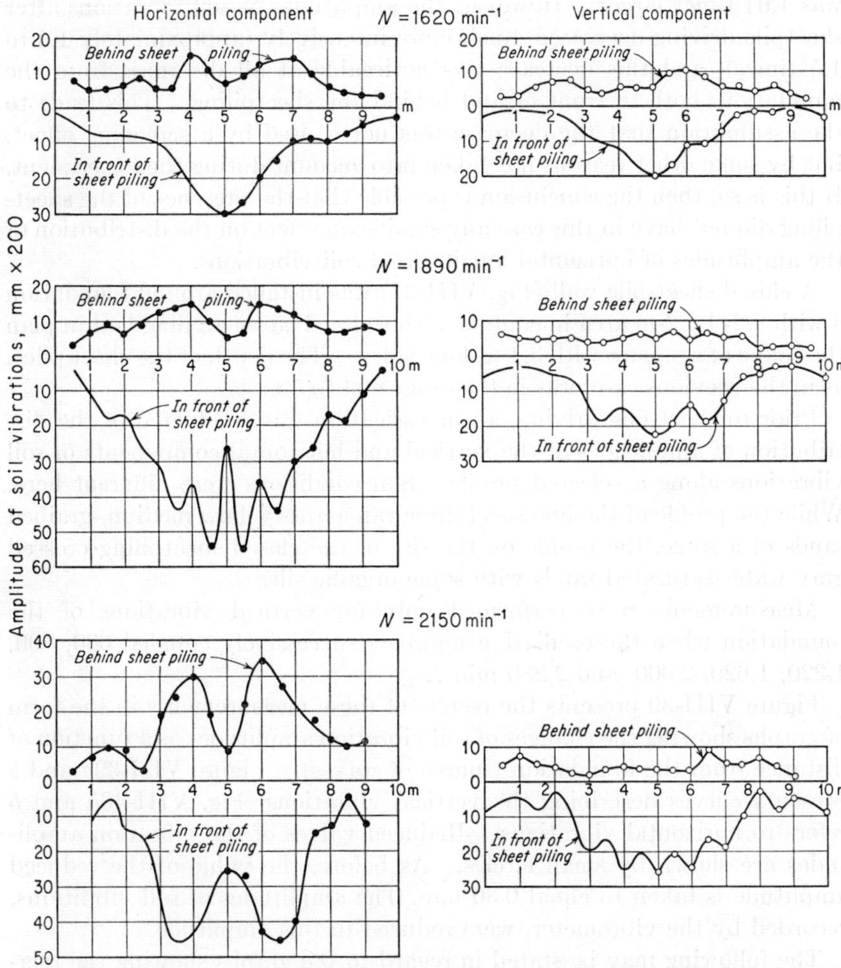

Fig. VIII-38. Variation of vibration amplitudes in front of and behind the sheet-pile line (Fig. VIII-37) at different oscillation frequencies of foundation 7.

measurements, are plotted as small full circles in Fig. VIII-36. These results indicate that the sheetpiling did not have any influence on the distribution of the vertical component of vibrations over the soil surface. The horizontal component of soil vibrations was found at all points to be somewhat smaller after than before sheet-pile driving.

Some influence on the decrease of horizontal soil vibration amplitudes was shown by the fact that the amplitudes of the horizontal component of transverse foundation vibrations after sheet-pile driving were somewhat smaller than those prior to driving; an average value of the amplitude prior to driving was 1.6 mm; after driving it was 1.57 mm; i.e., the former was 1.04 times larger. However, the amplitudes of soil vibrations after sheetpile driving decreased much more intensively (approximately 1.3 to 1.5 times), and this decrease was noticeable at all distances from the foundation, both in front of and behind the sheetpiling. This leads to the assumption that the decrease was not caused by a screening effect, but by some other reasons not taken into account during the experiment. If this is so, then the conclusion is possible that the presence of the sheet-piling did not have in this case any significant effect on the distribution of the amplitudes of horizontal longitudinal soil vibrations.

A closed sheet-pile wall (Fig. VIII-37) was installed around foundation 8 with a 2- by 2-m area in contact with soil. The sheetpiling had in plan the shape of a square with 3.4-m-long sides. The depth of the sheet piles, as in the previous case, was in the range of 4 to 5 m.

Prior to sheet-pile driving, an investigation was conducted of the distribution of amplitudes of the vertical and horizontal components of soil vibrations along a selected profile. Soil conditions were different here. While the profile of the line sheetpiling ran across yellow medium-grained sands of a ridge, the profile on the site of the closed sheetpiling crossed gray water-saturated sands with some organic silt.

Measurements were performed only for vertical vibrations of the foundation when the oscillation numbers successively equaled 600, 800, 1,220, 1,620, 2,000, and 2,280 min^{-1}.

Figure VIII-39 presents the results of these measurements in the form of graphs showing the changes of soil vibration amplitudes as a function of distance from the foundation (source of waves). Figure VIII-39*a* and *b* presents curves describing the vertical vibrations; Fig. VIII-40*a* and *b* refers to horizontal vibrations. Reduced values of the vibration amplitudes are shown by small circles. As before, the value of the reduced amplitude is taken to equal 0.30 mm. The amplitudes of soil vibrations, recorded by the vibrometer, were reduced to this amplitude.

The following may be stated in regard to the graphs showing the relationship between the amplitudes and distances from the foundation. As in the case of wave propagation from foundation 7 (when line sheetpiling was investigated), the amplitude of the vertical component of soil vibrations changes with distance from the foundation fairly monotonously, especially at low and medium frequencies. When $N = 2{,}000$ and $2{,}280\ \text{min}^{-1}$, the changes in vibration amplitudes are somewhat more complicated, probably due to the influence of wave interference.

Theoretical curves of changes in the amplitudes of the vertical component of soil vibrations, computed from Eq. (VIII-3-41)—which takes into account the absorption of wave energy—are shown by broken lines for $N = 600$, 800, 1,220, and 1,620 min^{-1}. It is seen from these graphs that theoretically obtained values of soil vibration amplitudes coincide fairly well with experimental values. As in the investigation discussed previously, the relationship between the horizontal component of soil vibrations and the distance from the foundation is much more complicated. The measurements performed did not provide sufficient data for generalizations concerning the nature of the decrease in amplitudes of the horizontal component with changes in distance from the source of waves.

After the measurements of the distribution of vibration amplitudes along a profile running from the foundation (source of waves) had been completed, this foundation was surrounded by timber sheetpiling of the type described above.

Then measurements of the distribution of amplitudes of the vertical and horizontal longitudinal components of soil vibrations were made along the same profile and for the same frequencies. Figure VIII-39 gives reduced values of the amplitudes of soil vibrations shown by means of small open circles joined by a continuous line.

The curves of the distribution of amplitudes of the vertical component of vibrations versus distance from the source of waves after sheet-pile driving, for all oscillation numbers, are not monotonous, with the exception of the smallest value investigated. These curves have a complicated shape with maximums and minimums located at different distances from the foundation. The first minimum lies, as a rule, directly behind the sheetpiling; then, at a distance of 1.5 to 2.5 m from the foundation, there comes a maximum. As the distance from the foundation grows, the amplitudes reach their smallest value as compared with all amplitude values along the investigated profile section. The location of this "principal minimum" depends on the frequency. The following relationship was established between the number of oscillations and the distance from the foundation of the "principal minimum."

N, min^{-1}	*Distance of "principal minimum" from foundation, m*
1,220	6.4
1,620	4.4
2,000	3.0
2,280	3.7

Hence, with an increase in the number of oscillations, there is a decrease in the distance from the foundation to the principal minimum.

At all frequencies, the amplitudes of the vertical component of soil vibrations decreased after the enclosing sheetpiling was installed. Table

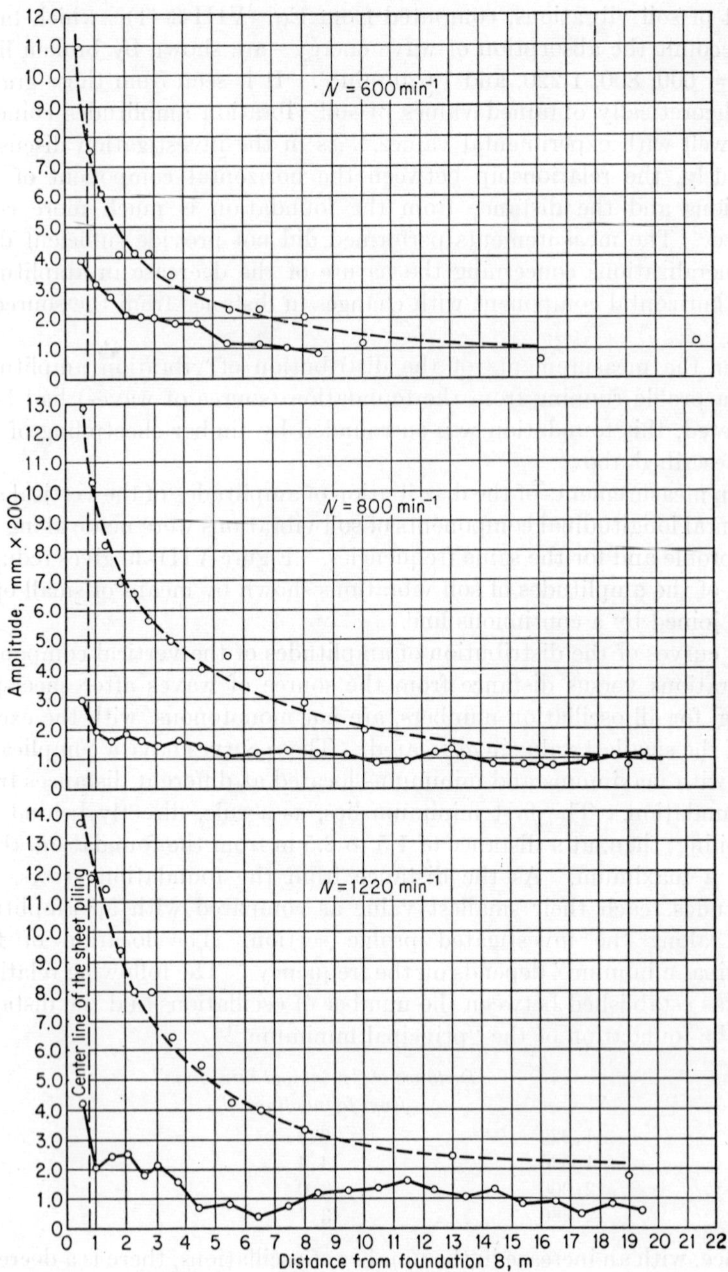

FIG. VIII-39*a*. Variation of vertical vibration amplitudes as a function of distance from vertically oscillating foundation 8 (see Fig. VIII-37) before (broken line) and after (full line) driving the sheet-pile enclosure.

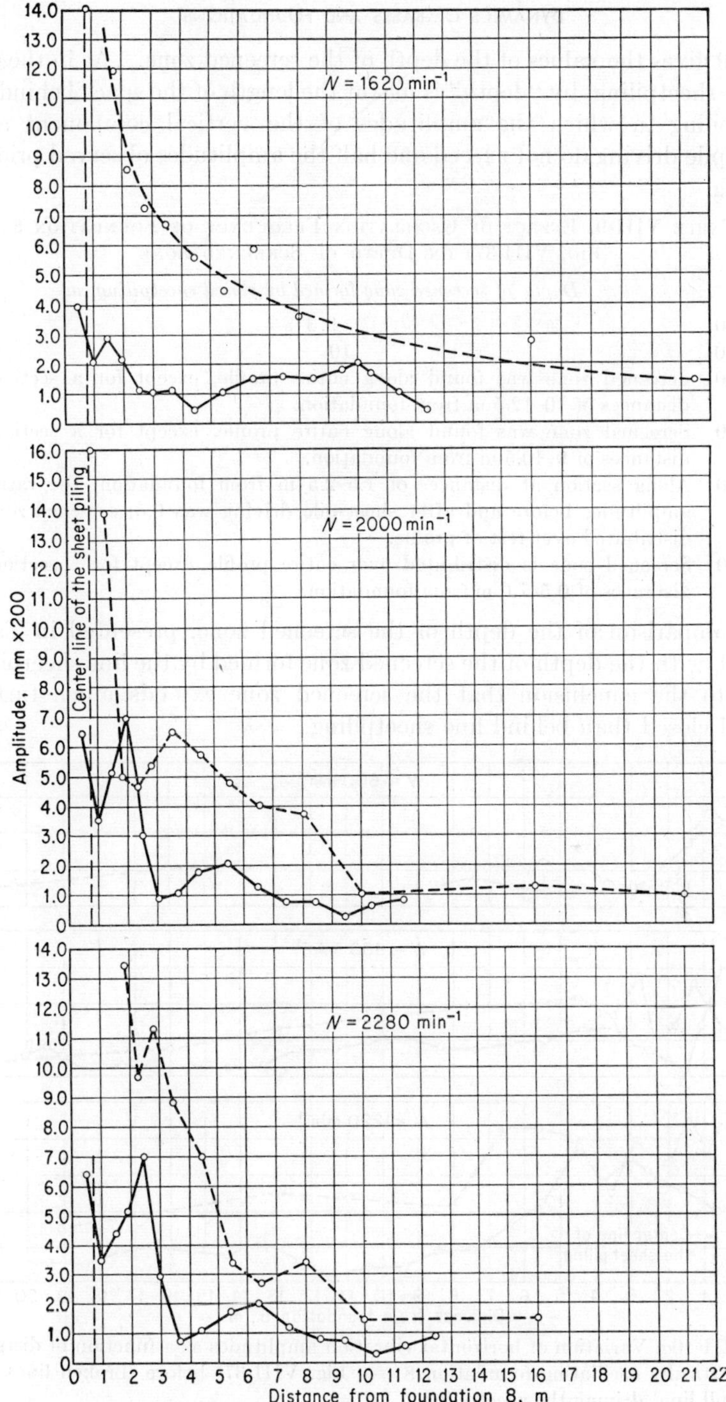

FIG. VIII-39*b*. Same as Fig. VIII-39*a*, but for other vibrator speeds.

VIII-9 gives the values of the depth of the screened zone. As in the case of line sheetpiling, by "depth" is meant the length of the space behind the sheetpiling in which the amplitudes of the vertical component after sheet-pile driving do not exceed one-half the amplitudes observed prior to driving.

TABLE VIII-9. EFFECT OF OSCILLATION FREQUENCY OF FOUNDATION 8 (FIG. VIII-37) ON DEPTH OF SCREENED ZONE

N, min^{-1}	*Depth of screened zone formed by closed sheetpiling, m*
600	1.3
800	10
1,200	Screened zone was found along entire profile, except for a section at distances of 10–12.5 m from foundation.
1,620	Screened zone was found along entire profile, except for a section at distances of 9–10.5 m from foundation.
2,000	Along section at distances of 1.5–2.5 m from foundation, the ratio of amplitudes before and after sheet-pile driving was 0.5; screened zone is distributed over rest of profile.
2,280	Screened zone is distributed over entire profile, except for a section at distances of 0.5–7.0 m from foundation.

A comparison of the depth of the screened zone, presented in Table VIII-9, with the depth of the screened zone formed by the line sheetpiling leads to the conclusion that the screened zone extends much further behind closed than behind line sheetpiling.

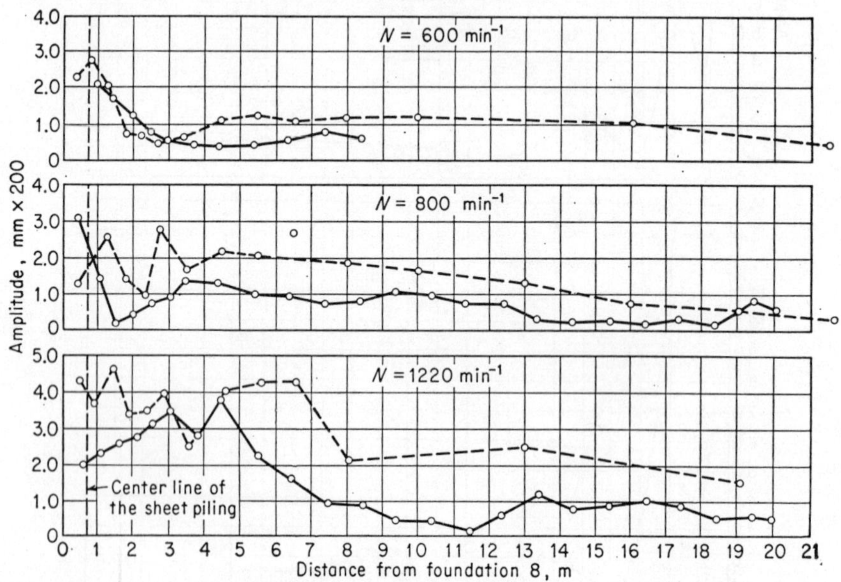

FIG. VIII-40*a*. Variation of horizontal vibration amplitudes as a function of distance from vertically oscillating foundation 8 (see Fig. VIII-37) before (broken line) and after (full line) driving the sheetpiling.

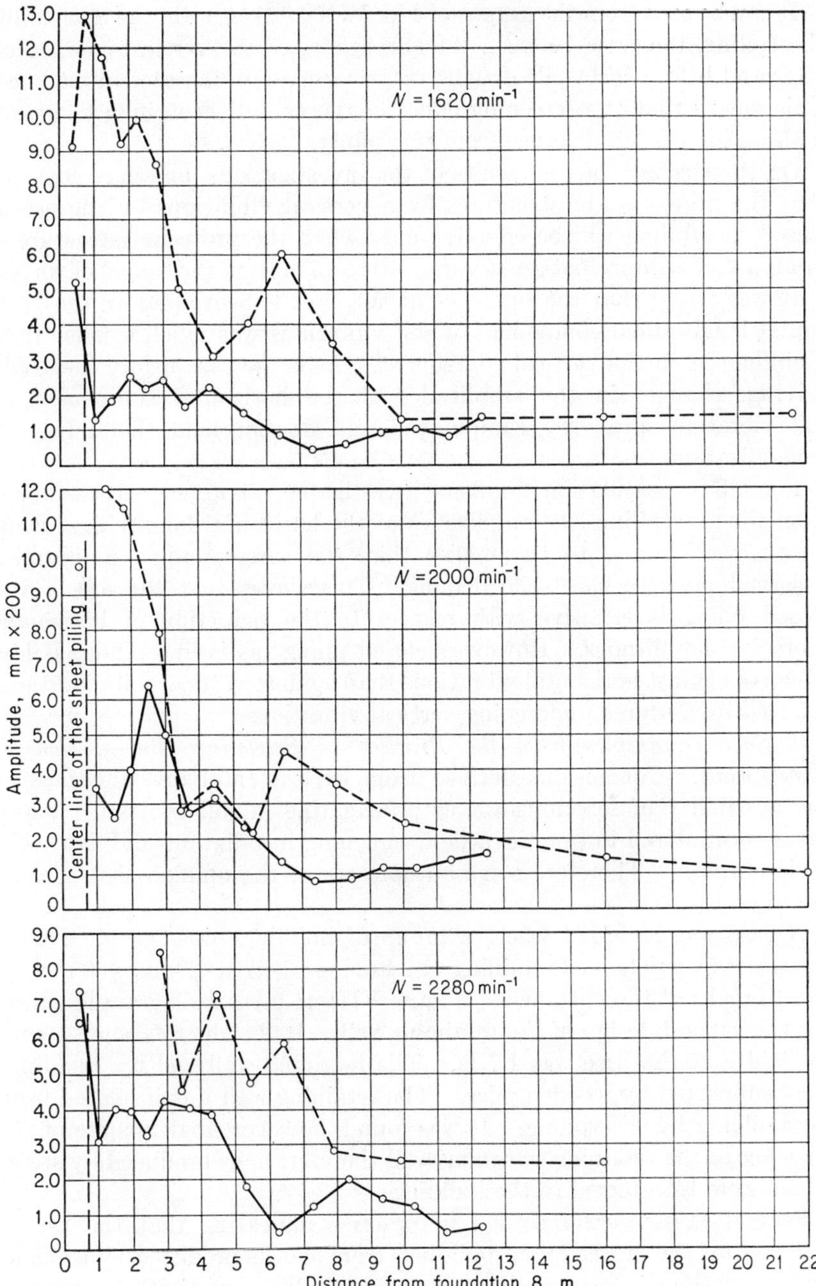

FIG. VIII-40*b*. Same as Fig. VIII-40*a*, but for other vibrator speeds.

It can be seen from the graphs of Fig. VIII-39 that, as in the case of line sheetpiling, the ratio between the amplitudes of the vertical component after and before sheet-pile driving increases with an increase in distance. This means that the screening effect of the closed sheetpiling decreases with an increase in distance from this piling.

On the strength of the results of the investigations, one may consider that the protection of structures from vertical vibrations by the use of closed sheetpiling will be effective only when the protected structure is located at a distance not greater than 10 to 15 m from the source of waves.

It was found that the influence of the closed sheetpiling on the horizontal longitudinal component of soil vibrations was much smaller than its influence on the vertical component. As is the case before sheet-pile driving, changes in the amplitudes of the horizontal component of soil vibrations are very complicated and do not lend themselves to generalization.

For all the oscillation numbers investigated, at almost all distances from the sheetpiling, the amplitudes of the horizontal longitudinal component were found to be smaller than the corresponding amplitudes observed prior to sheet-pile driving. This shows that the use of the closed piling is effective with respect to the reduction of horizontal vibration amplitudes. However, closed piling (as is line piling) is less effective against horizontal vibrations than against vertical soil vibrations induced by a source undergoing vertical vibrations.

d. Some Examples from the Practice of Screening Waves; General Conclusions. Conclusions derived from experimental investigations of the possibility of screening waves propagating in soil contradict many cases encountered in practice where screening installations did not fulfill their purpose. Therefore some authors are of the opinion that screens are of no use.

A case was recorded where a group of buildings was insulated from the street by a detached retaining wall having a length of some 400 m with short breaks at the entrances. Figure VIII-41 gives a diagram illustrating the general design of the retaining wall. Its height was some 4.3 m; the width at the base was 1.7 m. The retaining wall and the buildings were supported by wooden piles. The retaining wall was insulated from the building by sheetpiling. It was found, however, that in spite of the presence of the retaining screening wall the vibrations produced by street traffic were transferred to the buildings.

Several cases were recorded in practice indicating that the use of trenches for the protection of buildings from shocks produced by machinery foundations is less expedient than the employment of other screening devices. For example, there was a case in which residential buildings were separated from a plant by a river; in spite of this, it became necessary to insulate the machine shop of this plant because vibrations transferred from the shop to the residential buildings were too great and had to be decreased. In another case, a trench 7 m deep and 8.5 m wide was provided at a distance of 400 m from a plant in order to decrease shocks created by the operation of a 5-ton crushing hammer; however, no positive results were obtained.

Following are four cases in which the use of trenches did not have a noticeable effect on the decrease of vibrations beyond the trenches.

THE FIRST CASE. The necessity arose to decrease vibrations produced by horizontal reciprocating engines and transferred to adjacent buildings located at rather large distances from the source. For this purpose, a

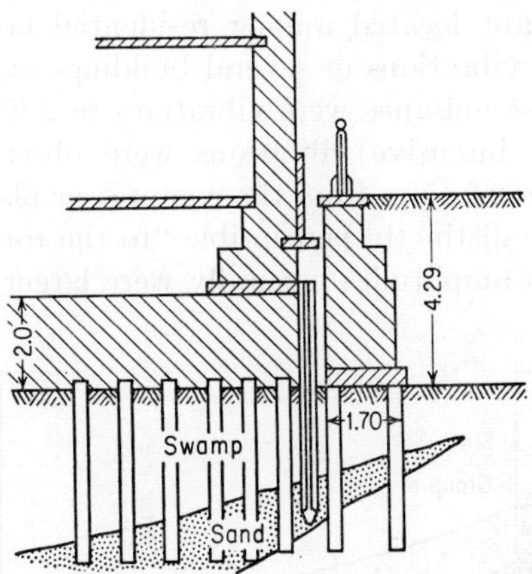

FIG. VIII-41. This retaining wall was not successful in preventing the transmission of traffic vibrations to an adjoining building.

trench of some 6 m depth was provided between the source of vibrations and the buildings; this, however, did not have any effect.

THE SECOND CASE. Here also no effect was obtained by providing a trench of about the same dimensions as above to protect buildings from vibrations produced by a horizontal reciprocating engine. The soil between the engine and buildings consisted of a mixture (fill) of sand and organic silt.

THE THIRD CASE. A deep and wide river was located between the foundation of a horizontal reciprocating engine and some buildings. In spite of this, these buildings were subjected to considerable vibrations. The soil was sandy. Trenches were installed, but were found to be of no use.

THE FOURTH CASE. A deep trench was located between a foundation under a vertical reciprocating engine and buildings subjected to vibrations. This trench, however, did not prevent the transmission of vibrations through soil which consisted mostly of organic silt. Thus the trench was of no use in this case either.

Let us discuss one more case of the use of screening installations at one of the plants in the U.S.S.R. in which again no positive results were obtained.

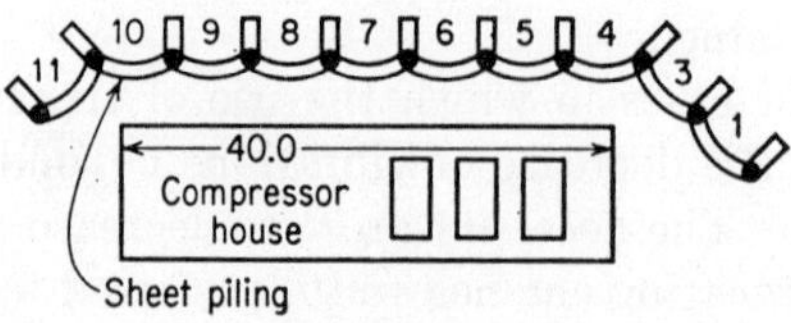

FIG. VIII-42. General plan of attempted screening of a compressor plant.

A compressor plant located among residential houses and auxiliary buildings produced vibrations of several buildings at different distances from it. Of greatest nuisance were vibrations in 2 five-story apartment houses. The most intensive vibrations were observed in a building located at a distance of 45 m from the compressor plant. In this building, vibrations were distinctly perceptible "to the touch." On the upper floors, the vibration amplitudes naturally were larger than on the lower

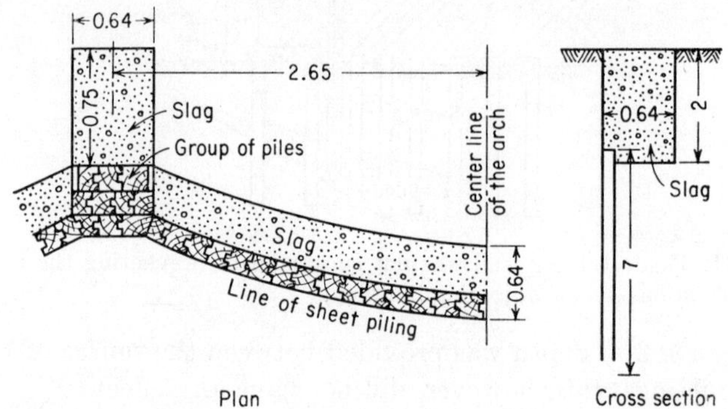

FIG. VIII-43. Details of attempted screening shown in Fig. VIII-42.

floors; they reached 0.3 mm on the fifth floor. An inspection of the outside surfaces of both buildings did not reveal any cracks or deformations.

In order to decrease the vibrations of the buildings, the decision was taken to provide a screen formed by a trench filled with cinders and by sheetpiling. A diagram of the screen as designed is shown in Fig. VIII-42; details of the design are given by Fig. VIII-43. The width of the trench filled with uncompacted cinders was 0.64 m. The sheetpiling was to consist of 7-m-long wooden piles of 16 by 16 cm cross section. The design provided 11 short arches supported by master pile clusters. Eight arches were to be located parallel to the longitudinal wall of the compressor plant nearest the vibrated buildings; one arch was to be located at the left-hand end, and two at the right-hand end of the wall. The pile cluster consisted of ten piles. In its wide section it had four piles; a trench 1 m long was provided behind. In the opinion of the authors of this design, the provision of separate arches with master pile clusters would contribute to an increase in the general rigidity of the screen; this would increase the screening effect of the installation.

This project was not fully completed, since after eight arches, designated in Fig. VIII-42 by the numbers 1 to 8, had been constructed, it was found that the finished section of the screen had no noticeable effect on the decrease of vibrations of the building. Since there was not much hope that the installation of the remaining three arches would change the situation, the decision was taken to stop construction of the screen. Thus again in this case, the sheetpiling did not prove its value.

In order to understand why in all the foregoing cases the use of screens was ineffective, it is necessary to remember that the damping of vibrations by a screen depends on the relationship between the smallest dimensions of the screen and the length of the propagating waves. At the same time, the wavelength depends on the velocity of wave propagation and the period of excitation. Wavelengths for different periods of the source and velocities of propagation are presented in Table VIII-10.

Table VIII-10. Wavelengths for Different Wave Velocities, Periods of Waves, and Numbers of Oscillations of Source of Waves

Number of oscillations of source, min^{-1}	Period of wave, sec	Length of wave, m, for velocity, m/sec, of:					Notes
		50.0	100	200	300	500	
50	1.2	60	120	240	360	600	Low-frequency machines (diesels, compressors, gas motors, etc.)
100	0.6	30	60	120	180	300	
200	0.3	15	30	60	90	150	
300	0.2	10	20	40	60	100	
500	0.12	6.00	12	24	36	60	
800	0.075	3.75	7.5	15	22.5	37.5	High-frequency machines (mainly turbodynamos and forge hammers)
1,000	0.060	3.00	6.0	13	18.0	30.0	
1,200	0.050	2.50	5.0	10	15.0	25.0	
1,500	0.040	2.0	4.0	8.0	12.0	20.0	

Knowing the wavelengths for different sources in different soils and taking the minimum ratio between screen dimensions and wavelength in the presence of which the effect of screening of vibrations is already noticeable, one may compute the smallest dimensions of the screen, as well as its smallest depth, at which the effect of screening will be felt.

Experience with the installation of trenches and sheetpiling in loessial soils showed that these devices had a screening effect when the ratio between the depth and the length of the propagating waves was about 0.3. This lower limit is approximately the same in other soils as well. It follows that in order to secure the screening of vibrations, it is necessary to make the depth of the screen not less than $0.3L$, where L is the length of the propagating waves.

The depth of a screening device should be counted from the depth of the source of waves. The maximum practical depth of the screening device is about 5 m, since machinery foundations are usually placed at a depth of the order of 3 m and more.

It appears that in most soils the velocity of propagation of surface waves is about 200 m/sec.

Taking into consideration both these factors, it may be held that in practice it is possible to damp vibrations with a number of oscillations not less than 600 to 700 min^{-1}. For example, in order to screen vibrations propagating from a diesel running at 200 rpm when the velocity of wave propagation in the soil is 200 m/sec, it would be necessary to install a screen to a depth of at least 17 to 25 m; when the velocity of wave propagation is some 300 m/sec, the depth of the screen should be increased to 27 to 30 m. It is hard in practice to construct screening devices at such depths.

Thus it is impossible in practice to screen waves propagating from low-frequency engines. This is the reason for the failures recorded in the practice of vibration damping by means of screens.

For engines of more than 600 to 700 rpm, it is possible to screen vibrations if the depth of the screening device equals 5 to 7 m.

Thus in most cases it is impossible to screen vibrations from such engines as diesels, piston compressors, and gas motors. Vibrations propagating from high-frequency engines, as well as from hammer foundations with fairly high natural frequencies, might be screened without too much difficulty. When the propagating vibrations have a whole spectrum of frequencies (for example, if vibrations are excited by traffic), not all frequencies will be screened, but mainly the higher frequencies.

Trenches may be employed as screening devices only if there is no ground water in the trench. If the ground-water level is close to the surface, then there is no use employing trenches as screening devices. In such cases, wooden or metallic sheetpiling might be used; if the piling is sufficiently long, then it is recommended to provide rigid junctions such as clusters of master piles.

Finally, if a sand of sufficient porosity is found on the site, then a screen may be formed by chemical or cement stabilization of a narrow strip of soil.

IX

EFFECT ON STRUCTURES OF WAVES FROM INDUSTRIAL SOURCES

IX-1. Classification of Industrial Sources of Elastic Waves in Soil

Some 30 to 40 years ago, no great practical importance was attached to the effects on structures of vibrations produced by seismic waves from industrial sources. At present these problems are very important, and it often happens that the normal work of a whole plant depends on their solution. For example, cases are on record in which shocks, produced by hammer foundations or by traffic and transferred to soil, led to failures in the work of molding shops or precision machines. Mining procedures employing explosives often had to be changed because of the harmful effects on structures of the seismic explosive waves. Several examples are cited in Art. IX-2 which show that waves propagating from low-frequency engines may lead to considerable vibrations of high structures.

Studies of vibrations of residential houses, caused by waves from the above sources, show that they have a very harmful physiological and psychological effect on the residents of such houses. As indicated in Art. IX-3, seismic waves from hammer foundations in some cases produce damage in forge shops.

Thus seismic waves from various industrial sources may have a harmful effect on structures, technological processes, and people. Therefore the study of the action of seismic waves from industrial sources on structures is of great practical importance. Equally important is the development of methods to decrease or eliminate the harmful influence of seismic waves.

With respect to their actions on buildings, industrial sources may be divided into the following main groups:

1. Machinery located within the buildings on floors of one or several stories, or supported by special constructions such as cantilevers tied to the main elements of the building. Machinery of this type includes machines in spinning and weaving mills, transmissions, printing presses, and others. The absence of strict synchronization and periodicity in work is a characteristic feature of these engines. They produce local vibrations of structural elements, e.g., of separate floor spans, of some columns, and of wall sections. These vibrations are not periodic and have a complicated nature due to the superposition of vibrations and shocks which are produced by many sources and which are often unstable in time. Vibrations produced by the type of engines under consideration are often transmitted through the soil to adjacent structures.

2. Powerful high-frequency engines, such as turbogenerators, turbocompressors, and turbofans. Such machines usually have only rotating parts of high static and dynamic stability running at more than 500 to 1,000 rpm. These machines may cause only local vibrations of some structural elements; the amplitudes of these vibrations are very small, not exceeding 0.05 to 0.10 mm. No cases of impermissible vibrations caused by such machines have been recorded in practice. Also no cases are known of vibrations of entire structures having been caused by the action of high-frequency engines, even where these engines were located close to the structures. Thus it can be held that high-frequency engines are safe from the point of view of vibrations of structures. This is partly explained by the fact that vibration amplitudes of high-frequency engine foundations are very small; therefore the amplitudes of soil vibrations are also small and even in areas adjacent to the foundation usually do not exceed 0.1 micron (0.0001 mm).

From the point of view of the safety of structures with respect to their reception of high-frequency waves, it is important that the natural vibration periods of structures be much lower than the working periods of high-frequency engines so that the phenomenon of resonance is prevented.

3. Automotive and railroad traffic exciting high-frequency vibrations with unstable regime and complicated spectrum. A relatively large vibration amplitude at the source is a specific feature of these vibrations. Therefore in spite of the high frequency of such vibrations, they may produce highly perceptible vibrations of some structural elements located close to roads with heavy traffic. Measures against such vibrations in structures comprise an important contemporary task of public interest in cities.

4. Low-frequency engines with periodic stable regime of work, including horizontal piston compressors, diesels, and reciprocating steam engines.

The speeds of these engines rarely exceed 500 rpm, and in the most powerful engines drop to 75 to 100 rpm. Engines of this type are more dangerous for structures than any other machines with a steady work regime. Cases are known in which foundations under such engines produced vibrations of structures located at distances of several hundred meters from the source. This is explained by the fact that the periods of natural vibrations of structures in most cases do not exceed the minimum periods of low-frequency engines; therefore resonance is possible, as a result of which vibration amplitudes may become very large. In antivibration work, cases of vibrations of structures caused by low-frequency engines are most frequent.

5. Forge hammers and other installations with impact actions on their foundations. The amplitudes of foundation vibrations under such units may reach 1.5 mm and more; the vibration amplitudes of soil close to a hammer foundation also may reach large values. Hammers do not exert a periodic action on foundations; therefore the vibrations of hammer foundations have an unstable character with respect to time. Waves propagating through soil from a hammer foundation in practice do not cause resonance with the natural frequencies of a building, especially since these foundations induce waves always higher in frequency than 500 periods per minute. Waves induced by hammer foundations may lead to dangerous shocks only in structures located in the immediate vicinity of these foundations.

Vibrations of hammers are dangerous because they lead to additional nonuniform settlements caused by waves propagating from their foundations. Vibration amplitudes of footings under walls or columns of structures located close to the source of vibrations are of the same order as the amplitudes of soil vibrations, which in the vicinity of hammer foundations may reach fairly high values. These vibrations of wall or column footings are characterized by fairly large amplitudes and result in dynamic pressure on the soil which may attain considerable magnitude and may act in addition to static pressure. Under certain conditions this dynamic pressure may cause considerable settlement of walls or columns, often leading to serious damage to a building (see Art. IX-3).

6. Explosions inducing waves, the nature of whose action on structures has much in common with waves propagating from foundations under impact-producing machines.

IX-2. Action on Structures of Seismic Waves Propagating from Foundations under Low-frequency Machinery

a. Resonance as the Main Cause of Vibrations of Structures. In most cases in which foundations under low-frequency machines cause vibrations of structures, the foundations themselves undergo vibrations at entirely permissible amplitudes. Seismograph measurements in areas close to the footings of walls or columns of a vibrating structure show that the amplitudes of soil vibrations are ten or a hundred times smaller than the amplitudes of vibration of the structure. Thus, the energy of waves which sometimes cause dangerous vibrations of structures is very small. In spite of this, amplitudes of structures receiving the waves may reach considerable values, sometimes several millimeters.

If the magnitude of excitement is small, the vibration amplitudes of a structure, and consequently the stresses therein, may reach large values only when the structure is in resonance with the excitation; then the damping forces acting in the structure are small in comparison with the elastic forces.

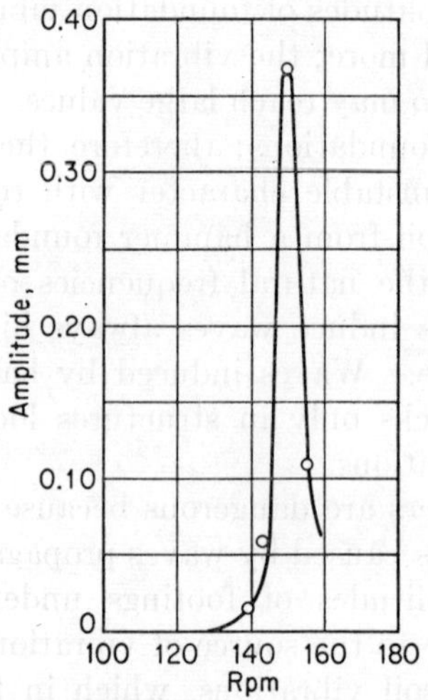

FIG. IX-1. Resonance curve of a 45-m-high structure.

Let us discuss two cases from practice which illustrate how structures are influenced by a coincidence of one of the natural frequencies of the structure with the frequency of the propagating waves.

THE FIRST CASE. Vibrations of the foundation under a steam engine of 650 hp caused vibrations of a 45-m-high structure. The vibration amplitude in the upper part of the structure was 0.37 mm; it decreased downward, and on the second floor the vibrations were so small that they could not be perceived.

The vibrations of the structure were measured within the range of 55 to 155 rpm. For this engine, the normal speed is 150 rpm. Figure IX-1 gives a curve of the relationship between the vibration amplitude of the structure and the speed of the engine, plotted on the basis of results of measurements. It is seen from this graph that the resonance curve of the structure has an extremely sharp peak with a maximum corresponding to the operating speed of the engine (150 rpm). If the speed increases to 155 rpm, i.e., becomes larger by 3 per cent than the speed at resonance, then the vibration amplitude is only 30 per cent of the amplitude corresponding to 150 rpm; if the speed decreases to 140 rpm (by 6.5 per cent), then the vibration amplitude is only 7 per cent of the vibration amplitude at resonance.

It is seen from Fig. IX-1 that in the case under consideration the structure was at resonance with waves propagating through soil from the engine foundation. In addition, it is seen from the graph that the resonance zone of vibrations is located within a very narrow range of frequencies whose values do not differ by more than 3 to 5 per cent from the resonance frequency of the structure.

Such a sharp resonance curve shows that the damping forces of the structure corresponding to the low frequencies of vibration are very small

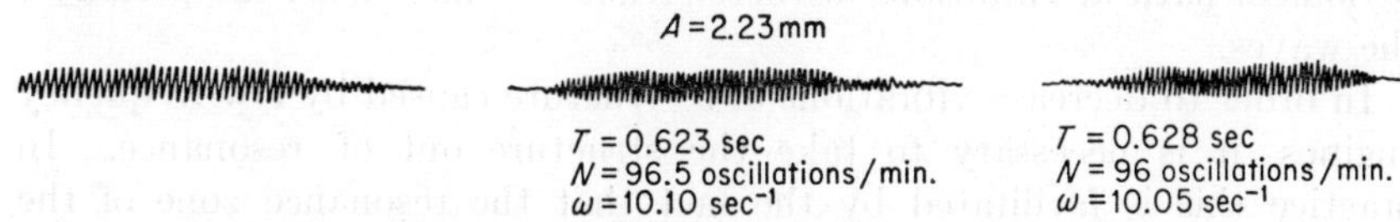

Fig. IX-2. Vibrograms of a 20-m-high structure during the operation of an adjoining compressor.

in comparison with the elastic forces. This is the reason why the structure reacts only to excitement having a period which does not differ much from its own.

The Second Case. The vibrations of a frame structure of 20 m height were caused by vibrating foundations under compressors operating at a maximum of 100 rpm. Figure IX-2 presents a diagram of vibrations of the structure when one compressor was operating at a speed of 97 rpm as measured by a tachometer. Actually, the speed of the compressor was not constant, but, due to some irregularity in speed typical of reciprocating engines, changed within a range of ± 3 to 4 per cent. The changes in vibration amplitudes of the structure, as recorded by the vibrograph, are explained by the fact that the indicated speed of the engine (97 rpm) was close to the lower natural frequency of the structure.

Due to the irregularity in the speed of the engine, the period of propagating waves at times approached the natural period of the structure and at times departed from it. Thus the structure at times was in resonance with waves propagating through the soil, and at times was not. The vibration amplitudes recorded by the vibrograph at times decreased to zero, then grew to values dangerous to the safety of the structure; this is explained by the fact that even 3 to 5 per cent changes in the period of excitement, by comparison with the period of natural vibrations of the structure, lead to an extremely sharp decrease in its vibration amplitudes. The time intervals during which vibrations were not observed at all alternated with intervals of intensive vibrations of the structure; these changes were irregular and the length of the intervals reached several minutes. When the speed of the compressor decreased to 90 rpm, the vibrations of the structure ceased completely.

Other similar cases of vibrations confirm that the resonance of structures with waves propagating from foundations under low-frequency engines is the main cause of their considerable vibrations. This is why vibrating foundations under low-frequency engines often do not affect structures located in the immediate vicinity, but induce considerable vibrations of structures located at a large distance. The latter structures get into resonance with waves propagating through soil from foundations under machines, while neighboring structures do not only because the periods of natural vibrations of these structures differ from the period of the waves.

In order to decrease vibrations of a structure caused by low-frequency engines, it is necessary to take the structure out of resonance. In practice this is facilitated by the fact that the resonance zone of the structure, corresponding to the low frequencies, lies within an extremely narrow range of frequencies located close to its frequency of natural vibrations. Therefore it suffices to change the ratio between the exciting frequencies and the frequencies of natural vibrations of structures by 3 to 5 per cent and the vibrations of the structure will completely die away.

Changing the operating speed of the engine is the easiest way to change the ratio between the frequencies of an engine and the natural frequencies of a structure. The maximum amplitude of vibrations of a structure under the most unfavorable conditions, and the decrease in the amplitude of its vibrations with changes in engine speed, may be established from the resonance curve of the structure obtained by means of a vibrograph for different speeds of the engine. It is simpler to record the resonance curve at the time of starting or stopping the engine. It is desirable to record diagrams of vibrations when the speed of the engine is somewhat greater than the operational speed.

The speed of the engine cannot always be changed, because this causes a change in the power of the engine and in the regime of its work. When this is inadmissible, it becomes necessary to change the natural frequency of the structure. This is generally done by increasing its rigidity. Here the condition should be satisfied that the natural frequency of the vibrating structure is larger than the operating speed of the engine. This may be checked by measuring the vibrations of the structure at the time of starting or stopping the engine. If possible, this check may also be performed by calculation. The increase in the rigidity of the structure must depend on the ratio between the frequency of natural vibrations of the structure and the frequency of the engine, i.e., on whether the first frequency is larger or smaller than the engine speed. For example, if it is established that the natural frequency of the vibrating structure is 3 per cent smaller than the operating frequency of the engine causing vibrations, then, in order to decrease the amplitude of vibrations of the structure, it is necessary to increase the rigidity of the structure so that its natural frequency becomes 7 per cent larger than the initial frequency. If, however, the natural frequency of the vibrating structure is 3 per cent larger than the operating frequency of the engine, then, in order to obtain the same effect, it suffices to increase the rigidity of the structure so that its natural frequency will be larger by 2 to 3 per cent.

In many cases calculations of natural frequencies of structures may be performed only with a high degree of approximation; therefore, in such cases the measures to change the natural frequency of the structure should provide for an increase in the frequency of its natural vibrations by some 5 to 10 per cent.

The selection of measures for changing the natural frequency of the vibrating structure depends on its design and on the equipment installed in it, as well as on the design of the foundation under the low-frequency engine.

We will now discuss several examples to illustrate the structural methods employed for decreasing vibrations.

b. Measures to Decrease Vibrations of a Power-station Wall Induced by Operating Saw Frames. Structures adjacent to a sawmill underwent noticeable vibrations created by waves propagating through soil from the foundations under saw frames. Three frames ran at 320 rpm and one at 300 rpm. Vibrations were noticeable even on the other side of a river of 8 m depth. The foundations underwent mainly vertical vibrations with amplitudes of some 0.25 mm. The horizontal component of rocking vibrations equaled only some 0.1 mm. Due to an inexact coincidence of the frame rotations, vibrations over the entire plant territory were of a pulsating type.

The foundations under the saw frames were located in a layer of water-saturated peaty soil. The soil conditions were such that they contributed intensively to the development of foundation vibrations and the propagation of vibrations. In spite of this, although the vibrations of structures were noticeable on the territory of the plant, they did not cause serious apprehensions. However, during the construction of a power station which had been designed as a wooden frame structure with brick filling, considerable transverse horizontal vibrations of a 20-m-long fireproof wall developed after it was erected to a height of about 4 m.

These vibrations reached such dimensions that it became dangerous to continue the work. A brick placed on its narrow side would fall off the wall, the set of the brick mortar was unreliable, and there was no assurance that the vibrations would not increase as the construction of the wall proceeded further, and that it would not lead to the collapse of the wall.

The newly erected wall was joined to the existing structure of the machine shop by interlocking masonary courses and was located at a distance of approximately 17 to 20 m from the nearest foundation under a saw frame. Measurements of vibrations showed that the wall vibrated almost exclusively in a horizontal (transverse) direction. The curve of the distribution of measured amplitudes along the center line of the wall was bent in such a way that the vibration amplitude was smallest near the wall edges, and the largest amplitude was found at a distance of

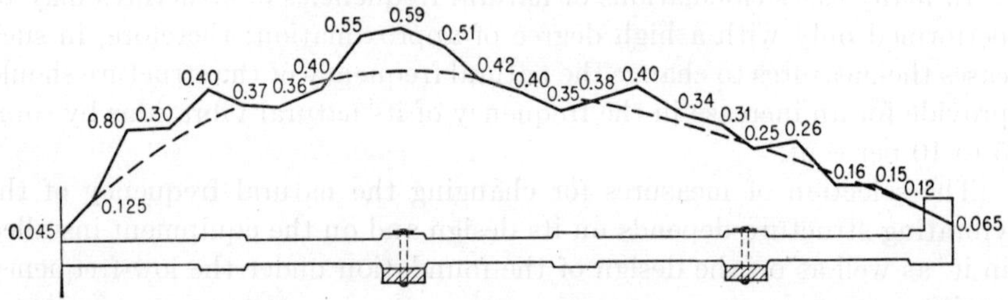

FIG. IX-3. Amplitudes (in millimeters) of transverse horizontal vibrations of a newly built brick wall, after it reached 4 m in height, as a result of resonance with a saw-frame engine operating 17 m away.

approximately 40 per cent of the total length from the end adjoining the existing machine shop.

The soil vibrations on the site of construction had been hardly perceptible, and the amplitude of vibrations had not exceeded several hundredths of a millimeter. Therefore it may be assumed that considerable wall vibrations, as indicated above, appeared only because the lowest natural frequency of the wall (corresponding to the shape of the wall vibrations as shown in Fig. IX-3) was close to the operating frequencies of the saw frames. Hence, in order to take the wall out of resonance, it was necessary to change its rigidity, which was done by providing a reinforced-concrete belt and counterforts (Fig. IX-4).

At first two abutments were installed which divided the wall into three uneven sections. However, these abutments did not exert any noticeable influence on the vibrations. Then a reinforced-concrete belt with double reinforcement was installed along the entire length of the wall. The construction of this belt proceeded at a time when the sawmill did not operate, and consequently the wall did not vibrate. Three to five days after the installation of the belt was completed, the sawmill resumed operations; the wall vibrations did not start again, and the construction was successfully completed.

c. Vibrations of a Boiler-shop Wall Induced by Operating Compressors. An almost analogous case of wall vibrations occurred due to the influence of a horizontal piston compressor with a capacity of 60 m^3/hr running at 167 rpm. Soil conditions were unfavorable; an upper layer of alluvial soil extended to a depth of 1.5 to 2 m and was underlaid by clays with some silt and sand and by loose sands. The ground-water level was at a

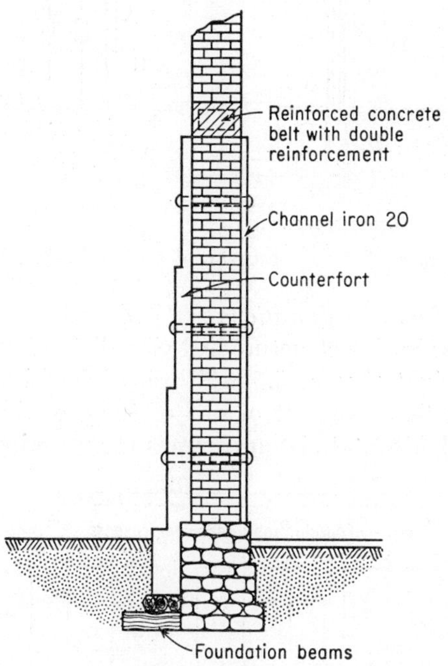

FIG. IX-4. Cross section of counterforts and reinforced-concrete belt used to eliminate the excessive wall vibrations shown in Fig. IX-3.

depth of some 3 m. Although the foundation vibrations were noticeable (especially horizontal vibrations in the direction of the piston motion), they were within the range of permissible values, since their amplitude did not exceed 0.2 mm. The observed vibrations did not exert any noticeable influence on machine operation or on the state of the foundation. In spite of this, considerable vibrations of several structures were observed in a zone with a radius up to 150 m. Extremely intensive vibrations of one of the walls of the boiler shop nearest the compressors were observed. This boiler shop was located at a distance of some 80 m from the compressor plant. In addition to this structure, several other structures located at various distances from the compressor plant also vibrated, but with smaller amplitudes. The maximum amplitude of wall vibrations reached 1.0 mm.

As in the foregoing case, a reinforced-concrete belt was cast along the entire wall in order to decrease the vibrations.

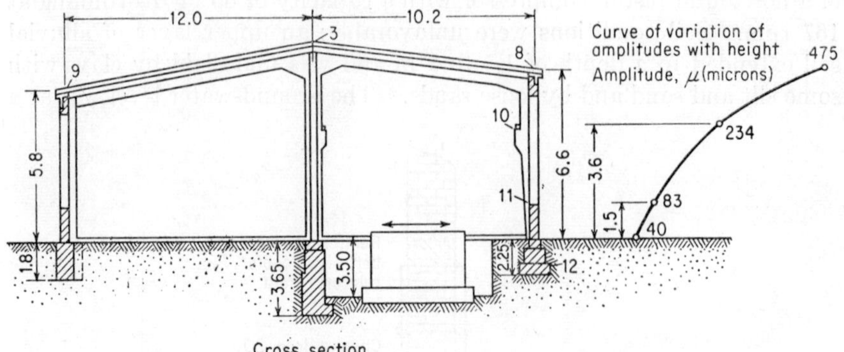

FIG. IX-5. Cross section of structure described in Art. IX-2-d.

d. Measures to Decrease Vibrations Induced by Waves from Compressor Foundations. A vibrating structure had plan dimensions of 33.0 by 22 m and an average height of some 5.5 m; it was divided by a brick wall into two sections. In the basement of one of these sections, five synchronously operating horizontal piston compressors were installed, each on a

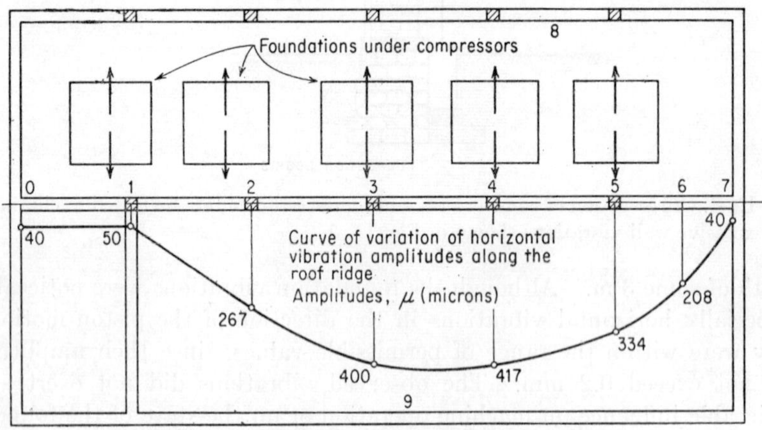

FIG. IX-6. Plan of structure described in Art. IX-2-d.

separate footing. Figures IX-5 and IX-6 show a cross section of the building and its schematic plan.

The vibrations of the building depended to a considerable degree on the season. In most cases of vibrations of buildings, and of sources of vibrations as well, an instability of vibration amplitudes is observed. This can be explained by minute variations in the elastic properties of materials which may cause minute changes in the natural frequencies and thus (especially under conditions of resonance) may cause changes in vibration amplitudes. If this is the case, then amplitudes which change with atmospheric conditions are an indirect indication of the fact that the structure vibrates under conditions of resonance. Changes in amplitudes of a structure's vibrations may depend on changes in soil moisture and on fluctuations in the ground-water level, especially in areas close to the foundation and the source of vibrations. An increase in soil moisture contributes to a decrease in the absorption by soil of the vibration energy; other conditions remaining equal, this leads to an increase in the transmission of wave energy to the structure.

An investigation of the structure's vibrations by a vibrograph provided the following data:

The structure underwent vibrations mostly in the direction of motion of the compressor pistons. In Fig. IX-6, the arrows show the main direction of wave propagation, perpendicular to the longitudinal walls of the structure; the walls are characterized by low rigidity in the transverse direction. Therefore, from the point of view of vibrations of the structure under consideration, the direction of motion of the compressor pistons (Fig. IX-6) is the least favorable, because the waves propagating from the compressor foundations exert maximum influence on the walls. This fact is not often taken into consideration when designing plants with engines inducing horizontal foundation vibrations (which may become the main cause of structural vibrations). If in the case under consideration the direction of piston motion was parallel to the longitudinal axis of the structure, then the wall vibrations would undoubtedly have had considerably smaller amplitude. In general, it is necessary to avoid as far as possible the installation of reciprocating engines so that the direction of piston motion is perpendicular to high walls elongated in plan and having no rigid sections.

Measurements showed that almost no vibrations were observed in the transverse walls of the structures when the external and internal longitudinal walls underwent vibrations with an amplitude reaching 0.475 mm. Points of measurements are indicated by numbers in Figs. IX-5 and IX-6. Vibrations were of a sinusoidal nature, and a superposition of components was observed at a frequency equaling the double frequency of machine rotation.

The curve of variation of vibration amplitudes with height is shown in Fig. IX-5, and the curve of variation of amplitudes along the roof ridge is shown in Fig. IX-6. It is seen from these curves that during the process of vibration the frames of the structure bend somewhat along their height.

Therefore the variation of amplitude with height has a curvilinear character. If the amplitudes had grown proportionally to height, the conclusion would have been possible that the frames do not bend during vibrations, but oscillate as a solid body rocking around a horizontal axis located in the plane of the foundation surface in contact with soil or somewhat below it.

The distribution of amplitudes along the length of the central wall of the building reveals that the presence of the transverse wall (Fig. IX-6) distorts the regularity in the distribution of amplitudes. There is almost no difference between the amplitudes at points 1 and 0. At other points, the distribution of amplitudes has a character which is usual for the vibrations of walls of buildings. The amplitudes increase toward the center of the wall and attain a maximum there; then they decrease toward the transverse wall; there they equal some 40 microns.

Thus it was established that the transverse wall decreases the amplitude of vibrations of the ridge at the place where the wall and the ridge are joined. Hence, the installation of a similar rigid brick or reinforced-concrete wall in the planes of other frames, for example, frame 4 or 3, would considerably decrease the amplitude of vibrations. There is no necessity to erect the wall along the whole height of the frame; it would suffice to install the wall in the upper part of the frame and support it with two or three columns.

In addition, vibration amplitudes of longitudinal walls of buildings could be decreased by the installation of two or three heavy counterforts outside the external walls. Finally, vibration amplitudes could be decreased considerably by the construction of extensions adjoining the longitudinal external walls of the vibrating building and properly tied to them. The latter method was employed to stop vibrations of the building shown in Figs. IX-5 and IX-6. After the construction of these extensions adjoining the building, the vibrations ceased.

e. Measures to Decrease Vibrations of a Refrigerator Plant Induced by Waves from a Compressor Foundation. The construction of the refrigerator plant proceeded as follows: first the building was erected, then the frame structure was placed within this building and separated from it by thermoinsulation. After the building was erected, it was found that one of its walls underwent considerable vibrations which caused cracks at several places. This wall was standing perpendicular to the direction of piston motion of the compressor located near it.

Attempts were made to decrease the wall vibrations by strengthening the foundation under the compressor and thus decreasing the amplitudes of foundation vibrations. These measures did not help at all, and operation of the compressor was temporarily interrupted. During this interruption, work on the construction of the frame structure continued.

As soon as each story of the frame was built, it was connected by ties with reinforced-concrete antiseismic belts installed in the wall at the level of each story of the frame (see Fig. IX-7). After three stories of the frame were constructed, the compressor was set into motion again. No vibrations of the wall were then observed.

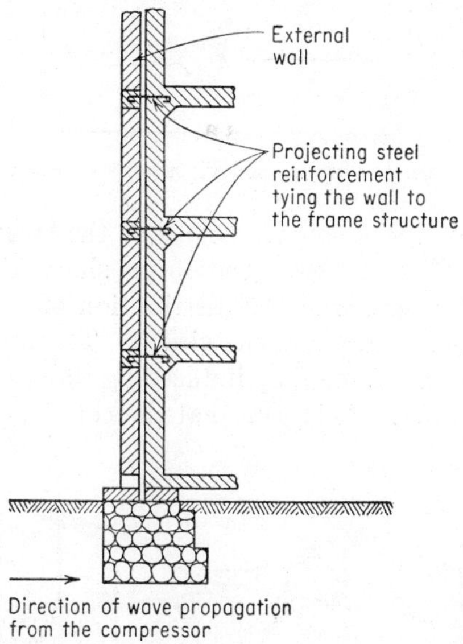

FIG. IX-7. Cross section of refrigerator-plant wall described in Art. IX-2-e.

In this case, the elimination of the vibrations is explained by the fact that the wall went out of resonance with waves propagating through the soil due to an increase in its rigidity after it was joined to the frame structure.

IX-3. The Action of Shocks Caused by Waves from Hammer Foundations

Unlike foundations under low-frequency machines with steady speeds, foundations under forge hammers and other machines with impact action are characterized by a considerable amplitude of vibrations and a relatively high frequency (70 to 120 $\sec^{-1}$). Soil vibrations induced by shocks on hammer foundations differ greatly from vibrations caused by waves from foundations under machines with a settled regime of work. Figure IX-8 shows a typical seismogram of vertical soil vibrations induced by a hammer operation; it indicates that, as in the case of hammer foundation vibrations, the soil vibrations are of an unsettled nature. Because of this, even if the frequency of the building vibrations coincides with the frequency of waves induced by the hammer foundation, the amplitudes of the building vibrations cannot reach considerable values if the amplitudes of soil vibrations on the site of the building are small. Therefore the amplitudes of forge-shop vibrations are usually of the same

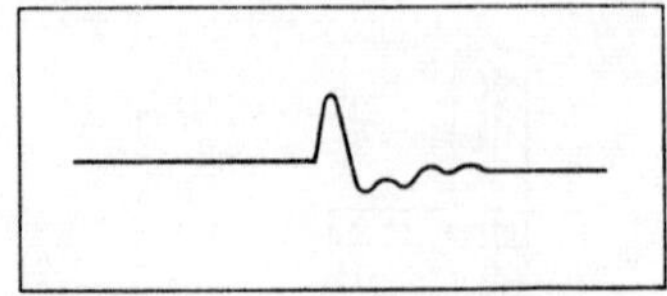

Fig. IX-8. Seismogram of vertical soil vibrations induced by hammer action.

order as the amplitudes of soil vibrations at the location of the footings under the walls or columns of the forge shop. Figure IX-9 shows several experimental curves of the distribution of amplitudes along the height of the columns and along the girder. These curves illustrate the nature of vibrations of a forge shop induced by waves from the foundation under a heavy hammer. It is seen that the columns undergo transverse

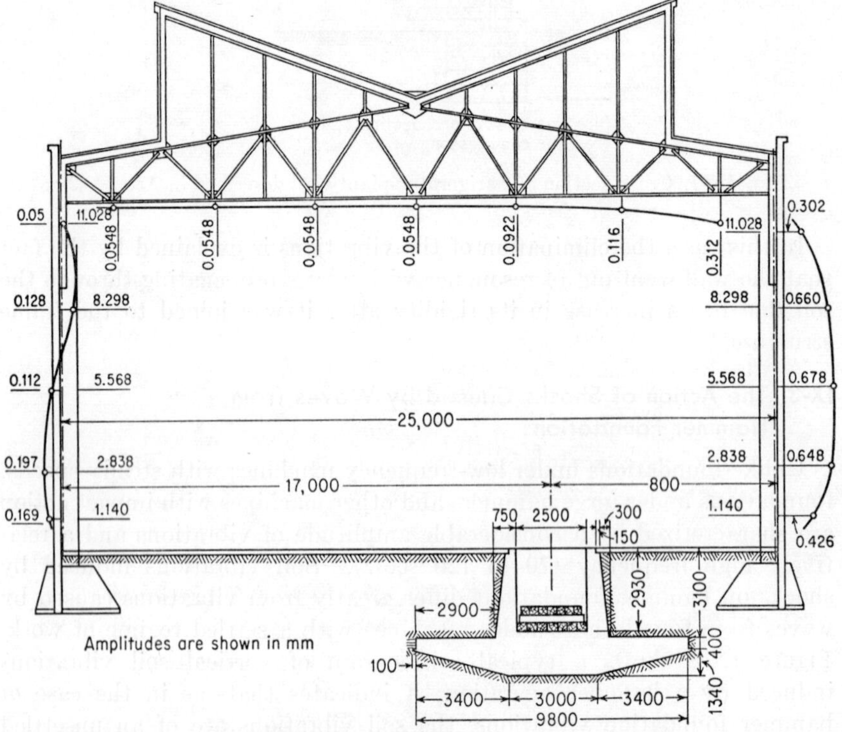

Fig. IX-9. Cross section of first forge-hammer shop described in Art. IX-3.

vibrations in bending, while the girders undergo mainly vertical vibrations. The vibrogram of Fig. IX-10 shows that the vibrations caused by waves from the hammer foundation are superimposed on the natural vibrations of the structure under consideration; the latter are damped to a comparatively slight degree.

Amplitudes of forge-shop vibrations are not large; therefore they are not dangerous with respect to the strength and stability of the structures.

In addition to the horizontal transverse vibrations, the shop columns, and consequently their footings, undergo vertical vibrations with an amplitude which often does not differ much from the amplitude of vertical vibrations of the hammer foundation. Therefore not only static, but dynamic loads as well, are transferred to the foundation bases. Since the acceleration of soil vibrations at small distances from the hammer

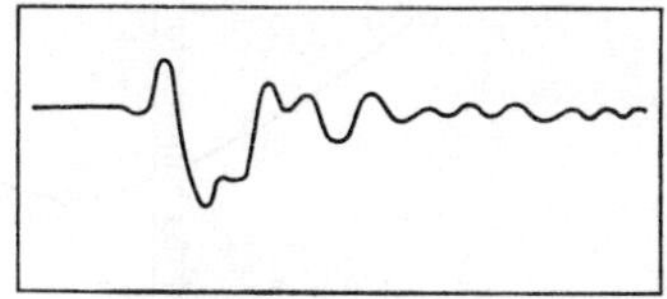

FIG. IX-10. Vibrogram of the frame structure in Fig. IX-9.

foundation may reach very high values (of the order of 0.7 to 1.0g), the real pressure on the base under columns will be 1.7 to 2 times higher than the static pressure. Therefore the footings under columns or wall sections subjected to impacts will undergo larger settlements than foundations subjected to the action of static loads only. Figure IX-11 shows graphs of the amplitudes of vertical vibrations and the residual settlements of the columns. It is seen from these graphs that the settlements are proportional to the amplitudes (or to the acceleration) of the vertical vibrations.

To illustrate the harmful effect of foundation settlements caused by the action of waves propagating from a hammer foundation, several cases are discussed below, derived from experience in the operation of forge shops.

Impacts transferred from a foundation under a 4.5-ton hammer were the reason for the complete destruction of a three-story building which was attached to a forge shop, but in which no manufacturing was done. The vibration amplitude of the hammer foundation was about 1.0 mm; the period of the waves propagating from the foundation was 0.075 to 0.070 sec. The brick auxiliary building was located at a distance of 6 m from the hammer foundation and was erected much later than the forge shop. It was supported by a continuous foundation built of cyclopean concrete; the pressure on soil was 1.75 to 2.0 kg/cm^2.

The foundation rested on fine-grained sands characterized by a higher than medium density; within this soil there occurred lenses of clay with some silt and sand. The ground-water level was at a depth of 4.0 m, i.e., approximately at the elevation of the hammer foundation surface in contact with soil.

The vibration amplitude of the foundation under the wall nearest the hammer foundation of the auxiliary building was 0.65 mm; hence, the acceleration of vertical vibrations of this wall was 0.54*g*.

The vibration amplitudes of the building foundations at different points varied from 0.05 to 0.65 mm; therefore the dynamic pressure on

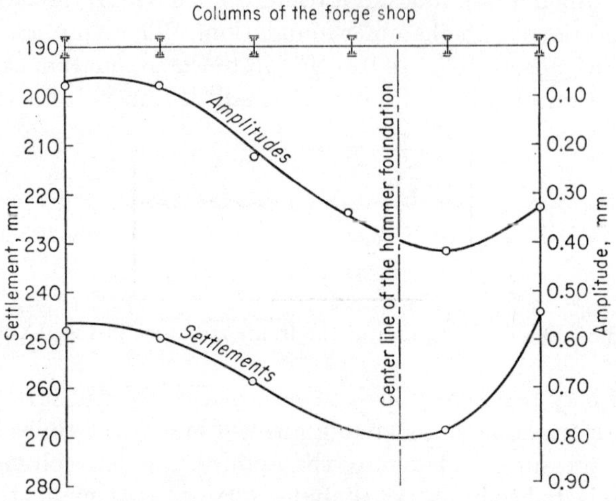

FIG. IX-11. Vibration amplitudes and settlement of footings under different columns of the forge-hammer shop shown in Fig. IX-9.

the soil was nonuniform and the foundations underwent considerable differential settlements resulting in the failure of the foundations and later of the walls.

There was another case in which shocks to a foundation under a 2.5-ton forge hammer were the cause of the destruction of the forge shop building. The foundation under the hammer underwent vertical vibrations with an amplitude of 0.55 to 0.60 mm; the period of these vibrations was 0.085 to 0.075 sec. The one-story brick forge shop rested on a continuous foundation built of cyclopean concrete; the static pressure on soil was about 2.0 to 2.5 kg/cm^2.

The foundations under the walls and hammer rested on gray and yellow fine-grained sands of medium density. The ground-water level was at a depth of 8.5 m.

Operation of the hammer for 200 hr caused differential settlements of the foundations under the walls of the forge shop and the formation of numerous cracks in the masonry of the walls.

In a third case, the source of vibrations was the foundation of a 3-ton forge hammer supported by 6-m-long timber piles. Fine-grained uniform sands of higher than medium density were found beneath the hammer foundations and the column foundations of the forge shop. Two groundwater horizons were found on the site of the shop: one was at a depth of 2.5 m, the other at a depth of 6 m. The pressure on the base of the columns of the forge shop was taken to equal 1.50 to 1.75 kg/cm^2.

The forge shop was a reinforced-concrete frame structure which consisted of columns tied together by continuous edge beams and crane-supporting beams. Soon after the 3-ton forge hammer was set in operation, a differential settlement of the shop columns nearest the foundation was noticed. This settlement led to the appearance of cracks in the reinforced-concrete structures of the shop, in the edge beams, in the crane-supporting beams, and in the masonry of the brick filling of the frame structure and the fireproof walls.

In all three cases, the development of a considerable differential settlement was caused by the combined action of static pressure and vibrations. The foundations rested on noncohesive soils—sands with somewhat higher than medium density.

Static investigations of such soils by means of load tests or in consolidometers show that they are characterized by a resistance much higher than that in cohesive soils (moist clays). Therefore sands of medium and higher than medium density present good bases for foundations transferring static loads only. However, if static loads and shocks (or vibrations) act simultaneously, sands and similar noncohesive soils may offer much less resistance to external loads. This was indicated in Chap. II and is illustrated by the above examples.

It was established in Chap. II that settlements caused by a simultaneous action of static loads and shocks (or vibrations) are proportional to the static pressure and inversely proportional to the coefficient of vibroviscosity. The value of this coefficient is inversely proportional to the acceleration of vibrations of soil particles at the point under consideration. Hence, the settlements of foundations subjected to vibrations and static loads are proportional to the accelerations.

Forge-hammer shops, similarly to other structures, are usually designed so that the static pressure on all columns or shop walls remains constant. However, the acceleration of vibrations depends essentially on the distance from the source of waves. Therefore columns or wall sections located near hammer foundations are subjected to higher accelerations of vibrations than building elements lying at larger distances from hammer foundations. This is the main reason for the differential settlements of foundations under columns, with all the consequences which follow.

In order to decrease the harmful influence of vibrations on the settlements of foundations under buildings, it is necessary not to assign uniform values of the static pressure on the foundation base. The values should depend on the distance of a column or a wall section from the source of waves. The smaller that distance, the lower should be the permissible bearing value. Thus permissible bearing values of soils beneath the foundations of a building undergoing vibrations induced by hammers should represent a certain function of the soil vibrations at the site of the foundation. The amplitude of soil vibrations, depending on the dimensions of the hammer foundation area in contact with soil; the amplitude and frequency of foundation vibrations; and the soil properties and distances from the foundation may be established similarly to the illustrative computation presented in Art. VIII-5.

Hence, in order to avoid differential settlements of foundations, it is necessary to assign permissible static pressures on the soil (depending on the amplitude of soil vibrations at the location of foundations under columns or walls of the shop) according to the directions presented in Chap. II.

REFERENCES

NOTES: 1. The reference number used in the original Russian text is given in parentheses at the end of each item of the present list; e.g., for item 1, (32).
2. The transliteration of Russian words follows the system adopted by the United States Board on Geographic Names.

1. Anikin, S. G.: Opredeleniye raschetnoy velichiny dinamicheskoy nagruzki, deystvuyushchey na fundamenty turbogeneratorov (Determination of Design Values of Dynamic Loads Acting on Foundations under Turbogenerators), *Sbornik informatsionnykh materialov Mosenergo,* part 1, 1946. (32)
2. Artbolevskiy, I. I., S. I. Artbolevskiy, and B. V. Edelshteyn: "Teoriya i metody uravnoveshivaniya shchekovykh drobilok" ("The Theory and Methods of Balancing Jaw Crushers"), ONTI, 1937. (35)
3. Barkan, D. D.: Eksperimentalnoye issledovaniye vibratsiy sploshnykh fundamentov, ustanovlennykh na svyaznykh gruntakh (An Experimental Investigation of Vibrations of Continuous Foundations Resting on Water-saturated Cohesive Soils), Symposium 4, VIOS, 1935. (25)
4. ———: O rasprostranenii ustanovivshikhsya kolebaniy v grunte (Distribution of Steady Vibrations in Soil), Symposium 4, VIOS, 1935. (40)
5. ———: "Raschët i proektirovaniye fundamentov pod mashiny s dinamicheskimi nagruzkami" ("Computations and Design of Foundations under Machinery with Dynamic Loads"), Moscow, 1938. (15)
6. ———: K voprosu o raschete i proektirovanii fundamentov pod turboagregaty (On the Question of Computations and Design of Foundations under Turbopower Systems), *Stroitel'naya Promyshlennost'*, no. 3, 1941. (31)
7. ———: O vybore glubiny zalozheniya istochnika i priyëmnika voln, rasprostranyayushchykhsya v grunte (On the Selection of the Depth of Placement of the Source and of the Receiver of Waves Propagated in Soil), *Zhurnal Tekhnicheskoy Fiziki,* vol. XI, no. 11, 1941. (37)
8. ———: Proektirovanie i raschet fundamentov pod kuznechnyye moloty i kopry v usloviyakh voyennogo vremeni (Design and Computations of Foundations for Forge Hammers and Large Drop Hammers under Wartime Conditions), *Stroitel'naya Promyshlennost'*, no. 6, 1942. (30)
9. ———: "Primeneniye vibrirovaniya pri ustroystve osnovaniy sooruzheniy"

("The Use of Vibrations for Treatment of Bases under Buildings"), Gosstroyizdat, Moscow, 1943. (16)

10. ———: Eksperimentalnyye issledovaniya ekranirovaniya voln, rasprostranyayushchikhsya v grunte (Experimental Investigations of Screening of Waves Propagated in Soil), *Inzhenernyy Sbornik Akademii Nauk SSSR*, Moscow-Leningrad, vol. II, no. 2, 1946. (51)
11. ———: Eksperimentalnyye issledovaniya kolebaniy fundamentov kuznechnykh molotov (Experimental Investigations of Vibrations of Foundations under Hammers), *NII, Sbornik trudov* no. 12, *Vibratsii sooruzheniy i fundamentov*, 1948. (29)
12. Deryagin, B.: O zatukhanii i dispersii seysmicheskikh voln (Damping and Dispersion of Seismic Waves), *Zhurnal Geofiziki*, no. 1–2, 1931. (38)
13. ———: Zatukhaniye seysmicheskikh i akusticheskikh voln i yego zavisimost ot chastoty (Damping of Seismic and Acoustic Waves and Its Dependence on Frequency), *Zhurnal Geofiziki*, no. 3–4, 1932. (38)
14. Die Anwendung dynamischer Baugrunduntersuchungen, *Veröffentlichungen des Instituts der Deutschen Forschungsgeselschaft für Bodenmechanik*, no. 4, 1936. (5)
15. Fillipov, A. P.: Vynuzhdennyye kolebaniya neogranichennoy plity lezhaschchey na uprugom poluprostranstve (Forced Vibrations of an Infinite Slab Resting on an Elastic Semi-infinite Space), *Prikladnaya Matematika i Mekhanika*, vol. IV, no. 2, 1940. (45)
16. Gerner: "Geologie und Bauwesen," H. 3, 1932. (10)
17. Gersevanov, N. M.: "*Dinamika gruntovoy massy*" ("Dynamics of Soil Mass"), Moscow, 1937. (4)
18. *Giproavtoprom* (Technical Division): "Rukovodstvo po proektirovaniyu i vozvedeniyu fundamentov pod kuznechnyye moloty i pressy" ("A Manual for Design and Erection of Foundations under Forge Hammers and Presses"), compiled by engineer A. N. Cheshkov; D. D. Barkan consulted; mimeographed edition, 1945. (28)
19. Gogoladze, V. G.: Uprugiye kolebaniya v srede s uprugim posledeystviyem (s nasledstvennost'yu) (Elastic Vibrations in a Medium Characterized by Elastic Recovery), *Trudy Seysmologicheskogo Instituta AN SSSR*, no. 109, 1941. (39)
20. Gorbunov-Posadov, M. I.: Osadki i davleniye pod zhestkimi pryamougol 'nymi fundamentnymi plitami (Settlements and Pressure under Rigid Rectangular Footings), *Stroitel'naya Promyshlennost'*, no. 8, 1940. (9)
21. Ivanov, N. N., and T. P. Ponomarev: "Stroitel'nyye svoystva gruntov" ("Engineering Properties of Soils"), Lengostransizdat, 1932. (1)
22. Ivanov, N. N., and M. D. Telegin: Uplotneniye dorozhnykh nasypey (The Compaction of Highway Embankments), *Novosti Dorozhnoy Tekhniki*, Symposium 8, Moscow, 1939. (17)
23. Kirillov, F. A., and S. V. Puchkov: Rasprostraneniye kolebaniya v grunte ot istochnika tipa impul'sa (The Propagation of Vibrations in Soil from a Source of the Impulse Type), *Trudy Seysmologicheskogo Instituta AN SSSR*, no. 59, 1935. (14)
24. Köhler, R.: *Z. Geophysik*, vol. VIII, no. 46, 1932. (43)

25. Korchinskiy, I. L.: Dinamicheskiye kharakteristiki drevesiny, betona, zhelezobetona (Dynamic Characteristics of Wood, Concrete, and Reinforced Concrete), *Symposium Dinamicheskiye Svoystva Stroitel'nykh Materialov*, Yu. A. Nilender, (ed.), Stroyizdat, 1940. (41)
26. Kupradze, V. D.: "Osnovnye zadachi matematicheskoy teorii diffraktsii (ustanovivshiyesya protsessy)" ("Basic Problems of the Mathematical Theory of Diffraction"), Moscow-Leningrad, 1935. (50)
27. Lamb, H.: On the Propagation of Tremors over the Surface of an Elastic Solid, *Phil. Trans. Roy. Soc. London*, series A, vol. 203, September, 1904. (44)
28. Leybenson, L. S.: "Kratkiy kurs teorii uprugosti" ("A Brief Outline of the Theory of Elasticity"), Chap. XIII, Moscow-Leningrad, 1942. (36)
29. Lur'ye, A. I.: Vliyaniye uprugosti grunta na chastoty sobstvennykh kolebaniy prostoy ramy (Effect of the Soil Elasticity on Frequencies of Natural Vibrations of a Simple Frame), *Symposium Vibratsii Fundamentov Ramnogo Tipa*, ONTI, Gosstroyizdat, 1933. (34)
30. ———: Vliyaniye uprugosti osnovaniya na chastoty svobodnykh kolebaniy turbofundamentov (Effect of Elasticity of the Base on Frequencies of Vibrations of Foundations under Turbodynamos), *Symposium Vibratsii Fundamentov Ramnogo Tipa*, ONTI, Gosstroyizdat, 1933. (33)
31. Lyav, A.: "Matematicheskaya teoriya uprugosti" ("Mathematical Theory of Elasticity"), Moscow, 1935. (13)
32. Pavliuk, N. P.: *Byulleten' Leningradskogo Instituta Sooruzheniy*, no. 15, 1931. (24)
33. ——— and A. D. Kondin: O pogashenii vibratsiy fundamentov pod mashiny (Damping of Vibrations of Foundations under Machinery), *Proekt i Standart*, no. 11, 1936. (26)
34. Pokrovskiy, G. I., and others: Novyye metody issledovaniya szhimaemosti i vnutrennego treniya v gruntakh (New Methods of Investigation of the Compressibility and Internal Friction in Soils), *Vestnik Voyenno-Inzhenernoy Akademii RKKA*, no. 6, 1934. (3)
35. Pol'shin, D. E.: "Ugol povorota fundamenta pri ekstsentrichnoy nagruzke yego" ("The Angle of Rotation of a Foundation when Loaded Eccentrically"). (12)
36. Ramspeck, A., and G. Schulze: *Veröffentlichungen des Instituts der Deutschen Forschungsgesellschaft für Bodenmechanik* (*Degebo*) *an der Technische Hochschule*, no. 6, 1936. (47)
37. Reissner, E.: Stationare axialsymmetrische, durch eine schütternde Masse, erregte Schwingungen eines homogenen elastischen Halbraumes, *Ingenieur Archiv*, December, 1936. (22)
38. Sadowsky, M.: Zweidimensionale Probleme der Elasticitätstheorie, *Z. Angew. Math. und Mech.*, Bd. 8, 1928. (7)
39. Schleicher: Zur Theorie der Baugrundes, *Der Bauingenieur*, Heft 48/49, 1926. (8)
40. Sezawa, K.: Dispersion of Elastic Waves, *Bull. Earthquake Research Inst., Tokyo Imperial University*, vol. 3, 1927. (46)
41. Shekhter, O. Ya.: Ob uchete inertsionnykh svoystv grunta pri raschete

vertikal'nykh vynuzhdennykh kolebaniy massivnykh fundamentov (Consideration of Inertial Properties of Soil in the Computations of Vertical Forced Vibrations of Massive Foundations), NII, *Symposium 12, Vibratsii Osnovaniy i Fundamentov*, Moscow, 1948. (23)

42. Sobolev, S. L.: O diffraktsii sfericheskikh uprugikh voln vblizi poverkhnosti sfery (Diffraction of Spherical Elastic Waves in the Vicinity of a Sphere Surface), *Trudy Seysmologicheskogo Instituta AN SSSR*, no. 7, 1930. (49)
43. ———: Teoriya diffraktsii ploskikh voln (Theory of Diffraction of Plane Waves), *Trudy Seysmologicheskogo Instituta AN SSSR*, no. 41, 1934. (48)
44. "Tekhnicheskiye usloviya na raschet i proektirovaniye fundamentov pod mashiny s dinamicheskimi nagruzkami (proekt)" ("Technical Rules for Computations and Design of Foundations under Machinery with Dynamic Loads"), Gosstroyizdat, 1939. (27)
45. Terzaghi, K.: "Stroitel'naya mekhanika grunta na osnove yego fizicheskikh svoystv" ("Soil Mechanics Based on Physical Properties of Soil"), translated from German, Gostekhizdat, Moscow-Leningrad, 1933. (2)
46. Tsytovich, N. A.: "Mekhanika Gruntov" ("Soil Mechanics"), Gosstroyizdat, Moscow, 1937. (6)
47. "Ukazaniya po proektirovaniyu i vozvedeniyu v usloviyakh voyennogo vremeni fundamentov pod mashiny s dinamicheskimi nagruzkami," U-56-42 ("Instructions for the Design and Erection of Foundations for Machinery with Dynamic Loads under Wartime Conditions," U-56-42), Gosstroyizdat, Moscow, 1942. (11)

NAME INDEX

SUBJECT INDEX